Conformation in Fibrous Proteins

MOLECULAR BIOLOGY

An International Series of Monographs and Textbooks

Editors: BERNARD HORECKER, NATHAN O. KAPLAN, JULIUS MARMUR, AND HAROLD A. SCHERAGA

A complete list of titles in this series appears at the end of this volume.

Conformation in Fibrous Proteins

AND RELATED SYNTHETIC POLYPEPTIDES

R. D. B. Fraser and T. P. MacRae

DIVISION OF PROTEIN CHEMISTRY
COMMONWEALTH SCIENTIFIC AND INDUSTRIAL RESEARCH ORGANIZATION
MELBOURNE, AUSTRALIA

ACADEMIC PRESS New York and London 1973

A Subsidiary of Harcourt Brace Jovanovich, Publishers

ACADEMIC PRESS, INC.
111 Fifth Avenue, New York, New York 10003

United Kingdom Edition published by
ACADEMIC PRESS, INC. (LONDON) LTD.
24/28 Oval Road, London NW1

Library of Congress Cataloging in Publication Data

Fraser, R   D   B
    Conformation in fibrous proteins.

    Bibliography: p.
        1.   Proteins.   2.   Conformational analysis.
I.   MacRae, T. P., joint author.   II.   Title.
[DNLM: 1.   Proteins.   QU55 C751 1973]
QP551.F8428   574.1'9245   72–9979
ISBN 0–12–266850–2

Contents

v

Chapter 2   **Electron Diffraction**

Chapter 3   **Electron Microscopy**

Chapter 4   **Optical Diffraction**

Chapter 10 **The Beta Conformation**

Chapter 11 **Poly(glycine) II, Poly(L-proline) II, and Related Conformations**

CONFORMATION IN FIBROUS PROTEINS

Chapter 15 Myofibrillar Proteins

Chapter 16 Keratins

Chapter 17 **Conformation in Other Fibrous Proteins**

Chapter 18 **Some General Observations on Conformation**

References

Preface

The elucidation of the nature of the regular secondary structures that occur in fibrous proteins has engaged the interest of structure analysts for more than half a century. The complexity of the problem, coupled with the paucity of readily interpretable data, has generated a large volume of literature concerned with speculations about possible conformations, and as the ability to perform complex calculations of predicted diffraction patterns and potential energies has advanced, it has become increasingly difficult to assess the relative merits of various claims and counterclaims. From the aura of finality that surrounds the description of many models the nonexpert could be excused for concluding that the search for the correct structure was over. Nothing could be further from the truth, and so far no model for the structure of a fibrous protein has been shown to account, in a quantitative manner, for the entire X-ray diffraction pattern. With the possible exception of the α- and β-forms of poly(L-alanine), the same is true of models for the structures of synthetic polypeptides.

Excellent surveys of the chemistry and solution properties of fibrous proteins and related synthetic polypeptides are already available (Bamford *et al.*, 1956; Katchalski *et al.*, 1964; Seifter and Gallop, 1964; Fasman, 1967a; Florkin and Stotz, 1968, 1971; Timasheff and Fasman, 1969) and the aim in preparing this monograph has been to provide a comprehensive and critical account of physical studies of these materials in the solid state. The monograph has been divided into three parts: in the first, an account is given of physical methods of determining conformation; in the second, relevant results from studies of synthetic polypeptides are discussed; and in the third, investigations of conformation in fibrous proteins are reviewed.

In the part dealing with methods of determining conformation (comprising Chapters 1–8) the extent of coverage has been determined by the availability elsewhere of adequate treatments and the extent to which the method has proved to be profitable in studies of fibrous proteins. Diffraction methods have provided much of the data upon which models of structures have been based and these data also provide the most searching test of correctness available at the present time. Particular emphasis has therefore been placed on the collection and interpretation of diffraction data.

The study of synthetic polypeptides has played an important role in the development of current ideas on the structure of fibrous proteins and selected topics are discussed in Chapters 9–12. In Chapter 9 data obtained on the α-helix are collected and these provide a starting point for the development of models for the k-m-e-f group of fibrous proteins. Similarly in Chapters 10 and 11 data are collected which are relevant to the β and collagen groups of fibrous proteins, respectively. In Chapter 12 evidence gained on the influence of side-chain composition on conformation is reviewed and compared with evidence obtained from studies of globular proteins.

In Chapters 13–17 the results of investigations of conformation in fibrous proteins are reviewed and discussed in terms of model structures. Consideration has been limited, in general, to the native, solid state and, with the exception of β-keratin, artificially induced conformations have not been dealt with. Similarly, studies of the conformational transitions that take place during muscular contraction have not been discussed since an adequate treatment of this topic would require an entire volume. In the concluding chapter, some general aspects of conformation are collected.

A large number of disciplines are spanned in the present monograph and where practicable we have adhered to the most widely used system of symbols for representing particular quantities, rather than defining a unique set of unfamiliar symbols. Inevitably some conflicts arise and the same symbols are used to denote different quantities. For example it has been recommended that the symbols ϕ and ψ be used to denote the torsion angles about the NC^{α} and $C^{\alpha}C'$ bonds, respectively (IUPAC-IUB Commission on Biochemical Nomenclature, 1970) and this conflicts with the almost universal use of ϕ as an angular measure in cylindrical polar coordinates. We believe that the meaning to be attached to a symbol will always be clear from the context. In the chapters dealing with methods of determining conformation, lists of the symbols used in the individual chapters are given. Polymers will be indicated by enclosing the repeating unit in square brackets and adding a subscript n. Thus for example poly(L-alanylglycine) would be abbreviated to $[Ala-Gly]_n$.

Acknowledgments

It is a pleasure to acknowledge the assistance of Mrs. Eunice Day who typed the manuscript, Mr. Ben Kowalski who prepared many of the diagrams, Mr. Royston Larkin who prepared many of the photographs, and Mrs. B. J. Adams who helped to proofread the manuscript and to prepare the index.

We are grateful to numerous colleagues for making available details of their work in advance of publication and to Drs. W. G. Crewther, A. Elliott, B. S. Harrap, and F. G. Lennox, Professor J. Lowy, Drs. A. Miller and D. A. D. Parry, Professor H. A. Scheraga, Mr. E. Suzuki, and Drs. P. A. Tulloch and E. F. Woods for their helpful comments and advice.

Abbreviations for Amino Acid Residues

The full chemical names of residues will sometimes be given but where their repeated use becomes cumbersome the following abbreviations will be used:

Abbreviation	*Amino acid*
Ala	L-alanine
Arg	L-arginine
Arg(HCl)	L-arginine hydrochloride
Asn	L-asparagine
Asp	L-aspartic acid
Asp(Bzl)	β-benzyl-L-aspartate
Asp(oClBzl)	β-o-chlorobenzyl-L-aspartate
Asp(mClBzl)	β-m-chlorobenzyl-L-aspartate
Asp(pClBzl)	β-p-chlorobenzyl-L-aspartate
Asp(Et)	β-ethyl-L-aspartate
Asp(Me)	β-methyl-L-aspartate
Asp(pMeBzl)	β-p-methylbenzyl-L-aspartate
Asp(PhEt)	β-phenethyl-L-aspartate
Asp(nPr)	β-n-propyl-L-aspartate
Asp(pIBzl)	β-p-iodobenzyl-L-aspartate
Cys	L-cysteine
Cys(Bzl)	S-benzyl-L-cysteine
Cys(BzlTh)	S-benzylthio-L-cysteine
Cys(Cbz)	S-carbobenzoxy-L-cysteine
Cys(CbzMe)	S-carbobenzoxymethyl-L-cysteine
Gln	L-glutamine
Glu	L-glutamic acid
Glu(Bzl)	γ-benzyl-L-glutamate
Glu(Et)	γ-ethyl-L-glutamate
Glu(isoAm)	γ-isoamyl-L-glutamate
Glu(Me)	γ-methyl-L-glutamate
Glu(Me₃Bzl)	γ-2, 4, 6-trimethylbenzyl-L-glutamate
Glu(Na)	L-glutamic acid, sodium salt

Abbreviation	*Amino acid*
Glu(oNO$_2$Bzl)	γ-o-nitrobenzyl-L-glutamate
Glu(mNO$_2$Bzl)	γ-m-nitrobenzyl-L-glutamate
Glu(pNO$_2$Bzl)	γ-p-nitrobenzyl-L-glutamate
Glu(NpMe)	γ-1-napthylmethyl-L-glutamate
Glu(pIBzl)	γ-p-iodobenzyl-L-glutamate
Gly	glycine
His	L-histidine
Hyp	L-hydroxyproline
Ile	L-*iso*leucine
Leu	L-leucine
Lys	L-lysine
Lys(Cbz)	ϵ-carbobenzoxy-L-lysine
Lys(HCl)	L-lysine hydrochloride
Met	L-methionine
Phe	L-phenylalanine
Pro	L-proline
Ser	L-serine
Ser(Ac)	O-acetyl-L-serine
Ser(Bzl)	O-benzyl-L-serine
Thr	L-threonine
Thr(Ac)	acetyl-L-threonine
Try	L-tryptophan
Tyr	L-tyrosine
Val	L-valine

METHODS OF DETERMINING CONFORMATION

LIST OF SYMBOLS FOR CHAPTER 1

λ	wavelength
$\bar{\lambda}$	weighted mean of K_{α} lines
x, y, z	Cartesian coordinates in real space
r, ϕ, z	cylindrical polar coordinates in real space
X, Y, Z	Cartesian coordinates in reciprocal space
R, ψ, Z	cylindrical polar coordinates in reciprocal space
$\mathbf{d}, \mathbf{D}$	vectors from origin in real and reciprocal space, respectively
a, b, c	primitive translations of space lattice
a^*, b^*, c^*	primitive translations of reciprocal lattice
h, k, l	Miller indices
$J_n(x)$	Bessel function of first kind of order n and argument x
ρ	electron density
ρ_m	electron density of continuous matrix
f	atomic scattering factor
σ_0	vector defining direction of incident radiation
σ	vector defining direction of scattered radiation
2θ	scattering angle
F	Fourier transform
I	intensity transform
$\langle\ \rangle_{\psi}$	average with respect to ψ
F^*	complex conjugate of Fourier transform
F_o, F_c	observed and calculated Fourier transforms, respectively
F_u	Fourier transform of "up" chain
F_d	Fourier transform of "down" chain
F_U	Fourier transform of repeating unit (structure factor)
I_U	intensity transform of repeating unit
F_L	Fourier transform of point lattice
I_L	intensity transform of point lattice
F_C	Fourier transform of crystal
I_C	intensity transform of crystal

F_s	Fourier transform of sheet
F_s	Fourier transform of uniform-electron-density solid
r	radius of helix
P	pitch of helix, Patterson function
r_0	radius of major helix in a coiled-coil
P_c	pitch of coiled-coil
t	unit twist of discontinuous helix
h	unit height of discontinuous helix (also used as a Miller index)
u	number of units in axial repeat of discontinuous helix
v	number of turns in axial repeat of discontinuous helix
α	rotation of helix about its axis
ω	inclination of helix axis to fiber axis
$C(\omega)$	orientation density function
$\mathbf{s, sr, s2, sr2}$	line groups
$G_{n,l}$	complex number describing contribution of Bessel function terms of order n to the lth layer line
$A_{n,l}, B_{n,l}$	real and imaginary parts of $G_{n,l}$, respectively
P_s	pitch of screw displacement
$\delta\phi$	displacement in azimuth about z axis
δz	displacement parallel to z axis
δs	displacement parallel to $\mathbf{D}$
B	temperature factor
$\mathcal{N}$	specification of rotation axis
β	tilt of fiber axis from normal to incident beam
μ	longitude of point on the sphere of reflection
χ	latitude of point on the sphere of reflection
y	distance on film measured parallel to meridian
x	distance on film measured perpendicular to y
ι	specimen–film distance in flat-film arrangement; cylinder radius for cylindrical film arrangement
$\mathcal{D}$	optical density
$\mathcal{I}_\mathrm{a}$	integrated intensity in an arc
$\mathcal{I}_\mathrm{s}$	integrated intensity per Å^{-1} in a layer-line streak
L_a	Lorentz factor for nonmeridional arcs
L_a'	Lorentz factor for meridional arcs
L_s	Lorentz factor for streaks
P	polarization factor
$\mathcal{R}'$	normalized mean deviation or residual
$\mathcal{R}$	normalized standard deviation
W	weight function
s	distance on film along the layer line

Chapter 1

X-Ray Diffraction

Most of the information that exists on the three-dimensional structure of molecules and assemblies of molecules has been obtained from diffraction studies. The amount of information which can be obtained in this way depends upon the degree of order that exists in the diffracting object. If the specimen is macroscopically crystalline, the arrangement of the atoms in the molecule can be determined with considerable precision; if the specimen is imperfectly crystalline, a limited amount of information can be obtained about the three-dimensional structure; and if the specimen is amorphous or is fluid, only a small amount of very general information of a statistical nature can be obtained.

Most fibrous proteins form partly ordered structures in their native state and diffraction studies offer the best prospect at present of gaining information about the detailed three-dimensional structures of these materials. The standard methods of crystal structure analysis (Wilson, 1970; Woolfson, 1970; Jeffery, 1971) are not directly applicable to the diffraction patterns obtained from fibrous proteins and special methods must be used. Introductory accounts have been given by Holmes and Blow (1965) and Wilson(1966), and a general review of the use of X-ray diffraction in the study of polymers has been given by Alexander (1969). In the present chapter the theory and technique of X-ray diffraction applicable to the study of conformation in fibrous materials are outlined. The theory and technique of electron diffraction will be discussed in Chapter 2.

A. THEORY OF X-RAY DIFFRACTION

The relative dimensions of X-ray wavelengths, diffracting objects, and recording apparatus are such that X-ray diffraction patterns are essentially of the Fraunhofer type. The most elegant method of formulating the theory of this type of diffraction is through the use of Fourier transforms and convolutions (James, 1954; Lipson and Taylor, 1958; Ramachandran and Srinivasan, 1970) and excellent introductions to these operations have been given by Jennison (1961) and Bracewell (1965).

I. GENERAL OUTLINE

a. *The Density Function*

The distribution of scattering matter in an object can be described by a function $\rho(\mathbf{d})$ which specifies the density of scattering matter at a point distant d from a suitable origin in the direction defined by the vector $\mathbf{d}$ (Fig. 1.1a). If there is a phase change on scattering, this can be allowed for by making $\rho(\mathbf{d})$ a complex quantity. In most instances the phase change is so small that the density function can be taken to be real. It is convenient to regard $\rho(\mathbf{d})$ as a function which extends throughout space but is zero outside the irradiated volume of the specimen.

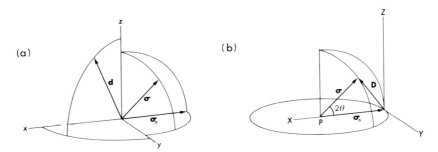

FIG. 1.1. (a) Coordinate system used to describe position in real space. **d** is a position vector and $\boldsymbol{\sigma}_0$ and $\boldsymbol{\sigma}$ are vectors of magnitude $1/\lambda$ which specify the directions of the incident and diffracted X-ray beams. (b) Coordinate system used to describe position in reciprocal space. **D** is a position vector and P is the center of the sphere of reflection. The angle between $\boldsymbol{\sigma}_0$ and $\boldsymbol{\sigma}$ is termed the scattering angle and is usually denoted by 2θ.

In the case of X-ray diffraction the magnitude of $\rho(\mathbf{d})$ is equal to the number of electrons per unit volume and is a continuous function. When dealing with scattering by atoms it is generally sufficient to regard the electrons associated with a particular atom as being concentrated at a point and $\rho(\mathbf{d})$ then becomes a discontinuous function. The effective scattering power associated with the point is termed the *atomic scattering factor f*, and is a function of both the scattering angle and the wavelength of the incident radiation. If a phase change occurs on scattering, f will be complex. Tables of atomic scattering factors are given by Ibers (1962).

Frequently $\rho(\mathbf{d})$ is periodic in one or more directions in space and the structure may be regarded as being built up by the continued repetition of a suitably chosen group of atoms.

b. *The Fourier Transform*

It is convenient to define the direction of the incident radiation by a vector σ_0 and to specify any particular scattering direction by a vector σ, both with magnitude $1/\lambda$ (Fig. 1.1a). It can be shown (see, for example, Lipson and Taylor, 1958) that if absorption and rescattering can be neglected, the amplitude and phase of the scattered wave, relative to an electron at the origin, are equal to the magnitude and phase, respectively, of the complex number

$$F(\sigma_0, \sigma) = \int \rho(\mathbf{d}) \exp[2\pi i (\sigma - \sigma_0) \cdot \mathbf{d}] \, dv \qquad (1.1)$$

where $(\sigma - \sigma_0) \cdot \mathbf{d}$ indicates a scalar product, dv is an element of volume, and the integral extends over all space. The complex number $F(\sigma_0, \sigma)$ is in fact the Fourier transform of $\rho(\mathbf{d})$ and it is convenient to define a vector $\mathbf{D} = \sigma - \sigma_0$ (Fig. 1.1b) so that Eq. (1.1) becomes

$$F(\mathbf{D}) = \int \rho(\mathbf{d}) \exp(2\pi i \mathbf{D} \cdot \mathbf{d}) \, dv \qquad (1.2)$$

This equation is the basis of diffraction theory.

For convenience the Fourier transform of a distribution is sometimes *normalized* so that it has a value of unity for $D = 0$. This is achieved by dividing the expression in Eq. (1.2) by $F(0)$, which is equal in magnitude to the number of electrons in the irradiated volume.

c. *Reciprocal Space*

The vector $\mathbf{D}$ in Eq. (1.2) has the dimension of reciprocal length and $F(\mathbf{D})$ may be regarded as a distribution in a *reciprocal space* in which position is defined by $\mathbf{D}$ (Fig. 1.1b). The distribution $F(\mathbf{D})$ specifies all the diffraction patterns which may be obtained for any combination of wavelength and direction of the incident radiation. The amplitude and phase of the radiation scattered for a particular combination of λ, σ_0, and σ can be calculated by evaluating $F(\mathbf{D})$, where $\mathbf{D}$ is obtained by constructing the so-called *sphere of reflection*. The sphere has a radius equal to $1/\lambda$ and is centered at a point P distant $1/\lambda$ from the origin in a direction opposite to that of the incident radiation (Fig. 1.1b). The vector $\mathbf{D}$ is the vector which joins the origin of reciprocal space to the point where the direction of the scattered radiation, laid off from P, intersects the sphere of reflection. The angle between σ_0 and σ is usually designated as 2θ and with this convention the magnitude of $\mathbf{D}$ is given by

$$D = 2 \sin \theta / \lambda \tag{1.3}$$

d. *The Intensity Transform*

The intensity of the scattered radiation is proportional to the square of the magnitude of $F(\mathbf{D})$, that is,

$$I(\mathbf{D}) = F(\mathbf{D})F^*(\mathbf{D}) \tag{1.4}$$

where $F^*(\mathbf{D})$ is the complex conjugate of $F(\mathbf{D})$. The distribution $I(\mathbf{D})$, which is conveniently referred to as the *intensity transform*, is of particular importance in the consideration of the diffraction pattern of a poly-crystalline specimen since the observed intensity is proportional to the appropriately averaged intensity transform of an individual crystallite.

The intensity transform represents the sum total of the observable diffraction data and the distribution $I(\mathbf{D})$ can be mapped by rotating the specimen so that all the accessible volume of the distribution intersects, in turn, the sphere of reflection (Fig. 1.1b). In the usual arrangement the direction of the incident radiation is fixed, so that the sphere of reflection remains stationary. As the specimen is rotated the intensity transform also rotates in such a manner that corresponding axes remain parallel.

In many specimens of fibrous proteins the axes of the polypeptide chains are aligned with a unique direction, for example, the z axis (Fig. 1.1a), but all possible orientations around this axis are present. This leads to a characteristic *fiber pattern* and the intensity transform is averaged by rotation about the Z axis (Fig. 1.1b). This cylindrically averaged intensity transform is usually denoted by $\langle I(\mathbf{D}) \rangle_\psi$, where ψ is an angle defining azimuth around the Z axis.

e. *Transforms of Periodic Structures*

If the density function is periodic in one dimension, for example, parallel to the z axis, $\rho(\mathbf{d})$ may be regarded as the convolution of a repeating unit ρ_U and a second function ρ_L consisting of a series of points spaced c apart along the z axis (Fig. 1.2). The Fourier transform

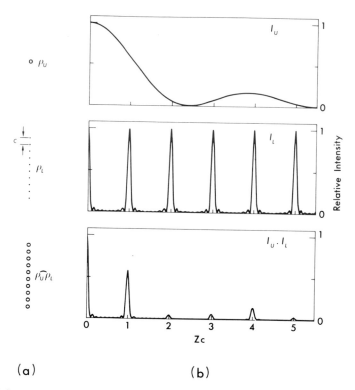

FIG. 1.2.   Example of the intensity transform of a periodic structure. (a) The density function ρ_U (top) specifies a repeating unit and the density function ρ_L (middle) specifies a lattice. The convolution of these two functions, shown in the lower part of the diagram, constitutes a periodic structure. (b) The intensity transform of this structure (lower curve) is the product of the intensity transform of the repeating unit (upper curve) and the intensity transform of the lattice (middle curve).

of a convolution of two functions is equal to the product of the Fourier transforms of the individual functions (Jennison, 1961) and this leads to the result

$$F(\mathbf{D}) = F_U(\mathbf{D})F_L(\mathbf{D}) \qquad (1.5)$$

where $F_U(\mathbf{D})$ is the Fourier transform of the repeating unit, often termed the structure factor, and $F_L(\mathbf{D})$ is the Fourier transform of the lattice of points. The corresponding intensity transform is, from Eq. (1.4),

$$I(\mathbf{D}) = I_U(\mathbf{D})I_L(\mathbf{D}) \qquad (1.6)$$

where $I_U(\mathbf{D}) = F_U(\mathbf{D})F_U^*(\mathbf{D})$ is the intensity transform of the repeating

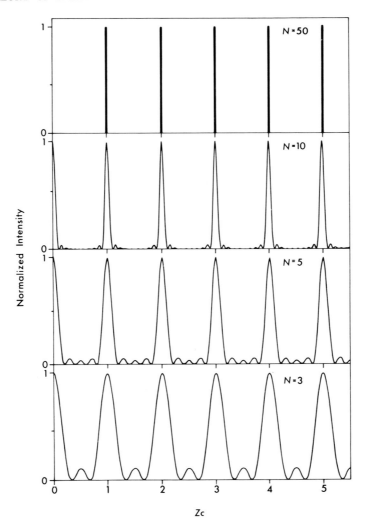

FIG. 1.3.   Intensity transforms or interference functions for point lattices containing different numbers N of points. The functions have been normalized to a value of unity at the origin.

unit and $I_L(\mathbf{D}) = F_L(\mathbf{D})F_L^*(\mathbf{D})$ is the intensity transform of the point lattice, sometimes called the *interference function*.

For a one-dimensional lattice the interference function has the form

$$I_L(Z) = [\sin^2(\pi NcZ)]/\sin^2(\pi cZ) \tag{1.7}$$

where N is the number of points in the lattice. This function is illustrated in Fig. 1.3 for various values of N. If N is very large, the interference function is effectively zero except over a set of plates with $Z = l/c$ (l an integer), where it has the value N^2. These are the *layer planes* and account for the characteristic *layer lines* which appear in the diffraction patterns of one-dimensionally periodic structures (Fig. 1.4b, lower). The distribution of intensity over the layer planes is continuous and equal to N^2 times the intensity scattered by the repeating unit.

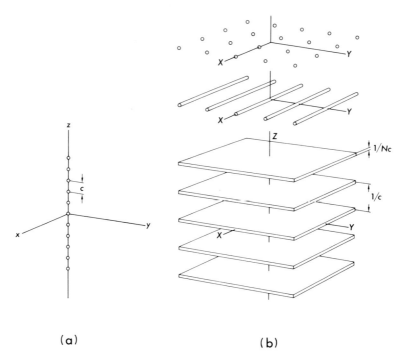

(a)   (b)

FIG. 1.4.   (a) One-dimensional point lattice along the z axis with repeat distance c. (b) The intensity transform, or interference function, has a very small value except over a set of planes perpendicular to the Z axis spaced $1/c$ apart. The effective thickness is inversely proportional to the number N of points in the lattice. In the case of a two-dimensional point lattice in the yz plane each plane in the intensity transform (lower) is broken up into a series of rods parallel to the X axis (middle). With a three-dimensional point lattice each plane is broken up into an array of small spheres (top).

When N is not very large the interference function has an appreciable value for values of $Z \neq l/c$ but most of the intensity is concentrated between the planes

$$Z = (l/c) \pm (1/2Nc) \qquad (1.8)$$

Thus the effect of a limited lattice size is to "broaden" the layer lines so that they have a "breadth" approximately equal to the reciprocal of the length of the lattice (Fig. 1.3). This effect is of particular importance in the study of fibrous proteins as the breadth of a reflection provides information about the size of the diffracting object. Detailed discussions of this topic have been given by James (1954) and Wilson (1962).

If like groups of atoms are arranged on a regular two-dimensional lattice, for example, in the plane $x = 0$, the interference function has the form of a set of lines parallel to the X axis in reciprocal space (Fig. 1.4b, middle) and if N is not very large, the lines are replaced by rods with a cross-sectional shape which is reciprocal to the shape of the lattice. That is, the effective diameter of the rod will be large parallel to a direction in which the lattice is small and vice versa.

In a perfect crystal, like groups of atoms are arranged on a regular three-dimensional lattice, and the interference function has the form of a point lattice in reciprocal space. This is termed the *reciprocal lattice* (Fig. 1.4b, upper). The points in the reciprocal lattice correspond to the "reflections" given by the crystal and a full discussion of the properties of real and reciprocal lattices may be found in James (1954) and in Buerger (1960).

In essence the regular repetition of groups of atoms on a lattice leads to a sampling of the Fourier transform of the group of atoms in a manner which depends upon the nature of the lattice. Thus the shapes and positions of intensity maxima in reciprocal space provide information about the nature of the lattice, while their magnitudes provide information about the nature of the repeating unit.

f. *Fourier Synthesis*

The Fourier transform $F(\mathbf{D})$ of the distribution $\rho(\mathbf{d})$ has the property (see for example Jennison, 1961) that

$$\rho(\mathbf{d}) = \int F(\mathbf{D}) \exp(-2\pi i \mathbf{D} \cdot \mathbf{d}) \, dV \qquad (1.9)$$

where dV is an element of volume in reciprocal space and the integration extends over the whole of reciprocal space. Thus the electron density distribution in the specimen can, in principle, be determined from the diffraction pattern. In practice two factors prevent the direct evaluation of $\rho(\mathbf{d})$ using Eq. (1.9). First, the observable effects of diffraction relate to the intensity transform $I(\mathbf{D})$, which gives information about the magnitude of $F(\mathbf{D})$ but not its phase. Second, all the regions of reciprocal space that are accessible to observation by rotating the specimen are contained in a sphere of radius $2/\lambda$ termed the *limiting sphere*, whereas the integration in Eq. (1.9) extends over the whole of reciprocal space.

Various means can be found for dealing with the second problem, but the so-called phase problem is a barrier to the direct application of Fourier synthesis to the observed data. If $F(\mathbf{D})$ in Eq. (1.9) is replaced by $I(\mathbf{D})$, as defined in Eq. (1.4), a direct synthesis can be carried out to give the so-called Patterson function $P(\mathbf{d})$. This function can be shown to correspond to the convolution of $\rho(\mathbf{d})$ and $\rho(-\mathbf{d})$ and in the case of a crystal provides important information about interatomic vectors (Buerger, 1959; Ramachandran and Srinivasan, 1970). A special type of cylindrical Patterson function, applicable to fiber patterns, has been described by MacGillavry and Bruins (1948); in this case $\langle I(\mathbf{D})\rangle_\psi$ rather than $I(\mathbf{D})$ is used in the synthesis.

g. *Coordinate Systems*

In some instances it is convenient to use Cartesian coordinates (x, y, z) and (X, Y, Z) to describe position in real and reciprocal space, respectively (Fig. 1.1). In this system the expression for the Fourier transform of an object becomes

$$F(X, Y, Z) = \int \rho(x, y, z) \exp[2\pi i(Xx + Yy + Zz)] \, dv \qquad (1.10)$$

When dealing with crystals it is more convenient to choose axes in real and reciprocal space which are parallel to the primitive translations of the real and reciprocal lattices, respectively. Position in real and reciprocal space may then be specified respectively by coordinates x, y, z and h, k, l measured parallel to these axes and expressed in terms of the corresponding primitive translations a, b, c and a^*, b^*, c^* of the respective lattices. It can be shown (see, for example, James, 1954) that the Fourier transform has the same form as in Eq. (1.10) except that X, Y, and Z are replaced by the Miller indices h, k, and l, respectively.

For many problems involving helical structures it is more convenient to use cylindrical polar coordinates (r, ϕ, z) in real space and (R, ψ, Z) in reciprocal space (Fig. 1.5). The scalar product $\mathbf{D} \cdot \mathbf{d}$ in Eq. (1.2) can be evaluated by resolving $\mathbf{D}$ and $\mathbf{d}$ into components parallel and perpendicular to Oz giving

$$\mathbf{D} \cdot \mathbf{d} = Rr \cos(\phi - \psi) + Zz \qquad (1.11)$$

The expression for the Fourier transform therefore becomes

$$F(R, \psi, Z) = \int \rho(r, \phi, z) \exp\{2\pi i[Rr \cos(\phi - \psi) + Zz]\} \, dv \qquad (1.12)$$

where $dv = r \, dr \, d\phi \, dz$.

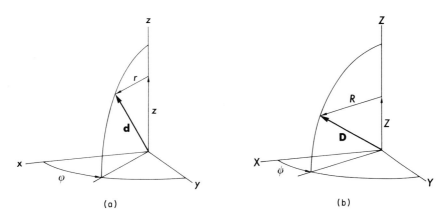

FIG. 1.5. Cylindrical polar coordinate systems used to describe position in (a) real and (b) reciprocal space.

II. DESCRIPTION OF HELICAL STRUCTURES

a. *Continuous Helices*

A helix can be defined as the locus of a point which satisfies the equations

$$r = \text{const} \tag{1.13a}$$

$$z = P\phi/2\pi \tag{1.13b}$$

where r, ϕ, z are cylindrical polar coordinates (Fig. 1.6). The coordinate system has been chosen so that the axis of the helix lies along Oz and the helix intercepts the plane $z = 0$ at $\phi = 0$. The corresponding description in Cartesian coordinates is

$$x = r \cos(2\pi z/P) \tag{1.14a}$$

$$y = r \sin(2\pi z/P) \tag{1.14b}$$

The constant r in Eqs. (1.13) and (1.14) represents the *radius* of the helix and P represents the *pitch* of the helix. The helix extends infinitely above and below the plane $z = 0$ and is periodic in z, as the portion of the helix contained between the planes $z = 0$ and $z = P$ is repeated precisely between $z = P$ and $2P$, $2P$ and $3P$, etc.

If the helix passes through the plane $z = 0$ at $\phi = \phi_0$ rather than at $\phi = 0$, Eq. (1.13) becomes

$$r = \text{const} \tag{1.15a}$$

$$z = P(\phi - \phi_0)/2\pi \tag{1.15b}$$

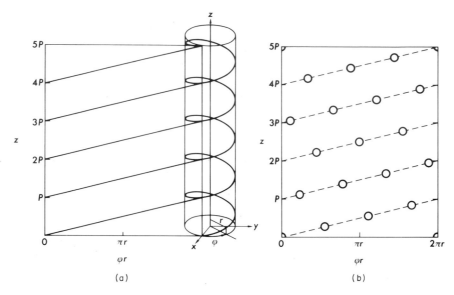

FIG. 1.6. (a) Illustration of the construction of a radial projection of a right-handed continuous helix. (b) Radial projection of a right-handed discontinuous helix with 18 units in five turns ($u = 18$, $v = 5$).

Continuous helices are enantiomorphic since reflection, for example, in the plane $z = 0$ in Fig. 1.6a produces a second continuous helix which cannot be superposed on the first by any combination of translations and rotations. The helix illustrated in Fig. 1.6a is conventionally referred to as being *right handed* and corresponds to a positive value of P in Eq. (1.15). Negative values of P give helices of opposite screw sense that are termed *left handed*.

The angle between the tangent to the helix at any point and the plane $z = 0$ is generally termed the *pitch angle* and its complement is sometimes referred to as the *tilt angle* (Lang, 1956a).

b. *Radial Projections*

In discussing helical conformations it is convenient to use a *radial projection* to illustrate the interrelationship of various features (Klug *et al.*, 1958). This is obtained by projecting the features along radial lines onto a cylindrical surface which is coaxial with the z axis. If this surface is imagined to be cut along a line parallel to the z axis corresponding to $\phi = 0$ and opened out flat, the result is as shown in Fig. 1.6a. The projection has the following properties:

(1) Axial displacements are preserved.

(2) Equal angular displacements in the structure give equal horizontal displacements in the projection.

(3) A continuous helix projects as a series of straight-line segments (Fig. 1.6a).

(4) With the convention used here, in which the outside of the opened-out cylindrical surface faces upward, lines of positive slope correspond to right-hand helices and lines of negative slope to left-hand helices.

(5) If the radius of the cylinder is chosen to be the same as that of a particular helix within the structure, both pitch angle and distance measured along the helix are preserved.

c. *Discontinuous Helices*

A discontinuous helix can be defined as an infinite set of points which lie on a continuous helix and are separated by a constant axial translation h (Fig. 1.6b). The points are related by a *screw axis* which may be specified in many equivalent ways. The simplest is probably in terms of the axial rise or *unit height* h and angular separation or *unit twist* $t = 2\pi h/P$ between axially consecutive equivalent points (IUPAC–IUB Commission on Biochemical Nomenclature, 1970). Another method, which is used in the development of helix diffraction theory, is to express h/P as a rational fraction u/v, where u and v are integers (Cochran *et al.*, 1952). The screw axis can then be specified by saying that u equivalent points occur in v complete turns. When P is negative, i.e., the helix is left handed, it is convenient to regard u as being positive and v negative.

The description of a discontinuous helix in terms of h, P, u, and v is not unique. For example, the right-handed discontinuous helix illustrated in Fig. 1.6b could equally well be described as $h = 1.5$ Å, $u = 18$, and

$$(1) \quad v = 5, \qquad P = 5.40 \text{ Å},$$

$$(2) \quad v = 23, \qquad P = 1.17 \text{ Å}, \qquad \text{etc.}$$

or

$$(3) \quad v = -13, \qquad P = -2.08 \text{ Å},$$

$$(4) \quad v = -31, \qquad P = -0.87 \text{ Å}, \qquad \text{etc.}$$

Cochran *et al.* (1952) used the term *primitive helix* to denote the choice leading to a minimum value for the magnitude of the unit twist $t = 2\pi v/u$. In the present example this would be description 1. Bear (1955b) applied the term primitive to describe 1 and 3 and used the term *genetic helix* for description 1. Klug *et al.* (1958) have used the term *basic helix* for description 1 and this will be used in the present treatment.

d. *Symmetry in Helical Structures*

Helical structures may contain more than one helical array of equivalent points and Klug *et al.* (1958) have classified the possible relationships between the arrays in enantiomorphic structures such as fibrous proteins. If all the points in all the arrays are equivalent, then the structure must belong to one of four *line groups*. The simplest, designated by **s**, involves only a screw axis, while the others involve in addition either a parallel $\mathcal{N}$-fold rotation (**sr**) or a perpendicular dyad axis (**s2**) or both (**sr2**). These are illustrated in Fig. 1.7. For structures belonging

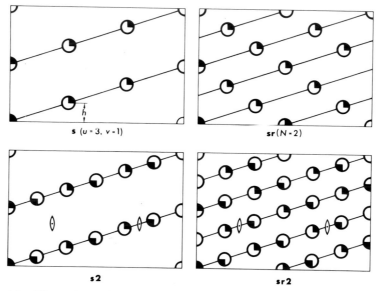

FIG. 1.7. Illustrations of line groups for enantiomorphic helical structures. Line group **s** has only a screw axis (not necessarily integral), **sr** has in addition a rotation axis parallel to the helix axis. Line group **s2** has, in addition to a screw axis, a horizontal dyad, while **sr2** contains both a rotation axis and a horizontal dyad.

to line group **s** there are *u* equivalent points in the repeat distance *c* and $\mathcal{N}u$, 2*u*, and 2$\mathcal{N}u$ in structures belonging to line groups **sr**, **s2**, and **sr2**, respectively.

e. *Coiled-Coils*

Crick (1953a) has given a detailed description of a conformation termed the *coiled-coil* in which a simple helix (the *minor helix*) is distorted in a regular manner so that its axis follows a helical path (the *major helix*). The parameters used by Crick to describe a continuous coiled-coil were the radius r_1 of the minor helix, the radius r_0 of the major helix, and the repeat distance c, in which it was supposed that the major helix made exactly N_0 turns while the minor helix made exactly N_1 turns in an internal frame of reference that rotates at the same rate as the major helix. If the convention is adhered to that N_0 and N_1 are positive for right-handed helices, the number of turns made by the minor helix in the repeat distance is $N_0 + N_1$. An example of a continuous coiled-coil is illustrated in Fig. 1.8b.

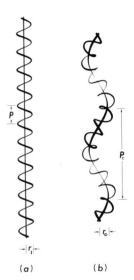

FIG. 1.8. (a) A continuous helix. (b) Coiled-coil obtained by a helical distortion of (a). P_c is the pitch and r_0 is the radius of the major helix.

(a)   (b)

When a discontinuous helix is distorted into a coiled-coil an additional parameter M is required to specify the number of points in the repeat distance c which may now may be a multiple of that of the corresponding continuous coiled-coil. The points of the original discontinuous helix are no longer equivalent, although groups of them will be equivalent if M and N_1 have a common factor. If C is the highest common factor of M and N_1, the discontinuous coiled-coil may be thought of as a simple discontinuous helix with M/C units, each containing C points, in N_0 turns. This greatly simplifies the computation of diffraction patterns.

Alternative methods of describing coiled-coils have been discussed by Lang (1956a), Ramachandran (1960), and Dickerson (1964).

III. Diffraction by Helical Structures

The Fourier transforms of helical structures have certain characteristic features and an understanding of these features is a prerequisite for the interpretation of diffraction patterns obtained from fibrous proteins. The theory of diffraction by helical structures was first given by Cochran *et al.* (1952) and considerably amplified by Klug *et al.* (1958).

a. *Continuous Helix*

The expression for the Fourier transform of a distribution of scattering material $\rho(r, \phi, z)$ was shown earlier (Section A.I.g) to be

$$F(R, \psi, Z) = \int \rho(r, \phi, z) \exp\{2\pi i[Rr \cos(\phi - \psi) + Zz]\} \, dv \qquad (1.16)$$

and in the case of an infinite, uniform, continuous helix we can replace $\rho(r, \phi, z) \, dv$ by a constant multiplied by $d\phi$ and obtain a line integral rather than a volume integral.

The helix is periodic in z, repeating exactly after a distance P, and so the transform will be zero everywhere except over the set of planes for which $Z = n/P$, where n is an integer. From Eq. (1.15) we have $z = P\phi/2\pi$ and so the product Zz in Eq. (1.16) becomes $n\phi/2\pi$. The normalized transform of the helix over the plane $Z = n/P$ is therefore

$$F(R, \psi, n/P) = (1/2\pi) \int_0^{2\pi} \exp[2\pi iRr \cos(\phi - \psi) + in\phi] \, d\phi \qquad (1.17)$$

Substituting $\theta = (\phi - \psi)$ and making use of the identity

$$2\pi i^n J_n(x) = \int_0^{2\pi} \exp(ix \cos \theta) \exp(in\theta) \, d\theta \qquad (1.18)$$

where $J_n(x)$ is a Bessel function of the first kind of order n, we obtain the result

$$F(R, \psi, n/P) = J_n(2\pi Rr) \exp[in(\psi + \tfrac{1}{2}\pi)] \qquad (1.19)$$

The transform is continuous over the layer planes and the magnitude is independent of ψ. The phase varies through n cycles between $\psi = 0$ and 2π. For $\psi = -\tfrac{1}{2}\pi$ the exponential term in Eq. (1.19) is unity and the nature of the transform for this value of ψ is illustrated in Figs. 1.9a and 1.9b. Since the magnitude is independent of ψ the normalized intensity transform is simply

$$I(R, n/P) = J_n^2(2\pi Rr) \qquad (1.20)$$

The function $J_n{}^2(2\pi Rr)$ is illustrated in Figs. 1.10a and 1.10b for values of $n = 0$–20. As n increases, the principal maximum occurs at progressively greater values of $2\pi Rr$, leading to the formation of a characteristic cross centered at the origin of the diffraction pattern.

These maxima are due to constructive interference from consecutive turns of the helix (Figs. 1.11a and 1.12a). For large values of n the principal maximum of $J_n(2\pi Rr)$ occurs at a value of $2\pi Rr = n + 0.8n^{1/3}$ (Jahnke and Emde, 1945) and so the angle δ in Fig. 1.12b approaches

$$\delta = \tan^{-1}\left[\frac{(n + 0.8n^{1/3})/2\pi r}{n/P}\right] = \tan^{-1}\left[\frac{P}{2\pi r}(1 + 0.8n^{-2/3})\right] \quad (1.21)$$

The dashed line shows the limiting value of $\delta = \tan^{-1}(P/2\pi r)$ as $n \to \infty$, which is parallel to the normal to the set of planes through consecutive turns of the helix (Fig. 1.11a).

b. *Discontinuous Helix*

The Fourier transform of a discontinuous helix can be derived by a conventional structure factor calculation (Tanaka and Naya, 1969) but the original derivation given by Cochran *et al.* (1952) is more instructive.

A discontinuous helix (Fig. 1.11c) may be thought of as the product of two functions, the first being zero everywhere except on a continuous helix of pitch P and radius r (Fig. 1.11a), the second being zero everywhere except on a set of planes perpendicular to Oz spaced h apart (Fig. 1.11b). The nonzero values of both functions are unity. The transform of the continuous helix is zero except on a set of planes with $Z = n/P$ (Fig. 1.12a) and the transform of the second function is simply a set of points spaced $1/h$ apart along the Z axis (Fig. 1.12b). The convolution of the two transforms reproduces the transform of the continuous helix at each of these points, giving the pattern shown diagrammatically in Fig. 1.12c. The transform of the discontinuous helix is therefore zero except over a set of planes given by

$$Z = (m/h) + (n/P) \quad (1.22)$$

where m and n are integers.

For a discontinuous helix with u points in v turns of the basic helix the structure repeats exactly after a distance $c = uh = vP$ and so the set of planes can be described in terms of the layer-line index $l\,(=cZ)$ by

$$l = um + vn \quad (1.23)$$

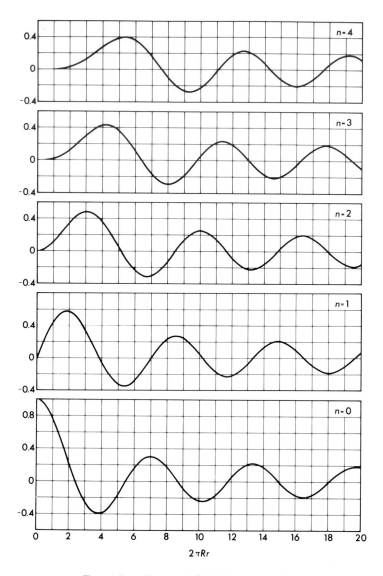

FIG. 1.9a.   Values of $J_n(2\pi Rr)$ for $n = 0$–4.

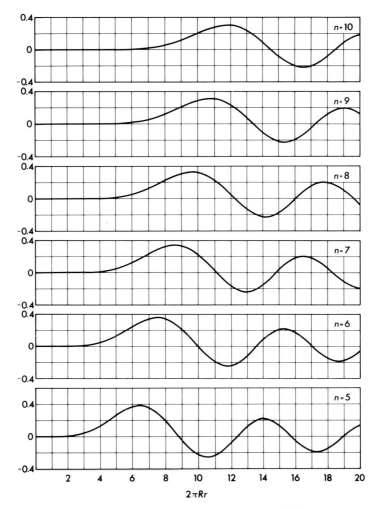

FIG. 1.9b.   Values of $J_n(2\pi Rr)$ for $n = 5$–10.

The transform of the discontinuous helix is obtained by summing the contributions from all the branches created by the convolution operation (Fig. 1.12). The normalized transform is therefore

$$F(R, \psi, l/c) = \sum_n J_n(2\pi Rr) \exp[in(\psi + \tfrac{1}{2}\pi)] \qquad (1.24)$$

where the summation extends over all values of n that satisfy Eq. (1.23). This expression is applicable to both right- and left-handed helices

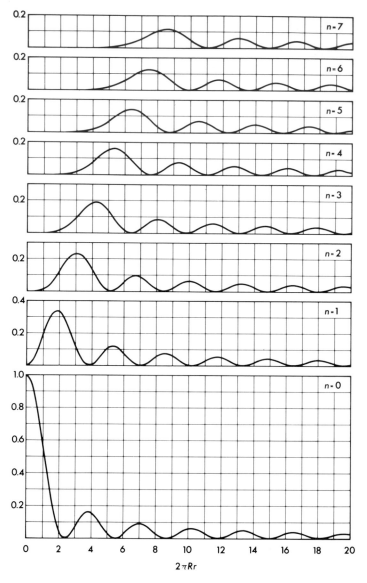

FIG. 1.10a.   Values of $J_n{}^2(2\pi Rr)$ for $n = 0$–7.

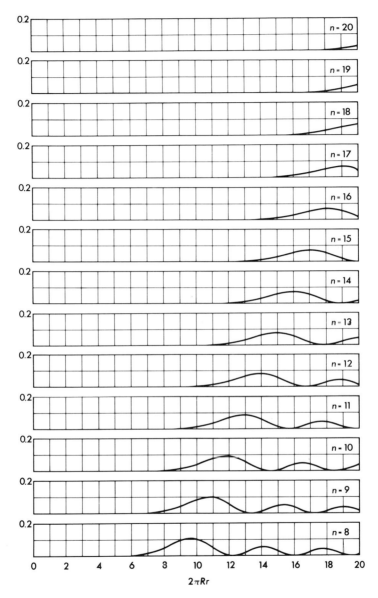

Fig. 1.10b.   Values of $J_n{}^2(2\pi Rr)$ for $n = 8$–20.

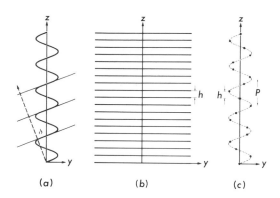

FIG. 1.11.   (a) Continuous helix; (b) set of planes perpendicular to the z axis spaced h apart; (c) discontinuous helix, indicated by dots, obtained by multiplying together the density functions for the continuous helix and the set of planes.

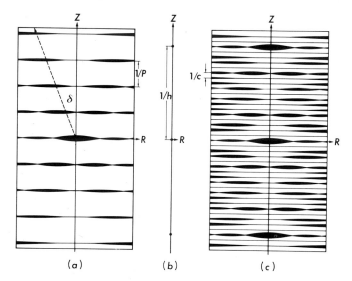

FIG. 1.12.   (a) Fourier transform of continuous helix shown in Fig. 1.11a; the broken line indicates the direction of the normal to the set of planes passing through successive turns of the helix; (b) transform of the set of planes shown in Fig. 1.11b; (c) convolution of the two transforms which gives the transform of the discontinuous helix shown in Fig. 1.11c.

provided the convention of taking v negative for left-handed helices is observed.

In the continuous helix the magnitude of the transform was independent of ψ but in general this will not be true for a discontinuous helix, since the various terms included in the summation in Eq. (1.24) for a particular layer line will oscillate at different rates with respect to ψ and the magnitude will be periodic with respect to this variable.

The cylindrically averaged intensity transform of the discontinuous helix, obtained by averaging FF^* over ψ, has the very simple form (Franklin and Klug, 1955)

$$\langle I(R, l/c)\rangle_\psi = \sum_n J_n^2(2\pi Rr) \tag{1.25}$$

where the summation extends over values of n that satisfy Eq. (1.23). The functions $J_n^2(2\pi Rr)$ are given in Figs. 1.10a and 1.10b and the intensity transforms for various discontinuous helices are illustrated diagrammatically in Fig. 1.13.

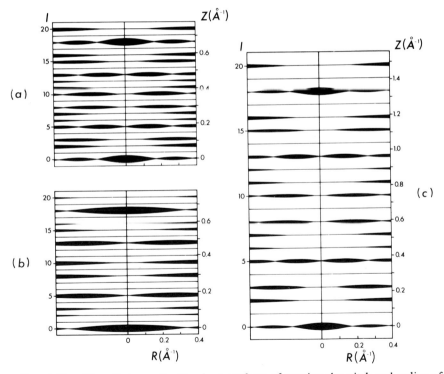

FIG. 1.13. The effects on the Fourier transform of varying the pitch and radius of a discontinuous helix with 18 units in five turns ($u = 18$, $v = 5$). (a) Transform of helix with pitch $P = 5.4$ Å and radius $r = 2.3$ Å; (b) transform of helix with pitch P and radius $r/2$; (c) transform of helix with pitch $P/2$ and radius r.

c. *Effect of Symmetry on the Transform*

If a set of $\mathcal{N}$ discontinuous helices with u points in v turns are related by an $\mathcal{N}$-fold rotation axis parallel to Oz (line groups **sr** or **sr2**), the period will be reduced to $c/\mathcal{N}$ if $\mathcal{N}$ happens to be a factor of u. In this case the transform is confined to layer lines (indexed on a repeat distance c) with indices that are multiples of $\mathcal{N}$.

This situation causes some difficulty in the diagnosis of helical parameters from the diffraction pattern since the characteristic cross centered at the origin (Fig. 1.13) is formed by Bessel functions of order 0, $\mathcal{N}$, $2\mathcal{N}$, $3\mathcal{N}$, etc., rather than 0, 1, 2, 3, etc., and attempts to deduce the radius of the helix from the positions of the maxima in a simple helix will lead to a result approximately $\mathcal{N}$ times too small.

In general $\mathcal{N}$ is not a factor of u and the normalized transform is

$$F(R, \psi, l/c) = (1/\mathcal{N}) \sum_{k=0}^{\mathcal{N}-1} \sum_n J_n(2\pi Rr) \exp\{in[(\psi + \tfrac{1}{2}\pi) - (2\pi k/\mathcal{N})]\} \quad (1.26)$$

The summations with respect to k yield terms of the form

$$[1 - \exp(-2\pi in)]/[1 - \exp(-2\pi in/\mathcal{N})] \quad (1.27)$$

which have the value zero unless n is a multiple of $\mathcal{N}$ and this imposes restrictions additional to that in Eq. (1.23) on Bessel functions that are to be included in the summation given in Eq. (1.24).

A perpendicular diad does not produce systematic absences in layer lines but affects the phase of the transform in a predictable manner (Klug *et al.*, 1958).

d. *Structure Factor Calculations*

In an actual structure the repeating unit is not a point but a group of atoms which is referred to as the asymmetric unit. Corresponding atoms in different units lie on helices having the same pitch P but different radii. In addition, the assumption, implicit in the foregoing treatment, that a point occurred at $(r, 0, 0)$ is no longer valid and Eq. (1.24) must be modified to include a summation over the atoms of a reference unit. The normalized transform of a helical set of points one of which occurs at (r, ϕ, z) is readily shown to be

$$F(R, \psi, l/c) = \sum_n J_n(2\pi Rr) \exp\{i[n(\psi + \tfrac{1}{2}\pi) - n\phi + (2\pi zl/c)]\} \quad (1.28)$$

(Cochran *et al.*, 1952) and when the atomic scattering factor f is introduced the expression for the structure factor per asymmetric unit for line group **s** becomes

$$F(R, \psi, l/c) = \sum_j \sum_n f_j J_n(2\pi R r_j) \exp\{i[n(\psi + \tfrac{1}{2}\pi) - n\phi_j + (2\pi l z_j/c)]\} \quad (1.29)$$

where the first summation extends over the atoms of one asymmetric unit and the second over values of n that satisfy Eq. (1.23). Klug *et al.* (1958) pointed out that Eq. (1.29) could be rewritten as

$$F(R, \psi, l/c) = \sum_n G_{n,l}(R) \exp[in(\psi + \tfrac{1}{2}\pi)] \quad (1.30)$$

where

$$G_{n,l} = \sum_j f_j J_n(2\pi R r_j) \exp\{i[-n\phi_j + (2\pi l z_j/c)]\} \quad (1.31)$$

is a complex number which is independent of ψ. The real and imaginary parts of $G_{n,l}$ are

$$A_{n,l} = \sum_j f_j J_n(2\pi R r_j) \cos[-n\phi_j + (2\pi l z_j/c)] \quad (1.32a)$$

$$B_{n,l} = \sum_j f_j J_n(2\pi R r_j) \sin[-n\phi_j + (2\pi l z_j/c)] \quad (1.32b)$$

The sine and cosine terms in Eq. (1.32) resemble the corresponding parts of the structure factor for the reflection with indices $-n, l$ from a two-dimensional rectangular lattice with a unit cell with sides $b = 2\pi r_j$ and c, and atoms at points defined by $r_j\phi_j$ and z_j (Henry and Lonsdale, 1952). This concept is of considerable value in the analysis of diffraction by fibrous proteins and provides a pictorial expression of the significance of the selection rule in Eq. (1.23). The $(-n, l)$ plot is in fact reciprocal to the reiterated radial projection (Fig. 1.14).

The cylindrically averaged intensity transform for an isolated helical molecule, obtained by averaging $FF^*(R, \psi, l/c)$ over ψ, is (Franklin and Klug, 1955)

$$\langle I(R, l/c) \rangle_\psi = \sum_n G_{n,l} G_{n,l}^* \quad (1.33)$$

where the summation extends over values of n that satisfy Eq. (1.23).

e. *Coiled-Coils*

Expressions for the Fourier transforms of both continuous and discontinuous coiled-coils were derived by Crick (1953a) and alternative derivations have been discussed by Lang (1956a) and Ramachandran (1960). Originally the theory was thought to be exact but Pardon (1967) pointed out that the expression obtained by Crick for the Fourier

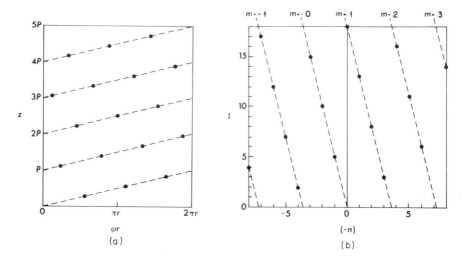

FIG. 1.14.   (a) Radial projection of a right-handed discontinuous helix with 18 points in five turns ($u = 18$, $v = 5$). (b) Lattice which is reciprocal to the reiterated radial projection. The positions of the points in this lattice correspond to the values of n and l that satisfy Eq. (1.23), that is, that satisfy $l = 18m + 5n$. The $(-n, l)$ plot can thus be used to predict the orders of the Bessel functions that contribute to each layer line.

transform of a continuous coiled-coil is incorrect, since the line density was not uniform. This also introduces errors into the original expression for the transform of the discontinuous coiled-coil. Pardon (1967) has given an approximate method for correcting Crick's formulas and has suggested that in some circumstances the correction may be small. Fortunately in the cases so far encountered in fibrous proteins M and N_1 (Section A.II.e) have had a common factor C and Fourier transforms can be calculated (Fraser *et al.*, 1964a) by generating the coordinates of the M/C atoms in the asymmetric unit from formulas given by Crick (1953a) and then using the expression given in Eq. (1.29) for a discontinuous helix. This gives an exact result.

The transform is zero except on layer lines for which

$$Z = l/c = (m/h) + (n/P_{\mathrm{c}}) \tag{1.34}$$

where m and n are integers, h is the unit height of the asymmetric set of M/C atoms, and P_{c} is the pitch of the major helix.

f. *Diffraction by Assemblies of Helices*

In some instances helices occur in regular three-dimensional crystals and then the Fourier transform is no longer continuous along the layer lines but is sampled at the reciprocal lattice points. The conventional approach via the concept of a unit cell and a space lattice (see, for example, Wilson, 1970; Woolfson, 1970; Jeffery, 1971) may be profitably employed in analyzing the diffraction from such an assembly.

In other cases the helices occur in bundles or fibrils and the transform, although still continuous along the layer lines, will be influenced by interhelical or *external* interference. An expression for the Fourier transform of the fibril can be derived (Fraser *et al.*, 1964b) from the treatment given by Cochran *et al.* (1952). If F_u is the normalized transform of a helix having some arbitrary rotation about Oz and displacement parallel to Oz, the transform of this helix can be calculated from Eq. (1.29). Each helix in the assembly can be generated from the reference helix by a rotation α_k around Oz, a displacement z_k parallel to Oz, and a displacement normal to Oz of magnitude r_k and azimuth ϕ_k. The normalized transform of the assembly can readily be shown to be

$$F(R, \psi, l/c) = (1/N) \sum_{k=1}^{N} F_u(R, \psi - \alpha_k, l/c)$$

$$\times \exp\{2\pi i[Rr_k \cos(\psi - \phi_k) + (lz_k/c)]\} \tag{1.35}$$

where the summation extends over the N helices in the assembly and does not, of course, include the imaginary reference helix.

If, as sometimes happens, only one Bessel function makes a significant contribution to each layer line, as for example with the α-helix, the intensity transform of the assembly averaged with respect to ψ can be shown to be

$$\langle I(R, l/c)\rangle_\psi = (1/N^2)[|\,G_{n,l}(R)|^2/|\,G_{0,0}(0)|^2]$$

$$\times \sum_j \sum_k J_0(2\pi Rr_{jk}) \cos[-n\alpha_{jk} + (2\pi lz_{jk}/c)] \tag{1.36}$$

where $G_{n,l}(R)$ is defined in Eq. (1.31), $\alpha_{jk} = \alpha_j - \alpha_k$, $z_{jk} = z_j - z_k$, and r_{jk} is the distance measured perpendicular to Oz between the axes of helices j and k.

To illustrate the application of Eq. (1.36) to the calculation of the effect of interhelical interference, consider the case of a pair of coaxial helices (i.e., $r_{jj} = r_{jk} = r_{kk} = r_{kj} = 0$) related by a twofold rotation axis parallel to Oz. We have $\alpha_j = 0$, $\alpha_k = \pi$ (i.e., $\alpha_{jj} = \alpha_{kk} = 0$, $\alpha_{jk} = -\pi$, $\alpha_{kj} = \pi$), and $z_j = z_k = 0$ (i.e., $z_{jj} = z_{kk} = z_{jk} = z_{kj} = 0$). If only one Bessel function contributes to each layer line, we obtain from Eq. (1.36)

$$\langle I(R, l/c)\rangle_\psi = \tfrac{1}{4}[|\,G_{n,l}(R)|^2/|\,G_{0,0}(0)|^2][1 + 1 + \cos(n\pi) + \cos(-n\pi)] \tag{1.37}$$

which is zero unless n is even. Hence interhelical interference in this
instance leads to elimination of diffraction on certain layer lines. This
confirms the result obtained earlier in Section A.III.c.

Helical chain molecules in fibrous proteins have polarity so that
a molecule rotated through π about a horizontal axis cannot be superimposed on an unrotated molecule by any combination of translations
and rotation about Oz. When assemblies contain mixtures of chains
with opposite polarities, often referred to as "up" and "down" chains,
Eq. (1.36) must be modified. If the translation and rotation of the
reference "up" chain are chosen suitably, the reference "down" chain
can be generated by rotating the "up" through π about the direction
$\phi = z = 0$. The normalized transform of the down chain is then

$$F_\text{d}(R, \psi, l/c) = F_\text{u}(R, -\psi, -l/c) \tag{1.38}$$

The expression for the Fourier transform of the assembly is obtained
by summing the contributions from the two types of chain using an
appropriate modification of Eq. (1.35).

The equatorial layer plane ($l = 0$) is frequently used to obtain information about the lateral arrangement of the helices in an assembly. If
$F_\text{u}(R, \psi)$ is the normalized Fourier transform of the helix for $l = 0$, the
normalized transform of the assembly becomes, from Eq. (1.35),

$$F(R, \psi) = (1/N) \sum_{k=1}^{N} F_\text{u}(R, \psi - \alpha_k) \exp[2\pi i R r_k \cos(\psi - \phi_k)] \tag{1.39}$$

and the normalized intensity transform is given by

$$I(R, \psi) = (1/N^2) \sum_{j=1}^{N} \sum_{k=1}^{N} F_\text{u}(R, \psi - \alpha_j) F_\text{u}{}^*(R, \psi - \alpha_k)$$

$$\times \exp\{2\pi i R[r_j \cos(\psi - \phi_j) - r_k \cos(\psi - \phi_k)]\} \tag{1.40}$$

If all values of α_j and α_k are equally probable, Eq. (1.40) simplifies to

$$I(R, \psi) = (1/N) F_\text{u}(R, \psi) F_\text{u}{}^*(R, \psi) + (1/N^2) \sum_{j} \sum_{k \neq j} \langle F_\text{u}(R) \rangle_\psi \langle F_\text{u}{}^*(R) \rangle_\psi$$

$$\times \exp[2\pi i R r_{jk} \cos(\psi - \phi_{jk})] \tag{1.41}$$

where r_{jk} is, as before, the distance measured perpendicular to Oz
between the axes of the helices j and k and ϕ_{jk} defines the direction of

the vector $\mathbf{r}_j - \mathbf{r}_k$. The cylindrically averaged intensity transform for $l = 0$ becomes (Tyson and Woods, 1964)

$$\langle I(R) \rangle_\psi = (1/N)\langle F_u(R)\,F_u^*(R)\rangle_\psi$$
$$+ (1/N^2)\langle F_u(R)\rangle_\psi \,\langle F_u^*(R)\rangle_\psi \sum_j \sum_{k \neq j} J_0(2\pi R r_{jk}) \qquad (1.42)$$

This can be recast (Vainshtein, 1966) to give

$$\langle I(R) \rangle_\psi = (1/N^2)\langle F_u(R)\,F_u^*(R)\rangle_\psi \sum_{j=1}^{N} \sum_{k=1}^{N} J_0(2\pi R r_{jk})$$
$$+ (1/N)[\langle F_u(R)\,F_u^*(R)\rangle_\psi - \langle F_u(R)\rangle_\psi \,\langle F_u^*(R)\rangle_\psi] \qquad (1.43)$$

The first term is the normalized product of the cylindrically averaged intensity transform of an individual helix and a term which corresponds to the cylindrically averaged intensity transform of the point lattice formed by the intersections of the axes of the helices with the plane $z = 0$. The numerator of the second term is the difference between the average intensity and the square of the average amplitude for an individual helix. From Eq. (1.30) and Eq. (1.33) this is simply

$$\langle F_u(R)\,F_u^*(R)\rangle_\psi - \langle F_u(R)\rangle_\psi \,\langle F_u^*(R)\rangle_\psi = \sum_{n \neq 0} G_{n,0}(R)\,G_{n,0}^*(R) \qquad (1.44)$$

where $n = \pm u, \pm 2u, \pm 3u$, etc.; u being the number of units in the repeat distance c of the helix.

If the Fourier transform of an individual helix is dominated on the equator by Bessel functions with $n = 0$, the second term in Eq. (1.43) vanishes and the expression reduces to that derived by Oster and Riley (1952) for the cylindrically averaged intensity transform for assemblies of cylindrically symmetric scattering units. Tyson and Woods (1964) and Vainshtein (1966) have pointed out that the second term in Eq. (1.43) has been overlooked in a number of instances (Burge, 1959, 1961, 1963; Wilson, 1963).

IV. Effects of Disorder

In formulating the theory of diffraction no account has been taken of the various types of imperfections that are present in naturally occurring aggregates of fibrous proteins. These have a profound effect on the intensity transform and must be taken into consideration when interpreting diffraction patterns. An exhaustive treatment of the various types of disorder encountered in assemblies of chain molecules has been given by Vainshtein (1966).

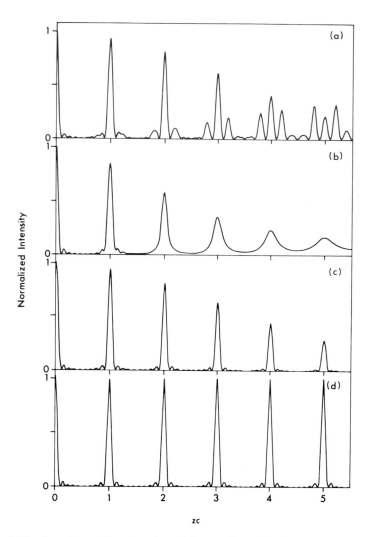

Normalized Intensity

zc

FIG. 1.15. Interference functions for a finite one-dimensional point lattice along the z axis with a repeat distance c and ten points. In (a) the lattice is subjected to a periodic distortion parallel to the z axis of period $5c$ and amplitude $0.05c$; in (b) the lattice is paracrystalline with random cumulative displacements parallel to the z axis with a root mean square amplitude of $0.05c$; in (c) the lattice is subject to random displacement parallel to the z axis about the lattice points with a root mean square amplitude of $0.05c$; (d) shows the interference function for the undistorted lattice.

a. *Random Displacements from Lattice*

When the atoms in a periodic structure are subject to random displacements from their idealized positions the value of $I(\mathbf{D})$ is decreased by an amount which increases with D (Fig. 1.15c) and a diffuse distribution without sharp maxima is superimposed on the pattern. If the directions of the displacements are random and the mean square amplitudes for each type of atom in the repeating unit are equal, the intensity transform becomes

$$I(\mathbf{D}) = F(\mathbf{D})F^*(\mathbf{D}) \exp(-\tfrac{1}{2}BD^2) \tag{1.45}$$

where B is related to the mean square of the displacement δs parallel to $\mathbf{D}$ by

$$B = 8\pi^2 \langle \delta s^2 \rangle \tag{1.46}$$

Values of B in the range 5–15 Å² are commonly encountered in fibrous structures corresponding to values of $\langle \delta s^2 \rangle^{1/2}$ from 0.25 to 0.44 Å.

The parameter B in Eq. (1.45) is usually termed the *temperature factor* even though it includes displacements due to other factors. In many instances the assumption that the displacements are isotropic is not justified and the value of B will be different for different directions in reciprocal space. This situation occurs, for example, in the pleated-sheet structures (Fraser *et al.*, 1969a). The assumption that all atoms have the same mean square displacement may also be inadequate. For example in homopolypeptides the main-chain atoms are usually highly ordered while the side chains may be disordered. In this case individual atoms or groups of atoms must be allotted individual temperature factors and this is conveniently achieved by combining the temperature factor with the atomic scattering factor to give

$$f(D) = f_0(D) \exp(-\tfrac{1}{4}BD^2) \tag{1.47}$$

where f_0 is the scattering factor for a stationary atom.

b. *Random Distortion of Lattice*

In the previous section it was assumed that the atomic displacements occurred around fixed lattice positions and this leads to the prediction that the intensities of the interference maxima will be attenuated with increasing D but the breadth of the maxima will not be affected (Fig. 1.15c). In some observed diffraction patterns it is evident that the breadth of the interference maxima increases with increasing D and this behavior is characteristic of the situation where the space lattice is imperfect and subject to random distortions between adjacent cells. The effects are cumulative so that the root mean square displacement of an atom from its idealized position increases with distance from the origin.

This type of lattice is sometimes described as "ideally paracrystalline" and the effect on the diffraction pattern has been investigated theoretically in considerable detail (Hosemann, 1951; Hosemann and Bagchi, 1962; Vainshtein, 1966). The effect on the interference function for a one-dimensional lattice is illustrated in Fig. 1.15b (Zernike and Prins, 1927).

c. Regular Distortion of Lattice

Regular, or periodic, distortion of a lattice produces additional maxima in the interference function and characteristically each maximum in the interference function is surrounded by a set of equispaced satellite or "ghost" maxima (James, 1954; Wooster, 1962). The separation of the satellite maxima is equal to the reciprocal of the distortion period. The discontinuous helix, discussed earlier, is a special case of a periodically distorted one-dimensional lattice. In this case the distortion period is equal to the pitch of the helix.

Helices themselves are frequently distorted in a periodic manner and the coiled-coil is in fact a special type of distorted helix. The effect of a periodic distortion on the Fourier transform of a discontinuous helix has been considered by Crick (1953a), Lang (1956a), Johnson (1959), and Holmes (Caspar and Holmes, 1969). In the special case of a discontinuous helix with a periodic axial distortion the Fourier transform is zero except over planes that satisfy the condition

$$Z = (m/h) + (n/P) + (s/P_d) \qquad (1.48)$$

where m, n, and s are integers and P_d is the period of the distortion. When this is compared with Eq. (1.22) it is seen that each layer line of the undistorted helix $(s = 0)$ is surrounded by a set of "ghost" layer lines corresponding to $s = \pm 1$, ± 2, etc. (Fig. 1.15a). The expression for the Fourier transform of a discontinuous helix in which the points are subject to sinusoidal axial displacements is

$$F(R, \psi, Z) = \sum_n \sum_s J_n(2\pi Rr)\{\exp[in(\psi + \tfrac{1}{2}\pi)]\} \, i^s J_s(2\pi Z \, \delta z) \qquad (1.49)$$

where δz is the maximum amplitude and the summations extend over values of n and s that satisfy Eq. (1.48). This may be compared with the transform for an undistorted discontinuous helix given in Eq. (1.24). If δz is small, then the ghost layer lines will be very weak for small Z since the value of $J_s(2\pi Z \, \delta z)$ is very small for small arguments if $s \neq 0$.

As Z increases, the intensity on the ghost layer lines will increase and the intensity on the original layer line will be reduced (Fig. 1.15a).

The expression in Eq. (1.49) is applicable to a discontinuous helix with a point at $(r, 0, 0)$ and a node in the distortion function at $z = 0$. The expression for the general case is given in Caspar and Holmes (1969). An important difference between the idealized case of diffraction by a helix of points and that by actual structures is that destructive interference between the atoms of the repeating unit leads to values of $\langle I(R, Z)\rangle_\psi$ which are, in general, very much less than $I(0, 0)$. Thus although $J_s(2\pi Z\,\delta z)$, $s \neq 0$, may be very small for small Z and δz, the origin set of ghosts with $m = n = 0$ and $s = \pm 1$, ± 2, etc. may be visible.

d. Disorder in Assemblies of Helices

In assemblies of helices the intrahelical bonding is generally stronger and more specific than the interhelical bonding. As a result a special type of disorder is encountered in which the helices are subject to random axial displacements or rotations without significant disturbance of the helical conformation. The effects of such disorder on the intensity transform of the assembly have been discussed by Klug and Franklin (1958), Clark and Muus (1962), Chiba et al (1966), Vainshtein (1966), and Tanaka and Naya (1969).

In the case of regular three-dimensional crystals containing helices the layer planes in the intensity transform of the helix are sampled by the interference function of the crystal lattice and the diffraction pattern consists of discrete reflections. Disorder involving the rotation or displacements of entire helices leads to the appearance of a continuous distribution of intensity over some or all of the layer planes so that "layer-line streaks" appear in the diffraction pattern.

The effects of various idealized types of disorder on the cylindrically averaged intensity transform of the crystal have been investigated by Clark and Muus (1962) and Tanaka and Naya (1969). The effect on the discrete reflections $\langle I_C(R, l/c)\rangle_\psi$ is as follows; expressions for the continuous distribution over the layer planes are given by Tanaka and Naya (1969).

(i) *Small Angular Displacements.* If the helices are supposed to be subject to small random displacements $\delta\phi$ about their idealized position, then

$$\langle I(R, l/c)\rangle_\psi = \langle I_C(R, l/c)\rangle_\psi \exp[-n^2\langle(\delta\phi)^2\rangle]   \qquad (1.50)$$

where it has been assumed that only one Bessel function of order n makes an appreciable contribution to reflections lying on the circle

$(R, l/c)$. Thus discrete reflections that are dominated by zeroth-order Bessel functions will not be affected while other discrete reflections will be reduced by an amount which depends both on the square of the order of the Bessel function and on the mean square angular displacement.

(ii) *Random Angular Displacements.* If $\delta\phi$ is uniformly distributed in the range 0–2π, the intensity transform becomes

$$\langle I(R, l/c)\rangle_\psi = \begin{cases} \langle I_C(R, l/c)\rangle_\psi & \text{if } n = 0 & (1.51a) \\ 0 & \text{if no } n = 0 & (1.51b) \end{cases}$$

where it is assumed in Eq. (1.51a) that the reflections on the circle $(R, l/c)$ are dominated by a single Bessel function.

(iii) *Small Axial Displacements.* If the helices are subject to small random axial displacements δz without rotation, the intensity transform becomes

$$\langle I(R, l/c)\rangle_\psi = \langle I_C(R, l/c)\rangle_\psi \exp[-(2\pi l/c)^2 \langle(\delta z)^2\rangle] \qquad (1.52)$$

The intensities of the discrete reflections on the equator $(l = 0)$ will not be affected but on other layer planes the intensities will be reduced by a factor which depends both on the Z coordinate of the layer plane and on the mean square displacement.

(iv) *Random Axial Displacement.* If the helices are subject to random axial translations which are uniformly distributed in the range $0 < \delta z < c$, the intensity transform becomes

$$\langle I(R, l/c)\rangle_\psi = \begin{cases} \langle I_C(R, l/c)\rangle_\psi, & l = 0 & (1.53a) \\ 0, & l \neq 0 & (1.53b) \end{cases}$$

Thus all discrete reflections disappear except on the equator.

(v) *Small Screw Displacements.* If the helices are subject to small screw displacements $(\delta\phi, \delta z)$ of pitch P_s, then

$$\delta\phi = (2\pi/P_s)\,\delta z \qquad (1.54)$$

and the intensity transform for the discrete reflections becomes

$$\langle I(R, l/c)\rangle_\psi = \langle I_C(R, l/c)\rangle_\psi \exp\{-[(lP_s/c) - n]^2 \langle(\delta\phi)^2\rangle\} \qquad (1.55)$$

where it is again assumed that the intensity on the circle $(R, l/c)$ is dominated by contributions from Bessel functions of order n. If $n = 0$, this reduces to Eq. (1.52) and the effects are similar to those caused by small axial translations. For equatorial reflections ($l = 0$) the expression reduces to Eq. (1.50) and the effects are similar to those caused by small rotations.

On other layer planes the intensity of the discrete reflections will be reduced except when

$$n = lP_s/c \tag{1.56}$$

This property provides a means of diagnosing screw disorder (Klug and Franklin, 1958).

e. *Disorders in Chain Direction*

Helical polypeptide chains inevitably have a polarity because the sequence of atoms in the main chain runs $-NH-CHR-CO-$ in one direction and $-CO-CHR-NH-$ in the other. The effect on interhelical interactions of reversing the chain direction may, however, be small and a common type of disorder is that in which the axes of the helices occupy regular positions in an assembly or on a lattice but chain direction is random.

The effects on the intensity transform of a crystal are similar to those discussed in the previous section in that the discrete reflections are overlaid by continuous scattering which extends over the layer planes (Elliott and Malcolm, 1958; Vainshtein, 1966; Takeda *et al.*, 1970). If all the "up" chains have the same orientation and z translation relative to the lattice and similarly for the "down" chains, the intensity transform for the crystal will be

$$I(R, \psi, l/c) = [\tfrac{1}{2}(F_u + F_d)][\tfrac{1}{2}(F_u^* + F_d^*)] I_L(R, \psi, l/c)$$
$$+ N[\tfrac{1}{2}(F_u - F_d)][\tfrac{1}{2}(F_u^* - F_d^*)] \tag{1.57}$$

where F_u and F_d are, respectively, the Fourier transforms of an up chain and a down chain at the origin, N is the number of unit cells in the crystal, and I_L is the interference function (Section A.I.e). The first term corresponds to the discrete reflections from a crystal composed of chains which have a density function equal to the mean of that for up and for down chains. The second term corresponds to a continuous distribution of intensity over the layer planes and depends upon the difference between the density functions for up and for down chains.

In some circumstances the chain direction may not be entirely random, for example, in the pleated-sheet structures (Pauling and Corey, 1951g, 1953a) the probability p that adjacent chains have opposite sense may be very different from the probability $1 - p$ that they have the same sense. The case $p = 0$ leads to the parallel-chain pleated sheet and the case $p = 1$ to the antiparallel-chain pleated sheet (Chapter 10). The nature of the intensity transform for intermediate values of p has been investigated by Fraser *et al.* (1969a).

f. *Disorders in Sheet Packing*

In the antiparallel-chain pleated-sheet structure (Chapter 10) there is a regular alternation in chain direction in the sheet so that the repeat distance perpendicular to the chain axes in the plane of the sheet is double the interchain distance. When a number of sheets pack together to form a crystallite (Fig. 1.16) the difference between energies of packing

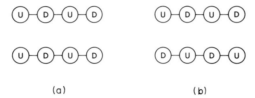

(a)   (b)

FIG. 1.16.   The *b*-axis projection of two types of packing for antiparallel-chain pleated sheets. U symbolizes the projection of an "up" chain and D that of a "down" chain.

arrangements (a) and (b) may be very small. This leads to disorder in the packing and if the number of sheets is N and the probabilities of arrangements (a) and (b) are equal, the normalized Fourier transform of the crystal at the reciprocal lattice point (h, k, l) of the idealized structure becomes

$$F_c(hkl) = (1/N) \sum_{j=1}^{N} F_s(hkl) \exp\{2\pi i[\mu_j(\tfrac{1}{2}ha) + (j-1)c]\} \qquad (1.58)$$

where F_s is the normalized Fourier transform of a sheet and μ_j may be zero or one with equal probability. If N is large, $F_c(hkl)$ becomes vanishingly small unless h is even and the transform corresponds to that which would be obtained from a crystal containing pleated sheets with chains having a density function equal to the mean of those for "up" and for "down" chains (Arnott *et al.*, 1967). Reflections with even h are not affected by the disorder but a very weak, continuous distribution of intensity will appear along lines parallel to the c^* axis passing through the points $(hk0)$.

If N is small, the continuous distribution becomes more important and the intensity at the reciprocal lattice points with odd h will no longer be negligible. It can be shown that the intensity at these points is proportional to $1/N$ (Fraser *et al.*, 1969a).

g. *Disorientation*

In most specimens with fiber-type orientation the axes of the assemblies or crystallites are not strictly parallel to the fiber axis. The distribution of directions is conveniently described by an orientation density function $C(\omega)$ such that the fraction of chains with directions inclined to the fiber axis at angles between ω and $\omega + d\omega$ is

$$2\pi C(\omega) \sin \omega \, d\omega \qquad (1.59)$$

Various idealized forms for $C(\omega)$ are described in Chapter 5, Section B.II.d.

The effect of the imperfect orientation on the observed diffraction pattern can be found by forming the convolution of the intensity transform for a single assembly or crystallite with the function $C(\omega)$. Discrete reflections appear as arcs of constant D value, while continuous distributions on layer planes appear as streaks with characteristic shapes (James, 1954; Vainshtein, 1966). Methods of correcting the observed data for the effects of disorientation are discussed in Section C.II.

B. EXPERIMENTAL METHODS

In a few instances the X-ray diffraction patterns obtained from fibrous proteins are sufficiently well developed for conventional single-crystal methods to be used and these are fully described in standard texts (Klug and Alexander, 1954; Henry and Lonsdale, 1952; Kasper and Lonsdale, 1959; MacGillavry and Rieck, 1962; Jeffery, 1971). More generally the recording and processing of the patterns are complicated by the presence of fine detail, which is difficult to resolve, and by various types of imperfections in orientation and crystallinity. Recent progress in dealing with these complications is not well documented and will therefore be discussed in some detail. A useful introductory account has been given by Holmes and Blow (1965) and a comprehensive guide to the literature up to 1955 is available (Peiser *et al.*, 1955). Very little use has been made of diffractometer techniques and in the present account attention will be restricted to photographic methods of recording.

In discussing X-ray diffraction it is customary to use the terms low-, medium-, and high-angle to denote different regions of the pattern. These regions are not precisely defined and in the present context the term low-angle will be used to denote the range of scattering angles, using Cu K_α radiation, that correspond to spacings greater than about 20 Å. Similarly medium-angle will be used to denote the range corresponding to spacings of about 5–50 Å, and the term high-angle to denote the range corresponding to spacings less than about 10 Å.

I. X-RAY GENERATORS

a. *Choice of Wavelength*

In Section A.I.f it was noted that the regions of reciprocal space that are accessible to observation are contained within a sphere of radius $2/\lambda$, where λ is the wavelength of the radiation used to produce the diffraction pattern. Many fibrous proteins yield information, particularly along the Z axis, to distances around 1 Å^{-1} and the maximum wavelength which may be used to record such information is about 2 Å. In practice it is inconvenient to work with scattering angles 2θ near 180° and this consideration sets a somewhat lower value for the maximum. The copper K_α lines ($\bar{\lambda} = 1.5418$ Å) or molybdenum K_α lines ($\bar{\lambda} = 0.7107$ Å) are normally used for such "high-angle" studies. Copper is a particularly suitable anode material because of its excellent thermal conductivity, and Cu K_α radiation is widely used for high-, medium-, and low-angle studies. The scattering angle 2θ and hence the physical size of the diffraction pattern may be increased by using longer wavelengths and the chromium K_α lines ($\bar{\lambda} = 2.2909$ Å) have been used for this purpose. The chief disadvantages of using wavelengths longer than about 2 Å are lower source brightness, increased specimen absorption, and increased air scattering.

If specimens contain heavy atoms, these may fluoresce and emit soft X rays. Fluorescence is particularly troublesome when the atomic number of the target material is two to four units greater than the atomic number of any of the heavy atoms in the specimen. The problem is usually overcome by using an alternative target material (Klug and Alexander, 1954).

b. *Sealed-Off Tubes*

Permanently evacuated X-ray tubes with focal dimensions about 10×0.4 mm are available commercially and these are convenient for use with pinhole cameras for medium- and high-angle studies. Tubes of this type were also used in early measurements of low-angle patterns but the low specific brightness led to inconveniently long exposure times.

c. *Demountable Tubes*

In order to obtain the best results with focusing cameras, small sources of very high specific brightness are required but these can only be obtained at the expense of rapid deterioration of the target surface. Various types of demountable tubes have been designed in which the filament and target can be replaced at regular intervals (Ehrenberg and Spear, 1951). In the Hilger Y33 microfocus generator (Rank Precision Industries, London) the tube may be operated at 45 kV and 0.4 mA tube current with a copper anode and a 0.04-mm-diameter spot focus. At the usual take off angle of 6° this gives a projected dimension of around 0.004 × 0.04 mm. In an alternative mode the tube is operated at 55 kV and 2 mA with a line focus of about 1.0 × 0.1 mm, giving a projected area of around 0.1 × 0.1 mm. Similar microfocus generators are produced by the Japan Electron Optics Laboratory, Tokyo (JMX series).

Most fibrous protein structures are weak diffractors and the time required to record a high-resolution, low-angle pattern using the type of generator just described may be several weeks in unfavorable cases. An appreciable reduction in exposure time can be effected by the use of a rotating anode generator (Huxley and Brown, 1967). In the GX3 generator manufactured by Elliott Automation Radar Systems Ltd., Borehamwood, England a focal area of about 1.0 × 0.1 mm may be loaded with about five times the power of a stationary anode (Lowy and Vibert, 1969). Higher loadings may be obtained by increasing the speed of rotation. Rotating anode tubes are considerably more complex and difficult to maintain than stationary anode tubes.

d. *Monochromatization*

The radiation emitted by an X-ray tube contains spectral lines characteristic of the target material together with a continuum of "white" X radiation. The strongest spectral lines are the K_α group and these can be isolated by using filters which selectively absorb the K_β lines and the white radiation (Roberts and Parrish, 1962). Absorption filters are adequate for most high-angle studies but a higher degree of monochromatization is desirable with hydrated specimens and for low-angle studies. This can be achieved either by reflecting the radiation from a polished surface close to the critical angle for K_α radiation or by collecting a diffracted beam from a suitable crystal (Bragg reflection).

Descriptions of various methods of producing monochromatic radiation have been given by Brindley (1955), Roberts and Parrish (1962), Herbstein *et al.* (1967), and Witz (1969). The latter review is particularly relevant to the experimental conditions encountered in recording diffraction patterns from fibrous proteins.

II. X-Ray Cameras

An X-ray camera is a device used to record the diffraction pattern produced when a specimen is irradiated with an X-ray beam. The design of a camera involves a compromise among angular resolution, which must be sufficient to reproduce all the detail faithfully, magnification, which must be sufficient to ensure accuracy in measurements of the positions and intensities of diffracted beams, and exposure time, which must be practicable.

a. *Collimating Systems*

The simplest arrangement is the "pinhole" camera, in which a narrow beam of X rays is defined by a pair of circular apertures and a third, coaxial aperture is placed between the second defining aperture and the specimen to limit the angular spread of the parasitic scatter from the defining apertures (Fig. 1.17a). Criteria for achieving the best compromise among speed, resolving power, and the minimum scattering angle at which reflections can be recorded have been discussed by Bolduan and Bear (1949), Huxley (1953), and Gerasimov (1970).

Pinhole collimation is useful for medium- and high-angle studies at moderate resolution and, although high resolution may be achieved by reducing the dimensions of the defining apertures, the exposure times very soon become unreasonably long.

Cylindrical cameras with pinhole collimation have been used widely in high-angle studies and Langridge *et al.* (1960) have given details of a flat-film camera used for medium- and high-angle studies.

Lead glass capillaries may be used as collimators (Hirsch, 1955) and provide beams of higher intensity than a pinhole collimator of the same dimensions due to the reflection that occurs at the walls of the capillary when the angle of incidence exceeds the critical angle for total reflection. Cameras of this type are useful for medium- and high-angle studies but the divergence of the reflected beams renders them unsuitable for use at low angles.

In an effort to overcome the long exposure times which result when pinhole cameras are used for low-angle studies, slit collimation has sometimes been used (Bear, 1944a; Huxley, 1953; Riley, 1955; Kratky, 1967). The disadvantage of this procedure is that detail parallel to the length of the slit is smeared out and the resulting patterns are unsuitable for intensity measurements.

b. *Focusing Cameras*

Although any desired degree of angular resolution can be achieved with pinhole collimation, the intensity transmitted, even using micro-

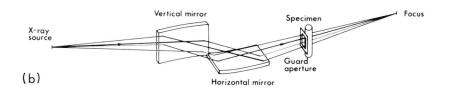

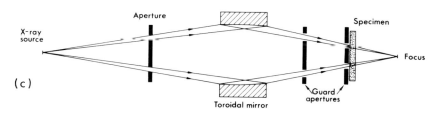

FIG. 1.17. Optical arrangements used to obtain X-ray diffraction patterns. (a) Pinhole collimation (Bolduan and Bear, 1949). (b) Focusing mirror system using elastically deformed optical flats (Franks, 1955, 1958). A similar arrangement is used with curved-crystal monochromators except that the reflectors are oriented so that the central ray lies in a principal plane of the surface. (c) Focusing mirror system using a toroidal reflector (Elliott, 1965) and annular apertures. In each case the angular widths of the beams have been exaggerated for clarity.

focus X-ray generators, is inconveniently low for the study of periodicities greater than about 50 Å. Many of the problems encountered with pinhole collimators can be overcome by using focusing cameras, which collect a divergent beam of radiation from the source and concentrate it into a fine spot or line either by total reflection from a curved mirror or by Bragg reflection from a curved crystal. The relative merits of these two systems have been discussed by Witz (1969).

(i) *Cylindrical Mirrors.* In the arrangement described by Franks (1955, 1958) two rectangular optical flats mounted at right angles are positioned so that total reflection close to the critical angle for the K_α lines occurs at both surfaces (Fig. 1.17b). If the flats are then elastically deformed into appropriate concave cylindrical shapes, the beam is brought to a focus in the plane of the film. When used in conjunction with a microfocus X-ray source about 0.04 mm in diameter this type of camera provides the most generally useful experimental arrangement for low-angle studies. The projected area of the source is about 0.004×0.04 mm and the image formed by the camera is slightly bigger.

The focal surface is curved but the depth of focus is considerable due to the low angular aperture, and a flat film is usually adequate. The energy transmitted is limited by the critical angle for the type of glass used, being approximately 0.22° for Pyrex. This angle can be increased to a value of around 0.5° by coating the glass with a thin, evaporated gold film but the reflectivity is substantially lower and the net gain in intensity when the angle of incidence is adjusted for maximum reflected energy is only about 40% per mirror. In addition there is an undesirable increase in parasitic scatter due to the granularity of the evaporated gold. A more satisfactory method is to use a glass with a very high content of lead oxide, for example, a "double extra dense flint glass" gives a comparable increase in reflected energy without any increase in parasitic scatter.

If the mirrors are adjusted so that reflection occurs close to the critical angle for the K_α lines, the K_β lines and the white radiation are almost completely eliminated from the focused beam. If adjusted for maximum energy, a small amount of K_β radiation will be present but this can readily be eliminated by using a suitable filter. The off-axis aberrations are severe for curved mirrors used at grazing incidence (Ehrenberg, 1949; Cosslett and Nixon, 1960) and attempts to use longer mirrors or larger sources or to depart appreciably from the symmetric configuration (Fig. 1.17b) produce undesirable spreading of the image.

If the guard slits (Fig. 1.17b) are well made and carefully adjusted, the parasitic scatter can be reduced to a low level and this is a particularly attractive feature when working with weak diffractors. Until recently most of the mirror cameras in use were custom built but commercially produced units are now available (Elliott Automation Radar Systems Ltd., Borehamwood, England).

(ii) *Toroidal Mirrors.* Elliott (1965) has described a camera suitable for use with Cu K_α radiation in which focusing is achieved by reflection from a gold-plated toroidal surface (Fig. 1.17c). The angular aperture of the reflector is much higher than that of the Franks camera and the intensity is correspondingly greater. The optical qualities of the toroidal reflector are poor and aberrations are severe even at unit magnification. Little reduction in the area of the focus is obtained by using projected source areas less than 0.1×0.1 mm and this is the size generally employed. A toroidal mirror camera based on Elliott's design is available commercially (from Elliott Automation Radar Systems Ltd.

The toroidal mirror camera is very useful for medium- and high-angle studies but suffers from two major disadvantages associated with the high convergence of the conical beam. First, the depth of focus is very small and since the focal surface is spherical a satisfactory focus can be obtained only in the immediate vicinity of an arc on a circular band of film. Second, the beam at the specimen position has the form of an annular ring and this requires a sometimes inconveniently large specimen volume over which uniform orientation must be present. It is essential to fill the beam, otherwise exposure time is increased and the high level of parasitic scatter from the gold-coated mirror becomes troublesome.

The range of angles over which reflections can be recorded is limited by parasitic scatter from the toroid. Some improvement in the minimum angle at which observations can be made is possible, at the sacrifice of intensity, by limiting the area of the toroid that is illuminated (Elliott and Lowy, 1970).

(iii) *Crystal Monochromators.* When large specimens are available advantage can be taken of the high angular apertures attainable with crystal monochromators (Witz, 1969). The optical arrangement used to obtain a point focus with a pair of crystals is similar to that shown in Fig. 1.17b. A very successful combination in which a cylindrical gold-surfaced mirror is used to illuminate a curved-crystal monochromator has been described by Huxley and Brown (1967).

An important advantage of using one or more crystal monochromators is the high degree of spectral purity attained so that very weak reflections can be recorded without interference from the usual background caused by traces of white radiation diffracted by the specimen.

c. *Elimination of Air Scatter*

Scattering of the main beam by air leads to an undesirable background which is particularly troublesome at low angles. This can be reduced to an insignificant level by evacuating the optical path between the specimen and the film to a pressure of about 0.1 Torr. Alternatively the air can be displaced by flushing the camera with hydrogen or helium. An incidental benefit is that a more intense beam is obtained and from this point of view it is advantageous to remove the air from the entire optical path.

III. Specimen Preparation

The selection and preparation of specimens for X-ray diffraction studies comprise an art which can only be learned by experience, but the guiding principles are simple. The amount of information which may be obtained from a fiber-type diffraction pattern is closely related to the dispersion of the molecular axes about the fiber axis in the irradiated volume. Dispersion over a cone with a semiangle of about 10°, for example, makes detailed interpretation almost impossible.

In some instances fibrous proteins occur naturally in a highly oriented form and no difficulty is encountered in preparing specimens of sufficient dimensions for use with toroidal focusing elements or curved-crystal monochromators. More generally the volumes over which good orientation persist are comparatively small and it becomes advantageous to use a Franks camera or a pinhole camera in which illuminated areas of about 0.1×0.1 mm are sufficient. Orientation can sometimes be improved by physical manipulation or chemical treatment (Table 1.1a) and it is generally advantageous to maintain the specimen in its native hydrated state. When the material is finely divided in its natural state, oriented aggregates must be prepared (Table 1.1b).

Most fibrous proteins are weak diffractors and it is important to use specimens which fill the beam and are close to the optimum thickness, otherwise parasitic scatter from the collimating or focusing system will build up a background density on the film which obscures weak reflections. For native proteins a thickness corresponding to 0.5–1.0 mm of dried protein is desirable. Thinner specimens will be needed when heavy atoms have been added to the protein.

The preparation of tactoids and oriented films and fibers of soluble fibrous proteins or fibrous protein derivatives is more difficult and accounts of various procedures that have been used may be found in the references listed in Table 1.1c. General accounts of methods used to prepare oriented specimens of various polymers have been given by Fraser (1960) and Elliott (1967, 1969).

IV. Data Collection

a. *Specimen Tilting*

In Section A.I.c the concept of the sphere of reflection was introduced; with fiber-type orientation the diffraction pattern corresponds to the intersection of this sphere with the rotated intensity transform. If the specimen is mounted with the fiber axis normal to the X-ray beam (Fig. 1.18a) there will be a range of R on all layer lines except the

TABLE 1.1

Examples of Methods Used to Prepare Specimens for X-Ray Diffraction Studies

a. Improvement of Orientation or Crystallinity

Material	Method	Reference
Bombyx mori fibroin	Doubly oriented by rolling	Herzog and Jancke (1929)
		Marsh *et al.* (1955b)
Chrysopa silk	Orientation increased by stretching in water or urea	Geddes *et al.* (1968)
β-Keratin	Oriented by stretching in steam	Astbury and Woods (1933)
	Doubly oriented by pressing in steam	Astbury and Sisson (1935)
	Doubly oriented by stretching and pressing in steam	Fraser and MacRae (1962a)
Collagen fibrils	Orientation increased by holding under tension	Cowan *et al.* (1955b)
		Miller and Wray (1971)
Paramyosin filaments	Crystallinity increased by drying and soaking in aqueous acetone	Elliott *et al.* (1968a)

b. Composite Specimens

Material	Reference
Whelk egg capsule	Flower *et al.* (1969)
Bacterial flagella	Burge and Draper (1971)
Silk fibers	Warwicker (1960a)
Bacteriophage	Marvin (1966)
TMV	Gregory and Holmes (1965)
Keratin cells	Woods (1938)

c. Oriented Specimens from Soluble Materials

Material	Method	Reference
Bombyx mori fibroin	Stretching and rolling contents of silk gland	Kratky (1929)
Fibrin, fibrinogen	Shearing of sol	Stryer *et al.* (1963)
Light meromyosin Fr. 1	Manipulation of precipitate	Szent-Györgyi *et al.* (1960)
Tropomyosin	Cast films stretched in formic acid vapor	Miller (1965)
	Fibers drawn from precipitate	Caspar *et al.* (1969)
F-Actin	Fibers drawn from precipitate	Cohen and Hanson (1956)
	Shearing of sol	Spencer (1969)
α-Keratose	Stretching precipitate	Happey (1955)
Feather keratin	Fibers drawn from precipitate	Rougvie (1954)
	Cast film	Fraser and MacRae (1963)
	Stretching precipitate	Burke (1969)

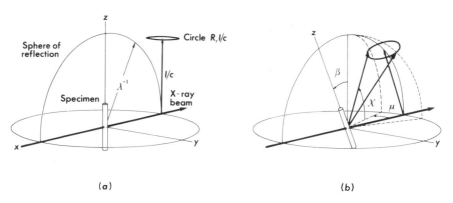

(a)   (b)

FIG. 1.18.   (a) If a specimen with fiber orientation is mounted with the fiber axis perpendicular to the direction of the incident beam, there are regions of reciprocal space for which no information is recorded. (b) By tilting the specimen through the appropriate angle, given by Eq. (1.60), additional information on specific layer lines can be obtained.

equator, for which no data will be recorded. In order to collect these data for a particular layer line it is necessary to tilt the fiber axis through an angle β (Fig. 1.18b) such that

$$\sin \beta = \lambda l/2c \qquad (1.60)$$

where λ is the X-ray wavelength, l is the layer-line index, and c is the axial period. The intensity transform is thereby rotated so that the sphere intersects the Z axis at the point $(0, l/c)$.

When focusing cameras employing highly convergent beams are used with tilted specimens special provision must be made to keep the pattern in focus since the focal surface rotates with the specimen position. Focus can be maintained along a circular arc on the film (Fig. 1.19ci) by using a Rowland mounting (Elliott, 1968).

b. *Recording*

By a fortunate combination of circumstances the relationship between optical density and exposure for X-ray films is almost linear over a considerable range. For the films commonly used for diffraction studies the linearity extends to optical densities of about 1.0–1.5 above background. The usable density range can be extended to 2.5 provided appropriate corrections for nonlinearity are applied (Matthews *et al.*, 1972). The corresponding range of intensity which may be recorded is not sufficient for most studies and a pack of films is used. With Cu K_α radiation each film absorbs an appreciable fraction of the radiation passing through it and so a wide range of intensities can be recorded within the linear range of one of the films in the pack. Precautions which must be observed in processing X-ray films and methods of scaling data from different films are given in Klug and Alexander (1954), MacGillavry and Rieck (1962), and Jeffery (1971).

The emulsions used for making X-ray film are extremely grainy and this causes considerable difficulty in microphotometry. The grain noise virtually dictates the specimen–film distance which must be used to prevent the graininess from interfering with the resolution of the camera. Some improvement can be obtained by using fine-grain X-ray film but the increase in exposure time is out of proportion to the small gain in resolution.

C. PROCESSING OF OBSERVED DATA

I. RECIPROCAL LATTICE COORDINATES

For a given value of specimen tilt β and direction of incidence of the X-ray beam (Fig. 1.18b) each point on the film corresponds to a particular direction of the vector σ. It is convenient to define the direction of σ by angles μ and χ (Fig. 1.18b) which correspond to a longitude and latitude on the sphere of reflection. The magnitude of the reciprocal lattice vector D, given by $D = 2 \sin \theta / \lambda$ (Section A.I.c), is related to these angles by

$$D = (1/\lambda)[2(1 - \cos \mu \cos \chi)]^{1/2} \tag{1.61}$$

and the reciprocal lattice coordinates are given by

$$Z = (1/\lambda)[\sin \beta (1 - \cos \mu \cos \chi) + \cos \beta \sin \chi] \tag{1.62}$$

$$R = (D^2 \quad Z^2)^{1/2} \tag{1.63}$$

$$\sin \psi = \sin \mu \cos \chi / \lambda R \tag{1.64}$$

The relationship between position on the film and the reciprocal lattice coordinates depends upon the type of camera used but expressions can readily be obtained to relate a set of Cartesian coordinates on the flattened-out film to the angles μ and χ. The reciprocal lattice coordinates can then be obtained from Eqs. (1.62)–(1.64). It is convenient to choose the point where the undeflected beam intercepts the film as origin and to measure y parallel to the meridional direction and x perpendicular to this direction. The relationship between x and y and the longitude and latitude of the diffracted beam are as follows for the most common arrangements (Fig. 1.19).

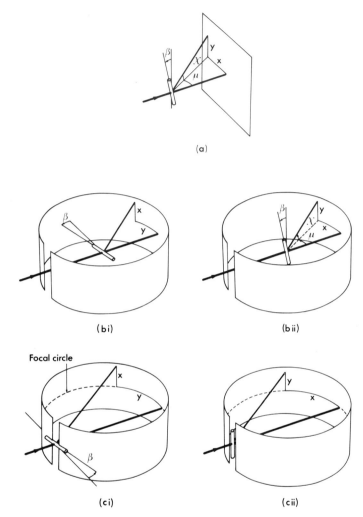

FIG. 1.19. Experimental arrangements used to record diffraction patterns photo-graphically. (a) Flat film. (b) Pinhole collimation used with cylindrical film. (c) Cylindrical film used with focusing cameras. In (i) the specimen should ideally be deformed to lie in the cylindrical surface and for tilted specimens ($\beta \neq 0$) the complete specimen/cylinder assembly must be rotated about the specimen and displaced in the direction of the beam to restore focus. This operation can be carried out automatically by use of a Rowland mounting (Elliott, 1968).

a. *Flat Film*

$$\tan \mu = x/\imath, \qquad \tan \chi = y(\imath^2 + x^2)^{-1/2} \qquad (1.65)$$

where $\imath$ is the specimen–film distance.

b. *Cylindrical Film (Pinhole Cameras)*

  (i) *Fiber Axis Normal to Cylinder Axis*

$$\sin \mu = (x/\imath)[(x/\imath)^2 + \cos^2(y/\imath)]^{-1/2} \qquad (1.66)$$

$$\sin \chi = \sin(y/\imath)[1 + (x/\imath)^2]^{-1/2} \qquad (1.67)$$

where $\imath$ is the radius of the cylinder.

  (ii) *Fiber Axis, Cylinder Axis, and Incident Beam Coplanar*

$$\mu = x/\imath, \qquad \tan \chi = y/\imath \qquad (1.68)$$

where $\imath$ is the radius of the cylinder

c. *Cylindrical Film (Focusing Cameras)*

  (i) *Fiber Axis Normal to Cylinder Axis*

$$\sin \mu = \frac{x}{2\imath}\left[\left(\frac{x}{2\imath}\right)^2 + \cos^2\left(\frac{y}{2\imath}\right)\cos^2\left(\frac{y}{2\imath} - \beta\right)\right]^{-1/2} \qquad (1.69)$$

$$\sin \chi = \sin\left(\frac{y}{2\imath}\right)\cos\left(\frac{y}{2\imath} - \beta\right)\left[\left(\frac{x}{2\imath}\right)^2 + \cos^2\left(\frac{y}{2\imath} - \beta\right)\right]^{-1/2}$$

$$(1.70)$$

where $\imath$ is the radius of the cylinder.

  (ii) *Fiber Axis Parallel to Cylinder Axis*

$$\mu = x/2\imath, \qquad \tan \chi = y/[2\imath \cos(x/2\imath)] \qquad (1.71)$$

where $\imath$ is the radius of the cylinder.

II. INTENSITY MEASUREMENTS

In early studies the intensities of diffracted beams were usually estimated visually by comparing the optical density at the center of each spot on the film with a standard set of graded optical densities. Various corrections are necessary to relate this peak intensity to the value of $I(\mathbf{D})$ at the reciprocal lattice point (see, for example, Franklin and Gosling, 1953; MacGillavry and Rieck, 1962; Jeffery, 1971). Certain of these corrections are difficult to apply to fiber patterns (Cella *et al.*, 1970) and the measurement of integrated intensities is to be preferred. The collection of suitable intensity data from a series of one-dimensional microdensitometer scans is a time-consuming process and some use has been made of automatic means for mapping equal-intensity contours (Miller *et al.*, 1964; Parsons *et al.*, 1965; Milledge and Graeme-Barber, 1973; Elliott, 1970). A more convenient method is to record the data in digital form using a two-dimensional scanning microdensitometer (see, for example, Matthews *et al.*, 1972). The digital data can be used to map equal-intensity contours if desired but are also ideally suited to automatic processing. Suitable instruments are marketed by Tech/Ops Instruments, Burlington, Mass. (Scandig) and by Optronics International Inc., Chelmsford, Mass. (Photoscan).

a. *Integrated Intensities of Arcs*

A quantity $\mathscr{I}_a(hkl)$, which is a measure of the integrated intensity of the arc corresponding to the reciprocal lattice point *hkl*, can be obtained by integrating the optical density above background over the region on the film occupied by the arc. Thus

$$\mathscr{I}_a(hkl) = \int_{\text{arc}} \mathscr{D}(x, y)\, dx\, dy \tag{1.72}$$

where x and y are the film coordinates defined in Section C.I and $\mathscr{D}$ is optical density. If x and y are measured in millimeters, the result will be in (optical density) × (area in mm²).

When a two-dimensional scanning microdensitometer is used $\mathscr{I}_a$ can be evaluated by numerical integration (Werner, 1970; Nockolds and Kretsinger, 1970). When a one-dimensional scanning instrument is used traces are usually taken radially across the center of the arc and tangentially along its length. The value of $\mathscr{I}_a$ is then taken as the product of the areas beneath the two traces, after subtraction of background, divided by the peak height (Marvin *et al.*, 1961). In order to reduce the labor involved in measuring tangential traces, some workers have taken $\mathscr{I}_a$ as being proportional to the product of the area under the radial trace multiplied by a calculated arc length (for example, Langridge *et al.*, 1960; Bradbury *et al.*, 1965; Yonath and Traub, 1969). This is less satisfactory, since the expressions used for arc length have only been very approximate (Cella *et al.*, 1970). The notion (Vainshtein, 1964; Burge and Draper, 1971) that integrated intensities can be estimated by increasing the measuring aperture to include the full width of a reflection is erroneous (see, for example, Jones, 1952).

b. *Layer-Line Streaks*

The quantity of interest in the case of continuous layer-line distributions is the integrated intensity $\mathscr{I}_s$ per Å⁻¹ increase in R, defined by

$$\mathscr{I}_s(R, l/c) = \lim_{\delta R \to 0} (1/\delta R) \left[\int\!\!\int_{R \pm \frac{1}{2}\delta R} \mathscr{D}(x, y)\, dx\, dy \right] \tag{1.73}$$

where the integration is confined to intensity due to the *l*th layer line. When digitized data from a two-dimensional microdensitometer scan are available the value of $\mathscr{I}_s$ can be calculated by numerical integration over the appropriate ranges of x and y. Otherwise $\mathscr{I}_s$ can be approximated, except in the immediate vicinity of the meridian, by taking traces at right angles to the layer-line streak and multiplying the area beneath the trace by $d\mathscr{s}/dR$, where $\mathscr{s}$ is the distance along the layer line. If x, y, and $\mathscr{s}$ are expressed in millimeters, the result will be in (optical density) $\times$ (area in mm²) per Å⁻¹. In the vicinity of the meridian the effect of disorientation is to spread the intersection of the circle $(R, l/c)$ with the sphere of reflection into an arc which makes an acute angle with the layer line of a perfectly aligned specimen. In this case it is necessary to take traces along the tangent to the line of constant R where it crosses the position of the layer line for a perfectly aligned specimen. The area under the trace must then be multiplied by $(d\mathscr{s}/dR) \sin \gamma$, where γ is the angle between the direction of the trace and the direction of $\mathscr{s}$.

c. Correction of Observed Intensities

(i) *Crystalline Patterns.* Fiber-type orientation leads to the smearing of the intensity associated with reciprocal lattice points into rings in reciprocal space (Fig. 1.18) and the magnitude of the Fourier transform $|F(hkl)|$ at the reciprocal lattice point hkl is related to the integrated intensity $\mathscr{I}_a(R, l/c)$ of the corresponding arc in the diffraction pattern by

$$|F(hkl)|^2 = k_a \mathscr{I}_a(R, l/c)/m \mathrm{L}_a \mathrm{P} \tag{1.74}$$

where k_a is a constant, m is the number of reciprocal lattice points with cylindrical polar coordinates $(R, l/c)$, L_a is the Lorentz factor for arcs, and P is a polarization factor. The Lorentz factor is a measure of the probability that a particular crystallite will be suitably oriented for it to contribute to the integrated intensity. If all the intensity associated with the reciprocal lattice point (hkl) is confined to a circular annulus $R(hkl) \pm \frac{1}{2}\delta R$ and $Z(hkl) \pm \frac{1}{2}\delta Z$ and if the disorientation is not too great, the probability that the reciprocal lattice point lies within the sphere of reflection is, from Fig. 1.20, equal to

$$\frac{\delta(1/\lambda) \, \delta R \, \delta Z}{\mathbf{n} \cdot \mathbf{v}} \frac{1}{2\pi R \, \delta R \, \delta Z} \tag{1.75}$$

where $\mathbf{n}$ is a unit vector normal to R and Z, and $\mathbf{v}$ is the normal to the sphere of reflection, which is assumed to have a radial thickness of $\delta(1/\lambda)$. Evaluation of the scalar product $\mathbf{n} \cdot \mathbf{v}$ (Arnott, 1965; Cella *et al.*, 1970) and substitution in Eq. (1.75) leads to the expression

$$\mathrm{L}_a = \{\lambda^2(R^2 + Z^2)[\cos^2\beta + \lambda Z \sin \beta - \tfrac{1}{4}\lambda^2(R^2 + Z^2)] - \lambda^2 Z^2\}^{-1/2} \tag{1.76}$$

where the term $\delta(1/\lambda)/2\pi$ has been omitted. For small and medium angles of diffraction $L_a \simeq (\lambda R)^{-1}$ and this approximation has frequently been used.

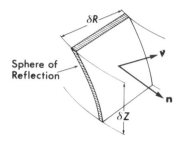

Fig. 1.20.

Provided that the camera does not polarize the incident radiation, the term P can be shown (see, for example, James, 1954) to be given by

$$P = 1 - \tfrac{1}{2}\lambda^2(R^2 + Z^2) + \tfrac{1}{8}\lambda^4(R^2 + Z^2)^2 \qquad (1.77)$$

Under certain conditions the use of crystal monochromators can affect the value of P that should be used (Witz, 1969).

It is tacitly assumed in Eq. (1.74) that all the reciprocal lattice points that lie on the circle $(R, l/c)$ have the same value of $|F|$, but this is not always true. In these cases Eq. (1.74) yields the mean value of $|F|^2$ for the set of reciprocal lattice points.

The expression given for L_a in Eq. (1.76) is not appropriate for $R = 0$ and values of $|F(00l)|$ must be obtained from a series of photographs taken with the specimen tilted so as to bring each of the $00l$ reciprocal lattice points in turn into the reflecting position (Section B.IV.a). After the films have been scaled to a common basis the integrated intensities of the meridional arcs are used to calculate $|F(00l)|$ by means of the expression

$$|F(00l)|^2 = k_a' \mathscr{I}_a(0, l/c)/L_a'P \qquad (1.78)$$

The effect of disorientation on a crystalline pattern is to produce an intensity transform in which the reciprocal lattice points on the Z axis are smeared over spherical caps of area proportional to Z^2. When the fiber is tilted so that the point $(0, l/c)$ intersects the sphere of reflection the peak intensity will be proportional to $Z^{-2}(\mathbf{n} \cdot \mathbf{v})^{-1}$ and the integrated intensity to $Z^{-1}(\mathbf{n} \cdot \mathbf{v})^{-1}$, giving

$$L_a' = (\lambda^2 Z^2 - \tfrac{1}{4}\lambda^4 Z^4)^{-1/2} \qquad (1.79)$$

The value of the ratio k_a/k_a' depends upon the nature of the distribution function $C(\omega)$ (Section A.IV.g) and must be regarded as an adjustable parameter in the refinement process discussed later.

(ii) *Layer-Line Streaks.* When a pattern consists of layer-line streaks the quantity of interest is the value of the cylindrically averaged intensity transform $\langle I(R, l/c)\rangle_\psi$ (Section A.I.d). As before, allowance must be made for the probability that a molecule is in the reflecting position, which in this case is proportional to $(\mathbf{n} \cdot \mathbf{v})^{-1}$. The factor $(2\pi R)^{-1}$ in Eq. (1.75) is not required since the intensity transform has already been averaged with respect to ψ. Thus we have

$$\langle I(R, l/c)\rangle_\psi = k_s \mathscr{I}_s(R, l/c)/L_s P \qquad (1.80)$$

where $L_s = \lambda R L_a$. For small and medium angles of diffraction $L_s \simeq 1$ and has often been neglected.

(iii) *Mixed Streaks and Arcs.* In many patterns the inner region contains arcs due to lattice reflections while the outer portions consist of continuous layer-line distributions. It is usually convenient to treat the lattice reflections as samples of the continuous rotated intensity transform $\langle I(R, l/c)\rangle_\psi$ and it can be shown (Marvin *et al.*, 1961; Yonath and Traub, 1969) that if Eq. (1.80) is used for streaks, then the corresponding expression for arcs, on the same scale, is

$$\langle I(R, l/c)\rangle_\psi = [k_s \mathscr{I}_a(R, l/c)/L_s P](2\pi R A/m) \qquad (1.81)$$

where A is the area of the c-axis projection of the unit cell, and m as before is the number of reciprocal lattice points with cylindrical polar coordinates $(R, l/c)$.

It should be noted that the expression given in Eq. (1.81) is based on the assumption that the mean value of the samples of $I(R, \psi, l/c)$ taken at the reciprocal lattice points lying on the circle (R, Z) gives an adequate representation of the ψ-average of I. This will not always be true.

(iv) *Low-Angle Patterns.* In the foregoing discussion it has been assumed that the primary cause of the arcing of meridional reflections was disorientation and this is generally true of the medium- and high-angle patterns obtained from crystalline polymer specimens. The situation is different, however, in the case of low-angle patterns since the natural breadth of the meridional reflections in a direction perpendicular to the Z axis may be an appreciable fraction of c^{-1}. It is usually inconvenient or impracticable to tilt the specimen for each individual meridional reflection and expressions have been derived for correcting the measured peak intensities to give estimates of $|F(l/c)|^2$ (Sikorski and Woods, 1960; Blaurock and Worthington, 1966).

(v) *Absorption.* As the incident beam traverses the specimen, its intensity is reduced by absorption, as are the diffracted beams before they leave the specimen. The absorption is a function of wavelength, specimen shape, and scattering angle, and means of calculating correction factors have been discussed by Lipson (1959) and by Buerger (1960). The specific case of a cylindrical specimen irradiated with a microbeam has been considered (Skertchly, 1957; Coyle and Schroeder, 1971; Coyle, 1972). In most work to date absorption corrections have generally been ignored in view of the low accuracy of the intensity measurements.

D. REFINEMENT OF MODEL STRUCTURES

Apart from the calculation of cylindrical Patterson functions (Section A.I.f), little direct interpretation of X-ray patterns obtained from fibrous proteins is possible and the data must be used instead to develop, test, and refine model structures. In the case of fiber patterns obtained from materials containing crystallites with three-dimensional order the determination of unit cell dimensions and space group follows conventional lines (Klug and Alexander, 1954; Kasper and Lonsdale, 1959; Jeffery, 1971) and least-squares optimization (Chapter 7) of the cell parameters may usefully be employed. Indexing of reflections is greatly simplified if doubly oriented specimens can be prepared. In the absence of any preferred orientation a "powder pattern" is obtained and the deduction of the correct unit cell is difficult. Unit cells derived on the basis of a few powder rings cannot be regarded as reliable.

The derivation of helical parameters from fiber patterns is based on the theory outlined in Sections A.III and A.IV. In some cases the layer-line translations can be indexed on a finite axial repeat of structure c and the pattern of meridional and near-meridional reflections can be used to determine the parameters u and v (Section A.II.c). In other cases the unit height h must be obtained from the meridional reflections and the pitch P from the local distribution of near-meridional reflections around each meridional reflection. Systematic methods of determining helical parameters have been discussed by Mitsui (1966, 1970). The main complications in the interpretation of diffraction patterns from helical structures arise from the presence of actual or pseudo rotation axes of symmetry (Section A.III.c), and from the presence of periodic distortions, which give rise to arcs and streaks in positions that are forbidden for an undistorted helix (Section A.IV.c). In the case of meridional reflections the intensities of these additional reflections may be comparable with those in the normal positions.

I. CRITERIA FOR REFINEMENT

a. *Crystalline Patterns*

The traditional measure used by crystallographers for judging goodness of fit between the observed values of the magnitude of the Fourier transform at the reciprocal lattice points $|F_o|$ and the values calculated for a model structure $|F_c|$ is the normalized mean deviation, or residual, given by

$$\mathscr{R}' = \left(\sum_i |\,|F_o|_i - |F_c|_i|\right)\bigg/\sum_i |F_o|_i \qquad (1.82)$$

(Kasper and Lonsdale, 1959; Buerger, 1960). A more convenient measure for least squares methods and for statistical analysis is the normalized standard deviation

$$\mathscr{R} = \left\{\left[\sum_i W_i(|F_o|_i - |F_c|_i)^2\right]\bigg/\sum_i W_i |F_o|_i^2\right\}^{1/2} \qquad (1.83)$$

where W_i is the weight to be attached to the observation of the ith reflection. The choice of W_i has an important bearing on the model which is selected as best fitting the observed data. Ideally W_i should be given a value proportional to the reciprocal of the variance of the value of the ith reflection. Methods of choosing W_i have been given by Cruickshank (1965), Arnott and Wonacott (1966a), and Bradbury *et al.* (1965).

It is sometimes more convenient to minimize the standard deviation of the intensities, in which case $|F_o|$ and $|F_c|$ are replaced by $|F_o|^2$ and $|F_c|^2$ in Eq. (1.83).

b. *Layer-Line Streaks*

When the diffraction pattern consists of continuous layer-line streaks it is customary to measure the goodness of fit by the subjective assessment of plots of the observed and calculated values of $\langle I(R, l/c)\rangle_\psi$. This is inconvenient from the point of view of automated refinement and is also susceptible to unconscious bias on the part of the experimenter. One possibility would be to use a quantity

$$\mathscr{R} = \left[\sum_l \int_{R_1}^{R_2} W(R, l)[\Delta\langle I(R, l/c)\rangle_\psi]^2\, dR\bigg/\sum_l \int_{R_1}^{R_2} W(R, l)[\langle I(R, l/c)\rangle_\psi]^2\, dR\right]^{1/2} \qquad (1.84)$$

equal to the normalized root mean square value of the deviation $\Delta \langle I(R, l/c) \rangle_\psi$ between observed and calculated values of $\langle I(R, l/c) \rangle_\psi$, where $W(R, l)$ is a weighting function and R_1 and R_2 define the range over which data were collected for the particular layer line.

II. Number of Parameters

The extent to which $\mathscr{R}$ may be reduced depends upon the number of independent parameters that are needed to define the model. For example, a model in which bond lengths and angles are regarded as variables will give a lower value of $\mathscr{R}$ than the same model with fixed bond lengths and angles. Procedures for assessing whether the decrease in $\mathscr{R}$ resulting from an increase in the number of adjustable parameters is statistically significant have been outlined by Hamilton (1965).

When least squares refinement is used the successful convergence onto the optimum values of the parameters is unlikely unless the number of intensity data exceeds the number of parameters by a sufficient margin. The difficulty encountered with fibrous protein models is that this condition is not generally satisfied. It is essential therefore to reduce the number of independent parameters by introducing the maximum number of constraints on the optimization (Arnott and Wonacott, 1966a; Arnott, 1968). In particular, bond lengths and angles and torsional angles about bonds are held fixed as far as possible at values determined from studies of model compounds. The method used to introduce these constraints is discussed in Chapter 7.

III. Refinement of Pseudocells

In a number of fibrous protein structures the main chain has a locally regular conformation with a comparatively short period whereas the pattern of side chains does not repeat in the distance over which this regular main-chain conformation persists. The problem therefore arises of finding a method to represent the contribution of the side chains to the Fourier transform of the contents of the pseudocell. A similar problem is encountered in representing the contribution of unordered water to the diffraction pattern of a hydrated specimen.

Several approaches have been used (Langridge *et al.*, 1960; Fraser *et al.*, 1965e, 1971a; O'Brien and MacEwan, 1970) based on the property that the Fourier transform F of an object immersed in a matrix of uniform electron density ρ_m is given by

$$F = F_c - \rho_m F_s \qquad (1.85)$$

where F_c is the transform calculated in the usual way and F_s is the transform of the volume not occupied by the continuum assuming it to have a uniform density of unity. To a first approximation the side chains or water can be assumed to be equivalent to a uniform electron distribution of appropriate density. A more accurate representation can be obtained by making ρ_m a function of D to allow for fall in atomic scattering factors with increasing scattering angle.

E. SUMMARY

Particular attention has been given to X-ray diffraction because it is the only method that is inherently able to provide sufficient information to enable the detailed conformation of fibrous proteins to be elucidated. However, the information is garbled by the effects of imperfect crystallinity and disorientation and the possibility of producing three-dimensional electron density maps at atomic resolution is remote. Recourse must therefore be made to trial-and-error methods in which plausible model structures, arrived at from a consideration of all the available physical and chemical data, are refined on the basis of the observed X-ray diffraction pattern. Co-refinement with energy calculations (Chapter 6) and infrared data (Chapter 5) offers attractive possibilities.

The main disadvantages of X-ray diffraction as a method of studying conformation are the large specimens required to give reasonable exposure times, which virtually precludes investigations at the cell level, and the insensitivity of the diffraction pattern to contributions from less-well-ordered regions, which constitute an appreciable proportion of many fibrous protein structures.

LIST OF SYMBOLS FOR CHAPTER 2

λ wavelength
V accelerating potential
f atomic scattering factor for X rays
f_e atomic scattering factor for electrons
D magnitude of reciprocal space vector
t crystallite thickness
N number of atoms in unit cell

Chapter 2

Electron Diffraction

Despite its potential advantages over X-ray diffraction, very little use has been made of electron diffraction in studies of conformation in fibrous proteins. Low-angle patterns have been recorded from protein specimens containing heavy atoms (Mahl and Weitsch, 1960; Murray and Ferrier, 1968; Ferrier, 1969; Glaeser and Thomas, 1969; Dobb, 1970) but these have been inferior to diffraction patterns obtained with X rays. This may be attributed to rapid deterioration of the specimen through radiation damage and it seems likely that the patterns which have been recorded are attributable to the heavy-atom framework rather than the protein.

In a recent study of the problem (Tulloch, 1971) it has been demonstrated, however, that with appropriate precautions both low- and high-angle diffraction patterns can be recorded from a wide variety of fibrous proteins without heavy-atom staining. This being so, the inherent advantages of electron diffraction, which stem mainly from the very small specimen volumes required, are likely to be increasingly exploited in future studies. In diffraction patterns obtained with X rays the arcs in fiber patterns result from the superposition of reflections from a large number of individual crystallites with a distribution of orientations. Because the irradiated volume is so much smaller in the case of electrons, an appreciable sharpening of the pattern would be expected in favorable cases and this is in fact realized (Fig. 2.1).

Excellent general accounts of the theory and technique of electron diffraction are available (Heidenreich, 1964; Vainshtein, 1964; Alderson and Halliday, 1965; Cowley, 1968; Rymer, 1970) and in the present chapter attention will be restricted to aspects that are relevant to studies of fibrous proteins.

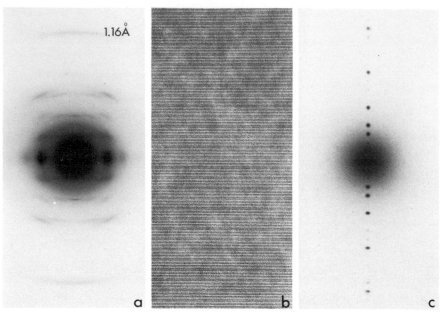

FIG. 2.1. Examples of electron diffraction patterns obtained from fibrous protein specimens (Tulloch, 1971). (a) High-angle pattern obtained from a longitudinal section about 800 Å in thickness cut from silk gut (*Bombyx mori*). (b) Tactoid of oyster muscle tropomyosin stained with uranyl acetate. The banding pattern has a period of about 380 Å. (c) Low-angle pattern obtained from (b). The central diffuse scatter in the diffraction patterns is due to inelastically scattered electrons.

A. THEORY OF ELECTRON DIFFRACTION

I. INTERACTION OF ELECTRONS WITH MATTER

a. *Wavelength*

It is the wave character of electrons which leads to the phenomenon of electron diffraction by matter and the wavelength associated with a moving electron is a function of its velocity. For a beam accelerated through a potential V, measured in volts, the wavelength, in angstroms, is given by

$$\lambda = 12.27(V + 0.978 \cdot 10^{-6} V^2)^{-1/2} \tag{2.1}$$

With an accelerating potential of 100 kV, for example, the wavelength given by Eq. (2.1) is 0.037 Å. This is very much smaller than the wavelengths commonly employed in X-ray studies and the sphere of reflection (Chapter 1, Section A.I.c) has a correspondingly greater radius. The collection of diffraction data from fiber patterns is thus greatly simplified since specimen tilting (Chapter 1, Section B.IV.a) is generally unnecessary.

b. *Elastic and Inelastic Scattering*

An electron beam is modified during its passage through a specimen by both elastic and inelastic interactions. Electrons that suffer inelastic collisions undergo a change in velocity, and energy is transferred to the specimen. The velocity change involves a change in wavelength and hence destroys the potentiality for producing interference effects with the electrons that emerge with unmodified velocities.

Electrons that only undergo elastic collisions emerge from the specimen with unmodified velocities and are capable of producing diffraction patterns similar to those obtained with monochromatic electromagnetic radiation. The inelastically scattered electrons give rise to an intense background which is concentrated in the forward direction and decreases with scattering angle (Fig. 2.1).

The cross sections of atoms for elastic and inelastic scattering are of the same order of magnitude and both decrease with increasing accelerating potential. It is generally believed that the ratio of elastic to inelastic cross section increases with increasing accelerating potential, so that the proportion of diffuse scatter is reduced as the potential is increased (Heidenreich, 1964).

The ratio of elastically to inelastically scattered electrons emerging from a specimen depends on the accelerating potential and on the nature and thickness of the specimen. An illustrative example is shown in Fig. 2.2 (Crick and Misell, 1971), where the calculated ratios at 100 kV for simulated protein specimens of different thicknesses are compared. The proportion of elastically scattered electrons rises to a maximum at a thickness of about 750 Å and then decreases slowly with increasing thickness. In contrast the proportion of inelastically scattered electrons shows a monotonic increase with thickness. The maximum in the curve for elastic scattering is very broad and a specimen with half the optimum thickness gives 0.83 of the maximum yield. When specimen thickness can be controlled accurately there are advantages in working at thicknesses somewhat less than that for the maximum yield of elastically scattered electrons, since a significant improvement in the elastic/inelastic ratio (Fig. 2.2) can be obtained with very little sacrifice in yield.

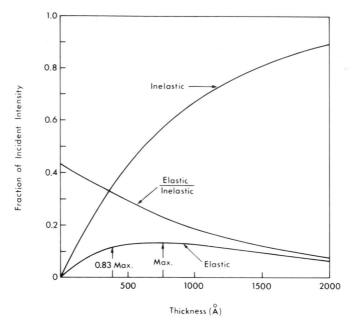

FIG. 2.2. Comparison of the proportions of elastically and inelastically scattered electrons from a carbon specimen irradiated with 100-keV electrons. Calculated for a density of 1.3 g cm⁻³ from the data given by Crick and Misell (1971).

The nature of the background produced by inelastically scattered electrons varies with specimen thickness as the half-width of the angular distribution increases with increasing specimen thickness (Crick and Misell, 1971).

c. *Radiation Damage*

The cross sections of atoms for inelastic scattering of electrons are several orders of magnitude greater than those for X rays and structural changes take place very rapidly in organic materials under the usual conditions of observation in the electron microscope (Stenn and Bahr, 1970). In addition to damage by heating, the electron bombardment induces ionization and radical formation and these are believed to be the chief causes of structural changes within the specimen. According to Breedlove and Trammell (1970) the cross section for ionization decreases less rapidly with accelerating potential than the cross section for elastic scattering. It might be concluded that a reduction in accelerating potential would enable more useful information to be obtained from a specimen before it was destroyed. According to other authors, however, the reverse is true (Kobayashi and Sakaoku, 1965; Thomas *et al.*, 1970; Parsons, 1970).

Quantitative measurements on poly(ethylene) show that the lattice spacings in this material change as the radiation dose increases (Fischer, 1964) and similar effects are to be expected in other polymers. Insofar as heating is involved in structural changes during irradiation, the provision of facilities for cooling the specimen should prolong the useful life of the specimen (Watanabe *et al.*, 1960; Tulloch, 1971).

II. Elastic Scattering

a. *Atomic Scattering Factors*

The periodic distribution of electrical potential in a crystal produces diffraction effects in the elastically scattered electrons which are superficially similar to those produced by the electron density distribution on a beam of X rays (Chapter 1, Section A.I).

For most purposes a point-atom approximation is sufficient and an atomic scattering factor for electrons f_e is used which is related to the atomic scattering factor f for X rays by the expression (MacGillavry and Rieck, 1962)

$$f_e(D) = 0.095736[Z - f(D)]/D^2, \qquad D \neq 0 \tag{2.2}$$

where D is the magnitude of the reciprocal space vector $\mathbf{D}$ (Chapter 1, Section A.I.c) and Z is the atomic number. This expression applies to neutral, isolated atoms and considerable changes in the effective value of f_e are brought about, particularly for small values of D, by ionization and chemical bonding. Values of f_e have been tabulated by Doyle and Turner (1968). The modifications to f_e that result from chemical bonding have been investigated by Harada and Kashiwase (1962).

For values of $D \leqslant 0.5$ Å^{-1} the values of f_e for C, N, and O atoms are in reverse order to the factors for X rays. Also, the disparity between the scattering factors for H atoms and those for C, N, and O atoms is less marked than for X rays. When computing the Fourier transform of the electron density distribution of model structures for comparison with observed X-ray patterns the contribution of H atoms is usually ignored, but when computing the Fourier transform of the electrical potential the H atoms cannot be ignored.

b. *The Kinematic Approximation*

In Chapter 1 X-ray diffraction was interpreted on the basis of a single-scattering approximation in which it was assumed that a scattered beam could pass through the crystallite without further interaction. The cross section of atoms for X-ray scattering with $\lambda \sim 1.5$ Å and the dimensions of crystallites in molecular crystals are such that this approximation is usually valid, but this is not generally true for electron diffraction. The cross section of atoms for elastic scattering of electrons is sufficiently great that multiple scattering must be taken into account. The intensity distribution in the diffraction pattern of a material in which appreciable multiple scattering has occurred within a single crystallite is virtually uninterpretable in the case of unit cells as large as those found in fibrous proteins and it is only when diffraction patterns are obtained under conditions where the effects of multiple scattering may be neglected that they are likely to be of value for structure analysis. Under these conditions the so-called kinematic approximation is valid and the interpretation of intensities closely follows that given for X rays in Chapter 1 except that $\rho(\mathbf{d})$ represents electrical potential rather than electron density (Vainshtein, 1964).

The prediction of the conditions under which the kinematic theory of scattering is valid is extremely difficult. According to Vainshtein (1964) an estimate of the value of the crystallite dimension parallel to the beam at which the kinematic approximation breaks down, in the case of unstained biological materials, is given by

$$t \sim 10 \sqrt{N}/\lambda \tag{2.3}$$

where N is the number of atoms in the unit cell and λ is given by Eq. (2.1). The transition from kinematic to dynamic scattering with increasing crystallite size is not uniform over the diffraction pattern and occurs first at low angles. From Eq. (2.3) it follows that an increase in accelerating potential or an increase in complexity of the structure favors kinematic scattering.

The complexity of fibrous protein structures and their poor crystallinity favor kinematic scattering and it is likely that the approximation will be valid in many cases for these materials. However, the expression given in Eq. (2.3), which is based on a gross oversimplification of the problem, can only be regarded as yielding an order of magnitude for the thickness at which the kinematic approximation breaks down.

B. EXPERIMENTAL TECHNIQUES

I. DIFFRACTION EQUIPMENT

Most modern electron microscopes provide facilities for electron diffraction studies but the requirements for recording the fine detail present in the rather weak diffraction patterns yielded by fibrous proteins are fairly stringent, and desirable features of the microscope are summarized in the present section.

The optimum specimen thickness with regard to the yield of elastically scattered electrons (Fig. 2.2) increases with accelerating potential. Provision for operation up to 100 kV or greater facilitates the preparation of specimens of optimum thickness over areas several microns square.

Various electron optical arrangements have been used for diffraction studies and their relative merits have been discussed by Ferrier (1969). The most generally useful method for fibrous proteins is the selected-area microdiffraction mode in which provision is made for visual selection of the specimen area to be irradiated. The area is usually delineated by means of an aperture in the object plane of the intermediate lens. The selection of the area must of course be performed at very low intensity to avoid, as far as possible, damage to the specimen. Photographic recording of the image of the irradiated area for survey purposes presents no problems since this may be done subsequent to the recording of the diffraction pattern.

In order to provide adequate separation of orders in specimens with axial periods up to about 1000 Å, an effective camera length of several meters is required when operating at 100 kV. A goniometer stage is required to tilt the specimen so as to bring the required portion of reciprocal space into the reflecting position (Chapter 1, Section A.I.c), and facilities must be provided to orient the specimen relative to anisodiametric selecting apertures.

Since specimen heating plays some part in radiation damage (Stenn and Bahr, 1970), a specimen cooling device may be advantageous but the extent of reduction in damage which may be achieved by this method has not been evaluated quantitatively.

II. SPECIMEN PREPARATION

Methods of preparing fibrous protein specimens for examination by electron diffraction are essentially similar to those used for transmission electron microscopy (Chapter 3) except that additional precautions must usually be observed if the native conformation is to be preserved. Provided that suitable embedding methods can be devised, it has been found (Tulloch, 1971) that the most generally useful method is sectioning (Glauert and Phillips, 1965) while in suitable cases mechanical cleavage and disintegration may be used. Care is needed when ultrasonic disintegration is used in order to avoid cellulosic contamination (Millward, 1969). Other methods, which have been used with synthetic polypeptides, include the preparation of collapsed surface films (Malcolm, 1968a,c, 1970) and thin single crystals (Keith *et al.*, 1969a,b; Padden *et al.*, 1969) and the stretching of cast films (Parsons and Martius, 1964; Vainshtein and Tatarinova, 1967a).

In order to avoid scattering by a supporting medium, it is convenient to use perforated plastic films (Fukami and Adachi, 1965; Bradley, 1965a) in which the hole sizes are a few microns in diameter.

As with other biological materials, the crystallinity of fibrous proteins is usually adversely affected by drying and the dehydration which occurs in the electron microscope is thus undesirable. Various attempts have been made to overcome this problem by the use of special chambers in which a moist environment can be maintained around the specimen (Stoyanova and Mikhailovskii, 1959; Dupouy *et al.*, 1960; Parsons, 1968, 1970). These have not so far been used routinely and the advantage gained must be weighed against the added specimen damage caused by reaction with radicals created by the interaction between the electron beam and the water.

III. PROCESSING OF OBSERVED DATA

The processing of photographically recorded electron diffraction patterns follows similar lines to those for X-ray diffraction patterns (Chapter 1). The response of photographic emulsions to electrons is such that optical density is proportional to exposure over a useful range (Valentine, 1966) but multiple exposures are usually necessary to encompass the wide range of intensities present in a single diffraction pattern.

The effective camera length is usually determined experimentally by recording the diffraction pattern of a calibrating substance (Alderson and Halliday, 1965; Ferrier, 1969). Sometimes it is convenient to evaporate a material such as gold or thallous chloride directly on to the specimen. The image of the specimen and the image of the diffraction pattern are rotated about the beam by different amounts during the imaging process and if they are to be correlated the extent of this difference must also be determined experimentally.

As mentioned earlier, the inelastically scattered electrons give rise to a background which decreases with increasing scattering angle. This is particularly troublesome when intensity measurements are to be made in a radial direction as the required profile is superimposed on a rapidly varying background.

C. SUMMARY

Electron diffraction studies have considerable potential in the determination of conformation in fibrous proteins. The main advantage over X-ray diffraction is the very much greater atomic scattering cross section so that the optimum thickness for specimens is of the order of a few hundred angstroms with 100-keV electrons. Coupled with the fact that selected areas of about 0.001×0.001 mm at the specimen can be used, the possibility arises of obtaining much higher degrees of orientation over volumes of these dimensions, with a corresponding improvement in the resolution of fine detail in the diffraction pattern. Other advantages are the very short photographic exposures required and the large radius of the sphere of reflection. Dynamical effects, which generally complicate the interpretation of intensities in electron diffraction patterns, are unlikely to be troublesome in the case of most fibrous protein specimens. When extensive crystallites of considerable perfection are encountered, for example, in certain insect silks (Chapter 13), conditions must be chosen to reduce dynamical effects to a minimum.

The main difficulties to be overcome relate to the very rapid destruction of conformation which occurs during irradiation. The development of high-resolution image intensifiers should materially assist in this regard. The requirements for specimen preparation are more stringent than with X-ray diffraction because the background due to inelastically scattered electrons increases with specimen thickness and soon becomes objectionably strong. The development of energy analyzers capable of simultaneous two-dimensional operation would be of assistance in overcoming this problem. The presently available raster-scan analyzers are of limited value due to the relatively large dose of radiation received by the specimen during the scanning process.

LIST OF SYMBOLS FOR CHAPTER 3

n order of Bessel function

l Miller index

ρ density of scattering material

σ density of scattering material in x-axis projection of ρ

x, y, z Cartesian coordinates in specimen

X, Y, Z Cartesian coordinates in reciprocal space

R, ψ, Z cylindrical polar coordinates in reciprocal space

F Fourier transform

$G_{n,l}$ complex number describing contribution of Bessel function terms of order n to the lth layer line

$g_{n,l}$ Fourier component of scattering density

Chapter 3

Electron Microscopy

At the present time electron microscopy is the only method available which offers the prospect of imaging fibrous protein molecules directly. Excellent accounts of the technique are available (Heidenreich, 1964; Kay, 1965; Slayter, 1969; Cosslett, 1971; Huxley and Klug, 1971) and in the present chapter attention will be restricted to the potentialities and limitations in the study of fibrous proteins. To date, transmission electron microscopy has provided information on the morphology of protein molecules, on their mode of aggregation, and on the distribution of chemical reactivity in the molecule. The instrumental resolution of many present-day electron microscopes is about 2 Å when operated under ideal conditions but a number of factors preclude the realization of this limit with proteins and other biological materials. The main difficulties relate to contrast, to radiation damage, and to dehydration.

A. CONTRAST

If a protein molecule is to be visualized, some form of contrast must be present between the image of the molecule and the image of its immediate surroundings. If the angular distributions of electrons scattered by the molecule and its surroundings are sufficiently different, adequate contrast can be introduced by inserting a limiting aperture in the back focal plane of the objective lens. In this way true amplitude contrast is obtained and when the image is accurately focused detail may be interpreted directly, except near the limit of resolution where

71

lens aberrations introduce phase contrast effects (Heidenreich, 1964; Erickson and Klug, 1971; Hanszen, 1971).

Unfortunately, the differences in density and composition between different parts of an assembly of fibrous protein molecules will not, in general, be sufficient to produce a useful amplitude contrast and artificial methods of increasing the contrast between a molecule and its surroundings must be used. This may take the form of a selective deposition of atoms of high atomic number leading to pronounced differences between the angular distributions of the electrons in beams which emerge from different parts of the specimen. This necessity of using artificial means of enhancing contrast with the ever-present possibility of artefactual changes in the specimen is the major shortcoming of electron microscopy as applied to protein structure. A second difficulty, which is partly philosophical, is that the information obtained relates to the heavy-atom distribution and only indirectly to the protein.

Additional contrast may also be obtained by defocusing the image so that phase contrast effects are introduced. Underfocus contrast enhancement is employed widely in the examination of biological specimens but care is necessary to avoid the introduction of spurious detail into the image (Ruska, 1966; Thon, 1966a,b; Sjöstrand, 1967; Haydon, 1968, 1969; Erickson and Klug, 1971). Erickson and Klug (1971) have made a quantitative study of the problem and formulated criteria for selecting the optimum degree of underfocus.

Another method of enhancing contrast, applicable to specimens with periodic structures, is to deflect the electron beam so that a particular diffracted beam passes along the optical axis of the microscope and to exclude the main beam by a suitably chosen objective aperture (Heidenreich, 1964; Glaeser and Thomas, 1969). The diffracting object or objects then appear bright while the remainder of the specimen is dark. This technique of dark-field imaging is potentially valuable for studying the distribution of regular conformation in fibrous protein structures.

Several methods of dark-field imaging applicable to aperiodic structures have been described. The central beam of undeflected electrons may be removed by means of a "contrast stop" (Dupouy et al., 1966; Johnson and Parsons, 1969) or by tilting the beam so that the undeflected electrons do not enter the objective aperture (Heidenreich, 1964). In an alternative method an annular condenser aperture is used and the diameter of the objective aperture is chosen so as to exclude unscattered electrons (Hall, 1948; Dupouy et al., 1969; Dubochet et al., 1971).

Dark-field electron microscopy has been used to image individual biological macromolecules (Ottensmeyer, 1969; Dubochet et al., 1971) but the information obtainable is severely limited by the granular background arising from inhomogeneities in the supporting membrane.

B. RADIATION DAMAGE

The damage caused to organic materials by irradiation with electrons was discussed briefly in Chapter 2 (Section A.I.c). The available evidence (Glaeser and Thomas, 1969; Stenn and Bahr, 1970) suggests that the native conformation of a protein molecule is completely destroyed under the conditions normally used to obtain transmission electron micrographs. The local disordering of the protein chains presumably also disrupts the local distribution of heavy-atom stains used to enhance contrast and explains why no interpretable detail is visible below a resolution of around 10–20 Å.

The loss of native conformation is accompanied by the disappearance of the high-angle diffraction pattern, which thus provides a useful check on the extent of radiation damage. It has been shown (Tulloch, 1971) that under favorable conditions electron micrographs can be obtained with preservation of the high-angle pattern and in these cases the detail can be interpreted in terms of the native conformation. An incidental advantage of routine monitoring of the high-angle diffraction pattern is that damage due to fixation, staining, or sectioning can also be detected.

Radiation damage can be reduced by the use of image intensifiers but with currently available equipment some loss of resolution is usually entailed.

C. SPECIMEN PREPARATION

Methods available for the preparation of specimens for transmission electron microscopy include sectioning (Glauert and Phillips, 1965; Sjöstrand, 1967), mechanical disintegration, and various means of mounting particulate specimens (Bradley, 1965b; Slayter, 1969; Dubochet et al., 1971). Enhancement of contrast can be achieved by shadow casting and positive or negative staining with heavy atoms (Bradley, 1965c; Glauert, 1965; Horne, 1956a,b; Slayter, 1969).

Shadow casting provides an important means of studying the morphology of individual molecules but the resolution attainable is limited by the grain size of the evaporated metal. With present methods this places a limitation of about 20 Å on this technique. A further limitation is due to surface irregularities in the supporting film but these may be overcome, albeit with some difficulty, by the use of a shadow-transfer technique devised by Hall (1956). Particles to be examined are applied as aerosols to freshly cleaved mica, which is extremely smooth, and the shadowed film is transferred to a supporting medium.

Staining for negative contrast also provides information about the morphology of molecules and aggregates of molecules and has proved particularly valuable in the study of filamentous structures from muscle (Huxley, 1963; Moore et al., 1970) and from keratins (Whitmore, 1972). Under suitable conditions the granularity in the stain can be reduced to a very low level and background due to inhomogeneities in the supporting film becomes noticeable. This can be eliminated by using perforated carbon films (Huxley and Zubay, 1960).

Shadow casting and the use of negative contrast have proved to be useful methods of investigating the morphology of fibrous protein molecules and assemblies. Positive staining, as a probe of the detailed internal structure of protein molecules, has proved, except in certain isolated cases, to be largely unsuccessful. This appears to be due to the difficulty of attaching sufficient concentrations of heavy atoms to produce significant contrast. Silver, mercury, and uranyl salts, osmium tetroxide, and phosphotungstates have all been used as positive stains for various fibrous proteins but the precise nature of the binding mechanism is unknown and it seems likely that part of the observed effect may be due to negative rather than positive contrast effects.

D. INTERPRETATION OF MICROGRAPHS

Much of the value of electron microscopy as a tool for the investigation of the structure of fibrous proteins depends upon the interpretation of fine detail in the image and a discussion of the many factors involved would be beyond the scope of the present treatment. Estimates of the resolution attainable in practice with biological specimens vary from 10 to 20 Å and valuable accounts of the requirements for realizing this limit have been given by Heidenreich (1964), Ruska (1966), and Haydon (1968, 1969).

The information contained in an electron micrograph can frequently be transformed into a more useful form by means of optical or computer processing and useful reviews have been given by Klug (1971), Nathan (1971), Horne and Markham (1972), Klug and Crowther (1972), Lake (1972a), Ottensmeyer et al. (1972) and Thompson (1972). Aspects of interpretation that are of special interest in the study of fibrous proteins are discussed in the following sections.

I. LINEAR MEASUREMENTS

The accuracy of length measurement attainable using the electron microscope is extremely poor due to the combination of a number of factors. Instrumental calibration is difficult and an accuracy of $\pm 5\%$ is probably all that is generally achieved. In addition, specimen dimensions undergo changes during fixation, embedding, staining, and irradiation. These changes may be small but their direction and magnitude cannot be assessed without reference to some other technique such as X-ray diffraction. Measurements of dimensions based on shadow casting are also subject to systematic errors which must be determined empirically (Hall, 1960; Slayter, 1969). Rowe and Rowe (1970) have considered the relationship between the distribution of shadow shapes and the shape of the shadowed object.

II. PARAMETERS OF PERIODIC STRUCTURES

Klug and Berger (1964) pointed out that the optical diffraction pattern of an electron micrograph provides a convenient means of analyzing periodicities in the image. These patterns can readily be obtained using an optical diffractometer and the procedure is described in Chapter 4. By means of the theory outlined in Chapter 1 the optical diffraction pattern can be used to identify the line or space group to which the periodic structure belongs and to estimate the values of the helical or unit cell parameters (see, for example, Klug and Berger, 1964; Hitchborn and Hills, 1967; Miller, 1968; Markham, 1968; Berger, 1969; Sternlieb and Berger, 1969; Burton, 1970; Lake and Slayter, 1970; Moore *et al.*, 1970; Champness, 1971; Mikhailov, 1971).

Elliott *et al.* (1968b) have described the use of a convolution camera (Hosemann and Bagchi, 1962) for the analysis of periodicities in electron micrographs. The camera produces a Patterson-like function and thus in general complicates rather than simplifies interpretation of periodicities. It was claimed that the convolution camera was capable of revealing periodicities which could not be detected by the optical diffraction method of Klug and Berger (1964).

III. IMAGE AVERAGING

Electron micrographs of regular structures such as helices or crystals are in general poor representations of the original material, due, first, to various types of distortion introduced during the preparation and recording of the image and, second, to a combination of irregular staining and various types of damage which causes originally like volumes to be modified in a random fashion. These random differences can be regarded as noise and McLachlan (1958) pointed out that by combining a number of images of originally like volumes, reinforcement of genuine detail would occur while the noise would tend to average out.

This principle has been applied to micrographs of a variety of biological materials (Fraser *et al.*, 1962b; Markham *et al.*, 1963, 1964; Valentine, 1964; Hitchborn and Hills, 1967; Labaw and Rossmann, 1969; Burton, 1970; Fraser and Millward, 1970; Erickson and Klug, 1971).

Markham *et al.* (1963) have described a method of averaging applicable to objects with rotational symmetry, and this was later elaborated by Crowther and Amos (1971). The method is useful when the symmetry is known, but attempts to *determine* the symmetry by this means are fraught with difficulty (Agrawal *et al.*, 1965; Norman, 1966; Frisch, 1969; Friedman, 1970).

When translational periodicities are present in a micrograph an averaged image of the repeating unit can be obtained by superposing a number of images (McLachlan, 1958). In the procedure described by Markham *et al.* (1964) a mechanical device was used to photograph an electron micrograph. Multiple exposures were made on a single sheet of film and the micrograph was translated a distance equal to the period between each exposure. The method is laborious, however, and requires an independent series of experiments to determine the direction and magnitude of the periodicity. Warren and Hicks (1971) have described an alternative method, employing multiple light sources, that enables the averaged image to be viewed directly.

Finch *et al.* (1967) showed that a reduction in noise in micrographs of periodic objects could be obtained by filtering the optical diffraction pattern of the micrograph, and techniques for obtaining a number of different types of averaged image by this means have been described by Fraser and Millward (1970). The advantages of the optical diffraction method are its flexibility and the fact that the magnitude and directions of periodicities do not need to be determined explicitly. Techniques used to obtain and filter optical diffraction patterns of electron micrographs are described in Chapter 4.

Although optical processing of images is simple, it is difficult to make it completely quantitative and the technique is gradually being replaced by exact numerical processing of digitized data obtained from the micrograph with a scanning microdensitometer. An important advantage of the latter method is that both the magnitude and the phase of the Fourier transform are readily available. The phase information is particularly valuable for assessing the degree of preservation of symmetric structures and for locating symmetry axes (Finch and Klug, 1971; Klug, 1971).

IV. THREE-DIMENSIONAL RECONSTRUCTION

The depth of field in the electron microscope is such that all the axial levels in the specimen are in focus at the final image and the superposition of detail at various levels complicates interpretation. With filaments having helical symmetry some resolution of the detail at the upper and lower levels can be obtained by filtering an optical diffraction pattern of the micrograph (Klug and DeRosier, 1966). The method is based on the fact that when a discontinuous helix is projected onto a plane parallel to its axis the distribution of projected points in the vicinity of the projected axis resembles that of two superimposed two-dimensional lattices, one corresponding to the "front" of the helix and the other corresponding to the "back" of the helix. If the helical parameters of the filament are known, the $(-n, l)$ plot (Fig. 1.14) can be used to predict regions in the optical diffraction pattern which contain contributions primarily from, say, the "front" of the filament. If these are isolated by filtering and are recombined using the methods outlined in Chapter 4, the resulting image will be freed from overlapping detail originating from the "back" of the filament.

Although this type of image processing has proved useful in a number of instances (Klug and DeRosier, 1966; Kiselev *et al.*, 1968; Kiselev and Klug, 1969), the method is open to the criticism that subjective decisions are required, and an incorrect choice of helical parameters produces an erroneous result which is not readily recognizable as such.

A more sophisticated and objective method, capable of providing a complete three-dimensional reconstruction of the scattering density in a specimen, has been described by DeRosier and Klug (1968). The method exploits the fact that a transmission electron micrograph is essentially a projection of the scattering density in the direction of the optical axis of the microscope. A series of such projections, obtained with varying orientations of the specimen, contains information about the three-dimensional distribution of scattering density and two methods for extracting this information have been described.

In the so-called algebraic reconstruction technique (Gordon *et al.*, 1970; Frieder and Herman, 1971; Herman and Rowland, 1971) optical density data from projections taken over a limited range of angles are used to construct a set of linear equations involving the scattering power at a grid of points throughout the specimen. These equations are then used as a basis for reconstructing the three-dimensional distribution of scattering power. The reliability of the result is difficult to assess (Crowther and Klug, 1971) and for the present, at least, methods based on a completely representative series of projections (Crowther *et al.*,

1970a,b; DeRosier and Moore, 1970; Crowther, 1971; Gilbert, 1972a,b) would appear to be more generally useful.

The algebraic reconstruction method relates the optical density data directly to the distribution of scattering density but the inversion property of the Fourier transform (Chapter 1, Section A.I.f) can usefully be employed in this procedure (DeRosier and Klug, 1968). If the scattering density ρ in an object is referred to a set of Cartesian coordinates x, y, and z with the x axis parallel to the optical axis of the microscope, the projected density $\sigma(y, z)$, which can be derived from a transmission micrograph, is given by

$$\sigma(y, z) = \int \rho(x, y, z)\, dx \tag{3.1}$$

The Fourier transform of the object (Chapter 1, Section A.I.g) is given by

$$F(X, Y, Z) = \iiint_{\text{object}} \rho(x, y, z)\, \exp[2\pi i(Xx + Yy + Zz)]\, dx\, dy\, dz \tag{3.2}$$

and it follows from Eq. (3.1) that

$$F(0, Y, Z) = \iint_{\text{projection}} \sigma(y, z)\, \exp[2\pi i(Yy + Zz)]\, dy\, dz \tag{3.3}$$

Thus the value of the Fourier transform over the plane $X = 0$ can be evaluated by numerical integration from the x axis projection by means of Eq. (3.3).

Each different projection of the object yields information about a plane in the Fourier transform which passes through the point $X = Y = Z = 0$ (a central section) and by a suitable choice of the number and directions of different projections the Fourier transform can be mapped to any required degree of accuracy (Crowther et al., 1970a,b). The density of scattering matter at any point in the object is then given by

$$\rho(x, y, z) = \iiint_{\text{transform}} F(X, Y, Z)\, \exp[-2\pi i(Xx + Yy + Zz)]\, dX\, dY\, dZ \tag{3.4}$$

In practice Eq. (3.4) is replaced by a summation over a series of elements of constant volume $\delta V = \delta X\, \delta Y\, \delta Z$ and the appropriate value of F must be obtained by interpolation from the measured point values.

In the case of helical structures a single micrograph contains a series of projections of the repeating unit at equispaced azimuths around the helix axis and such structures are particularly suited to the technique of three-dimensional reconstruction. A simplified procedure can be used (DeRosier and Moore, 1970) based on the theory of diffraction by helices (Chapter 1, Section A.III.d). If only one Bessel function contributes to each layer line in the Fourier transform of a helix, Eq. (1.30) becomes

$$F(R, \psi, l/c) = G_{n,l}(R) \exp[in(\psi + \tfrac{1}{2}\pi)] \tag{3.5}$$

where $G_{n,l}$ is defined in Eq. (1.31).

Application of Eq. (3.3) yields information about the Fourier transform over the half-planes $\psi = 0$ $(Y > 0)$ and $\psi = \pi$ $(Y < 0)$ and two estimates of $G_{n,l}$ can be obtained given by

$$G_{n,l}(R) = F(R, 0, l/c) \exp(- in\pi/2) \tag{3.6}$$

$$G_{n,l}(R) = F(R, \pi, l/c) \exp(in\pi/2) \tag{3.7}$$

The reconstruction is performed by using the inverse Fourier–Bessel transformation (Klug et al., 1958)

$$\rho(r, \phi, z) = \sum_l g_{n,l}(r) \exp\{i[n\phi - (2\pi l z/c)]\} \tag{3.8}$$

where

$$g_{n,l}(r) = 2\pi \int_{\text{layer line}} G_{n,l}(R) J_n(2\pi R r) R \, dR \tag{3.9}$$

This method has been applied successfully to the reconstruction of F-actin filaments (Chapter 15). The resolution attainable is limited by the restriction that measurements of F must be limited to a range of R for which only one Bessel function contributes to each layer line. Gilbert (1972a,b) has described an equivalent "direct" method, and Lake (1972b) has extended the Fourier method to sectioned helical structures.

E. SUMMARY

At present the method of transmission electron microscopy cannot be applied directly to fibrous protein specimens because the contrast between different portions of the specimen is too low. A second limitation is the very rapid deterioration of the specimen due to damage by the electron beam. A number of means are available for increasing specimen contrast by the addition of heavy atoms. In certain cases the heavy-atom deposits are sufficiently stable to the electron beam for micrographs to be recorded, even though the protein is disorganized. When used in conjunction with image-averaging and three-dimensional reconstruction techniques micrographs so obtained provide information about the morphology and the mode of aggregation of fibrous proteins.

The accuracy of the method with respect to the measurement of linear dimensions is inherently low and the presently achieved resolution (of the heavy-atom deposits) is probably only about 10 Å even in the most favorable cases.

LIST OF SYMBOLS FOR CHAPTER 4

r, ϕ, z	cylindrical polar coordinates in real space
R, ψ, Z	cylindrical polar coordinates in reciprocal space
n_p	number of projections
n_B	number of Bessel functions
u	number of units in the repeat of a helix
$\mathbf{s}, \mathbf{sr}$	line groups
$\mathcal{N}$	specification of rotation axis
U	lowest common multiple of $\mathcal{N}$ and u
n	order of Bessel function
l	Miller index
$[x]$	value of x rounded up to next highest integer
u, v, w	images
x, y, z	points
n	number of points

Chapter 4

Optical Diffraction

The idea of using optical diffraction to simulate X-ray diffraction was first suggested by Bragg (1939), and subsequently Lipson and co-workers (Lipson and Taylor, 1958; Taylor and Lipson, 1964; Lipson, 1972) developed the method to the stage where the optical diffractometer could be used as an analog computer for X-ray studies. Their methods have proved to be particularly valuable in the initial stages of the development of model structures for synthetic polypeptides and fibrous proteins (Elliott and Malcolm, 1958; Fraser and MacRae, 1959a, 1961a,b; Tomlin and Ericson, 1960; Rich and Crick, 1961; Bradbury *et al.*, 1962a) and in studies of the effects of disorder on fiber diffraction patterns (Hosemann, 1962; Hosemann and Bagchi, 1962; Predecki and Statton, 1965; Mukhopadhyay and Taylor, 1971; Taylor, 1972).

Optical diffractometry has also proved to be a powerful means of analyzing electron micrographs of periodic objects (Klug and Berger, 1964; Klug and DeRosier, 1966; Fraser and Millward, 1970; Lake, 1972a). Techniques are available for performing Fourier syntheses (Taylor and Lipson, 1964; Harburn, 1972) using an optical diffractometer but this process can be carried out more simply and precisely using a digital computer.

A. EXPERIMENTAL ARRANGEMENT

Accounts of experimental procedures for obtaining optical diffraction patterns have been given by Taylor and Lipson (1964), Markham (1968), and DeRosier and Klug (1972) and a typical arrangement is shown in Fig. 4.1. The source is a 5.0-mW helium–neon laser operated in the

83

TEM_{00} mode and the beam is condensed onto a pinhole about 50 μm in diameter by means of a microscope objective of sufficient magnifying power to give uniform illumination over the maximum area of specimen

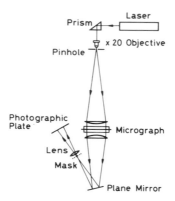

FIG. 4.1. Experimental arrangement used for filtering the optical diffraction pattern of an electron micrograph and reconstructing the filtered image. The diffraction pattern can be recorded by placing a photographic plate at the position marked "Mask" (Fraser and Millward, 1970).

to be used. The diameter of the beam at the pinhole is appreciably less than 50 μm and the purpose of the pinhole is merely to provide a reference point for alignment.

The specimen is placed in the parallel beam between two lenses. These have a focal length of 150 cm and the system is designed for minimum spherical aberration. The optical diffraction pattern of the specimen is recorded at the focal plane of the second lens. If the diffraction pattern is to be filtered, a suitable mask is placed in the focal plane of the second lens and the image of the specimen is reconstructed by means of a camera (Fig. 4.1). It is advantageous to treat all the lenses used in the diffractometer with a coating designed to reduce reflection.

The instrument illustrated in Fig. 4.1 is a modified version of a commercially available unit (R. B. Pullin and Co. Ltd, Brentford, England) and operates in the vertical position. This is preferable to a horizontal system (Wyckoff et al., 1957; Markham, 1968) because convection currents can cause a time-dependent distortion of the diffraction pattern in the latter arrangement unless special precautions are taken.

An alternative optical arrangement has been described by Elliott and Robertson (1955) in which a single parabolic mirror is used. The apparatus is extremely simple but suffers from the practical difficulty that the specimen and the diffraction pattern are separated vertically by a considerable distance.

B. TESTS OF MODEL STRUCTURES

When the optical diffractometer is used to test and refine model structures the positions of the atoms in a chosen projection are calculated and a mask is prepared in which the atoms are represented by circular holes or by annuli in an opaque screen. No entirely satisfactory method exists of simulating atomic scattering factors and most investigators have used circular holes of area proportional to the number of electrons in the atom. The factors involved in the choice of area per electron have been discussed by Taylor and Lipson (1964). It is convenient to plot the atomic positions on a large scale and to use a pantograph punch or drill to prepare a mask of the required size.

The complexity of structure that can be investigated is limited by overlap of the holes representing atoms at different levels along the axis of projection. A limited amount of overlap can be accommodated by punching a single hole with an area proportional to the sum of the electrons in the overlapping atoms.

Other techniques which have been used to prepare masks include etching (Hooper *et al.*, 1955; Taylor and Lipson, 1964) and photographic reduction (Bragg, 1942; Hooper *et al.*, 1955; Wyckoff *et al.*, 1957). Photographic techniques are particularly valuable for investigating diffraction by continuous distributions of scattering material.

The optical diffraction pattern of a projection of a structure gives information about the distribution of intensity over a plane through the origin (a central section) of the intensity transform (Chapter 1, Section A.I). In order to simulate a fiber pattern, several projections must be made normal to the fiber axis and the optical diffraction patterns combined photographically. The number of sections which must be used depends upon the maximum value of the radial reciprocal space coordinate R which is to be recorded and on the symmetry of the structure.

The problem of combining the patterns from a series of projections has been considered by Stokes (1955). For a structure of maximum radius r without rotational or screw symmetry about the fiber axis the number of projections required to give an adequate representation to a particular value of R was shown to be

$$n_{\rm p} = [4\pi Rr] + 3 \qquad (4.1)$$

where $[4\pi Rr]$ indicates a rounding up to the nearest integer. The projections are made at equispaced intervals of ϕ given by $\Delta\phi = 2\pi/n_p$.

With helical structures belonging to line group **s** (Chapter 1, Section A.II.d) the intensity transform has a u-fold rotation axis, where u is the number of units in the repeating distance (Cochran *et al.*, 1952). The number of projections required is thereby reduced and Klug *et al.* (1958) showed that n_B projections are required, uniformly spaced in the angular range $2\pi/u$, where n_B is the maximum number of Bessel functions contributing to any layer line in the intensity transform. This number can be estimated from the expression

$$n_B = [(4\pi Rr + 4)/u] \tag{4.2}$$

where [ ] indicates a rounding up to the nearest integer. If the helical structure has an $\mathcal{N}$-fold rotation axis (line group **sr**) the projections must be uniformly spaced in the angular range $2\pi/U$, where U is the lowest common multiple of u and $\mathcal{N}$.

The optical diffractometer was originally used to test model structures but it is difficult to make the method quantitative. The representation of atomic scattering factors is poor, the photographic process and subsequent densitometry are tedious, and systematic errors due to uneven illumination and imperfections in optical alignment are always present. As a result, quantitative assessment of model structures is usually carried out nowadays by digital computer. The main purpose for which the optical diffractometer is useful in this context is to explore the effects of the various types of disorder which are present in fiber specimens (Elliott and Malcolm, 1958; Fraser and MacRae, 1961a,b; Hosemann and Bagchi, 1962; Mukhopadhyay and Taylor, 1971).

C. OPTICAL PROCESSING OF ELECTRON MICROGRAPHS

Optical diffraction patterns can be obtained directly from the photographic plate used to record electron miscroscope images but it is usually more convenient to reproduce the image in reversed contrast on a second plate. At the same time the magnification can be adjusted to suit the characteristics of the diffractometer. In order to reduce the optical path differences through the photographic plate due to surface irregularities it is desirable to mount the plate between a pair of optical flats with a capillary film of a suitable immersion medium filling the gaps between the plate and the flats (Bragg and Stokes, 1945). Cedarwood oil is suitable for this purpose.

I. ANALYSIS OF PERIODIC STRUCTURES

The Fraunhofer diffraction pattern of an electron micrograph, provided by the optical diffractometer, is a decomposition of the image into its periodic components. Each discrete spot indicates the existence of a periodicity and the sphere of reflection (Chapter 1, Section A.I.c) is sufficiently flat for the direction and magnitude of the periodicity to be determined directly from the optical diffraction pattern. If a line is drawn from the spot to the center of the pattern, the length is inversely proportional to the period and the direction is parallel to that of the periodicity. The constant of proportionality can be determined by recording the diffraction pattern of a mask with a series of holes punched in it at regular intervals along a line.

Analysis of optical diffraction patterns obtained from electron micrographs of helical structures follows the lines discussed in Chapter 1 (Section A.III) but it must be remembered that some distortion might be present in the preparation and different portions of the structure might be stained differently. These factors combine to destroy the precise helical symmetry. These and other imperfections combine to produce a background of "noise" superimposed on the pattern of discrete spots or lines.

A series of rings, lines, or spots emanating from the origin of the optical diffraction pattern will always be present due to the fact that the selected area of the micrograph has finite dimensions. If overlap with a region of interest occurs, the shapes and positions of the subsidiary maxima around the origin can be adjusted by varying the shape and size of the illuminated area of the micrograph.

II. THREE-DIMENSIONAL RECONSTRUCTION BY FILTERING

The technique devised by Klug and DeRosier (1966) for separating the images of the "back" and "front" from the projection of helical structures was discussed in Chapter 3, Section D.IV. An optical diffraction pattern is recorded from a selected area of a plate prepared from the micrograph and the Bessel functions contributing to each layer line are identified. This, of course, presupposes that the helical parameters have been correctly determined. The $(-n, l)$ plot (Chapter 1, Section A.III.d) can be used to identify maxima originating from the same side of the helix and a mask is cut to pass only these regions of the transform (Fig. 4.2). The central cross is common to both sides and the amplitude must be reduced to one-half. This is conveniently achieved by using a fine mesh screen with about 50% transmission. When an image is reconstructed from the filtered diffraction pattern, using the arrangement in Fig. 4.1, it approximates the projection of one side of the structure (Fig. 4.2). A detailed account of the procedure has been given by DeRosier and Klug (1972).

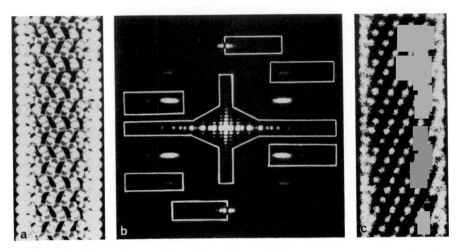

FIG. 4.2. (a) Positive replica of a photographic transparency representing the orthogonal projection of a helical structure onto a plane through the axis. (b) Optical diffraction pattern of projection. (c) Filtered image of projection obtained by admitting only those regions of the diffraction pattern indicated by boxes. From Klug and DeRosier (1966).

Although the results obtained with negatively stained specimens with prominent surface features are very striking, the method is not particularly suited to the study of internal structure since the assumption that maxima in the diffraction pattern can be wholly ascribed to the "back" or "front" of the structure will not always be valid.

III. IMAGE AVERAGING BY FILTERING

The optical diffractometer can be used as a convenient means of obtaining an averaged image of the repeating unit in periodic structures (Fraser and Millward, 1970). The method is based upon the fact that the averaged image is contained in the convolution of the image of the periodic structure with the image of a point lattice on which the repeating units are arrayed. The process is illustrated in Fig. 4.3a, in which the image of the periodic structure contains three repeating units u, v, w all differing slightly in appearance. A second image x, y, z is the point lattice on which u, v, and w are situated, and is conveniently termed the averaging function.

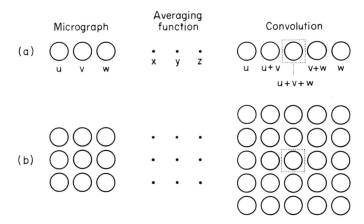

FIG. 4.3. Illustrations of the convolution operation. An arbitrary point in a micrograph of a periodic structure is moved successively to each of the points in the averaging function and the results superimposed. From Fraser and Millward (1970).

The convolution of the micrograph and the averaging function is obtained by choosing some arbitrary point in the micrograph, for example, the center of v, and adding together the result of successively transferring v to x, y, and z. As shown in Fig. 4.3a, the central image in the convolution is a summation of u, v, and w, that is, an average of the three repeating units in the original micrograph. An identical process can be applied to a two-dimensional array and in Figure 4.3b the central image in the convolution is the summation, or average, of the nine images in the micrograph. The averaging function does not necessarily have to have the same number of points as there are repeating units in the selected area of the micrograph.

The convolution operation can be carried out by forming the diffraction pattern of the micrograph and filtering it with a mask which represents the transform of the averaging function. The two Fourier transforms are thereby multiplied together and if the transmitted light is resynthesized to give an image, using the arrangement shown in Fig. 4.1, this will be the convolution of the micrograph and the averaging function.

The masks used to approximate the Fourier transform of the averaging function can be prepared by photographing the diffraction pattern of the micrograph and enlarging this by a factor equal to the reduction ratio of any pantograph available. The Fourier transform of the required averaging function can be drawn directly onto the enlargement and the required mask prepared with the pantograph using a punch with a suitable diameter.

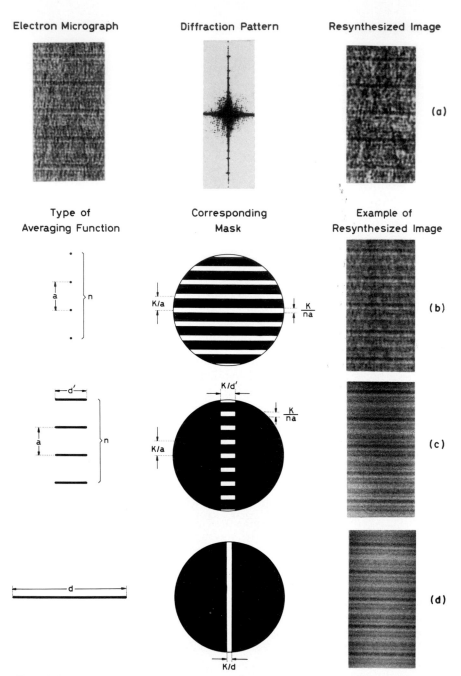

FIG. 4.4. Illustrations of the use of optical filtering to obtain an averaged image of the repeating unit in a one-dimensionally periodic structure (paracrystal of tropomyosin). In (a) the image of the electron micrograph is reconstructed without filtering, in (b) a point lattice is used as an averaging function, in (c) a combination of these two types of average is used, while in (d) the image is continuously averaged perpendicular to the repeat direction. From Fraser and Millward (1970).

Several types of averaging functions can be applied to the micrograph depending on the type of regularity which is present. The simplest, which is applicable to structures such as helices and other one-dimensionally periodic structures, consist of a linear array of points (Fig. 4.4b). A reduction in image "noise" by a factor of about $\sqrt{n}$ would be expected, where n is the number of points in the averaging function. In practice cumulative defects in the actual lattice cause the positions of repeating units separated by greater than a certain number of lattice positions to be uncorrelated. The value of n used must be less than this number.

The same principles apply when two-dimensional regularity is present (Fig. 4.5). Linear averaging functions (Fig. 4.4c) and periodic linear averaging functions can also be used (Fig. 4.4d).

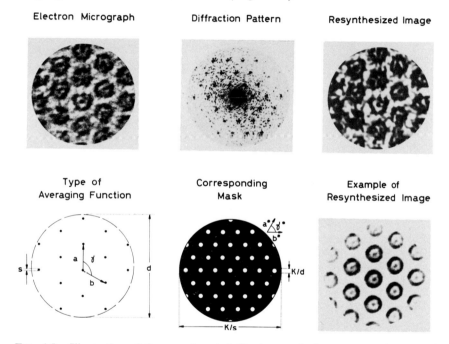

FIG. 4.5. Illustration of the use of optical filtering to obtain an averaged image of the repeating unit of a two-dimensionally periodic structure (α-keratin). From Fraser and Millward (1970).

D. SUMMARY

The use of optical diffraction to test model structures is limited by the inadequate representation of atomic scattering factors and the difficulty of obtaining quantitative results. It has considerable value, however, in semiquantitative studies of the diffraction patterns of continuous distributions of scattering matter and in investigations of the effects of disorder on fiber patterns.

The optical diffraction pattern of an electron micrograph provides information about a central section of the Fourier transform of the specimen and is useful for analyzing periodicities. By the use of filtering and reconstruction techniques image averaging can be carried out and some resolution of detail in the direction of projection is possible. For quantitative studies it is preferable to record the optical density data in a digital form, using a two-dimensional scanning microdensitometer, and to compute the Fourier transform. Transforms and resynthesized images can be presented photographically if desired by means of suitable digital–analog devices.

LIST OF SYMBOLS FOR CHAPTER 5

a, b, c	Cartesian coordinates
x, y, z	Cartesian coordinates
α, ψ	polar angle and azimuth
θ, ϕ	polar angle and azimuth
f_θ, f_ϕ	orientation parameters for amide-group axes
f	orientation parameter for fiber specimens
C_a, C_b, C_c	orientation density functions
$\langle \, \rangle$	average value
$\mathscr{A}$	observed absorbance or optical density
ν	frequency
ν_0	unperturbed frequency of amide-group vibration
$\nu_\parallel, \nu_\perp$	vibrational frequencies of ordered structures
ν_m	frequency of band maximum
A_m	absorbance at band maximum
$\Delta\nu_{1/2}$	bandwidth at half-height
B	baseline function
g	fraction of Gauss function in a band shape
γ	asymmetry parameter in a band shape
D	interaction coefficient
k_x, k_y, k_z	integrated absorption coefficients or intensities
$\mathbf{M}_a, \mathbf{M}_b, \mathbf{M}_c$	components of transition moment
R_0	dichroic ratio with perfect order
R	dichroic ratio with disorder present
G	constraint function in optimization process
P	parameter in optimization process
W	weighting function
S	least squares function
v	extension ratio in Kratky rodlet function

Chapter 5

Infrared Spectrophotometry

Infrared absorption spectra yield information about the nature and frequencies of molecular vibrations and the magnitudes and spatial directions of the associated transition moments. Molecular vibrations are sensitive to the conformation of the polypeptide chain and to the pattern of interactions between neighboring chains and their study thus provides a powerful means of investigating molecular structure. General accounts of the use of infrared spectra in investigations of conformation in polypeptides and proteins are available (Bamford *et al.*, 1956; Fraser, 1960; Elliott, 1969; Susi, 1969; Fraser and Suzuki, 1970a) and in the present chapter attention will be restricted to the quantitative interpretation of spectra. Experimental procedures, together with precautions which must be observed in data collection for quantitative studies, have been described in detail by Fraser and Suzuki (1970a).

A. VIBRATIONAL MODES OF FIBROUS PROTEINS

Any vibration of a polyatomic molecule can be analyzed into a linear combination of independent or *normal* modes of vibration and the number of such modes is $3N - 6$, where N is the number of atoms in the molecule (Herzberg, 1945). When a normal mode of vibration is excited by the absorption of a quantum of radiation of appropriate energy all the atoms in the molecule oscillate about their equilibrium positions with the same frequency. However, there will be some modes that involve significant amplitudes of vibration only in a particular grouping

95

of atoms, and it has been established empirically that the frequency of vibration differs only slightly in different molecules containing the same grouping. Most of the studies of fibrous proteins carried out so far concern these so-called "group vibrations." General accounts of group vibrations have been given by Bellamy (1958, 1968), Miller and Willis (1961), Nakanishi (1962), Flett (1963), Colthup et al. (1964), Brand and Eglinton (1965), and Yamaguchi (1965) and group vibrations in protein side chains have been discussed specifically by Fraser (1960), Bendit (1967, 1968a), Susi (1969), and Fraser and Suzuki (1970a).

I. AMIDE-GROUP VIBRATIONS

Investigations of the nature of the amide-group vibrations have been described in detail by Miyazawa (1962, 1963, 1967), Elliott (1969), and Susi (1969). Although the vibrations are highly localized, none is capable

TABLE 5.1

Amide-Group Vibrations[a]

Designation		Approximate frequency[b] (cm^{-1})	Description
Amide	A	3300	Result of Fermi resonance between the first excited
	B	3100	state of the N—H stretching mode and the second
			excited state of the amide II vibration[c]
Amide	I	1650	In-plane modes[d]; the potential energy distributions are
	II	1550	such that none is capable of simple description
	III	1300	although amide I approximates to a C=O
	IV	625	stretching mode
Amide	V	650	Out-of-plane modes[e]; amide V involves an N—H
	VI	550	bending motion, amide VI a C=O bending motion,
	VII	200	and amide VII a torsional motion about the C—N bond
		4600	Combination of amide A and amide III[f]
		4860	Combination of amide A and amide II[g]
		4970	Combination of amide A and amide I[f]

[a] Trans configuration.
[b] The frequencies are sensitive to environment.
[c] Badger and Pullin (1954), Miyazawa (1960b, c), Tsuboi (1964).
[d] Miyazawa et al. (1956, 1958), Miyazawa (1962, 1963, 1967).
[e] Kessler and Sutherland (1953), Miyazawa (1961b, 1963, 1967).
[f] Fraser (1956a).
[g] Glatt and Ellis (1948).

of simple description in terms of individual bond-stretching or -bending modes. The vibrations have been given arbitrary designations and their main characteristics are summarized in Table 5.1. Absorption bands associated with binary combinations of the amide-group vibrations have also been used to study conformation in fibrous proteins and the assignments are given in Table 5.1.

Information on the directions of the transition moments associated with the in-plane amide-group vibrations has been obtained by studying simple model compounds (Sandeman, 1955; Abbott and Elliott, 1956; Bradbury and Elliott, 1963a) and the results are summarized in Table 5.2.

TABLE 5.2

Transition Moment Directions Associated with Certain Amide-Group Vibrations

Material	Transition moment direction[a]			Reference
	Amide A (deg)	Amide I (deg)	Amide II (deg)	
N, N'-Diacetylhexa-methylene diamine	+8	+17	+68, +77 (double)	Sandeman (1955)
Acetanilide	+11 or −8	+22 (±2)	+72 or −57	Abbott and Elliott (1956)
N-Methyl acetamide	+8 or +28	+15 to +25	+73 or −37	Bradbury and Elliott (1963a)
Silk fibroin	—	+19	—	Suzuki (1967)
β-Keratin	—	+19	—	Fraser and Suzuki (1970c)

[a] Measured relative to the direction of the CO bond:

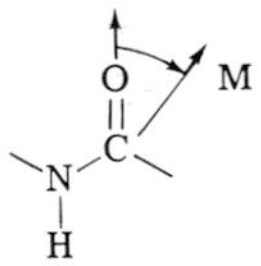

The term "transition moment direction" is somewhat misleading. All the values are arbitrarily quoted in the range $0 \pm 90°$, since transition moments inclined at θ and $\theta + 180°$ cannot be distinguished by dichroism measurements with linearly polarized radiation.

The transition moments associated with the out-of-plane modes (V, VI, and VII) are predicted to be perpendicular to the plane of the amide group. In polypeptides the environment of the amide group will be different from that in crystals of model compounds and so the transition moment directions may be somewhat different. Estimates of the transition moment direction in an isolated amide group for the amide I vibration have been obtained from silk and from β-keratin and these are included in Table 5.2. The transition moment directions associated with the combination modes listed in Table 5.1 are not known precisely but some preliminary limits have been set (Fraser, 1956a).

In fibrous proteins and synthetic polypeptides the vibrations in different amide groups are coupled through the main chain and through the formation of hydrogen bonds. The form of the coupling affects both the frequencies of the normal modes of the assembly and the directions of the associated transition moments and thus provides a basis for investigating the conformation and orientation of the main chain.

II. EFFECTS OF CONFORMATION

In considering the main-chain vibrations of fibrous protein molecules it is convenient to distinguish between regions of "regular" conformation, in which the amide groups of the main chain are related by symmetry operations, and regions of irregular conformations, in which they are not so related. Individual molecules generally contain both types of conformation. The terms random, amorphous, unordered, and disordered have also been used to describe these regions of irregular conformation. None is entirely satisfactory since the irregular portion of the molecule may be arranged in a specific conformation that is repeated in all the other molecules in the structure.

To a first approximation the regular and irregular portions of the main chain in an individual molecule can be treated separately and the coupling between them neglected. Thus it is possible to consider the observed main-chain absorption bands as arising from a linear combination of contributions from regions of regular and irregular conformation.

The effects of a regular conformation on the vibrational modes of a polymer are generally approached through a consideration of the modes of an isolated chain of infinite length. This greatly simplifies the problem since it can be shown that the only vibrations that are potentially infrared- or Raman-active are those in which corresponding atoms in successive axial repeats of structure vibrate in phase. The number of such vibrations is $3ru - 4$, where r is the number of atoms in a monomer unit and u is the number of units in the axial repeat. The nature of these normal vibrations and the associated transition moments can be analyzed by a procedure described by Liang *et al.* (1956), Liang and Krimm (1956), Krimm (1960a), and Elliott (1969).

a. *Isolated Helical Chain*

In the case of a helical conformation it can be shown (Higgs, 1953) that the only modes that are infrared-active are those for which the atomic motions in successive units are related by a constant phase difference δ given by

$$\delta = 0, \pm t \tag{5.1}$$

where t is the unit twist (Chapter 1, Section A.II.c). The transition moment associated with the mode for which $\delta = 0$ is parallel to the helix axis and the transition moments for the modes with $\delta = \pm t$ are mutually perpendicular and normal to the helix axis.

The various localized vibrations of the amide group have counterparts in the normal modes of an infinite helical polypeptide chain. Miyazawa (1960a) has shown that the vibrational frequencies are given approximately by

$$\nu(\delta) = \nu_0 + \sum_j D_j \cos(j\delta) \tag{5.2}$$

where ν_0 is the unperturbed frequency of the group vibration, D_j is an interaction coefficient which depends on the magnitude of the coupling between the reference unit and the jth unit along the chain, and δ is the phase angle between motions in neighboring units, which is restricted to the values given in Eq. (5.1).

In the case of the α-helix, significant interactions are to be expected between the reference residue and the adjacent residue ($j = 1$) through the common alpha carbon atom and between the reference residue and the residue to which a hydrogen bond is formed ($j = 3$). The interactions for $j = -1$ and -3 are implied by the presence of interactions for $j = 1$ and 3 and need not be explicitly included. Each group vibration therefore gives rise to three infrared-active vibrations in an infinite α-helix with frequencies

$$\nu(0)_\parallel = \nu_0 + D_1 + D_3 \tag{5.3}$$

$$\nu(\pm t)_\perp = \nu_0 + D_1 \cos(\pm t) + D_3 \cos(\pm 3t) \tag{5.4}$$

The vibrations corresponding to $\delta = \pm t$ have the same frequency and are said to be degenerate.

For an isolated chain in a regular β conformation $t = \pi$ and the only interaction of consequence will be for $j = 1$; thus from Eq. (5.2)

$$\nu(0)_\parallel = \nu_0 + D_1 \tag{5.5}$$

$$\nu(\pm \pi)_\perp = \nu_0 - D_1 \tag{5.6}$$

The magnitude of D_1 in Eqs. (5.5) and (5.6) will be different from that in Eqs. (5.3) and (5.4) since the spatial relationship between successive residues in the α-helix is different from that in the β conformation.

b. *Interchain Interactions*

In the α-helical conformation the interactions between amide groups are intrachain and the frequencies given by Eqs. (5.3) and (5.4) would be expected to be a good approximation even when the helix was surrounded by other chains. The situation is quite different with the more extended conformations found in polyglycine II and β-sheets since the vibrations of the amide groups in adjacent chains are coupled through interchain hydrogen bonds. In these and similar cases an interchain interaction term must be introduced into Eq. (5.2) giving

$$\nu(\delta, \delta') = \nu_0 + \sum_j D_j \cos(j\delta) + D_1' \cos \delta' \qquad (5.7)$$

where the coefficient D_1' is a measure of the interchain interaction and δ' is the phase angle between motions in the two coupled amide groups (Miyazawa, 1960a). The permissible values of δ' are determined by the fact that the only potentially infrared-active modes of assemblies of helices are those in which the motions of corresponding atoms in different unit cells are in phase (Liang *et al.*, 1956). Krimm and Abe (1972) have suggested that Eq. (5.7) may not be adequate to account for the amide I components of pleated sheet structures and proposed that an additional interchain term should be included to take account of transition dipole coupling.

In the parallel-chain pleated-sheet β conformation (Pauling and Corey, 1951g, 1953a) the repeat of structure contains only one chain (Fig. 5.1) and it follows that the only permissible value of δ' is zero (Miyazawa, 1960a). Thus the observable vibrational frequencies are given by

$$\nu_\parallel(0, 0) = \nu_0 + D_1 + D_1' \qquad (5.8)$$

$$\nu_\perp(\pi, 0) = \nu_0 - D_1 + D_1' \qquad (5.9)$$

where it has been assumed that only nearest neighbor interactions are significant.

The directions of the transition moments associated with these vibrations can be understood by considering the way in which the contributions from the two amide groups in the repeat of structure are combined (Fig. 5.1). For convenience an axis b is chosen parallel to the chain axis, an axis a in the plane of the sheet normal to b, and an axis c normal to both a and b. If $\mathbf{M}$ is the contribution of residue 1 to the transition

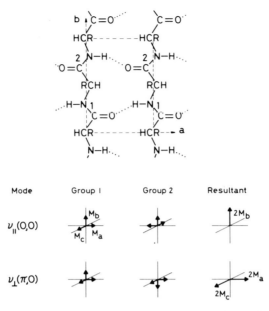

FIG. 5.1. Arrangement of amide groups in the parallel-chain pleated-sheet β conformation (Pauling and Corey, 1953a). Two phase combinations between the motions in residues 1 and 2 are possible (Miyazawa, 1960a), one leading to a resultant transition moment parallel to the b axis, the other to a transition moment perpendicular to the b axis. The transition moment $\mathbf{M}$ has been taken in an arbitrary direction. From Fraser and Suzuki (1970a).

moment, this can be resolved into components $\mathbf{M}_a$, $\mathbf{M}_b$, and $\mathbf{M}_c$. When combined with the corresponding components from residue 2 the result is $2\mathbf{M}_b$ for the $\nu_\parallel(0, 0)$ mode and $2(\mathbf{M}_a + \mathbf{M}_c)$ for the $\nu_\perp(\pi, 0)$ mode. If $\mathbf{M}$ makes an angle α with the b axis and its projection onto the ac plane makes an angle ψ with the c axis, the magnitudes of the transition moments become

$$2M_b = 2M \cos \alpha \tag{5.10}$$

for the $\nu_\parallel(0, 0)$ mode and

$$2M_a = 2M \sin \alpha \sin \psi \tag{5.11}$$

$$2M_c = 2M \sin \alpha \cos \psi \tag{5.12}$$

for the $\nu_\perp(\pi, 0)$ mode.

In the antiparallel-chain pleated sheet the repeat of structure contains four amide groups and the phase angle δ' relating the vibrational motions

in the two chains can assume values of zero or π (Miyazawa, 1960a).
The frequencies of the four modes of vibration are therefore given by

$$\nu(0, 0) = \nu_0 + D_1 + D_1' \tag{5.13}$$

$$\nu_\parallel(0, \pi) = \nu_0 + D_1 - D_1' \tag{5.14}$$

$$\nu_\perp(\pi, 0) = \nu_0 - D_1 + D_1' \tag{5.15}$$

$$\nu_\perp(\pi, \pi) = \nu_0 - D_1 - D_1' \tag{5.16}$$

and the individual contributions to the transition moment from the
four amide groups in the repeat of structure are depicted in Fig. 5.2.
The $\nu(0, 0)$ mode is infrared-inactive and the resultant transition
moments for the other modes are

$$\nu_\parallel(0, \pi) \qquad 4M_b = 4M \cos \alpha \tag{5.17}$$

$$\nu_\perp(\pi, 0) \qquad 4M_a = 4M \sin \alpha \sin \psi \tag{5.18}$$

$$\nu_\perp(\pi, \pi) \qquad 4M_c = 4M \sin \alpha \cos \psi \tag{5.19}$$

The transition moment directions can also be predicted directly from
a consideration of the symmetry of the pleated sheet (Miyazawa, 1967).

In the model proposed for the structure of polyglycine II by Crick
and Rich (1955), which is illustrated in Fig. 11.4, the unit twist t is
$\pm 2\pi/3$ and the unit cell contains only one chain. The permissible values
of δ are therefore zero and $\pm 2\pi/3$, while δ' is restricted to the value
zero; hence the observable frequencies are given by

$$\nu_\parallel(0) = \nu_0 + D_1 + D_1' \tag{5.20}$$

$$\nu_\perp(2\pi/3) = \nu_0 + D_1 \cos(2\pi/3) + D_1 \tag{5.21}$$

c. *Numerical Values of Amide Frequencies*

Early studies of the infrared spectra of fibrous proteins and synthetic
polypeptides (Ambrose and Elliott, 1951a,b,c; Elliott, 1953a; Bamford
et al., 1956) established that the frequencies of certain amide-group
vibrations were sensitive to conformation. In particular the frequencies
of the main components of the amide I and amide II bands were found
to be 25–30 cm^{-1} lower in the β form than in the α form. These frequency
differences were used as a basis for qualitative estimates of the contents
of the α and β forms in different materials but later it was realized that
unordered material gave rise to amide I and II bands with frequencies
close to those associated with the α form (Elliott *et al.*, 1958). Empirical
correlations of frequency and conformation have also been established

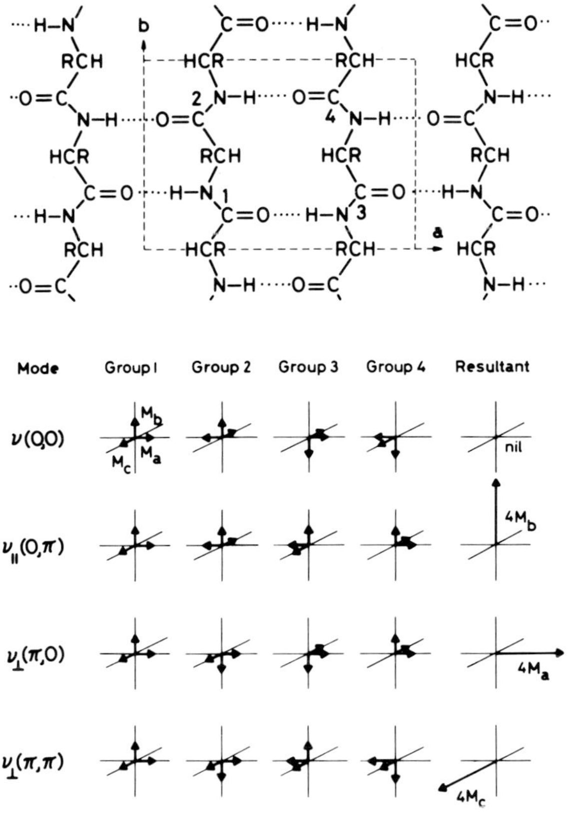

FIG. 5.2. Arrangement of amide groups in the antiparallel-chain pleated-sheet β conformation (Pauling and Corey, 1953a). Four phase combinations of the motions in residues 1, 2, 3, and 4 are possible (Miyazawa, 1960a); the $\nu(0, 0)$ mode is infrared-inactive, the $\nu_\parallel(0, \pi)$ mode has a transition moment parallel to the b axis, the $\nu_\perp(\pi, 0)$ has a transition moment parallel to the a axis, and the $\nu_\perp(\pi, \pi)$ mode has a transition moment perpendicular to the ab plane. The transition moment $\mathbf{M}$ has been taken in an arbitrary direction. From Fraser and Suzuki (1970a).

for the amide A band, which occurs at a frequency about 10 cm^{-1} lower in the β than in the α form (Elliott, 1954) and about 30 cm^{-1} higher in collagen than in the α form (Fraser, 1950b), and for the amide V band (Miyazawa, 1967; Miyazawa et al., 1967; Masuda et al., 1969). The frequency changes in the amide-group modes, associated with changes in conformation, have counterparts in the binary combination modes (Elliott et al., 1954; Fraser, 1960).

The analysis of the effects of conformation on the vibrational frequencies of the amide group, discussed in Sections A.II.a and A.II.b, offers the possibility of predicting the frequencies of the amide absorption bands for different conformations but in order to do this it is necessary to determine values for the unperturbed frequency ν_0 and for the interaction coefficients applicable to different conformations. No method exists at present for calculating these quantities but Miyazawa and Blout (1961) obtained estimates of their values for the amide I and II modes in various conformations by comparing spectra obtained from different materials. On the basis of these estimates a table correlating frequency and conformation was drawn up. This was extended by Krimm (1962), but Elliott and Bradbury (1962) obtained experimental evidence which suggested that the value of ν_0 varied from conformation to conformation and they concluded that the detailed analysis given by Krimm was not justified. Subsequent studies, particularly of spectra obtained from material containing the antiparallel-chain pleated sheet (Table 5.3), have confirmed their suspicion. It is clear that both the

TABLE 5.3

Estimates of ν_0, D_1, and D_1' for the Amide I Mode of the
Antiparallel-Chain Pleated Sheet[a]

Material	ν_0[b] (cm^{-1})	D_1 (cm^{-1})	D_1' (cm^{-1})	Reference
Poly(glycine)	1659	16	−11	Bradbury and Elliott (1963a)
Poly(L-alanine)	1665	—	—	Elliott (1954)
Poly(L-alanylglycine)	1666	—	—	Fraser et al. (1965a)
Poly(γ-methyl-L-glutamate)	1660	—	—	Elliott (1953a)
Poly(L-serine)	1658	23	−14	Koenig and Sutton (1971)
Poly(β-propyl-L-aspartate)	1670	20	−13	Bradbury and Elliott (1963a)
Poly(β-propyl-L-aspartate)[c]	1670	13	−19	—
Anaphe moloneyi silk	1664	—	—	Elliott and Malcolm (1956a)
Bombyx mori silk	1667	—	—	Fraser et al. (1965a)
Stretched porcupine quill	1662	7	−27	Fraser and Suzuki (1970c)
Feather rachis	1663	—	—	Fraser and Suzuki (1965)
Range	1658	7	−11	
	1670	23	−27	
Original estimate[d]	1658	8	−18	Miyazawa and Blout (1961)

[a] ν_0 is the unperturbed frequency, D_1 the intrachain interaction constant, and D_1' the interchain interaction constant.

[b] Rounded to nearest wave number.

[c] Recalculated on basis of assignment of weak perpendicular band at 1676 cm^{-1} to $\nu_\perp(\pi, \pi)$ of antiparallel-chain pleated sheet.

[d] Based on a comparison of frequencies in poly(glycine) I and nylon 66.

unperturbed frequency and the interaction coefficients vary from material to material for the amide I vibration and the situation with the amide II vibration is probably worse (Elliott and Bradbury, 1962; Bradbury and Elliott, 1963b). In these circumstances the tables of predicted frequencies that have been given (Miyazawa and Blout, 1961; Krimm, 1962) must be treated with circumspection.

The amide I vibration has been more thoroughly studied than other vibrations. The vibrational frequencies are sensitive to conformation (Ambrose and Elliott, 1951a,b,c; Elliott, 1953a, 1969; Bamford et al., 1956) and the perturbation treatment given by Miyazawa (1960a) has enabled measurements of dichroism in this region to be used for quantitative structure analysis in synthetic polypeptides (Tsuboi, 1962) and in fibrous proteins (Suzuki, 1967; Fraser et al., 1969a, 1971a; Fraser and Suzuki, 1970c). In the antiparallel-chain pleated sheet the $v_\perp(\pi, 0)$ and $v_\parallel(0, \pi)$ modes have frequencies of about 1635 and 1700 cm^{-1}, respectively, and are well separated from the $v_\parallel(0)$ and $v_\perp(t)$ modes of the α-helix, which occur near 1655 cm^{-1} and differ by only about 2 cm^{-1}. The $v_\perp(\pi, \pi)$ mode of the antiparallel-chain pleated sheet has been tentatively identified at 1683 cm^{-1} in β-keratin (Fraser and Suzuki, 1970c) and at 1650 cm^{-1} in poly(L-serine) (Koenig and Sutton, 1971). Neither of these frequencies agrees with that predicted by Miyazawa and Blout (1961), who estimated a value of 1668 cm^{-1} for this mode. The infrared-inactive mode $v(0, 0)$ has been observed by Koenig and Sutton (1971) in the Raman spectrum of poly(L-serine).

d. *Effects of Finite Chain Length*

In the treatment of the vibrational modes of helical chains given in Section A.II.a it was assumed that the chain was infinite and the extent to which this treatment is applicable to actual, finite structures must be considered. Liang et al. (1956) estimate that if a polymer contains ten or more repeating units, any differences from the spectra of longer polymers are due only to the end groups. A quantitative treatment of this problem has been given by Zbinden (1964) and the principal effect of a finite number of repeating units was predicted to be the appearance of satellite bands on the high-frequency side of the main absorption band. The width of the envelope was predicted to contract with increasing chain length.

Individual satellite bands associated with finite chain length have not so far been observed in spectra from polypeptides but effects related to crystallite size have been noted in spectra obtained from *Bombyx mori* silk fibroin (Suzuki, 1967; Fraser and Suzuki, 1970c). In this material the crystallites are elongated parallel to the fiber axis and it was found that the half-bandwidth of the $\nu_\parallel(0, \pi)$ amide I component was appreciably less than that of the $\nu_\perp(\pi, 0)$ component. According to the treatment given by Zbinden (1964), the band shapes would be expected to be asymmetric, but the bands observed in silk fibroin are symmetric in shape and no satisfactory theoretical treatment has so far been given for this effect.

e. *Irregular Conformations*

In the sections of chain that have an irregular conformation, the nature of the coupling between adjacent groups along the main chain will be variable, as will the relative orientations of hydrogen-bonded groups. This leads to a broadening of the absorption band and it has generally been assumed that this broadening takes place symmetrically about the unperturbed frequency ν_0. This assumption is questionable and probably accounts for some of the difficulties encountered in predicting the frequencies of regular conformations using values of ν_0 so derived.

B. DICHROISM

Measurements of infrared dichroism have played an important part in the development of models for the structure of fibrous proteins but the full potential of the method has not generally been realized because of the difficulties encountered in the quantitative measurement and interpretation of dichroism. The precautions that must be observed in measuring dichroism have been described in detail by Fraser and Suzuki (1970a) and procedures for analyzing the spectrum into individual absorption bands are discussed in Section C. In the present section the relationship between dichroism, and conformation and orientation will be considered.

I. OPTICAL PROPERTIES OF ANISOTROPIC SPECIMENS

The optical properties of anisotropic absorbing media can be specified by a dielectric tensor and a conductivity tensor (Born and Wolf, 1965). The conductivity tensor defines the absorption properties and can be investigated by using polarized radiation. The mode of propagation of radiation through absorbing crystals of low symmetry for an arbitrary direction of incidence and polarization is complex (Ward, 1955; Susi, 1961) and criteria for selecting suitable directions have been given by Newman and Halford (1950). These are that eigenvectors of the dielectric and conductivity tensors be parallel and that the radiation be incident normally on a crystal face containing two principal axes of the dielectric ellipsoid and polarized parallel to one of them. Under these conditions the radiation is propagated without refraction and remains plane polarized.

In crystals of low symmetry the directions of the eigenvectors of the dielectric and conductivity tensors vary with frequency and may also vary independently. It follows that spectra obtained from crystal mosaics and partially oriented microcrystals will be difficult to interpret (Newman and Halford, 1950) and Susi (1969) has stressed that such difficulties may be encountered in the interpretation of infrared dichroism in fibrous proteins. However, Elliott (1969) has pointed out that the dimensions of the crystallites in polymeric materials are very much smaller than the wavelengths of infrared radiation and so it will be the orientation density function (Section B.II), rather than the unit cell type of the crystallites, which determines the optical properties.

The orientation density functions encountered in fibrous protein specimens always have three mutually perpendicular mirror planes and so the principal axes of the dielectric and conductivity tensors are coincident and fixed by symmetry (Born and Wolf, 1965). In these circumstances the three principal absorption coefficients can readily be determined (Ambrose *et al.*, 1949a; Zbinden, 1964; Koenig *et al.*, 1967) and the difficulties foreseen by Susi (1969) do not eventuate in practice. The principal absorption coefficients completely define the absorption properties of the specimen at a particular frequency and their variation with frequency comprises the information that must be related to the orientation and conformation of the protein chains.

II. ORIENTATION DENSITY FUNCTIONS

The preferred orientation of the polypeptide chain in an anisotropic specimen can be described by means of orientation density functions. A convenient type of function for the interpretation of infrared absorption measurements is obtained by choosing a set of three mutually perpendicular axes Oa, Ob, and Oc in a like manner in each residue and describing the spatial density of axes in terms of the spherical polar coordinates θ and ϕ (Fig. 5.3). If the specimen contains a mixture of conformations, it may be necessary to consider them individually if the frequencies of the absorption bands under consideration are conformation-sensitive.

Many types of preferred orientation are possible and various classifications have been suggested (Sisson, 1936; Heffelfinger and Burton, 1960; Zbinden, 1964). Unfortunately, no uniform terminology exists at present but two main types of preferred orientation can be distinguished. In the first the density functions are symmetric with respect to rotation about a unique direction and the X-ray diffraction pattern obtained

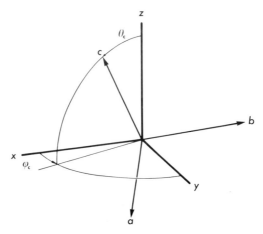

FIG. 5.3.   Coordinate systems used to describe orientation in partially ordered specimens. In doubly oriented specimens x, y, and z are chosen so that they are parallel to the symmetry axes of the orientation density functions. These are also the principal axes of the dielectric and conductivity tensors. In fiber-type orientation z is the fiber axis and x and y are equivalent. The axes a, b, and c define the orientation of an amide group or a region of regular structure. The direction of each of these axes is specified by a polar angle θ and an azimuth ϕ.

when the beam is directed perpendicular to this axis is of the fiber type. It is convenient to refer to the unique axis as the fiber axis even though the specimen may be in the form of a sheet or film. In the second type of preferred orientation the density functions have three mutually perpendicular mirror planes and the specimens are said to exhibit double orientation. Specimens with fiber-type orientation exhibit optical properties which resemble those of an optically uniaxial crystal, while doubly oriented specimens have properties resembling those of optically biaxial crystals of the orthorhombic system.

a. *Doubly Oriented Specimens*

It is convenient to refer orientation density functions for doubly oriented specimens to the three mutually perpendicular symmetry axes of the distribution and to describe the distribution of a axes by a density function $C_a(\theta, \phi)$ such that the fraction of axes in the angular ranges θ and $\theta + d\theta$ and ϕ and $\phi + d\phi$ is $C_a(\theta, \phi) \sin \theta \, d\theta \, d\phi$. Similar functions can be defined for the distributions of the b and c axes. From the definition of C_a it follows that

$$\int_0^{2\pi} \int_0^{\pi} C_a(\theta, \phi) \sin \theta \, d\theta \, d\phi = 1 \tag{5.22}$$

and likewise for C_b and C_c. It is often convenient to choose the orientation of the axes a, b, and c in the amide group in such a manner that the functions C_a, C_b, and C_c have maxima or minima in the direction of the symmetry axes as illustrated in Fig. 5.4a. Since the three axes a, b,

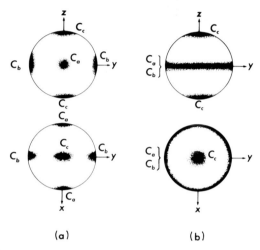

(a)   (b)

FIG. 5.4. Diagrammatic representation of the orientation density functions for the a, b, and c axes in (a) a doubly oriented specimen and (b) a specimen with fiber orientation.

and c have a constant spatial relationship, it follows that C_a, C_b, and C_c are not independent. A further property dictated by the symmetry is that

$$C_a(\theta, \phi) = C_a(\theta, -\phi) = C_a(-\theta, \pm\phi) = C_a(\theta, \pi \pm \phi) = C_a(-\theta, \pi \pm \phi) \tag{5.23}$$

Stein (1961, 1964) has described a method of characterizing doubly oriented specimens by means of a pair of orientation parameters for each of the axes a, b, and c. They are defined by

$$f_\theta = \tfrac{1}{2}(3\langle \cos^2 \theta \rangle - 1) \tag{5.24}$$

$$f_\phi = 2\langle \cos^2 \phi \rangle - 1 \tag{5.25}$$

where $\langle\cos^2\theta\rangle$ and $\langle\cos^2\phi\rangle$ are the mean values of $\cos^2\theta$ and $\cos^2\phi$, respectively, and are given by

$$\langle\cos^2\theta\rangle = \int_0^{2\pi}\int_0^\pi C(\theta,\phi)\cos^2\theta\sin\theta\,d\theta\,d\phi \qquad (5.26)$$

$$\langle\cos^2\phi\rangle = \int_0^{2\pi}\int_0^\pi C(\theta,\phi)\cos^2\phi\sin\theta\,d\theta\,d\phi \qquad (5.27)$$

For the complete description of the orientation texture in a specimen three pairs of orientation parameters are needed for each type of regular conformation present and three pairs for the regions of irregular conformation. In favorable cases these parameters can be determined from X-ray diffraction studies (see, for example, Alexander, 1969) and can be related to the principal infrared absorption coefficients (Stein, 1964; Desper and Stein, 1966).

Doubly oriented specimens are of considerable value in the assignment of infrared absorption bands and in the indexing of X-ray diffraction patterns. They are less valuable for quantitative studies since the orientation density functions are difficult to determine experimentally.

b. *Fiber Orientation*

With fiber orientation the density functions for the a, b, and c axes are rotationally symmetric about the fiber axis so that the parameter f_ϕ defined in Eq. (5.25) is zero for all three axes. The quantities required for the interpretation of dichroism (Section B.III.a) are $\langle\cos^2\theta_a\rangle$, $\langle\cos^2\theta_b\rangle$, and $\langle\cos^2\theta_c\rangle$, where

$$\langle\cos^2\theta\rangle = 2\pi\int_0^\pi C(\theta)\cos^2\theta\sin\theta\,d\theta \qquad (5.28)$$

Since the a, b, and c axes are mutually perpendicular, the three quantities are related by the expression

$$\langle\cos^2\theta_a\rangle + \langle\cos^2\theta_b\rangle + \langle\cos^2\theta_c\rangle = 1 \qquad (5.29)$$

In many cases there is a special direction, usually the chain axis, about which all azimuthal orientations of the amide group are equally probable. If this direction is chosen as the c axis, then

$$\langle\cos^2\theta_a\rangle = \langle\cos^2\theta_b\rangle = \tfrac{1}{2}(1 - \langle\cos^2\theta_c\rangle) \qquad (5.30)$$

and the orientation can be characterized by a single parameter f, equal to f_θ for the c axis, given by

$$f = \tfrac{1}{2}(3\langle\cos^2\theta_c\rangle - 1) \qquad (5.31)$$

It can be shown (Fraser, 1958a) that the absorption properties are equivalent to those of a specimen in which a fraction f of the c axes are aligned parallel to the fiber axis and a fraction $(1 - f)$ are randomly distributed.

In certain specimens, such as cast films, the c axes may be preferentially aligned parallel to the plane of the film while the density function is symmetric about the normal. This is sometimes referred to as uniplanar orientation and is a special case of fiber orientation in which $-\frac{1}{2} \leqslant f \leqslant 0$.

c. Determination of Orientation Parameters from X-Ray Studies

In specimens exhibiting fiber orientation the chain axes generally have a preferred orientation parallel to the fiber axis, and if the c axis is chosen to be parallel to the chain axis, the density function C_c will have maxima for $\theta = 0$ and π (Fig. 5.4b). Information on C_c can be obtained from the distribution of intensity in a meridional X-ray reflection and the quantity $\langle \cos^2 \theta_c \rangle$ in Eq. (5.28) evaluated by numerical integration. Similar measurements on equatorial reflections yield values for $\langle \cos^2 \theta_a \rangle$ and $\langle \cos^2 \theta_b \rangle$ (Hermans, 1946; Stein, 1958). Alternatively, C_a, C_b, and C_c can be fitted to a theoretical model by the variation of one or more adjustable parameters. The value of $\langle \cos^2 \theta_c \rangle$ can then be calculated directly from the parameter values (Warwicker and Ellis, 1965).

In most specimens the X-ray reflections from the individual crystallites are appreciably sharper in the meridional direction than in the equatorial direction and the most suitable arcs, in the observed pattern, for determining orientation density functions are equatorials. If $\langle \cos^2 \theta_a \rangle$ and $\langle \cos^2 \theta_b \rangle$ can be determined in this way, $\langle \cos^2 \theta_c \rangle$ can be obtained (Hermans, 1946) from the relationship

$$\langle \cos^2 \theta_c \rangle = 1 - (\langle \cos^2 \theta_a \rangle + \langle \cos^2 \theta_b \rangle) \tag{5.32}$$

If all orientations around the c axis are equally probable, the density functions C_a and C_b are identical and a single measurement suffices since

$$\langle \cos^2 \theta_c \rangle = 1 - 2\langle \cos^2 \theta_a \rangle = 1 - 2\langle \cos^2 \theta_b \rangle \tag{5.33}$$

In the latter case it can be shown that the density functions C_a, C_b, and C_c are related by the expression

$$C_a(\theta) = C_b(\theta) = \frac{2}{\pi} \int_{(\pi/2)-\theta}^{\pi/2} \frac{C_c(\xi) \sin \xi \, d\xi}{(\sin^2 \theta - \cos^2 \xi)^{1/2}} \tag{5.34}$$

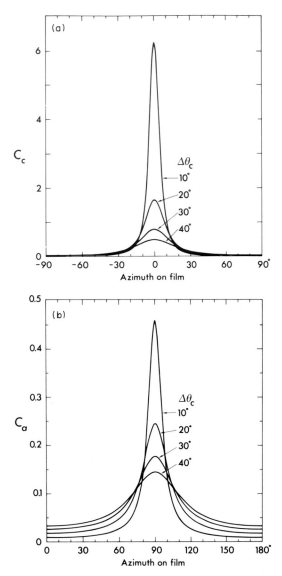

FIG. 5.5. (a) Orientation density function C_c for the c axes in the Kratky rodlet model. (b) Orientation density function C_a for the a or b axes. The c axis is assumed to be preferentially aligned parallel to the direction of stretching and all orientations of the a and b axes around c are assumed to be equally probable. For small and medium scattering angles these correspond to the azimuthal distributions of scattered intensity for (a) meridional arcs and (b) equatorial arcs. $\Delta\theta_c$ is the azimuthal width of the meridional reflection at half-height.

and if a theoretical model is assumed for C_c, the corresponding models for C_a and C_b can be derived in terms of the same adjustable parameters. The optimized values of these parameters can then be used to calculate $\langle \cos^2 \theta_c \rangle$. Methods of determining orientation parameters from dichroism studies are discussed in Section B.III.b.

d. Orientation Density Models

Various models for orientation density functions have been suggested to represent C_c in the case of fiber specimens where all orientations of the a and b axes about the c axis are equally probable and these have been reviewed by Fraser (1958a). The most realistic so far devised is that due to Kratky (1933) which describes the preferred orientation introduced by stretching an array of randomly oriented rodlets embedded in an amorphous matrix. The properties of this system have been investigated in detail (Kuhn and Grün, 1942; Fraser, 1958a; Zbinden, 1964). The orientation density is related to the extension ratio v by the expression

$$C_c(v, \theta) = v^3/\{4\pi[1 + (v^3 - 1)\sin^2 \theta]^{3/2}\} \tag{5.35}$$

and is illustrated, for various values of v, in Fig. 5.5. The corresponding functions for C_a and C_b obtained by carrying out the integration in Eq. (5.34) are

$$C_a(v, \theta) = C_b(v, \theta) = \frac{E[(1 - v^{-3})^{1/2} \sin \theta]}{2\pi^2 v^{3/2}[1 - (1 - v^{-3})\sin^2 \theta]} \tag{5.36}$$

where $E[x]$ is the complete elliptic integral of the second kind of modulus x (Warwicker and Ellis, 1965). The value of $\langle \cos^2 \theta_c \rangle$ for this model in terms of v is (Fraser, 1958a)

$$\langle \cos^2 \theta_c \rangle = \frac{v^3}{v^3 - 1} \left[1 - \frac{\tan^{-1}(v^3 - 1)^{1/2}}{(v^3 - 1)^{1/2}} \right] \tag{5.37}$$

and the relationship between the orientation parameter f defined in Eq. (5.31) and the extension ratio v is depicted in Fig. 5.6a. A rapid method of estimating $\langle \cos^2 \theta_c \rangle$, assuming that X-ray reflections have an azimuthal profile given by Eq. (5.35) or (5.36) as appropriate, is to measure the angular width at half-height. The predicted relationship between $\Delta\theta_{1/2}$ and the orientation parameter f for meridional and equatorial reflections is given in Fig. 5.6b.

Other model functions which have been used to represent C_c are a mean angle of disorientation $\bar{\theta}$ (Fraser, 1953a) and a uniform distribu-

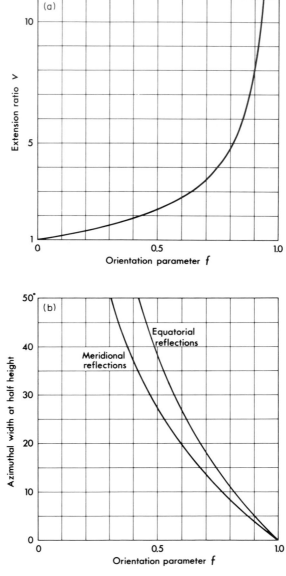

FIG. 5.6. (a) Relationship between extension ratio v and orientation parameter f for the Kratky rodlet model. (b) Angular width at half-height of meridional and equatorial X-ray diffraction arcs as a function of the orientation parameter f for the Kratky rodlet model.

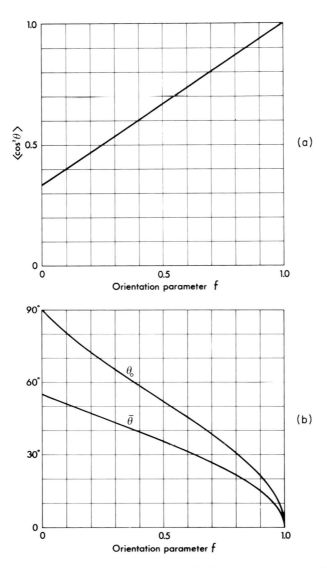

FIG. 5.7. (a) Relationship between $\langle \cos^2 \theta \rangle$ and orientation parameter f. (b) Relationship between orientation parameter f and the angles $\bar{\theta}$ and θ_0 defined in Eqs. (5.38) and (5.39) respectively.

tion of c axes in the range $0 \leqslant \theta \leqslant \theta_0$ (Trotter and Brown, 1956; Sandeman and Keller, 1956). The corresponding values of $\langle \cos^2 \theta_c \rangle$ and the orientation parameter f are given by (Fraser, 1958a)

$$\langle \cos^2 \theta_c \rangle = \cos^2 \bar{\theta} \tag{5.38a}$$

$$f = \tfrac{1}{2}(3 \cos^2 \bar{\theta} - 1) \tag{5.38b}$$

$$\langle \cos^2 \theta_c \rangle = \tfrac{1}{3}(1 + \cos \theta_0 + \cos^2 \theta_0) \tag{5.39a}$$

$$f = \tfrac{1}{2}(\cos \theta_0)(1 + \cos \theta_0) \tag{5.39b}$$

These relationships are illustrated in Fig. 5.7b.

III. INTERPRETATION OF ABSORPTION COEFFICIENTS

The absorption properties of a polymer specimen in which the distribution functions have three mutually perpendicular mirror planes of symmetry can be specified by means of the three principal absorption coefficients. The values of these absorption coefficients for a particular frequency will represent a summation of contributions from the regions of regular conformation, of which there may be more than one type, and from the regions of irregular conformation.

In order to carry out a quantitative analysis, the observed spectrum must be resolved into individual absorption bands and the integrated absorbance estimated for each band. Methods of carrying out this analysis are discussed in Section C. In this section the symbols k_x, k_y, and k_z will be used to denote the integrated absorption coefficients, also termed the integrated intensities, obtained by such an analysis. These quantities can be related directly to the magnitude of the transition moment (Pauling and Wilson, 1935; Herzberg, 1945; Wilson et al., 1955; Brown, 1958; Overend, 1963).

If the transition moment associated with a particular vibration has components $\mathbf{M}_a$, $\mathbf{M}_b$, $\mathbf{M}_c$ relative to the amide-group axes, then the integrated absorption coefficients are given by (Fig. 5.3)

$$k_z = K[M_a{}^2\langle \cos^2 \theta_a \rangle + M_b{}^2\langle \cos^2 \theta_b \rangle + M_c{}^2\langle \cos^2 \theta_c \rangle] \tag{5.40}$$

$$k_x = K[M_a{}^2\langle \sin^2 \theta_a \cos^2 \phi_a \rangle + M_b{}^2\langle \sin^2 \theta_b \cos^2 \phi_b \rangle + M_c{}^2\langle \sin^2 \theta_c \cos^2 \phi_c \rangle] \tag{5.41}$$

$$k_y = K[M_a{}^2\langle \sin^2 \theta_a \sin^2 \phi_a \rangle + M_b{}^2\langle \sin^2 \theta_b \sin^2 \phi_b \rangle + M_c{}^2\langle \sin^2 \theta_c \sin^2 \phi_c \rangle] \tag{5.42}$$

where K is a constant of proportionality. Equations (5.40)–(5.42) form the basis for the interpretation of infrared dichroism in partially oriented specimens.

a. Fiber Orientation

With fiber orientation all values of ϕ are equally probable so that $\langle \sin^2 \theta_a \cos^2 \phi_a \rangle = \frac{1}{2} \langle \sin^2 \theta_a \rangle$, etc., and if all orientations about a particular axis, say c, are equally probable, then $\langle \sin^2 \theta_a \rangle = \langle \sin^2 \theta_b \rangle$. Thus from Eqs. (5.33) and (5.40)–(5.42) we obtain the result, for this commonly encountered situation, that

$$k_z = K[M_c^2 \langle \cos^2 \theta_c \rangle + \tfrac{1}{2}(M_a^2 + M_b^2)(1 - \langle \cos^2 \theta_c \rangle)] \qquad (5.43)$$

$$k_x = k_y = \tfrac{1}{2}K[M_c^2(1 - \langle \cos^2 \theta_c \rangle) + \tfrac{1}{2}(M_a^2 + M_b^2)(1 + \langle \cos^2 \theta_c \rangle)] \qquad (5.44)$$

which relates the principal integrated absorption coefficients to the transition moment direction in the amide group and to $\langle \cos^2 \theta_c \rangle$. If the transition moment direction in the amide group makes an angle α with the c axis, Eqs. (5.43) and (5.44) can be rewritten in terms of α and the orientation parameter f, defined in Eq. (5.31), as follows:

$$k_z = KM^2[f \cos^2 \alpha + \tfrac{1}{3}(1 - f)] \qquad (5.45)$$

$$k_x = k_y = KM^2[\tfrac{1}{2}f \sin^2 \alpha + \tfrac{1}{3}(1 - f)] \qquad (5.46)$$

The dichroic ratio R of a specimen with uniaxial optical properties is generally defined as

$$R = k_z/k_x \qquad (5.47)$$

and so for the case just considered the result

$$R = [f \cos^2 \alpha + \tfrac{1}{3}(1 - f)]/[\tfrac{1}{2}f \sin^2 \alpha + \tfrac{1}{3}(1 - f)] \qquad (5.48)$$

is obtained. With helical molecules the c axis is generally parallel to the helix axis and so α is restricted to values of zero and $\frac{1}{2}\pi$ (Sec-A.II.a). For the case $\alpha = 0$, Eq. (5.48) reduces to

$$R = (1 + 2f)/(1 - f) \qquad (5.49)$$

and for $\alpha = \frac{1}{2}\pi$, to

$$R = 2(1 - f)/(2 + f) \qquad (5.50)$$

The relationship between R and f for various values of α is illustrated in Fig. 5.8.

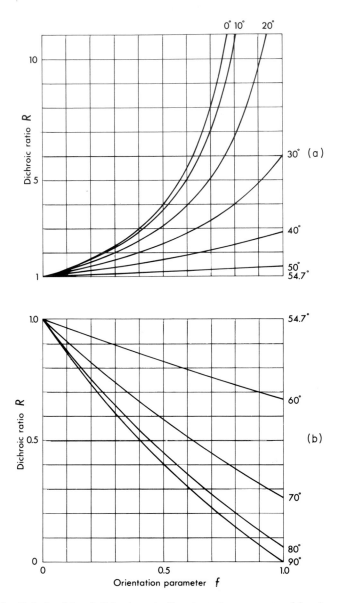

FIG. 5.8.   Relationship of dichroic ratio R, orientation parameter f for the c axis, and the inclination of the transition moment direction α to the c axis for fiber orientation. The c axis is assumed to be chosen so that it is aligned with the fiber axis in a perfectly ordered specimen ($f = 1$). (a) $1 \leqslant R \leqslant 11$, ($b$) $0 \leqslant R \leqslant 1$.

b. *Determination of Orientation Parameter from Dichroism Studies*

The form of the orientation density function C_c cannot be determined from infrared absorption measurements but it follows from Eq. (5.48) that if α is known, then the orientation parameter f can be determined from the observed dichroic ratio R. If all the c axes were perfectly aligned parallel to the fiber axis, the value of f would be unity and the dichroic ratio would be, from Eq. (5.48),

$$R_0 = 2 \cot^2 \alpha \qquad (5.51)$$

Thus if α is known, the value of the orientation parameter can be obtained from the expression (Fraser, 1958a)

$$f = [(R - 1)(R_0 + 2)]/[(R_0 - 1)(R + 2)] \qquad (5.52)$$

Even if α is unknown, the observed dichroic ratios can be used to set limits on the value of f (Fraser, 1958b). Other aspects of the relationship between dichroic ratio and transition moment direction have been discussed by Beer (1956), Krimm (1960a,b), Liang (1964), Zbinden (1964), Chirgadze (1965), and Rill (1972).

C. INTENSITY MEASUREMENTS

A considerable overlap of absorption band envelopes occurs in the infrared spectra of proteins and as mentioned in Section B conclusions drawn from measurements of peak heights can be grossly misleading. A typical example of such a situation is illustrated in Fig. 5.9, which shows the spectrum of isotropic silk fibroin. The ratios of the integrated absorption coefficients for the various components of the amide I bands are completely different from the ratios of the peak heights (Table 5.4).

Various approximate methods of resolving overlapping bands have been suggested (Tubomura, 1956; Kakiuti *et al.*, 1967) but none is adequate for the complex spectra given by fibrous proteins. Analog "curve resolvers" have also been used to analyze spectra by generating a series of functions electrically and adjusting the function parameters sequentially to give the best visual fit to the observed curve. The method, however, suffers from the disadvantage that curve-fitting by sequential variation of the parameters is both slow and inefficient. The use of digital methods is greatly to be preferred since least squares optimization can be used (Chapter 7) and a precise measure of goodness of fit obtained (Pitha and Jones, 1966; Fraser and Suzuki, 1966, 1970a,b, 1973).

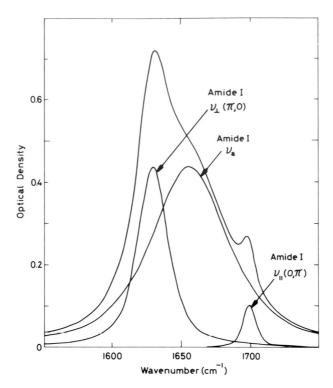

FIG. 5.9.   Spectrum of isotropic silk fibroin (Fraser and Suzuki, 1970a). The contributions of the individual components of the amide I vibration are indicated and it is clear that gross errors would be involved if apparent peak heights were used as a guide to integrated intensity.

TABLE 5.4

An Illustration of the Errors Introduced by the Use of Peak Heights Instead of
Integrated Intensities

Band	ν_m (cm^{-1})	$\Delta\nu_{1/2}$ (cm^{-1})	A_m	Relative area[a]	Relative peak height[a]	Error
Amide I ν_a [b]	1656	66.5	0.438	1.00	1.00	0
Amide I $\nu_\perp(\pi, 0)$	1630	25.9	0.437	0.39	1.00	157%
Amide I $\nu_\parallel(0, \pi)$	1699	11.3	0.102	0.04	0.23	475%

[a] Taken from the spectrum of isotropic silk fibroin given in Fig. 5.9.
[b] ν_a indicates amorphous phase.

I. FUNCTIONS FOR REPRESENTING BAND ENVELOPES

The shapes of infrared absorption bands have been discussed in detail by Seshadri and Jones (1963) and Young and Jones (1971). For liquid-phase systems the shape is determined primarily by collision broadening and would be expected to approximate to a Cauchy function:

$$A(\nu) = A_m/\{1 + [2(\nu - \nu_m)/\Delta\nu_{1/2}]^2\} \tag{5.53}$$

where A is absorbance, A_m is the peak absorbance, ν_m is the frequency of the peak, and $\Delta\nu_{1/2}$ is the bandwidth at half-height. In protein spectra broadening occurs for a variety of reasons and it has been found that bands arising from normal modes of regular conformations are often symmetric but intermediate in shape between that given by Eq. (5.53) and the Gaussian function

$$A(\nu) = A_m \exp\{-(\log_e 2)[2(\nu - \nu_m)/\Delta\nu_{1/2}]^2\} \tag{5.54}$$

Bands due to irregular conformations are often asymmetric.

a. *Symmetric Band Shapes*

A great variety of symmetric band shapes can be generated by combining Gaussian and Cauchy functions (Fraser and Suzuki, 1966, 1969; Pitha and Jones, 1966, 1967) but the most generally useful is a sum function in which the two components have the same bandwidth and are combined in the ratio g to $(1 - g)$:

$$A = A_m \left(g \exp \left\{ -(\log_e 2) \left[\frac{2(\nu - \nu_m)}{\Delta\nu_{1/2}} \right]^2 \right\} + \frac{1 - g}{1 + [2(\nu - \nu_m)/\Delta\nu_{1/2}]^2} \right) \tag{5.55}$$

The parameter g may be assumed to be the same for all bands or may be individually optimized for each band. The effect on the band shape of varying g is illustrated in Fig. 5.10. The optimum value of g for the amide I components of the pleated-sheet conformation in silk fibroin was found to be 0.3 (Suzuki, 1967).

The integrated absorption coefficient for band shapes that can be represented by Eq. (5.55) will be proportional to the area beneath the curve, which can be shown to be

$$\tfrac{1}{2}A_m(\Delta\nu_{1/2})[g(\pi/\log_e 2)^{1/2} + (1 - g)\pi] \tag{5.56}$$

An alternative approach, applicable to both symmetric and asymmetric band shapes, is discussed in Section C.I.c.

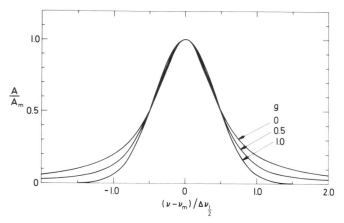

FIG. 5.10.   Gaussian ($g = 1$) and Cauchy ($g = 0$) band shapes. The Cauchy band shape has more pronounced "tails" and intermediate shapes can be produced by taking a linear combination of the two functions in the proportion g to $(1 - g)$, respectively. The bands in Fig. 5.9 have a shape close to that obtained by setting $g = 0.3$. Reprinted from Fraser and Suzuki, *Anal. Chem.* **41**, 37 (1969). Copyright 1969 by the American Chemical Society. Reprinted by permission of the copyright owner.

b. *Asymmetric Band Shapes*

Empirical fitting of asymmetric band shapes with analytical functions is not always successful and the standard shape method described in Section C.I.c is often to be preferred. Functions which have been used to represent bands with small amounts of asymmetry include the so-called "log normal" band shape

$$A = A_m \exp\left[-(\log_e 2)\left\{\frac{\log_e[1 + 2\gamma(\nu - \nu_m)/\Delta\nu_{1/2}]}{\gamma}\right\}^2\right], \quad \frac{2\gamma(\nu - \nu_m)}{\Delta\nu_{1/2}} > -1$$

(5.57)

$$A = 0, \quad \frac{2\gamma(\nu - \nu_m)}{\Delta\nu_{1/2}} \leqslant -1$$

(5.58)

The value of the parameter γ determines the degree of asymmetry. As $\gamma \to 0$, the band reduces to a symmetric Gaussian, while negative values of γ produce negatively skewed bands. Other properties of this function are discussed by Fraser and Suzuki (1969, 1970b).

Another method of introducing small amounts of asymmetry into a symmetric function is to multiply it by a factor

$$\{1 + \tanh[2\gamma(\nu - \nu_m)/\Delta\nu_{1/2}]\}$$

(5.59)

where γ is an adjustable parameter.

c. *Standard Shape Functions*

Band shapes determined experimentally, for example, from solution spectra of model compounds, can be used as models for fitting protein spectra. The standard shape is stored in digital form as a series of values of $V = A^s/A_m{}^s$ at equispaced intervals of $U = (\nu - \nu_m{}^s)/\Delta \nu^s_{1/2}$ where the superscript s denotes standard. Bands in the protein spectrum can be represented in terms of the standard function $V(U)$ as

$$A = A_m V[(\nu - \nu_m)/\Delta \nu_{1/2}] \tag{5.60}$$

This method of representing shapes for fitting observed data has been discussed by Keller *et al.* (1966), Anderson *et al.* (1970), Trombka and Schmadebeck (1970), and Fraser and Suzuki (1973).

II. OPTIMIZATION PROCEDURE

Details of procedures for fitting model functions to observed data by the least squares method have been given by Fraser and Suzuki (1973) together with computer programs for carrying out these operations. In applying these procedures to the analysis of overlapping absorption bands in infrared spectra the absorbance A is represented by the sum of a series of functions A_r representing absorption bands and a function B representing the baseline; thus the model function can be written

$$A(\nu) = \sum_{r=1}^{N_b} A_r(\nu) + B(\nu) \tag{5.61}$$

and the least squares function to be minimized (Chapter 7) becomes

$$S = \sum_{i=1}^{N} W_i[A(\nu_i) - \mathscr{A}(\nu_i)]^2 \tag{5.62}$$

where N_b is the number of bands, N is the number of data points, $\mathscr{A}(\nu)$ is the observed absorbance, and W_i is a weighting function (Chapter 7).

The band functions will be those given in Eq. (5.55) or (5.60) and a linear baseline is usually adequate, so that

$$B(\nu) = B_1 + [(\nu - \nu_1)(B_N - B_1)/(\nu_N - \nu_1)] \tag{5.63}$$

where B_1 and B_N are the baseline contributions to the observed absorbance at the extremes of the frequency range to be analyzed.

Initial estimates of B_1, B_N, and the parameters A_m, ν_m, $\Delta \nu_{1/2}$ (and g if applicable) for each band are obtained from the observed spectrum and their values optimized by an iterative procedure (Fraser and Suzuki,

1973). When polarized radiation is used the least squares function will consist of the sum of a set of expressions like the one in Eq. (5.62), one for each direction of polarization.

In order to hasten convergence to a solution, the number of independent parameters must be reduced to a minimum and this can be effected by applying as many constraints to the problem as possible (Chapter 7). When weak bands are present the corresponding parameters can be held fixed in the initial stages of the optimization by applying constraints of the type

$$G(P_j) = P_j - P_j' = 0 \qquad (5.64)$$

where P_j is a parameter and P_j' is the initial estimate of its value.

When sets of spectra obtained with polarized radiation are to be analyzed advantage can be taken of the fact that the parameters ν_m, $\Delta\nu_{1/2}$, and g will be the same for corresponding bands in different spectra. This constraint takes the form

$$G(P_j, P_k) = P_j - P_k = 0 \qquad (5.65)$$

where P_j and P_k are corresponding parameters in different spectra. Additional constraints can be applied to the integrated areas beneath various bands if the directions of the transition moments are known (Beer, 1956; Liang, 1964) through the relationship given in Eq. (5.52). Since the orientation parameter f must be same for all bands associated with a particular conformation, the constraints take the form of a set of relationships between the observed and calculated dichroic ratios for pairs of bands, given by

$$G(R^i, R^j, R_0^i, R_0^j) = \frac{(R^i - 1)(R_0^i + 2)}{(R^i + 2)(R_0^i - 1)} - \frac{(R^j - 1)(R_0^j + 2)}{(R^j + 2)(R_0^j - 1)} = 0 \qquad (5.66)$$

where R is the dichroic ratio expressed in terms of the function parameters, R_0 is defined in Eq. (5.51), and the superscripts refer to bands.

The main problems encountered in the quantitative analysis of spectra concern the representation of side-chain absorptions (Bendit, 1967, 1968a; Fraser and Suzuki, 1970a) and the frequent near-coincidence of the contributions from different regular conformations, from different modes of the same conformation, or from regular and irregular conformations. In some instances these difficulties can be reduced by partial deuteration or by analysis of the dichroism spectrum rather than the absorption spectrum.

The side-chain absorptions which are most troublesome in analyzing the amide A and B bands are due to the O—H stretching modes of serine, tyrosine, hydroxyproline, and un-ionized acidic groups, and the N—H stretching modes of arginine, lysine, tryptophan, and amides. The amide I and II bands are overlapped by absorptions due to carboxyl and side-chain amide groups and to the aromatic ring in tyrosine. Other side chains which absorb in this region, although to a lesser extent, include arginine and lysine.

LIST OF SYMBOLS FOR CHAPTER 6

b	bond length
τ	bond angle
ϕ	torsion angle about NC^α of main chain
ψ	torsion angle about $C^\alpha C'$ of main chain
ω	torsion angle about $C'N$ of main chain
χ	torsion angle in side chains
h	unit height in discontinuous helix
t	unit twist in discontinuous helix
x, y, z	Cartesian coordinates
$\mathbf{r}$	vector defining position
$\mathbf{T}$	transformation matrix
U	potential energy
U_ϕ , U_ψ , U_ω , U_χ	torsional potentials
U_{nb}	nonbonded interatomic potential
U_{es}	electrostatic potential
U_{hb}	hydrogen-bond potential
U_τ	covalent bond angle deformation potential
U_{b}	covalent bond length deformation potential
r_{ij}	internuclear distance
a_{ij} , b_{ij} , c_{ij} , d_{ij} , e_{ij}	parameters describing nonbonded interactions
q	partial charge on atom
D	dielectric constant
d^*, e^*	parameters describing hydrogen bond potential
k_τ	bending force constant
k_{b}	stretching force constant
v	atomic volume
v_{w}	volume excluded by water molecule
G	solvent free energy
F	conformational energy

Chapter 6

Analysis and Prediction of Conformation

Early attempts to devise conformations of the polypeptide chain which would be consistent with observed X-ray diffraction and other data from fibrous proteins were based on the use of scale atomic models. In these models the distances between covalently bonded atoms were fixed, as were the interbond angles, but no restriction was placed on torsional motion around bonds. In the space-filling type of model each atom was supposed to be surrounded by an impenetrable sphere of radius equal to the so-called van der Waals radius. These restrictions were not sufficient to significantly reduce the number of possible conformations and little progress was made. An important advance over earlier work was an attempt by Huggins (1943) to formulate a set of structural principles that might reasonably be expected to be obeyed and this was followed by a systematic search for conformations which satisfied these requirements. The concept that the main-chain portions of the residues should be equivalent led naturally to the idea of screw axes and helices and attention was also focused on the importance of hydrogen bonding and on close packing. Later Bragg *et al.* (1950) made an extensive survey of hydrogen-bonded helical conformations but restricted their attention to helices with integral numbers of residues per turn.

The first real success of the model building technique came when Pauling *et al.* (1951) described the α-helix conformation, which was shown to be consistent with many features of the X-ray diffraction patterns yielded by synthetic polypeptides in the α form. The stereochemical principles used to restrict the search for stable conformations were as follows.

127

(1) Residues should be structurally equivalent apart from the side chains.

(2) As a consequence of (1), only helical conformations were acceptable.

(3) The interatomic distances and bond angles were assumed to have values similar to those found in crystals of simple model compounds.

(4) The amide group was assumed to be planar due to the loss of resonance energy which would result from a departure from planarity.

(5) Each N—H group was required to form a hydrogen bond with a C=O group, with restrictions on the angle $\tau(H, N, O)$ and the distance $b(N, O)$.

Only two intramolecularly hydrogen-bonded helices were found that satisfied these requirements. These were the α helix with 18 residues in five turns of pitch 5.4 Å and the γ helix with 36 residues in seven turns of pitch 5.04 Å. Other possible helices were described by Donohue (1953).

All the regular secondary conformations so far recognized in fibrous proteins were first described in generic form during a period of intense activity in model building based on the structural principles enunciated by Pauling and Corey. These conformations were the α-helix (Pauling et al., 1951), the antiparallel-chain pleated sheets (Pauling and Corey, 1951g, 1953a), the poly(glycine) II helix (Crick and Rich, 1955), and the poly(proline) II helix (Cowan and McGavin, 1955).

Despite these successes the precise form of the regular polypeptide chain conformations which occur in fibrous proteins proved more elusive and insufficient data are available at present to establish whether the coiled-coil models for collagen (Ramachandran and Kartha, 1954, 1955a,b; Rich and Crick, 1955) and for α-fibrous proteins (Crick, 1952, 1953a,b; Pauling and Corey, 1953b) are correct in detail. Further refinement of these models depends upon the collection of more accurate physical data and improvements in the energy minimization procedure for the prediction of stable conformations.

The use of scale atomic models for the analysis and prediction of conformations has largely been replaced by automated procedures using high-speed digital computers. A given conformation can be analyzed very rapidly by this means for bad contacts and the effects of varying various parameters can be examined in a systematic manner. In addition there is the prospect of refining models on the basis of energy minimization procedures provided that the various potential functions can be accurately represented in an analytical form. Valuable surveys of the extensive literature on this topic have been given by Scheraga (1968a, 1971) and Ramachandran and Sasisekharan (1968).

In this chapter a critical review of the procedures currently available for conformation analysis and prediction is given.

A. DESCRIPTION OF CONFORMATION

I. NOMENCLATURE

In order to simplify the description of conformation a standard nomenclature has been proposed (IUPAC-IUB Commission on Biochemical Nomenclature, 1970) and the conventions relating to the mainchain atoms

$$-NH-\overset{|}{C}{}^\alpha H^\alpha -CO-$$

are illustrated in Fig. 6.1. Residues are numbered sequentially starting from the N terminus and atoms belonging to the ith residue are distinguished by the subscript i. The torsional rigidity about the

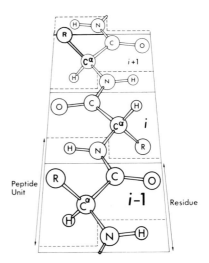

FIG. 6.1. Standard conventions for the description of residues (enclosed by full lines) and peptide units (enclosed by broken lines) recommended by the IUPAC-IUB Commission on Biochemical Nomenclature (1970).

$C_i N_{i+1}$ bond is high and in many cases the group of atoms $C_i^\alpha - C_i O_i - N_{i+1} H_{i+1} - C_{i+1}^\alpha$ (Fig. 6.1) behaves as a rigid unit. In some instances it is therefore convenient to choose a *peptide unit* $-CHR-CO-NH-$ rather than a residue $NH-CHR-CO$ as the repeating group. The length of the bond between atoms A and B is written $b(A, B)$ and the angle included between atoms A, B, and C is written $\tau(A, B, C)$.

Rotation about the main-chain bonds is specified by *torsion angles* ϕ_i about $N_i C_i^\alpha$, ψ_i about $C_i^\alpha C_i$, and ω_i about $C_i N_{i+1}$. The choice of origin for these angles is such that $\phi = \psi = \omega = 180°$ for all i corresponds to a fully extended chain (Fig. 6.1). Right-handed rotation is regarded as positive, so that when looking along a bond the group of atoms at the far end rotates clockwise relative to the group at the near end (Fig. 6.2). Torsion angles are measured in the range $-180°$ through zero to $+180°$.

Side-chain atoms are distinguished by the superscripts β, γ, δ, ϵ, ζ, η according to their sequence along the chain, and the bonds between the atoms in positions α and β, β and γ, γ and δ, etc. are numbered consecu-

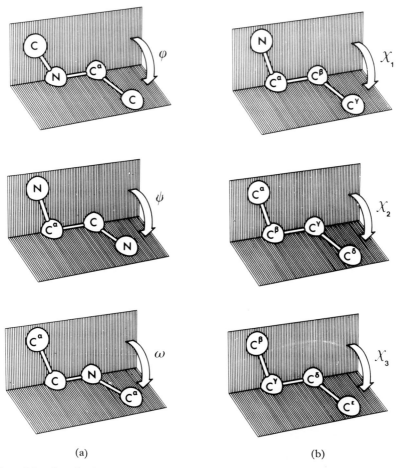

(a)   (b)

FIG. 6.2.   Standard convention for the description of torsion angles in (a) main chain and (b) side chains (IUPAC-IUB Commission on Biochemical Nomenclature, 1970).

tively 1, 2, 3, etc. When chain branching occurs the branches are numbered according to a set of rules given in the IUPAC-IUB publication and the atoms and bonds in different branches in corresponding sequential positions in the branches are distinguished by the addition of the branch numeral to their designation. For example, the carbon skeleton of isoleucine is written

$$C^\alpha \overset{1}{\text{---}} C^\beta \overset{2,1}{\diagup} C^{\gamma_1} \overset{3,1}{\text{---}} C^{\delta_1}$$
$$\underset{2,2}{\diagdown} C^{\gamma_2}$$

For unbranched chains the angle of rotation about the jth bond is denoted by χ_j and the choice of zero is such that the eclipsed conformation (Fig. 6.2) corresponds to $\chi = 0$. Right-handed rotation is again taken as positive. For the $C^\alpha - C^\beta$ bond χ_1 is taken as zero when the $C^\beta - C^\gamma$ bond eclipses the $N - C^\alpha$ bond. When chain branching occurs branch 1 is used to define zero rotation.

Hydrogen atoms are identified by a superscript indicating the heavier atom to which they are attached and when more than one hydrogen is attached to an atom they are distinguished by an additional numeral in the superscript chosen according to the rules given in the IUPAC-IUB publication.

The nomenclature recommended by the IUPAC-IUB Commission differs in several important respects from that used by early workers in the field and also from that recommended by Edsall *et al.* (1966).

II. GENERATION OF ATOMIC COORDINATES

The conformation of a polypeptide chain can be described by specifying the amino acid sequence together with the bond lengths b, bond angles τ, and torsion angles ϕ, ψ, ω, and χ. These *internal coordinates* are unsuitable for most analytical purposes and the problem arises of determining the coordinates of the individual atoms relative to some external frame of reference. It is often convenient to define this frame of reference by means of a group of three linked atoms, for example, $C_1^\alpha - C_1 - O_1$ (Fig. 6.3). The set of internal coordinates then defines a series of operations by means of which the external coordinates of any atom can be generated.

A general procedure for evaluating external coordinates has been given by McGuire *et al.* (1971). A local frame of reference is established in each residue and the coordinates of the constituent atoms in this frame of reference are calculated. These local coordinates are then transformed via the local frames in preceding residues to the chosen external frame of reference. Other accounts of procedures for generating external coordinates have been given by Brant and Flory (1965), Némethy and

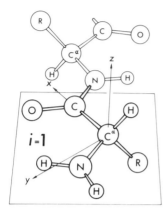

FIG. 6.3. Local frame of reference within a residue. The x axis is along $C^\alpha C'$, the y axis is in the plane $C^\alpha C'O$, and the z axis is perpendicular to this plane.

Scheraga (1965), Ooi *et al.* (1967), Ramachandran *et al.* (1966b), and Ramachandran and Sasisekharan (1968). It is important to remember that these treatments must be modified where necessary to take account of the changes in the definitions of the torsional angles recommended by the IUPAC-IUB Commission.

A convenient local frame of reference for the ith residue is similar to that illustrated in Fig. 6.3 except that the origin is at C_i^α. The C_i atom has coordinates $(b(C_i^\alpha, C_i), 0, 0)$ and the O_i atom lies in the plane $z = 0$ with positive x and y coordinates (Ooi *et al.*, 1967). If $\mathbf{r}_i$ is a vector defining the coordinates of an atom in the local frame of the ith residue, the vector $\mathbf{r}_{i-1}$ defining the coordinates of the same atom in the local frame of residue $i - 1$ is given by

$$\mathbf{r}_{i-1} = \mathbf{r}_{i-1}(C_i^\alpha) + \mathbf{T}_{i \to i-1}\mathbf{r}_i \qquad (6.1)$$

where $\mathbf{r}_{i-1}(C_i^\alpha)$ is the vector in the local frame of residue $i - 1$ which defines the position of C_i^α, and $\mathbf{T}_{i \to i-1}$ is a matrix which rotates the axes of local frame i until they are parallel to the corresponding axes of frame $i - 1$.

The matrix $\mathbf{T}_{i \to i-1}$ can be expressed as a product $\mathbf{T}_{\tau'}\mathbf{T}_\omega\mathbf{T}_{\tau''}\mathbf{T}_\phi\mathbf{T}_\tau\mathbf{T}_\psi$ (McGuire *et al.*, 1971) in which

$$\mathbf{T}_{\tau'} = \begin{bmatrix} -\cos \tau'_{i-1} & \sin \tau'_{i-1} & 0 \\ -\sin \tau'_{i-1} & -\cos \tau'_{i-1} & 0 \\ 0 & 0 & 1 \end{bmatrix} \qquad (6.2)$$

where

$$\tau'_{i-1} = \tau(C^{\alpha}_{i-1}, C'_{i-1}, N_i) \tag{6.3}$$

$$\mathbf{T}_\omega = \begin{bmatrix} 1 & 0 & 0 \\ 0 & \cos \omega_{i-1} & -\sin \omega_{i-1} \\ 0 & \sin \omega_{i-1} & \cos \omega_{i-1} \end{bmatrix} \tag{6.4}$$

$$\mathbf{T}_{\tau''} = \begin{bmatrix} -\cos \tau''_i & \sin \tau''_i & 0 \\ -\sin \tau''_i & -\cos \tau''_i & 0 \\ 0 & 0 & 1 \end{bmatrix} \tag{6.5}$$

where

$$\tau''_i = \tau(C'_{i-1}, N_i, C^{\alpha}_i) \tag{6.6}$$

$$\mathbf{T}_\phi = \begin{bmatrix} 1 & 0 & 0 \\ 0 & \cos \phi_i & -\sin \phi_i \\ 0 & \sin \phi_i & \cos \phi_i \end{bmatrix} \tag{6.7}$$

$$\mathbf{T}_\tau = \begin{bmatrix} -\cos \tau_i & \sin \tau_i & 0 \\ -\sin \tau_i & -\cos \tau_i & 0 \\ 0 & 0 & 1 \end{bmatrix} \tag{6.8}$$

where

$$\tau_i = \tau(N_i, C^{\alpha}_i, C'_i) \tag{6.9}$$

and

$$\mathbf{T}_\psi = \begin{bmatrix} 1 & 0 & 0 \\ 0 & \cos \psi_i & -\sin \psi_i \\ 0 & \sin \psi_i & \cos \psi_i \end{bmatrix} \tag{6.10}$$

and it can be shown that

$$\mathbf{r}_{i-1}(C^{\alpha}_i) = \begin{bmatrix} b - b' \cos \tau'_{i-1} + b''(\cos \tau'_{i-1} \cos \tau''_i - \sin \tau'_{i-1} \sin \tau''_i \cos \omega_{i-1}) \\ -b' \sin \tau'_{i-1} + b''(\sin \tau'_{i-1} \cos \tau''_i + \cos \tau'_{i-1} \sin \tau''_i \cos \omega_{i-1}) \\ -b'' \sin \tau''_i \sin \omega_{i-1} \end{bmatrix} \tag{6.11}$$

where $b = b(C^{\alpha}_{i-1}, C'_{i-1})$, $b' = b(C'_{i-1}, N_i)$, and $b'' = b(N_i, C^{\alpha}_i)$.

By repeated application of Eq. (6.1) the positions of all the atoms in the molecule can be referred to the local frame of the first residue, which then serves as an external frame of reference.

The problem of finding the coordinates of the atoms in a residue relative to its local frame of reference can be tackled in an analogous way to that discussed above (Ooi *et al.*, 1967; McGuire *et al.*, 1971). With regular helical structures it is convenient to use an external frame of reference in which one of the axes is coincident with the axis of the helix. Sugeta and Miyazawa (1967, 1968) describe a method whereby the local frame of reference in the repeating unit of a helical polymer can be transformed directly to this external frame of reference without recourse to the repeated application of the transformation in Eq. (6.1) necessary in the general case. The application of this method to polypeptides has been discussed by McGuire *et al.* (1971).

III. CONFORMATIONAL MAPS

In the analysis of conformations it is usually necessary to make a number of simplifying assumptions to reduce the magnitude of the problem. Frequently a property of the molecule is explored by fixing all the parameters bar two and then varying the values of these in a systematic way. It is convenient to present the results of such a study by means of a conformational *map* such as that shown in Fig. 6.4a.

The torsional rigidity of the peptide bond is sufficiently great for the group of atoms $C^{\alpha}-NH-CO-C^{\alpha}$ to be treated in many instances as a rigid planar unit. If $\tau(N, C^{\alpha}, C')$ is fixed, the conformation of the main chain is then completely defined by the set of torsional angles $\{\phi, \psi\}$. In regular structures where the residues (apart from the side chain) are equivalent all values of ϕ must be equal and all values of ψ must be equal. Any property of the set of helices so generated can be analyzed by systematically varying the values of ϕ and ψ, and the IUPAC-IUB Commission recommends that the results be presented with the conventions shown in Figs. 6.4a and 6.4b.

IV. HELICAL PARAMETERS

The helical parameters of regular conformations can be calculated directly from the internal coordinates by means of formulas given by Miyazawa (1961a) and Sugeta and Miyazawa (1967, 1968). If all residues (apart from side chains) are equivalent, the unit height h and unit twist t (Chapter 1, Section A.II.c) of the helix are related to the internal coordinates of the main chain by the expressions

$$\cos(\tfrac{1}{2}t) = \cos[\tfrac{1}{2}(\phi + \psi + \omega)] \sin(\tfrac{1}{2}\tau) \sin(\tfrac{1}{2}\tau') \sin(\tfrac{1}{2}\tau'')$$
$$-\cos[\tfrac{1}{2}(\phi - \psi + \omega)] \cos(\tfrac{1}{2}\tau) \cos(\tfrac{1}{2}\tau') \sin(\tfrac{1}{2}\tau'')$$
$$-\cos[\tfrac{1}{2}(\phi + \psi - \omega)] \sin(\tfrac{1}{2}\tau) \cos(\tfrac{1}{2}\tau') \cos(\tfrac{1}{2}\tau'')$$
$$-\cos[\tfrac{1}{2}(\phi - \psi - \omega)] \cos(\tfrac{1}{2}\tau) \sin(\tfrac{1}{2}\tau') \cos(\tfrac{1}{2}\tau'') \tag{6.12}$$

and

$$h \sin(\tfrac{1}{2}t) = (b + b' + b'') \sin[\tfrac{1}{2}(\phi + \psi + \omega)] \sin(\tfrac{1}{2}\tau) \sin(\tfrac{1}{2}\tau') \sin(\tfrac{1}{2}\tau'')$$

$$+(b - b' - b'') \sin[\tfrac{1}{2}(\phi - \psi + \omega)] \cos(\tfrac{1}{2}\tau) \cos(\tfrac{1}{2}\tau') \sin(\tfrac{1}{2}\tau'')$$

$$-(b - b' + b'') \sin[\tfrac{1}{2}(\phi + \psi - \omega)] \sin(\tfrac{1}{2}\tau) \cos(\tfrac{1}{2}\tau') \cos(\tfrac{1}{2}\tau'')$$

$$+(b + b' - b'') \sin[\tfrac{1}{2}(\phi - \psi - \omega)] \cos(\tfrac{1}{2}\tau) \sin(\tfrac{1}{2}\tau') \cos(\tfrac{1}{2}\tau'') \quad (6.13)$$

where $\tau = \tau(N, C^\alpha, C)$, $\tau' = \tau(C^\alpha, C, N)$, $\tau'' = \tau(C, N, C^\alpha)$, $b = b(C^\alpha, C)$, $b' = b(C, N)$, and $b'' = b(N, C^\alpha)$. If the values $\tau = 110°$, $\tau' = 114°$, $\tau'' = 123°$, $b = 1.53$ Å, $b' = 1.32$ Å, $b'' = 1.47$ Å, and $\omega = 180°$ suggested by Corey and Pauling (1953a) are inserted in Eqs. (6.12) and (6.13), they reduce to

$$\cos(\tfrac{1}{2}t) = -0.817 \sin[\tfrac{1}{2}(\phi + \psi)] + 0.045 \sin[\tfrac{1}{2}(\phi - \psi)] \quad (6.14)$$

$$h \sin(\tfrac{1}{2}t) = 2.966 \cos[\tfrac{1}{2}(\phi + \psi)] - 0.663 \cos[\tfrac{1}{2}(\phi - \psi)] \quad (6.15)$$

Values of the number of residues per turn $n = 360°/t$ and h corresponding to different values of ψ and ϕ have been tabulated (Ramachandran and Sasisekharan, 1968) and this information has also been presented in graphical form (Ramakrishnan, 1964). In the conformational maps shown in Fig. 6.4 h and t are plotted as functions of ϕ and ψ.

The point (180°, 180°) corresponds to the fully extended chain with $t = 180°$ and $h = 3.63$ Å. The contour $h = 0$ corresponds to conformations (all impossible for an infinite chain) in which the path is circular. Whenever the contours $h = 0$ or $t = 180°$ are crossed the screw sense of the helix changes to the opposite hand (Fig. 6.4). The positions on the map corresponding to left-handed and right-handed helices of the same type are symmetrically disposed about the origin, for example, the right-handed α-helix formed by poly(L-alanine) has coordinates $(-57°, -47°)$ and the corresponding left-handed helix has coordinates $(57°, 47°)$.

While a given pair of values for the internal coordinates ϕ and ψ uniquely define the values of h and t for a helical conformation, the reverse is not true and as can be seen from Fig. 6.4 there are, in general, two solutions (ϕ_1, ψ_1) and (ϕ_2, ψ_2) corresponding to each physically realizable combination of h and t. If the sign of t is not known, as, for example, in data obtained from fiber diffraction patterns, the four combinations of ϕ and ψ that must be considered as the solutions for h and $|t|$ are (ϕ_1, ψ_1), (ϕ_2, ψ_2), $(-\phi_1, -\psi_1)$, and $(-\phi_2, -\psi_2)$.

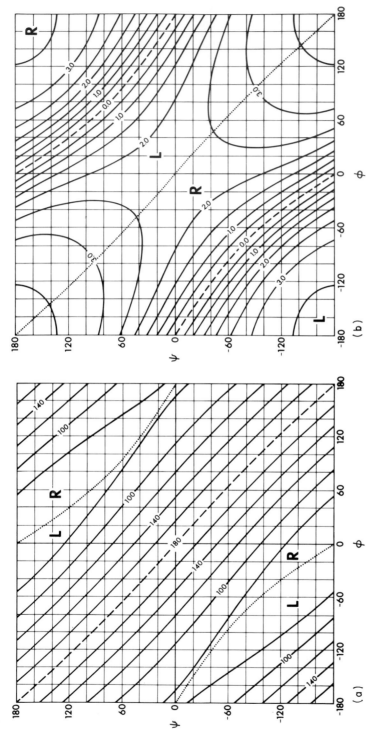

Fig. 6.4. (a) Variation of unit twist $|t|$, measured in degrees, with torsion angles ϕ and ψ for regular helical structures with the standard amide-group dimensions given in Table 6.1 and $\omega = 180°$. The screw sense of the helix changes when the contour $t = 180°$ (broken line) is crossed and when the dotted line is crossed. Contour interval, 20°; grid interval, 20°; calculated from Eq. (6.14). (b) Variation in unit height h, measured in Å, with torsion angles ϕ and ψ for regular helical structures with the standard amide-group dimensions given in Table 6.1 and $\omega = 180°$. The screw sense of the helix changes when the contour $h = 0$ (broken line) is crossed and when the dotted line is crossed. Contour interval, 0.5 Å; grid interval, 20°; calculated from Eqs. (6.14) and (6.15).

B. "ALLOWED" CONFORMATIONS

When scale atomic models are used to study steric hindrance each atom is allotted a van der Waals radius and a particular molecular conformation is regarded as being sterically "allowed" if all possible pairs of nonbonded atoms are separated by distances which exceed the sum of their respective van der Waals radii. It has long been recognized that this procedure represents an oversimplification and various improvements have been suggested, such as the use of non-spherical shapes for the atoms and the introduction of a limited amount of flexibility in certain bonds (Hartley and Robinson, 1952; Robinson and Ambrose, 1952; Corey and Pauling, 1953b; Koltun, 1965).

Such models proved to be adequate for preliminary investigations of regular conformations of the polypeptide main chain but are unsuitable for detailed studies of the large number of interatomic contacts that occur in proteins. With the advent of high-speed digital computers, however, it has become possible to investigate the interatomic contacts in a particular conformation in a precise and systematic manner. The problem of choosing bond lengths and angles, and criteria for allowed contact distances between nonbonded atoms, are just as acute as with scale atomic models, but in contrast the effects of varying these parameters can readily be investigated.

The use of van der Waals radii to determine whether a conformation is sterically allowed has come to be known as the "hard-sphere" approximation and the various studies that have been made using this method have been summarized by Ramachandran and Sasisekharan (1968) and Scheraga (1968a).

I. GEOMETRIC DATA

In order to generate a conformation it is necessary to specify a complete set of internal coordinates and the values used have mostly been obtained from structural studies of simple model compounds. The amide group is generally assumed to be planar and to have the geometry given in Table 6.1 (Corey and Pauling, 1953a). It should be noted, however, that Winkler and Dunitz (1971) have suggested that out-of-plane deformation may be possible at a very modest energy cost. The remaining parameters required to define the main-chain geometry are $\tau(N, C^\alpha, C)$, which is usually assumed to be 109.5° or 110°, $\tau(N, C^\alpha, H^\alpha)$, and $\tau(H^\alpha, C^\alpha, C)$. Comprehensive lists of values of bond distances and angles used for side chains have been given by Scheraga (1968a).

TABLE 6.1

"Standard" Dimensions of Planar Amide Groups

Dimension[a,b]	Å	Dimension	Trans[a] (deg)	Cis[c] (deg)
$b(N, C^\alpha)$	1.47	$\tau(C^\alpha, C', O)$	121	119
$b(C^\alpha, C')$	1.53	$\tau(C^\alpha, C', N)$	114	118
$b(C', N)$	1.32	$\tau(O, C', N)$	125	123
$b(C', O)$	1.24	$\tau(C', N, C^\alpha)$	123	126
$b(N, H)$	1.00	$\tau(C', N, H)$	123	121
		$\tau(H, N, C^\alpha)$	114	113

[a] Corey and Pauling (1953a).
[b] Leach et al. (1966a).
[c] Ramachandran and Sasisekharan (1968).

II. Van der Waals Contact Distances

The greatest weakness of the hard-sphere approximation is the difficulty of specifying the closest distance of approach which can be tolerated between two nonbonded atoms before rejecting the conformation as being sterically disallowed. The various tables of van der Waals radii given in the literature have been compiled from surveys of the average distance of approach observed in molecular crystals. However, in certain situations an approach several tenths of an angstrom closer than the sum of these average radii (Table 6.2a) have been observed. Ramachandran and co-workers have compiled tables of allowable distances for interatomic constants in which both normal and extreme limits are recognized (Table 6.2b). This distinction was not made by Scheraga and co-workers and their values for limiting contact distances are also given in Table 6.2b for comparison. The greatest discrepancies are in the minimum distances allowed for carbon–carbon and methylene group contacts. In a later study Gō and Scheraga (1970) used a set of contact distances designed for the detection of large steric overlaps, and the minimum allowable distances were generally somewhat smaller than the extreme values suggested by Ramachandran and co-workers (Table 6.2b). So far no attempt appears to have been made to incorporate an angular dependence into these minimum contact distances such as those used in some space-filling atomic models.

III. Steric Maps

The short-range contacts between main-chain atoms and between side-chain and main-chain atoms have been extensively investigated by mapping the sterically allowed conformations of the group of atoms illustrated in Fig. 6.5a. Scheraga and co-workers have suggested that

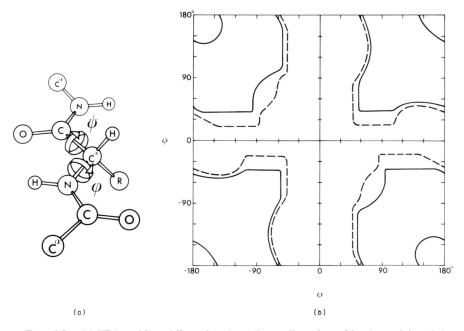

(a)   (b)

FIG. 6.5. (a) "Dipeptide unit" used to investigate allowed combinations of ϕ and ψ for different residues. (b) Steric map showing allowed combinations of ϕ and ψ around a glycyl residue (R $\equiv$ H) with $\tau(C^\alpha) = 110°$, standard amide geometry, and planar amide groups. The four regions enclosed by full lines correspond to combinations of ϕ and ψ permitted with "normal" contact distances (Rn in Table 6.2b) and the regions enclosed by broken lines correspond to combinations permitted with "extreme" contact distances (Re in Table 6.2b). Redrawn from Ramachandran and Sasisekharan (1968).

conformational maps of this type be called *steric maps*. Summaries of these investigations have been given by Ramachandran and Sasisekharan (1968) and Scheraga (1968a). In the studies carried out by Ramachandran and co-workers contacts between atoms separated by three or more bonds were included but Scheraga and co-workers omitted contacts between atoms separated by three bonds.

The steric map for the group of atoms in Fig. 6.5a for the case of a glycyl side chain (R $\equiv$ H), $\tau(N, C^\alpha, C) = 110°$, and planar amide groups ($\omega = 0$) is shown in Fig. 6.5b (Ramachandran and Sasisekharan, 1968). About 45% of the total area is allowed when the normal van der Waals radii are used and this increases to 61% when the extreme limits are used. In the steric map for an alanyl side chain (R $\equiv$ CH$_3$) these

TABLE 6.2a

"Observed" Mean van der Waals Radii[a]

Atom	H	C	N	O	S
Van der Waals radius (Å)	1.20	1.70	1.55	1.52	1.80

[a] Bondi (1964).

TABLE 6.2b

Selected Values of Allowed Interatomic Contact Distances (Å)[a]

	Ref.[b]	H	C	C′	C_{ar}	CH	N	O
H	S	2.0	—	2.4	2.7	2.7	2.3	2.2
	Rn	2.0	2.4	2.4	2.4	—	2.4	2.4
	Re	1.9	2.2	2.2	2.2	—	2.2	2.2
C	S	—	—	—	—	—	—	—
	Rn	2.4	3.0	3.0	3.0	3.2	2.9	2.8
	Re	2.2	2.9	2.9	2.9	3.0	2.8	2.7
C′	S	2.4	—	2.9	3.1	3.2	2.7	2.7
	Rn	2.4	3.0	3.0	3.0	3.2	2.9	2.8
	Re	2.2	2.9	2.9	2.9	3.0	2.8	2.7
C_{ar}	S	2.7	—	3.1	3.4	3.4	3.0	3.0
	Rn	2.4	3.0	3.0	3.0	3.2	2.9	2.8
	Re	2.2	2.9	2.9	2.9	3.0	2.8	2.7
CH	S	2.7	—	3.2	3.4	3.5	3.0	3.0
	Rn	—	3.2	3.2	3.2	3.2	—	—
	Re	—	3.0	3.0	3.0	3.0	—	—
N	S	2.3	—	2.7	3.0	3.0	2.6	2.6
	Rn	2.4	2.9	2.9	2.9	—	2.7	2.7
	Re	2.2	2.8	2.8	2.8	—	2.6	2.6
O	S	2.2	—	2.7	3.0	3.0	2.6	2.6
	Rn	2.4	2.8	2.8	2.8	—	2.7	2.7
	Re	2.2	2.7	2.7	2.7	—	2.6	2.6

[a] Identification of atoms: C, aliphatic carbon atom; C′, carbonyl carbon atom; C_{ar}, aromatic carbon atom; CH, includes CH_2 and CH_3.

[b] S, Values used by Scheraga and co-workers (Leach et al., 1966a). R, Values used by Ramachandran and co-workers (Ramachandran and Sasisekharan, 1968): n, normal values; e, extreme values.

allowed areas are drastically reduced and constitute only 7.5 and 22.5%, respectively, of the total area (Fig. 6.6a). Additional atoms in the side chain further reduce these areas. It has been found that allowed areas are very sensitive to the choice of $\tau(\text{N}, \text{C}^\alpha, \text{C})$ and ω (Ramakrishnan and Balasubramanian, 1972) and the steric map for an alanyl side chain with $\tau(\text{N}, \text{C}^\alpha, \text{C}) = 115°$ is shown in Fig. 6.6b.

Steric maps have also been calculated for a regular helical conformation of poly(L-alanine) and the presence of additional residues reduces the allowed area particularly in the vicinity of the α-helix (Fig. 6.7a). The allowed conformations in the region $(-100°, 100°)$ are closer to the fully extended conformation than the α-helix and are not influenced by the addition of further alanyl residues to the structure shown in Fig. 6.5b.

The formation of intramolecular hydrogen bonds in helical structures has also been investigated by the use of steric maps and the results are summarized in Fig. 6.7b. The regions in which satisfactory hydrogen bonds of the type $\text{N}_i-\text{H}_i\cdots\text{O}_{i-3}=\text{C}_{i-3}$, $\text{N}_i-\text{H}_i\cdots\text{O}_{i-4}=\text{C}_{i-4}$ (as in the α-helix), and $\text{N}_i-\text{H}_i\cdots\text{O}_{i-5}=\text{C}_{i-5}$ can be formed are indicated.

Steric maps give a clear picture of the important part played by steric hindrance in reducing the number of conformations possible for a polypeptide chain. A further drastic reduction occurs if regular intramolecular hydrogen bonds are required to be formed.

C. CALCULATION OF POTENTIAL ENERGIES

The assumption of a hard-sphere potential function between nonbonded atoms enables allowed helical conformations to be mapped in a relatively simple manner but this approach has the limitation that it gives no information on the relative stabilities of different conformations. A second difficulty is that the acceptance as an allowed conformation is sensitive, in some areas of the steric map, to the choice of van der Waals radii and to the values assumed for the bond angles, particularly $\tau(\text{N}, \text{C}^\alpha, \text{C})$.

The first attempt to obtain quantitative information about the relative intramolecular potential energies of different polypeptide chain conformations was that reported by De Santis et al. (1965), who calculated the sum of the pairwise nonbonded interactions for regular helical conformations of poly(L-alanine). Since that time various refinements to the procedure have been introduced by Scheraga and co-workers (Scheraga, 1968a, 1971), Ramachandran and co-workers (Ramachandran and Sasisekharan, 1968), Brant and Flory (1965), and Liquori and co-workers (Liquori, 1969). In this section methods of calculating intramolecular potential energy will be described but other factors, which will be discussed in Section D, must be taken into account in

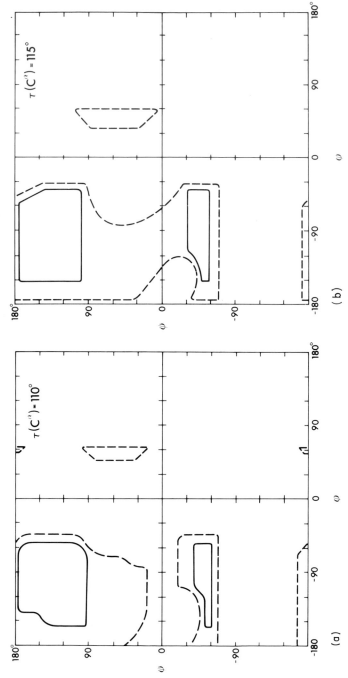

FIG. 6.6. (a) Steric map showing allowed conformations of ϕ and ψ around an alanyl residue (R $\equiv$ CH$_3$) with $\tau(\mathrm{C}^{\alpha}) = 110°$, standard amide geometry (Table 6.1), and planar amide groups. The two regions enclosed by full lines correspond to combinations of ϕ and ψ permitted with "normal" contact distances (Rn in Table 6.2b) and the regions enclosed by broken lines correspond to combinations permitted with "extreme" contact distances (Re in Table 6.2b). (b) As in (a) except $\tau(\mathrm{C}^{\alpha}) = 115°$, showing the considerable increase in permitted area for "extreme" contact distances. Redrawn from Ramachandran and Sasisekharan (1968).

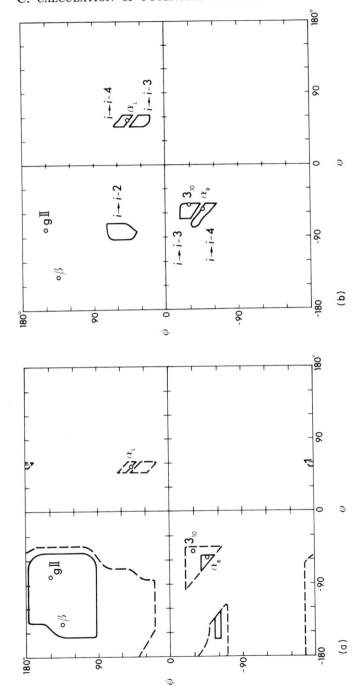

FIG. 6.7. (a) Steric map showing allowed combinations of ϕ and ψ in regular helical conformations of poly(L-alanine) with $\tau(C^\alpha) = 110°$, standard amide-group geometry (Table 6.1), and $\omega = 180°$. The regions enclosed by full lines correspond to combinations of ϕ and ψ permitted with "normal" contact distances (Rn in Table 6.2b) and the regions enclosed by broken lines correspond to combinations permitted with "extreme" contact distances (Re in Table 6.2b). The (ϕ, ψ) combinations corresponding to the left-handed α-helix (α_L), right-handed α-helix (α_R), 3_{10}-helix, poly(glycine) II (g II), and the anti-parallel-chain pleated sheet (β) are indicated. (b) Steric map showing combinations of ϕ and ψ in regular helical conformations for which satisfactory hydrogen bonds may be formed. Redrawn from Ramachandran and Sasisekharan (1968).

attempting to predict the most stable conformation in a given environment.

The contributions which must be included in the potential energy U arise from the following sources.

(1) Rotation about bonds (U_ϕ, U_ψ, U_ω, U_x).
(2) Interactions between nonbonded atoms (U_{nb}).
(3) Electrostatic interactions (U_{es}).
(4) Hydrogen bonding (U_{hb}).
(5) Distortion of the normal equilibrium interbond angles (U_τ) and bond lengths (U_b).

No completely satisfactory means exist for representing any of these contributions and a variety of empirical functions have been used. As a result, different groups of workers have obtained different estimates for the potential energy of the same conformation.

The empirical approach has been criticized by Pullman (1971), who comments:

> Praiseworthy as such attempts are, they suffer from two obvious drawbacks. In the first place, whatever the practical justification for the partitioning of the total potential energy into a series of components, the procedure involves necessarily an element of arbitrariness and, possibly, incompleteness. Secondly, the fundamental formulae and parameters used to define the various components are far from being well established and differ, often appreciably, from one author to another. A more rigorous deal may therefore be expected from a quantum mechanical approach [p. 319].

However, the feasibility of applying the quantum mechanical approach to large structures such as protein molecules remains to be established and in the interim an empirical approach offers the only practical alternative. The functions and procedures used in this method are described in the following sections.

I. TORSIONAL POTENTIALS

Torsional potentials exist such that there are preferred orientations of the groups of atoms attached to either end of a covalent bond. These potentials are considered to arise from two effects, the first due to an exchange interaction between the electrons in the bond with those in the neighboring bonds and the second due to van der Waals interaction between the two groups of atoms attached to the ends of the bond.

a. *Total Potential*

Quantitative information on the nature of torsional potentials has been obtained mainly from studies on simple compounds and one approach to the assignment of torsional potentials for rotation about side-chain bonds has been to assume that they are similar to those observed in analogous small molecules (Gibson and Scheraga, 1967). The side-chain rotational barriers used by these authors are listed in Table 6.3. When this method is used the nonbonded interactions between

TABLE 6.3

Total Barriers to Torsional Rotation about Side-Chain Bonds[a]

Bond number	Residue	χ Trans (deg)	χ Gauche (deg)	$U_{\text{gauche}} - U_{\text{trans}}$ (kcal mole^{-1})	$U_{\text{max}} - U_{\text{trans}}$ (kcal mole^{-1})
1	Aromatic	-60	60, 180	0.40	3.50
1	Branched	-60, 180	60	0.40	3.80
1	Seryl	-60	60, 180	0.20	1.00
1	Other	-60	60, 180	0.20	3.50
2	Leucyl, isoleucyl	-60, 180	60	0.75	3.50
2	Aromatic	—	—	—	0.00
2	Seryl	-60, 60, 180	—	—	2.00
2	Asparaginyl	-60, 60, 180	—	—	0.50
2	Aspartate	—	—	—	0.00
2	Other	180	-60, 60	0.75	3.50
3	Methionyl	180	-60, 60	0.40	2.00

[a] Gibson and Scheraga (1967).

side-chain atoms separated by three bonds must be omitted when the nonbonded term U_{nb} is being calculated.

b. *Intrinsic Component*

An alternative approach which has been used is to separate the torsional potential into exchange and nonbonded components and to incorporate the latter into the calculation of U_{nb}. The exchange, or intrinsic, component is generally assumed to be adequately represented by a function of the form

$$U = \tfrac{1}{2} U_0 (1 \pm \cos m\chi) \tag{6.16}$$

where U_0 is the barrier height and m is an integer which depends upon the rotational symmetry of the torsional potential. If the positive sign is used $\chi = 0$ corresponds to a maximum; if the negative sign is used $\chi = 0$ corresponds to a minimum. There is no general agreement about the precise values to be assigned to U_0 for the various situations met with in proteins and in some instances the values used have been no better than plausible guesses. A summary of the reasoning used to assign potentials for rotation about various bonds has been given by Scheraga (1968a). The forms of Eq. (6.16) used by Scheraga and co-workers are as follows (values in kcal mole^{-1}).

(i) *Rotation about* N—C$^\alpha$ (Scott and Scheraga, 1966):

$$U_\phi = 0.3(1 - \cos 3\phi) \tag{6.17}$$

(ii) *Rotation about* C$^\alpha$—C (Scott and Scheraga, 1966):

$$U_\psi = 0.1(1 + \cos 3\psi) \tag{6.18}$$

(iii) *Rotation about* C′—N (McGuire *et al.*, 1971):

$$U_\omega = 10(1 - \cos 2\omega) \tag{6.19}$$

which for small values of ω can be approximated by

$$U_\omega = 20(\Delta\omega)^2 \tag{6.20}$$

(iv) *Rotation about Bonds between Tetrahedral Carbon Atoms* (Ooi *et al.*, 1967):

$$U_x = 1.4(1 + \cos 3\chi) \tag{6.21}$$

(v) *Rotation about Side-Chain Ester Bonds* (Ooi *et al.*, 1967; Yan *et al.*, 1968):

$$\text{C} - \text{(COO)} \qquad U_x = 0.1(1 + \cos 3\chi) \tag{6.22}$$

$$\text{(C} = \text{O)} - \text{OC} \qquad U_x = 4.375(1 - \cos 2\chi) \tag{6.23}$$

$$\text{(COO)} - \text{C} \qquad U_x = 0 \tag{6.24}$$

(vi) C$^\beta$—C$^\gamma$ *in Tyrosine* (Ooi *et al.*, 1967):

$$U_x = 0.29(1 + \cos 6\chi) \tag{6.25}$$

(vii) CH_2—C_6H_5 *in Benzyl Groups* (Yan *et al.*, 1968):

$$U_x = 0.25(1 + \cos 6\chi) \tag{6.26}$$

(viii) C_6H_5—NO_2 *in p-Nitrobenzyl Group* (Yan *et al.*, 1968):

$$U_x = 3.9(1 - \cos 2\chi) \tag{6.27}$$

Ramachandran and Sasisekharan (1968) also suggest the use of Eqs. (6.19) and (6.20) to represent torsional potential about C—N. Chandrasekharan and Balasubramanian (1969) have used Eq. (6.21) with a barrier height of 2.0 kcal mole^{-1}, giving a value of 1.0 for $\frac{1}{2}U_0$ rather than 1.4, and have used $m = 2$ together with the positive choice of sign in Eq. (6.16) to represent the potential about C—S bonds. The barrier height was assumed to be 2.5 kcal mole^{-1}.

There is a divergence of opinion about the nature of torsional potential (i); Liquori and co-workers use

$$U_\phi = \tfrac{1}{2}U_\phi{}^0(1 + \cos 3\phi) \tag{6.28}$$

with minima corresponding to the maxima of the version used by Scheraga and co-workers. Brant and Flory (1965a) and Maigret et al. (1970) also favor the alternative form given in Eq. (6.28).

II. Nonbonded Interactions

The simplest representation of nonbonded interactions is the "hard-sphere" approximation discussed in Section B, in which it is assumed that there is zero interaction if atoms are separated by a distance greater than the sum of their van der Waals radii and infinite repulsion if the separation is less than this sum (Fig. 6.8). A more realistic representation of the nonbonded interactions can be obtained by using expressions containing a pair of functions, the first representing repulsive forces and the second attractive forces. The selection of suitable functions has been discussed by Ramachandran and Sasisekharan (1968) and Scheraga (1968a). The two representations commonly employed for the nonbonded potential U_{nb} between atoms i and j are

$$U_{nb} = (d_{ij}/r_{ij}^{12}) - (e_{ij}/r_{ij}^{6}) \tag{6.29}$$

and

$$U_{nb} = a_{ij}\exp(-b_{ij}r_{ij}) - (c_{ij}/r_{ij}^{6}) \tag{6.30}$$

where r_{ij} is the internuclear distance and a_{ij}, b_{ij}, c_{ij}, d_{ij}, and e_{ij} are constants which depend on the atom types. In both cases the first term represents a repulsion which decreases rapidly with internuclear distance and the second an attraction which decreases less rapidly with internuclear distance. As with torsional potentials, no general agreement exists on the best means of representing nonbonded interactions and some idea of the variation can be obtained from Fig. 6.8, which shows the dependence of nonbonded potential on internuclear distance computed by Venkatachalam and Ramachandran (1967) for various func-

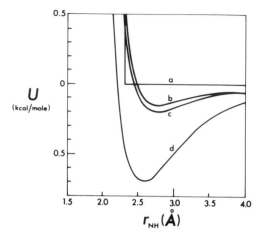

FIG. 6.8. Potential energy functions used to represent the nonbonded interaction between a hydrogen atom and a nitrogen atom. (a) Hard-sphere approximation with an allowed contact distance of 2.3Å (Table 6.2b); (b) Scott and Scheraga (1966) (Table 6.4b); (c) Ramachandran *et al.* (1966b) (Table 6.4a); (d) De Santis *et al.* (1965). Redrawn from Venkatachalam and Ramachandran (1967).

TABLE 6.4a

Parameter Values Used in the Potential Function for Nonbonded Interactions between Atoms Given in Eq. (6.29)[a]

	Parameter	H	C	N	O
H	$d \cdot 10^{-3}$	4.5	38	27	25
	e	47	128	125	124
	r_{min}	2.40	2.90	2.75	2.72
C	$d \cdot 10^{-3}$	38	286	216	205
	e	128	370	366	367
	r_{min}	2.90	3.40	3.25	3.22
N	$d \cdot 10^{-3}$	27	216	161	153
	e	125	366	363	365
	r_{min}	2.75	3.25	3.10	3.07
O	$d \cdot 10^{-3}$	25	205	153	145
	e	124	367	365	367
	r_{min}	2.72	3.22	3.07	3.04

[a] Scott and Scheraga (1966) and Ooi *et al.* (1967). Here r_{min} is the internuclear distance at which the potential function has a minimum value and is equal to the sum of the "observed mean" van der Waals radii given in Table 6.2a. Values of d in kcal mole^{-1} Å^{12}, e in kcal mole^{-1} Å^{6}, r_{min} in Å.

tions. In discussing this problem Scheraga (1968a) emphasizes that all these functions are based on limited data and assumptions that are difficult to substantiate.

The representation in Eq. (6.29) has been used extensively by Scheraga and co-workers and a list of the parameters employed is given in Table 6.4a. The representation in Eq. (6.30) has been employed by

TABLE 6.4b

Parameter Values Used in the Potential Function for Nonbonded Interactions between Atoms Given in Eq. (6.30)[a]

Interaction	Ramachandran and co-workers[b,c]			Liquori and co-workers[d]		
	$a \cdot 10^{-3}$	b	c	$a \cdot 10^{-3}$	b	c
$H \cdots H$	8.29	4.6	46.8	6.6	4.08	49.2
$H \cdots C$	59.7	4.6	127.0	31.4	4.20	121.1
$H \cdots C'$	77.9	4.6	165.8	31.4	4.20	121.1
$H \cdots CH_2$	149.0	4.6	226.9	—	—	—
$H \cdots N$	53.3	4.6	156.1	28.1	4.32	99.2
$H \cdots O$	38.3	4.6	124.1	28.1	4.32	99.2
$H \cdots S$	152	4.6	290.2	—	—	—
$C \cdots C$	559	4.6	363.2	237.0	4.32	297.8
$C \cdots C'$	716	4.6	465.3	237.0	4.32	297.8
$C \cdots CH_2$	1870	4.6	822.8	—	—	—
$C \cdots N$	472	4.6	445.1	212.1	4.44	244.0
$C \cdots O$	341	4.6	363.8	212.1	4.44	244.0
$C \cdots S$	1479	4.6	845.9	—	—	—
$C' \cdots C'$	924	4.6	600.2	237.0	4.32	297.8
$C' \cdots CH_2$	1870	4.6	822.8	—	—	—
$C' \cdots N$	605	4.6	571.4	212.1	4.44	244.0
$C' \cdots O$	433	4.6	461.9	212.1	4.44	244.0
$C' \cdots S$	1882	4.6	1076.0	—	—	—
$CH_2 \cdots CH_2$	3820	4.6	1128.0	—	—	—
$CH_2 \cdots N$	1210	4.6	783.6	—	—	—
$CH_2 \cdots O$	860	4.6	633.9	—	—	—
$N \cdots N$	404	4.6	547.1	186.4	4.55	200.0
$N \cdots O$	294	4.6	446.2	186.4	4.55	200.0
$N \cdots S$	1242	4.6	1038.0	—	—	—
$O \cdots O$	217	4.6	368.9	186.4	4.55	200.0
$O \cdots S$	905	4.6	855.2	—	—	—

[a] a in kcal mole^{-1}, b in Å^{-1}, and c in kcal mole^{-1} Å^6.
[b] Ramachandran and Sasisekharan (1968).
[c] Chandrasekharan and Balasubramanian (1969).
[d] Liquori (1969).

Liquori and co-workers and the values assumed for the parameters are given in Table 6.4b. A disadvantage of this function is that it tends to $-\infty$ rather than ∞ for very small values of r_{ij}.

III. ELECTROSTATIC INTERACTIONS

Electrostatic interactions occur between the electrical dipoles arising from charge separation in chemical bonds. The most widely used method of representing the electrostatic potential (Scott and Scheraga, 1966; Brant et al., 1967; Ooi et al., 1967) is to use the so-called monopole approximation in which partial charges are assigned to individual atoms so as to reproduce both the individual bond moments and also the overall moment. The electrostatic potential can then be calculated by considering the interaction between pairs of atoms. The interaction energy U_{es} between a pair of atoms with partial charges q_i and q_j has the form

$$U_{es} = 332 q_i q_j / D r_{ij} \qquad (6.31)$$

where the charges are expressed as fractions of the electronic charge, r_{ij} is the internuclear distance expressed in Å, D is the effective dielectric constant, and U_{es} is expressed in kcal mole^{-1} (Ooi et al., 1967).

The assignment of partial charges has been based on a consideration of the dipole moments of analogous simple compounds and the values used by Scheraga and co-workers and Brant et al. (1967) are given in Table 6.5. A consistent set of values for both main-chain atoms and side-chain atoms has also been obtained by calculation (Poland and Scheraga, 1967). The selection of a value for the effective dielectric constant poses some difficulty (Brant and Flory, 1965; Scheraga, 1968a; Ramachandran, 1969). The effective value depends upon the internuclear distance and increases rapidly with distance in the presence of a polar solvent. Brant and Flory (1965a) used a constant value of 3.5 for nonpolar solvents, whereas values ranging from one to four have been used by Scheraga and co-workers (Scott and Scheraga, 1966; Ooi et al., 1967; McGuire et al., 1971).

In an aqueous environment account must be taken of the increase in D with r_{ij} and this problem has been considered by Gibson and Scheraga (1967), who took

$$D = 3/g(r_{ij}) \qquad (6.32)$$

where

$$g(r) = \begin{cases} \dfrac{[1 - (r/r_0)^2]^4}{[1 - (r/r_0)^2]^4 + [\frac{1}{16}(r/r_0)^2]^4}, & 0 \leqslant r < r_0 \qquad (6.33) \\[2ex] 0, & r \geqslant r_0 \qquad (6.34) \end{cases}$$

TABLE 6.5

Partial Charges Used in Monopole Approximation of Dipole Interaction[a]

Group	Atom	A[b]	B[c]	C[d]	D[e]	E[f]
Amide	H	+0.271	+0.281	+0.204	+0.30	—
	C′	+0.450	+0.394	+0.318	+0.40	—
	N	−0.304	−0.281	−0.202	−0.30	—
	O	−0.417	−0.394	−0.422	−0.40	—
Net charge		0.000	0.000	−0.102	0.00	—
Amide (Pro, Hyp)	C′	—	—	+0.311	—	—
	N	—	—	+0.048	—	—
	O	—	—	−0.425	—	—
Net charge				+0.162		—
$C^\alpha H^\alpha - R$ (glycyl)	C^α	0	0	0.000	—	—
	$H^{\alpha 1}$	0	0	+0.051	—	—
	$H^{\alpha 2}$	0	0	+0.051	—	—
Net charge		0	0	+0.102	—	—
$C^\alpha H^\alpha - R$ (nonglycyl)[g]	C^α	0	0	min +0.037	—	—
	C^α	0	0	max +0.068	—	—
	H^α	0	0	+0.046	—	—
Net charge (Pro, Hypro)		0	0	−0.162	—	—
Net charge (others)		0	0	+0.102	—	—
$-CH_2-C\diagup^O_{OH}$ C (alkyl)	C (alkyl)	—	—	—	—	+0.035
	C (carbonyl)	—	—	—	—	+0.192
	O (carbonyl)	—	—	—	—	−0.278
	O (hydroxyl)	—	—	—	—	−0.252
	H (hydroxyl)	—	—	—	—	+0.303
Net charge		—	—	—	—	0.000

[a] Expressed as fractions of the electronic charge.
[b] Ooi et al. (1967).
[c] Brant et al. (1967).
[d] Poland and Scheraga (1967).
[e] Ramachandran and Sasisekharan (1968).
[f] Yan et al. (1968).
[g] Partial charges for the atoms of various R groups are given by Poland and Scheraga (1967).

and r_0 is the sum of the van der Waals radii (given in Table 6.6) plus 2 Å. This function has a value close to 3 for $r < 0.9r_0$ and becomes infinite for $r/r_0 = 1.0$. Srinivasan (1969) considered the general problem of the change in D with internuclear distance and suggested that the value for atoms separated by a distance equal to the sum of their van der Waals radii was about 2 in the interior of a solid protein and about 4.5 in aqueous solution. At twice this distance it was found that the effective dielectric constant had risen to almost its macroscopic value (about 81 in the case of water).

A feature of electrostatic interactions that has generally been neglected in calculations of potential energies is the fact that the charge distribution will vary with conformation. For example, the charges which should be assigned to the atoms of the amide group in the monopole approximation will be a function of the torsional angle ω (Yan *et al.*, 1970).

IV. HYDROGEN BONDS

Although hydrogen bonding plays an important part in determining conformation, the nature of the potential function for this type of interaction is only very imperfectly understood and a variety of empirical or semiempirical forms have been used (De Santis *et al.*, 1965; Scott and Scheraga, 1966; Ooi *et al.*, 1967; Poland and Scheraga, 1967; McGuire *et al.*, 1972). Discussions of the relative merits of these and other forms have been given by Ramachandran and Sasisekharan (1968), Scheraga (1968a), Chidambaram *et al.* (1970), and Balasubramanian *et al.* (1970). The simplest representation so far suggested is that of Poland and Scheraga (1967). The potential is regarded as being made up of two parts, the first being the normal sum of electrostatic and nonbonded interactions for all the interacting atoms less the nonbonded interaction between the donor and acceptor atoms, the second being a term

$$(d^*/r^{12}) - (e^*/r^6) \tag{6.35}$$

similar to that of the nonbonded potential function given in Eq. (6.29) except that d^* and e^* are chosen to reproduce the experimentally observed equilibrium distance and energy. The values suggested for $N-H\cdots O=C$ hydrogen bonds between amide groups were $d^* = 2423$ kcal mole^{-1} Å^{12} and $e^* = 56.08$ kcal mole^{-1} Å^6. These values will only be applicable if the functions used to calculate the electrostatic interactions are the same as those used by Poland and Scheraga (1967). A detailed consideration of this means of representing hydrogen bond potential functions has been made by McGuire *et al.* (1972).

V. DISTORTION POTENTIALS

In order to reduce the problem of computing conformational potential energies to manageable proportions, it is usual to assume a fixed set of values for the bond distances and angles. It is evident from crystallographic studies of molecular crystals that intermolecular interactions lead to minimum-energy states in which the geometry of the individual molecules is distorted compared with the minimum-energy conformation for an isolated molecule. In the large molecules of fibrous proteins it is likely that such effects will be important in long-range intramolecular interactions as well as in intermolecular interactions.

a. *Bond Stretching*

The potential function for small changes in bond length from the normal equilibrium value can be represented as

$$U_b = \tfrac{1}{2}k_b(\delta b)^2 \tag{6.36}$$

where k_b is the stretching force constant which may, in favorable cases, be established from studies of infrared spectra. Values of k_b are generally about 1000 kcal mole^{-1} Å^{-2} and so a change $\delta b = 0.1$ Å leads to an increase in energy of about 5 kcal mole^{-1}. It follows that variation in bond length is unlikely to be a significant variable in conformation analysis.

b. *Bond Angle Deformation*

The potential function for small changes in the normal equilibrium values of interbond angles can be represented by

$$U_\tau = \tfrac{1}{2}k_\tau(\delta\tau)^2 \tag{6.37}$$

where k_τ is the bending force constant and $\delta\tau$ is in radians. The values of k_τ for the main chain are of the order of 80 kcal mole^{-1} (Ramachandran and Sasisekharan, 1968) so that a change $\delta\tau = 5°$ leads to an increase in energy of about 0.3 kcal mole^{-1}. Except for a few isolated cases the potential for bond angle deformation has been omitted from energy calculations for polypeptide chains and this is a serious omission since there is abundant evidence that bond angle distortion is a common feature in molecular crystals. In proteins a small variation in bond angle may have a very significant effect on the interactions between atoms remote from the bond, due to the amplification of the perturbation by the intervening distance. Balasubramanian and Ramakrishnan (1972) have explored the effects of variation in the angle $\tau(N, C^\alpha, C')$ on the potential energy of a pair of peptide units.

Another difficulty which arises from the neglect of this potential is that cumulative effects of small distortions of bond angles, particularly along the main chain, may appreciably affect the position of one portion of the chain relative to another in folded structures.

The bending force constants deduced from infrared spectra are not particularly reliable and cannot be used directly in Eq. (6.37) since they include contributions from nonbonded and electrostatic interactions, which have already been allowed for in the terms U_{nb} and U_{es}, respectively.

VI. Computational Procedure

When it is wished to compare the intramolecular potential energies of different conformations of an isolated protein molecule the expression generally calculated is

$$U = U_\phi + U_\psi + U_\omega + U_\chi + U_{nb} + U_{es} + U_{hb} + U_b + U_\tau \qquad (6.38)$$

where the individual terms on the right-hand side represent summations over the molecule. In practice, bond lengths are usually regarded as fixed and U_b is therefore omitted. Side-chain ring structures are generally regarded as having a fixed geometry and torsional motion about the bonds in the ring is not considered. The amide group has also been taken as a fixed structure, with $\omega = 180°$, in many instances, and the term U_ω omitted. Nonbonded and electrostatic interactions between the atoms in structures of fixed geometry are not usually included in U.

When torsional potentials of the type suggested by Gibson and Scheraga (1967) are used (Section C.I.a) the term U_{nb} is calculated as a pairwise summation over atoms separated by more than three bonds, subject to the restriction mentioned for structures of fixed geometry. If the torsional potential is represented as the sum of nonbonded and exchange terms (Section C.I.b), nonbonded terms from atoms separated by three bonds are also included. Depending on the type of hydrogen bond potential which is assumed, nonbonded interactions between certain of the atoms in the N—H and C=O groups may have to be omitted. If the function proposed by Poland and Scheraga (Section C.IV) is used, the term corresponding to the H↔O interaction is replaced, for amide hydrogen and oxygen atom pairs, by the term given in Eq. (6.35). Nonbonded interactions decrease rapidly with internuclear distance and may be neglected for $r_{ij} \geqslant 7.0$ Å (Scheraga, 1968a).

The electrostatic term U_{es} is calculated as a pairwise summation over all partial charges subject to the restriction mentioned for fixed structures. Again, depending upon the type of hydrogen bond potential which is assumed, terms involving pairs in interacting amide groups may have to be omitted. This latter restriction does not apply if the hydrogen bonding function proposed by Poland and Scheraga is used. The electrostatic interactions decrease rather slowly with internuclear distance and care must be taken to ensure that sufficient terms are included when calculating the energies per residue of idealized conformations such as an infinite α-helix (Parry and Suzuki, 1969a).

As mentioned earlier, bond angles are usually regarded as fixed in order to reduce the number of variables and U_τ is omitted from Eq. (6.38). There is no rationale for this procedure and it would seem to be essential to include this term in calculations of the energies of, for example, coiled-coil conformations.

D. INTERMOLECULAR INTERACTIONS

In addition to the intramolecular potential energy U discussed in Section C, energy terms arising from intermolecular interactions must be considered.

I. SOLVENT INTERACTIONS

Fibrous proteins are synthesized in an aqueous environment and in most instances adopt their native conformation while fully hydrated. The interaction of proteins with water plays an important part in determining their conformation since the free energy of hydration favors conformations in which the polar groups are exposed and hydrophobic groups cluster so as to exclude water molecules.

The problem of calculating the solvent free energy has been discussed by Gibson and Schcraga (1967), who devised a method for approximating this term. It was assumed that unless two atoms approach each other to within a distance equal to the sum of their van der Waals radii plus the diameter of a water molecule, the solvent that is displaced does not contribute to the energy. A function

$$N_i = \sum_{i \neq j} g(r_{ij})(v_j/v_w) \tag{6.39}$$

was used to estimate the number of water molecules removed from nearest-neighbor contact with the ith atom by the approach of other atoms, where v_j is the volume (Bondi, 1964) of the jth atom, v_w is the volume excluded by a water molecule (taken as 30 Å^3), and $g(r)$ is the function given in Eqs. (6.33) and (6.34). The free-energy contribution arising from the removal of N_i solvent molecules from the ith atom was assumed to be

$$G_i = G_i{}^\circ N_i{}^\circ \{1 - \exp[-N_i/(N_i{}^\circ - N_i)]\} \tag{6.40}$$

where $N_i°$ is the maximum number of solvent molecules which can occur in the first shell around the particular type of atom and $G_i°$ is the free energy for the removal of one solvent molecule (Table 6.6). The total solvent free energy is then obtained by summing G_i over all atoms.

TABLE 6.6

Data for Evaluating Solvent Free Energies with Aqueous Environment

Atom of group		$N°$	$G°$ (kcal mole^{-1})	r^W (Å)	v/v_w
H		2	0.31	1.30	0.11
C	(aromatic)	2	0.11	1.80	0.18
CH	(aromatic)	3	0.11	1.90	0.28
CH	(aliphatic)	2	−0.13	1.95	0.22
CH$_2$	(aliphatic)	3	−0.13	1.95	0.34
CH$_3$	(aliphatic)	8	−0.13	1.95	0.46
N	(amide)	2	0.63	1.65	0.27
N$^+$	(imidazole)	3	3.30	1.65	0.27
N$^+$	(guanidine)	6	1.20	1.65	0.27
NH$_3^+$	(amine)	5	15.40	1.75	0.27
O	(carbonyl)	4	0.94	1.60	0.22
O	(hydroxyl)	6	0.84	1.60	0.27
O$^-$	(carboxyl)	5	4.80	1.60	0.22
S		6	−0.17	1.90	0.44

Identification of column headings: $N°$, Number of water molecules in first shell (Gibson and Scheraga, 1967). $G°$, Free energy for removing one solvent molecule (Gibson and Scheraga, 1967). r^W, Van der Waals radii used for calculating the value of r_0 in Eq. (6.33). For a pair of atoms i and j, $r_0 = r_i^W + r_j^W + 2.0$. The values of r^W are 0.1 Å greater than those given by Bondi (1964) and Brant and Flory (1965). v, Atomic or group volumes used in Eq. (6.39). v_w, Volume of water molecule.

II. OTHER INTERMOLECULAR INTERACTIONS

In their native state most fibrous proteins occur in ordered assemblies and intermolecular interactions must be included in calculations of the total energy. Terms must be included to allow for nonbonded, electrostatic, and hydrogen bond potentials, and for solvent free energy when the assembly is hydrated. The nonbonded interaction will, in general, be dominated by hydrogen–hydrogen terms and so it is particularly important to have an accurate representation of the potential for this particular interaction. Scheraga (1971) has suggested that a separate set of potential functions may be required which are chosen for accuracy at the generally longer distances involved in intermolecular interactions.

E. ENERGY MINIMIZATION

The mapping of the potential energy function U of a model for a fibrous protein structure is relatively straightforward since a large number of constraints may generally be imposed upon the conformational parameters on the basis of data obtained from physical studies. Scheraga and co-workers (Scheraga, 1971) have tackled the more complex problem of predicting the conformation of a protein molecule in solution by searching for minima of a "conformational energy function" F which is the sum of the potential energy function U and a second function V equal to the free energy for all interactions involving the solvent.

The conformational energy F may be considered as a function in n-dimensional space, where n is the number of independent variables in the expression for F. In general this function will have many minima, the lowest of which is referred to as the global minimum. At present the only known method of locating the global minimum is to find all the local minima and select the one of lowest energy. Methods of locating minima have been discussed by Gibson and Scheraga (1967), Ramachandran and Sasisekharan (1968), and Scheraga (1968a), and a global search algorithm has been described by Crippen and Scheraga (1971a).

The global minimum of the conformational energy function does not necessarily correspond to the conformation of lowest free energy, since small fluctuations about this conformation contribute a conformational entropy (Gibson and Scheraga, 1969a,b; Gō and Scheraga, 1969). The native conformation is, in fact, a macroscopic concept which embraces a collection of microscopic states in the vicinity of the minimum of F that has the lowest free energy. If the statistical weight of such a minimum is sufficiently great, the concept of a native conformation is a valid one. If there are a number of minima of comparable statistical weight, the concept is imprecise.

The prediction of conformation therefore requires the location not only of the global minimum of F but also all the local minima of comparable energy. The statistical weights of these minima must then be calculated. Methods of tackling these problems are discussed by Scheraga (1971) but it is doubtful whether the conformations of fibrous proteins could be predicted *ab initio* by this method at present, because of the inadequacies of the empirical potential functions used to calculate F and the difficulty of assigning statistical weights. However, it has already been demonstrated (Fraser *et al.*, 1969a) that if the minimum of greatest statistical weight can be located approximately by other methods, all that is required is to locate the precise minimum of F in order to find the native conformation.

F. SUMMARY

The procedures described in this chapter enable conformations to be described precisely in terms of internal coordinates and, by a transformation to an external frame of reference, the feasibility of the conformation can be tested by computing interatomic distances and searching for bad contacts. When used in conjunction with X-ray diffraction or other studies this method is likely to remain a valuable guide to the intuitive development of model structures.

A more ambitious application is the *ab initio* prediction of the conformation of lowest free energy and it seems unlikely that this will be possible in the near future. Calculations of conformational energy, however, seem likely to play a major role in the development and refinement of models for the conformation of fibrous proteins. Except in the case of fibrous proteins that form true crystals, the X-ray data are insufficient to deduce the full three-dimensional structure of the molecule, but if model structures can be co-refined on the basis of X-ray data and energy calculations together with data from other studies, this may be possible. The X-ray data provide a geometric framework within which the molecule must fit and this drastically reduces the volume of configurational space which must be explored in the search for minima of the configurational energy. When energy minima are located the corresponding structures may then be refined on the basis of a comparison of the observed X-ray intensity data with the calculated intensity transform.

The major difficulty in co-refining model structures on the basis of X-ray data and calculations of conformational energy is in establishing the relative weights to be attached to the two sets of data. The theory of X-ray diffraction is well established and the experimental errors may be assessed fairly accurately; in contrast, energy minimization in its present state is of unpredictable accuracy. The dilemma has been nicely stated by Flory [(1967), p. 345]: "Calculations of this nature, being mainly empirical in basis and approximate in numerical result, should be treated with circumspection. They cannot be regarded as being accurate in detail."

The key to this problem lies in the development of an internally consistent set of potential functions to represent both intermolecular and intramolecular potentials and the accurate description of protein–solvent interactions.

LIST OF SYMBOLS FOR CHAPTER 7

S	function to be minimized
$S_{\min}$	global minimum of S
P	parameter to be optimized
m	number of parameters to be optimized
W	weighting function
F	calculated value
Y	observed value
$\hat{\sigma}$	unbiased estimate of standard deviation
G	constraint function
r	number of constraints
n	number of data
λ	Lagrange undetermined multiplier
L	lower limit of parameter value
U	upper limit of parameter value

Chapter 7

Optimization Techniques

A frequently encountered problem in studies of conformation in fibrous proteins involves the optimization of the parameters of a model or function on the basis of certain criteria. For example, the potential energy of a conformation can be expressed as a function of the internal coordinates (Chapter 6) and the conformation of lowest potential energy found by optimizing these parameters for minimum potential energy. Again the intensity transform of a model structure can be expressed as a function of the atomic coordinates (Chapter 1) and these parameters must be optimized so as to minimize the discrepancy between the observed and calculated intensity transforms.

To date very little use has been made of systematic optimization techniques in studies of fibrous proteins and the purpose of this chapter is to draw attention to the need for objective optimization of parameters and to indicate the methods that are available for this purpose. Excellent accounts of the mathematical background of these methods are available (Cooper and Steinberg, 1970; Kowalik and Osborne, 1968; Hamilton, 1964).

Most optimizations involve the search for minima of a quantity S which may be expressed as a function of a set of m parameters P_1, P_2,..., P_m and has the property that the lowest minimum, termed the global minimum, occurs at the optimum values of the parameters. Various means of locating local minima of S are available but there is no general method for locating the global minimum directly.

161

A. DIRECT SEARCH METHODS

In direct search methods (Rosenbrock, 1960; Hooke and Jeeves, 1961; Powell, 1964; Nelder and Mead, 1965) S is evaluated for an initial set of parameter values and for one or more additional sets. The resulting values of S are then compared and used as a basis for selecting a better starting set. Repeated application of the algorithm is used to locate a local minimum. None of these methods requires the calculation of derivatives.

The efficiency of direct search methods depends upon such factors as the value of m and the functional dependence of S upon the parameters. The most generally useful method appears to be that due to Powell (Fletcher, 1965; Box, 1966; Duke and Gibb, 1967).

B. GRADIENT METHODS

The efficiency of the optimization process can be improved when information is available on both S and $(\partial S/\partial P_j)$ ($j = 1,..., m$), and the theory of a number of algorithms which use this information has been given by Kowalik and Osborne (1968).

Comparisons of various optimization procedures have been reported by Box (1966), Duke and Gibb (1967), Gibson and Scheraga (1967), Kowalik and Osborne (1968), Pearson (1969), and Warme et al. (1972). The most generally useful appears to be that of Davidon (1959) as modified by Fletcher and Powell (1963).

C. LEAST SQUARES METHODS

In many problems a model system which is to be fitted to n experimental data can be specified by a set of m parameters ($m < n$) and a widely used criterion for optimization is that the function

$$S = \sum_{i=1}^{n} W_i(F_i - Y_i)^2 \qquad (7.1)$$

should be a minimum, where Y_i is one of the n observed data, F_i is the calculated value, and W_i is the weight attached to the particular observation. This is the least squares criterion and the particular form of Eq. (7.1) enables special methods to be used to find minima of S.

If F_i is a linear function of the parameters, the set that minimizes S can be obtained by solving a set of simultaneous linear equations

(Hamilton, 1964; Kowalik and Osborne, 1968). In general, however, F_i has a nonlinear dependence on at least some of the parameters and iterative methods must be used to find the set of parameter values that minimizes S.

Direct search and gradient methods can be used for nonlinear least squares optimizations but the most generally useful is the Gauss or Gauss–Newton method (Kowalik and Osborne, 1968; Hamilton, 1964). Various refinements to the method have been suggested (Levenberg, 1944; Marquardt, 1963; Meiron, 1965; Pitha and Jones, 1966; Kowalik and Osborne, 1968) and detailed accounts of the procedure for carrying out least squares optimization have been given (Fraser and Suzuki, 1970b, 1973).

The choice of weighting function in Eq. (7.1) will clearly have an important bearing on the solution. Factors influencing the choice of W_i have been discussed by Hamilton (1964) and Blackburn (1970). In most cases W_i is assigned a value which is inversely proportional to the estimated variance of the observation. The choice of weighting function for the refinement of model structures on the basis of X-ray diffraction data has been discussed by Cruickshank et al. (1961), Cruickshank (1965), Bradbury et al. (1965), and Arnott and Wonacott (1966a). When analyzing infrared absorption spectra W_i is generally taken as unity if the range of optical densities in the spectrum is not too great (Chapter 5, Section C.II).

When S has been minimized the goodness of fit can be measured by means of the unbiased estimate of the standard deviation (Marquardt et al., 1961)

$$\hat{\sigma} = [S_{\min}/(n - m)]^{1/2} \tag{7.2}$$

where $S_{\min}$ is the minimum value of S. Unbiased estimates of the standard deviations of the parameter values can also be calculated (Hamilton, 1964; Luenberger and Dennis, 1966; Ederer, 1969).

Hypothesis testing can also be carried out on the basis of least squares optimization and when the model function F is known to be capable of representing error-free data exactly, precise tests can be made (Hamilton, 1964, 1965).

D. CONSTRAINED OPTIMIZATION

So far attention has been restricted to unconstrained optimization in which the parameters were independent and could take any value.

In many applications the optimization of the parameters must be carried out subject to certain equality constraints of the type

$$G_k(P_1,...,P_m) = 0, \qquad k = 1,...,r \tag{7.3}$$

For example if the value of P_1 was required to be equal to $P_2 + P_3$, the constraint would be expressed by $G = P_1 - P_2 - P_3 = 0$. Constrained optimization can be performed using the methods described earlier by optimizing the unconstrained function

$$S' = \sum_{i=1}^{n} W_i(F_i - Y_i)^2 + \sum_{k=1}^{r} \lambda_k G_k \tag{7.4}$$

where the λ_k are known as Lagrange's undetermined multipliers and are to be treated as independent variables.

The efficiency of optimization procedures decreases rapidly as the number of parameters increases and every constraint that can be introduced increases the probability of obtaining a solution. The use of Eq. (7.4) is particularly convenient since a problem such as the refinement of a model structure to fit X-ray diffraction data can be formulated in a completely general way and equality constraints introduced as required to improve the data/parameter ratio (Arnott and Wonacott, 1966a).

Another type of restriction which can be applied to parameters is the inequality constraint

$$L_j < P_j < U_j \tag{7.5}$$

where L_j and U_j are, respectively, lower and upper limits and this can be treated by a suitable transformation of the independent variable P_j. For example, the transformation

$$P_j = L_j + (U_j - L_j) \sin^2 P_j' \tag{7.6}$$

could be made and unconstrained optimization carried out with respect to P_j' (Kowalik and Osborne, 1968). An alternative approach is to introduce severe penalties on S when the forbidden region is transgressed.

E. SEARCHES FOR GLOBAL MINIMA

It was pointed out earlier that the function S may have many minima and there is no guarantee that the optimization procedures discussed so far will converge onto the global minimum. Although the problem is general, it has been found to be particularly acute in conformational energy minimizations (Scheraga, 1971). Another situation in which this problem is often encountered is when attempting to fit an incorrect model to observed data or a correct model to poor data. In the latter cases the general vicinity of the global minimum is often readily located but the final rate of convergence is slow and many shallow local minima exist in the vicinity of the global minimum.

No general procedure for locating the global minimum of S has been devised and the selection of a strategy in the search requires a certain amount of low cunning. The magnitude of the problem can quickly be assessed by choosing a number of widely separated starting points. If all lead to the same minimum, this is probably the global minimum. If all lead to different minima, empirical methods must be used to try and locate lower minima using these as starting points (Gibson and Scheraga, 1969a, 1970; Crippen and Scheraga, 1969, 1971a,b; Momany et al., 1969). A global search algorithm has been described by Crippen and Scheraga (1971a).

LIST OF SYMBOLS FOR CHAPTER 8

$[\alpha]$	specific rotation
λ	wavelength
λ_0 , a_0 , b_0	parameters
M	molecular weight
M_r	mean residue weight
n	refractive index
n_a , n_b	principal refractive indices
P	specific polarizability
P_a , P_b	principal specific polarizabilities
p_l , p_t	longitudinal and transverse bond polarizabilities
p_1 , p_2	longitudinal and transverse molecular polarizabilities
N	Avogadro's number
f	orientation function (see Chapter 5)
ρ	macroscopic density

Chapter 8

Other Methods of Studying Conformation

In the foregoing chapters methods which have proved to be especially valuable in the elucidation of conformation in fibrous proteins have been discussed in some detail. There are, in addition, a number of other methods which, although inherently incapable of yielding detailed information about conformation, nevertheless provide a means of measuring and detecting changes in a small number of molecular parameters. Concerning such methods, Schellman and Schellman (1964) remark: "The limitation is inescapable: If all the groups of a molecule contribute to a physical property, then the interpretation must invoke simple (and presumably oversimplified) models. If only a few groups contribute, then only a small part of the desired structure is under experimental attack" [p. 4]. However, many of these methods provide information which may be of assistance in devising models of fibrous protein structures and the most profitable, in this respect, will be reviewed briefly in this chapter.

Information on molecular shape and size can be obtained from studies of solutions of fibrous proteins by the techniques of sedimentation analysis (Coates, 1970), gel filtration (Andrews, 1970), viscosity measurement (Bradbury, 1970), light scattering (Timasheff and Townend, 1970), and low-angle X-ray scattering (Brumberger, 1967; Pilz, 1973). In addition, ultraviolet absorption (Gratzer, 1967), optical rotatory dispersion, and circular dichroism spectra (Yang, 1967, 1969; Beychok, 1967; Jirgensons, 1969; Tinoco and Cantor, 1970) exhibit effects which are associated with the spatial arrangement of the amide-group chromophores of the main chain and have been widely used to study conformation in solution. Nuclear magnetic resonance has also been used in studies of conformation and conformational change in solution (Metcalfe, 1970; Lui and Anderson, 1970; Roberts and Jardetzky, 1970; Sheard and Bradbury, 1971) but its main value in the study of fibrous proteins at present appears to be in investigations of the role of water in native structures (Clifford and Sheard, 1966; Berendsen, 1968; Cope, 1969; Dehl and Hoeve, 1969; Hazlewood *et al.*, 1969; Katz, 1970; Tait and Franks, 1971; Khanagov, 1971).

The foregoing methods of studying conformation in solution have been very adequately described in the references cited. In the following sections attention will be confined to certain methods which are applicable or potentially applicable to the study of solid fibrous protein structures.

A. ULTRAVIOLET ABSORPTION SPECTRA

The ultraviolet absorption spectra of proteins and synthetic poly-peptides have been studied in considerable detail (Donovan, 1969) and the absorption bands near 200 nm associated with the main-chain amide chromophore have been found to be conformation-sensitive (Gratzer, 1967). This sensitivity to conformational change is a result of differences in the nature of the interactions between electronic transitions in different amide groups.

Exciton splitting, resulting from the coupling of the $\pi_1 \rightarrow \pi^*$ transitions, has been observed in oriented films of poly(γ-methyl-L-glutamate) in the α-helical form by Gratzer *et al.* (1961) and Momii and Urry (1968) and the directions of polarization were found to be as predicted by Moffitt (1956). Exciton splitting has also been observed in the α form of poly(L-alanine) (Gratzer *et al.*, 1961; Momii and Urry, 1968; Bensing and Pysh, 1971) and in other regular structures such as the pleated-sheet β conformations (Pysh, 1966; Rosenheck and Sommer, 1967; Woody, 1969) and the polyproline I and II conformations (Rosenheck *et al.*, 1969). Measurements of this type on fibrous proteins are likely to yield information similar to that provided by observations of infrared dichroism. However, specimen preparation is more exacting, measurement of spectra is more difficult, band overlap is considerable, and the prospect of obtaining reliable quantitative estimates of the contents of different conformations in a specimen seem remote at the present time. An important feature of linear dichroism measurements, however, is the possibility of distinguishing between the parallel-chain and antiparallel-chain pleated sheets (Woody, 1969).

Hypochromism, resulting from interactions between electronic transitions of different energies in adjacent amide groups, has been observed in polypeptides in the α-helical form in solution (Imahori and Tanaka, 1959; Rosenheck and Doty, 1961; Tinoco et al., 1962; Goodman and Listowsky, 1962; Goodman et al., 1963c; Goodman and Rosen, 1964; McDiarmid and Doty, 1966) and used as a basis for estimating α-helix content of various proteins in solution (Rosenheck and Doty, 1961; Gratzer, 1967). The results obtained were not encouraging, however, and the possibility of applying this method to solid specimens of fibrous proteins seems slight in view of the difficulties of making quantitative intensity measurements on solid specimens.

B. OPTICAL ROTATORY DISPERSION
AND CIRCULAR DICHROISM

Extensive studies of the optical rotatory dispersion and circular dichroism of synthetic polypeptides in solution have shown that these properties, which are closely related (Moffitt and Moscowitz, 1959; Moscowitz, 1960), are sensitive to conformation and thus a potentially useful means of investigating protein structure. Valuable accounts of experimental methods and reviews of the topic have been given by Yang (1967, 1969), Beychok (1967), Jirgensons (1969), and Tinoco and Cantor (1970). Attempts have been made to assign characteristic spectra to different conformations (Moffitt and Yang, 1956; Pysh, 1966; Rosenheck and Sommer, 1967; Greenfield et al., 1967; Greenfield and Fasman, 1969; Woody, 1969; Chen and Yang, 1971; Saxena and Wetlaufer, 1971), thus providing a basis for estimating the contents of different conformations from observed spectra. There are, however, a number of theoretical and experimental problems which remain to be solved before the method can be made quantitative (Yang, 1969; Fasman et al., 1970; Jirgensons, 1970; Chen et al., 1972; Dalgleish, 1972; Madison and Schellman, 1972).

Measurements of optical rotatory dispersion in the visible and near-ultraviolet regions are generally fitted to an equation proposed by Moffitt and Yang (1956)

$$\frac{3}{n^2 + 2} \frac{M_r}{100} [\alpha] = a_0 \frac{\lambda_0^2}{\lambda^2 - \lambda_0^2} + b_0 \left(\frac{\lambda_0^2}{\lambda^2 - \lambda_0^2}\right)^2 \tag{8.1}$$

where $[\alpha]$ is the specific rotation, n is the refractive index of the medium, M_r is the mean residue weight, λ_0 is a constant which has a value close to 212 nm in many cases, and a_0 and b_0 are adjustable parameters. The value of a_0 has been found to be highly variable and to depend upon conformation, composition, and the molecular environment. The value of b_0, on the other hand, exhibits a strong dependence on the conformation of the main chain. Estimates of b_0 for synthetic polypeptides in the α-helical conformations yield values of about -630 deg cm^2 dmole^{-1} and the quantity $-100b_0/630$ has been used widely as an empirical estimate of α-helix content. The value so obtained can only be regarded as approximate, however, if the α-helix content is low or regular conformations other than the α-helix are present (Jirgensons, 1969; Yang, 1969). Chen and Yang (1971) have analyzed optical rotatory dispersion data for a number of globular proteins of known structure and suggest an alternative value of -580 deg cm^2 dmole^{-1} for the α-helix. In certain polymers the quantitative correlation between b_0 and α-helix content breaks down completely due to a contribution to b_0 from a regular arrangement of electronically coupled side-chain chromophores (Goodman et al., 1963b; Fraser et al., 1967b).

Attempts have been made to assign a value of b_0 to the β conformation (Fasman and Blout, 1960; Imahori, 1960; Wada et al., 1961; Bradbury et al., 1962b; Imahori and Yahara, 1964; Davidson et al., 1966; Iizuka and Yang, 1966; Sakar and Doty, 1966) but the results obtained so far are inconclusive. It seems certain, however, that the magnitude of b_0 is very much less for the β conformation than for the α-helix. From a study of globular proteins of known structure Chen and Yang (1971) suggest a value of 60 ± 30 deg cm^2 dmole^{-1} for the β conformation. Chirgadze et al. (1971) have pointed out that the optical rotatory dispersion of polypeptides and proteins in the near-infrared can be represented by an expression similar to that given in Eq. (8.1) provided that a constant term c_0 is added. The value of c_0 is conformation-sensitive and has zero value for the random-coil conformation. Its magnitude for the β conformation is an order of magnitude greater than for the α-helix and so the method has potentialities for the quantitative estimation of β contents.

Elliott et al. (1957b, 1958) have stressed the difficulties associated with the measurement of optical rotatory dispersion spectra of solids. These arise from birefringence due to local preferred orientation and to residual strain. Visible and near-ultraviolet spectra were obtained from synthetic polypeptides and from films cast from solubilized silk fibroin. The results, however, were less accurate than those obtained from solution studies. In the case of poly(γ-benzyl-L-glutamate) the b_0 values in the solid state were similar to those measured in solution, but with poly(β-benzyl-L-aspartate) anomalous results were obtained and this was attributed to a regular array of interacting side-chain chromophores (Bradbury *et al.*, 1959; Elliott *et al.*, 1962).

Conformation-dependent Cotton effects, associated with the electronic transitions discussed in Section A, are observed below 250 nm in the optical rotatory dispersion spectra of polypeptides in solution and these have been used as a basis for conformation analysis (see for example Greenfield *et al.*, 1967; Yang, 1969). The spectra of solid films of a number of synthetic polypeptides in this region have been reported by Fasman and Potter (1967). Very little work has been done on native fibrous protein structures, due to the difficulties of obtaining optically homogeneous specimens and estimating the mass per unit area (McGavin, 1964).

The measurement of circular dichroism spectra rather than optical rotatory dispersion spectra offers certain advantages and the two types of spectra can be related to one another by means of the Kronig–Kramers relationships (Moscowitz, 1960). Circular dichroism studies of polypeptides and proteins in solution have been reviewed by Beychok (1967), Jirgensons (1969), Yang (1969), and Tinoco and Cantor (1970). As noted earlier, attempts to use circular dichroism spectra for the estimation of the contents of different conformations in a molecule have not been completely successful. Measurements have been made on solid films of both synthetic polypeptides and globular proteins (Fasman *et al.*, 1965, 1970; Fasman and Potter, 1967; Stevens *et al.*, 1968) but very little work has been done on fibrous proteins. Spectra for powdered elastin have been reported by Mammi *et al.* (1970).

A practical problem encountered in obtaining spectra from oriented specimens is the presence of linear birefringence and dichroism, which make it difficult to extract the required circular dichroism spectrum (Tinoco and Cantor, 1970). There are also a number of unresolved problems in the interpretation of the circular dichroism spectra of solid films (Fasman *et al.*, 1970).

C. BIREFRINGENCE

The phenomenon of birefringence arises from anisotropy of polarizability, which is in turn related to molecular and textural anisotropy. General accounts of the theory of birefringence in polymer specimens and methods of measurement have been given by Hermans (1946), Bunn (1961), Morton and Hearle (1962), Stein (1964, 1969), and Wilkes (1971). Most naturally occurring fibrous protein structures are birefringent and the observed values are, in general, the sum of an intrinsic component, which is related to the anisotropy of polarizability of the individual molecules, and a second component, arising from the texture (Wiener, 1912; Schmitt, 1939), which is termed form birefringence.

An approximate value for the intrinsic component of the birefringence of a material of known structure can be calculated by assigning longitudinal and transverse polarizabilities to individual chemical bonds and summing their contributions to the polarizabilities parallel to the principal directions of the refractive index ellipsoid using the formula

$$P = \sum p_l \cos^2 \theta + \sum p_t \sin^2 \theta \tag{8.2}$$

where P is the intrinsic component of the specific polarizability, p_l and p_t are the longitudinal and transverse bond polarizabilities, respectively, and θ is the angle between the bond and the principal direction. The summation extends over unit volume, and provided the birefringence is small, the difference between the refractive indices for two principal directions can be shown to be (Stein, 1964)

$$n_a - n_b = \frac{2\pi(n^2 + 2)^2}{9n}(P_a - P_b) \tag{8.3}$$

where P_a and P_b are the corresponding specific polarizabilities and n is the mean value of the refractive index. Values of p_l and p_t for various bonds have been given by Wang (1939), Denbigh (1940), Bunn and Daubeny (1954), Bunn (1961), and Nagatoshi and Arakawa (1970). The bond polarizabilities given by these authors can only be regarded as average values and the true value is expected to vary with the electronic structure of the bond. Under these circumstances, birefringence values calculated using Eq. (8.3) can only be regarded as approximate and are likely to be seriously in error if the birefringence is small. Ingwall and Flory (1972) have investigated the anisotropy of the polarizability of amides and derived values for a glycyl peptide unit.

An alternative approach is to use Eq. (8.2) to calculate the polarizability p_1 parallel to the axis of an elongated molecule and a mean polarizability p_2 perpendicular to that axis and to use the concept of an orientation parameter f as defined in Chapter 5. In this case the intrinsic component of the macroscopic anisotropy of specific polarizability will be given by

$$P_a - P_b = (Nf\rho/M)(p_1 - p_2) \tag{8.4}$$

where N is Avogadro's number, M is the molecular weight, and ρ is the macroscopic density. If a number of phases are present, the right-hand side of Eq. (8.4) must be replaced by an appropriate summation. Since the intrinsic birefringence is a function of conformation and orientation, it provides a means of detecting changes in these molecular properties. A complication in quantitative studies is the existence of form birefringence, which arises from the anistropic shape of the molecular aggregates. This effect has been treated theoretically (Wiener, 1912, 1927) but the contribution to the observed birefringence must, in general, be determined experimentally (Hermans, 1946; Haly and Swanepoel, 1961; Stein, 1964; Cassim et al., 1968).

Measurements of birefringence provide a useful means of comparing the degree of orientation in different specimens of the same material and have also been used to detect changes in orientation and conformation in, for example, collagen fibers (Ramanathan and Nayudamma, 1962; Basu et al., 1962), muscle fibers (Fischer, 1947; Mommaerts, 1950; Takahashi et al., 1962), and keratin fibers (Fraser, 1953b; Feughelman et al., 1962; Feughelman, 1966, 1968).

D. LIGHT SCATTERING

Many fibrous proteins are laid down in ordered aggregates which extend over distances comparable with the wavelength of visible light and the accompanying fluctuations in polarizability give rise to light scattering. The angular distribution of the scattered light depends upon the nature of the fluctuations and two approaches to the interpretation of light-scattering results on solids have been used. In the first a model for the fluctuations is assumed, for example, optically anisotropic spheres embedded in an isotropic medium, and the observed data are used to test the model and to estimate values of the parameters describing the model. In the second approach the problem is dealt with in terms of a set of statistics which defines the correlation between fluctuations in polarizability in different parts of the specimen. Details of the experimental arrangement used to measure light scattering from solids and a discussion of the theory are given by Stein (1964).

Moritani et al. (1971) have recorded light scattering from several types of collagen films prepared from acid-soluble collagen and collagens solubilized by treatment with proteolytic enzymes. An optical model for the microscopic texture of collagen was assumed and changes in the light scattering with denaturation were interpreted in terms of this model. Although it is difficult to relate the optical model to the texture of the molecular aggregates in collagen, these experiments are important because they point the way to making similar measurements in vivo on fibrous protein assemblies such as those found in connective tissue and muscle.

E. RAMAN SPECTRA

Raman spectra arise from a scattering process in which a change takes place in the vibrational energy of the molecule. The frequency of the scattered light differs from that of the incident light by an amount corresponding to the change in vibrational energy, and information which is in many ways complementary to that given by infrared spectra can be obtained on the normal modes of vibration of a molecule (Colthup *et al.*, 1964; Koenig, 1971). Raman scattering is in general much weaker than the Rayleigh type of scattering, discussed in Section D, in which there is no change in frequency on scattering. Experimental aspects have been discussed by Hawes *et al.* (1966), Beattie (1967), Tobin (1968, 1971), Cornell and Koenig (1968), and Fanconi *et al.* (1969) and valuable reviews of the application of Raman spectra to polymers have been given by Hendra (1969) and Gall *et al.* (1971).

The selection rules for Raman activity of vibrations require that the polarizability must change during the vibration, and the intensity of the Raman band is a function of the magnitude of the change in polarizability. The selection rules for Raman scattering from helical molecules have been given by Higgs (1953) and a detailed treatment of the polarized spectra of poly(L-alanine) has been given by Fanconi *et al.* (1969). Partially ordered systems have been considered by Snyder (1971). The interpretation of polarized Raman spectra is more complex than with polarized infrared spectra and the results are less readily interpretable in terms of molecular conformation. However, observations of Raman spectra are of value in cases where vibrations are infrared-inactive, such as the $\nu(0, 0)$ modes of the antiparallel-chain pleated sheet (Chapter 5, Section A.II.b), and these modes, which are Raman-active, have been observed in the spectra of synthetic polypeptides (Koenig and Sutton, 1971).

Most vibrational modes of proteins are both infrared- and Raman-active and the relative intensities in the two types of spectra are markedly different. The Raman spectrum is not dominated by the main-chain amide bands, as is the case with the infrared spectrum, and side-chain bands are prominent. The aromatic side-chain groups of phenylalanine, tryptophan, and tyrosine give rise to sharp and intense bands and bands associated with the $-C-S-S-C-$ grouping are prominent in cystine-containing proteins (Lord and Yu, 1970a,b). Thus it is in the finer details of side-chain conformation that Raman spectra seem likely to yield information, not available from other methods, on the structure of fibrous proteins. The other property of Raman spectra which seems likely to be exploited in the future is that the bands associated with water are relatively very much less intense than in infrared spectra, thus permitting observations on hydrated materials. One of the problems which must be overcome before satisfactory Raman spectra can be obtained is the elimination of fluorescence and the reduction, as far as possible, of turbidity. These present difficult problems with naturally occurring fibrous protein structures and most of the studies of polypeptides so far reported (Fanconi et al., 1969; Koenig and Sutton, 1969, 1970, 1971; Rippon et al., 1970; Deveney et al., 1971) have been made on synthetic materials, from which fluorescent impurities can be removed by appropriate chemical treatment. Other problems encountered in obtaining spectra from solid films are overheating and photodegradation.

F. MECHANICAL AND THERMAL PROPERTIES

Since fibrous proteins have a primarily structural function in living tissue, the mechanical properties and thermal stability are topics of considerable interest and have been studied extensively. The classical investigations of Astbury and co-workers (Astbury and Street, 1931; Astbury and Woods, 1933; Astbury and Sisson, 1935) for example, provide a wealth of data against which model structures can be tested. While studies of this type still have a place in investigations of conformation, there is a growing realization that the direct correlation of detailed conformation with mechanical and thermal data is fraught with difficulties.

The value of mechanical studies and the difficulties associated with their interpretation in terms of molecular structure have been elegantly summarized by Chapman (1969a):

> A mechanical model of fiber structure may be found useful in a number of ways: firstly, merely as a convenient method of describing structure and predicting mechanical behavior in a fairly compact way; secondly, as a means of testing structures that are proposed on the basis of other work; and lastly, it may be possible from experimental results to infer, on the basis of a particular model, some general structural features. In the last case, however, the conclusions drawn concerning the structure are usually highly dependent on the particular model system chosen to describe the behavior .... Of necessity, the model must generally be an oversimplification of the actual structure, to allow any degree of mechanical analysis [p. 1102].

There are a number of instances, however, where the use of a simple model has served to give a very clear understanding of the mechanical properties in terms of the molecular texture, for example, in the filament-matrix model used to explain the behavior of the elastic moduli of α-keratin fibers on hydration (Feughelman, 1959).

The literature on the mechanical properties of fibrous proteins, their dependence on time, temperature, hydration, and other parameters, and attempts to interpret these properties in terms of molecular structure is voluminous. Valuable reviews of these topics and of experimental techniques have been given by Morton and Hearle (1962), Crewther *et al.* (1965), Harkness (1968), Chapman (1969b), Hearle and Greer (1970), and by various authors in the Encyclopedia of Polymer Science and Technology (Bikales, 1966–70).

Structural transitions occur when fibrous proteins and other polymers are heated or deformed and these can be studied by differential thermal analysis (Ke, 1964a,b; Morita, 1970). This technique has been applied to collagen (Witnauer and Wisnewski, 1964; Naghski *et al.*, 1966), silk (Schwenker and Dusenbury, 1960; Pande, 1965; Crighton and Happey, 1968), and α-keratin fibers (Schwenker and Dusenbury, 1960; Felix *et al.*, 1963; Menefee and Yee, 1965; Pande, 1965; Haly and Snaith, 1967; Crighton *et al.*, 1967; Crighton and Happey, 1968).

SYNTHETIC POLYPEPTIDES AS MODELS OF FIBROUS PROTEINS

Chapter 9

The Alpha Helix

A. INTRODUCTION

Following the discovery of the α-helix (Pauling and Corey, 1950, 1951a; Pauling *et al.*, 1951) it was soon established that the overall distribution of intensity in the diffraction patterns of certain synthetic polypeptides was consistent with the presence of such a helix (Pauling and Corey, 1951b; Perutz, 1951a; Cochran *et al.*, 1952; Bamford *et al.*, 1952, 1954). As originally described, the α-helix embraced two distinct structures for residues other than glycine, and for L-amino acids these correspond either to a right-handed or to a left-handed helix. Early studies of synthetic polypeptides (Yakel *et al.*, 1952; Brown and Trotter, 1956) and proteins (Riley and Arndt, 1952) were held to favor the left-handed alternative, but Huggins (1952) pointed out that the distance between the C_i^β and O_i atoms in the same residue was only 2.64 Å in the left-handed helix whereas no significant steric hindrance occurred in the right-handed alternative. It was concluded that the right-handed α-helix would be the more stable for amino acid residues with the L-configuration. This prediction has been largely borne out by subsequent studies. Confirmation of the occurrence of the α-helix in proteins was obtained in the 2-Å Fourier synthesis of myoglobin calculated by Kendrew *et al.* (1960) and the screw sense was found to be right-handed in all the helical sections of the molecule.

If values are assumed for the bond angles and distances in the amide group and for $\tau(N, C^\alpha, C')$, the formulas given by Sugeta and Miyazawa (1967) can be used to generate atomic coordinates for the α-helix relative

179

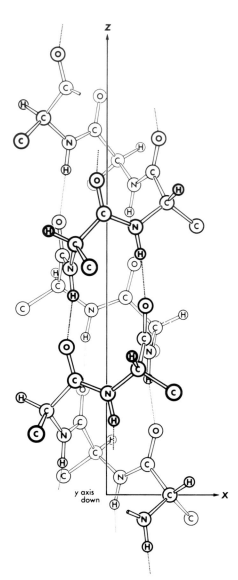

FIG. 9.1. The α-helical conformation of the polypeptide chain (Pauling and Corey, 1951a). Systematic hydrogen bonds of the type $N_iH_i\cdots O_{i-4}$ are formed.

to the helix axis for any permissible combination of unit height h and unit twist t. If the disposition of the bonds around C^α is assumed to be tetrahedral, that is, $\tau(N, C^\alpha, C') = 109.47°$, and the amide group is assumed to be planar and to have the "standard" dimensions given in Table 6.1, the coordinates of a "standard" α-helix with $h = 1.5$ Å and

$t = 100°$ can be calculated. This helix is illustrated in Fig. 9.1 and the coordinates are given in Table 9.1 (Parry and Suzuki, 1969a).

TABLE 9.1

Atomic Coordinates of an L-Residue in a Right-Handed α-Helix[a]

Atom	x (Å)	y (Å)	z (Å)	r (Å)	ϕ (deg)
N	1.360	−0.744	−0.873	1.550	−28.69
H	1.485	−0.559	−1.847	1.586	−20.63
C$^\alpha$	2.280	0	0	2.280	0
[b]H$^\alpha$	2.989	−0.699	0.467	3.069	−13.17
[b]C$^\beta$	3.049	1.029	−0.830	3.218	18.65
C$'$	1.480	0.718	1.089	1.645	25.88
O	1.765	0.579	2.288	1.858	18.15

[a] Unit height $h = 1.5$ Å, unit twist $t = 100°$, $\tau(N, C^\alpha, C') = 109.47°$, and the "standard" planar amide-group geometry, the dimensions of which are given in Table 6.1. Other helix parameters are: axial period $c = 27.0$ Å, pitch $P = 5.4$ Å, $\phi = −51.45°$, $\psi = −52.74°$, $\omega = 180°$, $u = 18$ residues in $v = 5$ turns.

[b] Calculated assuming a tetrahedral distribution of bonds around C^α and $b(C^\alpha, H^\alpha) = 1.1$ Å, $b(C^\alpha, C^\beta) = 1.53$ Å.

In this chapter information which has been obtained on the α-helix from studies of ordered aggregates of synthetic polypeptides is reviewed in Sections B and C and factors affecting the sense and geometry of the helix are considered in Section D. A discussion of the stability of the α-helix in relation to other conformations and the effects of composition and sequence on stability will be reserved for Chapter 12.

B. DIFFRACTION STUDIES

I. POLY(L-ALANINE)

Oriented specimens of [Ala]$_n$ in which the majority of the polymer has an α-helical conformation can be prepared by stretching spun fibers (Elliott, 1967). The X-ray diffraction pattern of such specimens was examined in detail by Brown and Trotter (1956) and a plot of the observed data in reciprocal space coordinates (Chapter 1, Section A.I.g) is shown in Fig. 9.2. The pattern consists mainly of sharp reflections characteristic of a three-dimensionally ordered structure (Chapter 1, Section A.I.e) but streaks are present on certain layer lines. The layer lines can be indexed on an axial period of $c = 70.3$ Å and the sharp

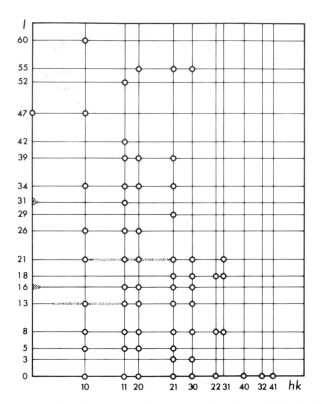

FIG. 9.2.   X-ray diffraction data obtained from the α form of [Ala]ₙ . The pattern consists of discrete reflections (circles) together with layer-line streaks. The discrete reflections can be indexed on a hexagonal unit cell with $a = 8.55$ Å and $c = 70.3$ Å. (Redrawn from Brown and Trotter, 1956.)

reflections fall on a set of vertical row lines which can be indexed on a hexagonal unit cell of side $a = 8.55$ Å.

No evidence for a larger cell could be found and since the cross-sectional area of the cell perpendicular to the fiber axis was close to that expected for a single chain in the α-helical conformation, Brown and Trotter attempted to interpret the pattern in terms of a crystallite in which the repeating unit was a section of α-helix of length 70.3 Å. This implied that all the chains in a crystallite had the same direction as regards the main-chain sequence $-NH-CHR-CO-$.

(i) *Helical Parameters.*   From the theory of diffraction by helical structures (Chapter 1, Section A.III) the α-helix would be expected to produce a pattern consisting of a set of layer lines spaced about $1/5.4$ Å^{-1} apart, centered on the origin, and similar sets centered on a series of meridional reflections spaced about $1/1.5\,\text{Å}^{-1}$ apart. The observed pattern can in fact be interpreted in these terms, which are summarized in the selection rule for Bessel function orders n contributing to a layer line (Chapter 1, Section A.III.b)

$$l = um + vn \qquad (9.1)$$

where l is the layer-line index, u is the number of residues and v is the number of turns in the repeat distance c, and m is an integer.

The only Bessel functions that can give rise to meridional reflections are those of order $n = 0$ (Fig. 1.10a) and since the meridional reflection of spacing about $1.5\,\text{Å}^{-1}$, corresponding to $m = \pm 1$, occurs on $l = 47$ (Fig. 9.2), the value of u must be 47. The first layer line of the origin set, corresponding to $m = 0$ and $n = \pm 1$, has an index $l = 13$, and so from Eq. (9.1), $v = \pm 13$. Thus it was concluded that a helix with $u = 47$ and $v = \pm 13$ was present in the α-form of $[\text{Ala}]_n$ and since $c = 70.3\,\text{Å}$ the unit height $h = 70.3/47 = 1.496\,\text{Å}$, the pitch $P = \pm 70.3/13 = \pm 5.41\,\text{Å}$, and the unit twist $t = \pm 360 \times 13/47 = \pm 99.6°$.

From the coordinates given in Table 9.1 it can be seen that the largest value of r for a nonhydrogen atom in the α-helical form of poly(L-alanine) is likely to be about 3.2 Å. The observed data extend to $R \sim 0.6\,\text{Å}^{-1}$ and so the maximum value of $2\pi Rr \sim 12$. From Eq. (1.29) and Fig. 1.10 it can be seen that Bessel functions of order $|n| > \sim 8$ are unlikely to make any significant contribution to the observed pattern and in fact the indices l of the layer lines on which diffraction is observed (Fig. 9.2) all correspond to values of $-7 \leqslant n \leqslant 7$ in Eq. (9.1).

(ii) *Helix Sense.* With the exception of the meridional reflections in the vicinity of $l = 16$ and 31, and the layer-line streaks, the overall form of the diffraction pattern is in substantial agreement with that predicted for a hexagonally packed arrangement of helices with $u = 47$ and $v = \pm 13$. However, the intensity distribution, except on the equator, was found to be in poor agreement with that calculated for either right-handed ($v = 13$) or left-handed ($v = -13$) α-helix models (Brown and Trotter, 1956). Of the two possibilities the left-handed helix was favored. It was also suggested that the unexpected meridional reflections could be explained in terms of distortion (Chapter 1, Section A.IV.c) of the methyl side-chain groups due to unfavorable interhelical contacts and the layer-line streaks were attributed to axial and azimuthal disorder (Chapter 1, Section A.IV.d).

Elliott and Malcolm (1956b, 1958) pointed out that the type of disorder involving random axial displacements and random azimuthal rotations was sterically impossible with a fixed interchain distance due to the bulkiness of the protruding methyl side-chain groups and suggested that a different type of disorder was present in which the chain direction, with respect to the sequence $-NH-CHR-CO\rightarrow$, was random (Chapter 1, Section A.IV.e). The intensities of the sharp reflections in this case depend upon the sum of the Fourier transforms of the "up" and "down" chains, while the layer-line streaks depend upon the difference between the two Fourier transforms. This packing model was explored by means of optical transforms using an optical diffractometer and it was found that within the limitations of accuracy imposed by this method the intensities of both the sharp reflections and the layer-line streaks could be explained by a crystal model in which right-handed α-helices were arranged with random chain sense.

An important feature of the studies by Elliott and Malcolm (1958) was the fact that the agreement between the observed and calculated intensities was significantly better for the right-handed than for the left-handed model, thus supporting Huggins's suggestion as regards their relative stabilities for L-amino acid residues.

(iii) *Departures from Helical Symmetry.* The reasons for the appearance in the observed diffraction pattern of "forbidden" meridional reflections has been investigated in detail by Elliott and Malcolm (1958) and Johnson (1959). Plausible explanations, which have been summarized by Elliott (1967), were given for these reflections in terms of regular distortions occasioned by intermolecular steric hindrance. Strictly speaking, the values for h and t calculated earlier from u and v and the axial repeat of structure c must be treated, in the absence of strict helical symmetry, as average values although the variation in value from residue to residue for the main-chain atoms is probably small in the case of $[Ala]_n$.

An unfortunate feature of the disorder in chain sense which appears to exist in crystallites of the α-helical form of $[Ala]_n$ is that a Fourier synthesis would yield an image of an "up" chain superposed on a "down" chain and so far no means has been devised for overcoming this difficulty.

(iv) *Refinement of the α-Helical Model for Poly(L-alanine).* Elliott and Malcolm (1958) attempted a limited empirical refinement of the atomic coordinates given by Brown and Trotter (1956) using optical diffraction methods and found that a better fit between calculated and observed intensities could be obtained if the radial coordinate of the oxygen atom was increased to 1.98 Å and that of the nitrogen atom decreased to 1.49 Å (cf. Table 9.1).

Arnott and co-workers (Arnott and Wonacott, 1966a,b; Arnott and Dover, 1967; Arnott, 1968) remeasured the diffraction pattern of $[Ala]_n$ and obtained quantitative intensity data on 61 reflections. With this amount of information it was found possible to perform a limited least squares refinement (Chapter 1, Section D; Chapter 7, Section C) in which the number of independent variables was reduced by incorporating various assumptions about the stereochemistry of the amide group. The refined values of the bond and torsion angles are listed in Table 9.2 and the atomic coordinates are given in Table 9.3. In both

TABLE 9.2

Bond and Torsion Angles in the α-Helical Conformation of Poly(L-alanine)[a]

Angle	$[Ala]_n$ (deg)	"Standard"[b] (deg)	Angle	$[Ala]_n$ (deg)
$\tau(C^\alpha, C', O)$	121.0	121	$\tau(N, C^\alpha, C')$	109.7
$\tau(C^\alpha, C', N)$	115.4	114	$\tau(N, C^\alpha, C^\beta)$	109.5
$\tau(O, C', N)$	123.6	125	$\tau(H^\alpha, C^\alpha, C')$	109.4
$\tau(C', N, C^\alpha)$	120.9	123	$\tau(H^\alpha, C^\alpha, C^\beta)$	109.5
$\tau(C', N, H)$	123.0	123	$\tau(C^\beta, C^\alpha, C')$	109.4
$\tau(H, N, C^\alpha)$	116.1	114	ϕ	-57.4
ω	-179.8	180	ψ	-47.5

[a] Arnott and Dover (1967).
[b] Corey and Pauling (1953a).

TABLE 9.3

Atomic Coordinates of a Residue in the α-Helical Conformation of Poly(L-alanine)[a]

Atom	x (Å)	y (Å)	z (Å)	r (Å)	ϕ (deg)
N	1.375	-0.711	-0.906	1.548	-27.35
H	1.459	-0.490	-1.878	1.539	-18.57
C^α	2.288	0	0	2.288	0
[b]H^α	2.929	-0.705	0.485	3.013	-13.54
[c]C^β	3.139	0.998	-0.808	3.294	17.63
C'	1.481	0.760	1.054	1.664	27.16
O	1.777	0.691	2.256	1.906	21.24

[a] Calculated from data given by Arnott and Dover (1967). Helix parameters: unit height $h = 1.495$ Å, unit twist $t = 99.57°$, repeat of structure $c = 70.3$ Å contains $u = 43$ residues in $v = 17$ turns of pitch $P = 5.41$ Å.
[b] $b(C^\alpha, H^\alpha) = 1.07$ Å.
[c] $b(C^\alpha, C^\beta) = 1.54$ Å.

cases these must be regarded as average values in view of the departures, mentioned earlier, from perfect helical symmetry. According to Arnott and Dover (1967), there are no intramolecular contacts in the refined model less than the "fully allowed" values given by Ramachandran *et al.* (1963) and the crystal packing parameters determined from the X-ray study were shown to be close to minima in the calculated intermolecular potential energy (Arnott and Wonacott, 1966b).

(v) *Single Crystals of Poly(L-alanine).* Padden and Keith (1965) examined the crystalline morphology of films of [Ala]$_n$ and found that platelets with a truncated hexagonal habit and sheaves of ribbonlike crystals were present. Selected-area electron diffraction patterns obtained from the sheaves resembled those obtained from the α form using X rays, and from the spatial orientation of the reflections it was concluded that the molecular chains were oriented transversely with respect to the long axis of the sheaves. The sheaves were found to consist of platelets about 150 Å thick stacked on edge and since this dimension is too small to accommodate a complete molecule in the α-helical conformation (Keith *et al.*, 1969a), it can be concluded that crystal reentrant folding must occur. Lamellar single crystals about 100 Å in thickness in which the chains fold back and forth on themselves have been prepared from a wide variety of synthetic polymers (Geil, 1963; Mandelkern, 1964, 1970; Keller, 1968).

II. Esters of Poly(L-glutamic acid)

a. *Poly(γ-methyl-L-glutamate)*

Early attempts to test the α-helix model centered on the diffraction pattern obtained from [Glu(Me)]$_n$ and the overall distribution of intensity was found to be consistent with this model (Cochran *et al.*, 1952; Bamford *et al.*, 1952; Yakel *et al.*, 1952). Diffraction patterns showing a high degree of orientation and crystallinity have been obtained from this polymer (Bamford *et al.*, 1953b, 1956), but so far it has not proved possible to develop a detailed model for the conformation of the side chain and so the conformation of the main chain has not been refined in the way that proved possible for the α form of [Ala]$_n$. A curious feature of the crystal structure of [Glu(Me)]$_n$ is that the density calculated from dimensions of the unit cell is less than the observed density (Bamford *et al.*, 1953b). This is the reverse of the situation normally encountered in polymeric materials where the presence of an amorphous phase of density lower than the crystalline phase leads to a macroscopic density which is less than that calculated for the crystal.

Polymorphism occurs in the α form of $[\text{Glu(Me)}]_n$: The crystalline form originally studied was hexagonal with a unit cell edge of $a = 11.95\,\text{Å}$ and an axial repeat of structure $c = 103\,\text{Å}$ corresponding to a helix with $u = 69$ and $v = \pm 19$ (Brown and Trotter, 1956). However, Brown (1956) obtained specimens, from a different sample by different conditions of preparation, in which the axial repeat was $c = 27\,\text{Å}$, corresponding to a helix with $u = 18$ and $v = \pm 5$. Both forms exhibit "forbidden" meridional reflections with spacings of 4.32 and 4.50 Å, indicating a departure from perfect helical symmetry, and as with $[\text{Ala}]_n$ this can be attributed to distortions arising from intermolecular contacts (Johnson, 1959; Elliott, 1967).

Electron diffraction patterns have been obtained from $[\text{Glu(Me)}]_n$ (Vainshtein and Tatarinova, 1967a) and the intensity data were used to compute a Fourier synthesis of the cylindrical average of the z-axis projection of the potential. The presence of a peak between $r = 1$ and $2\,\text{Å}$ was consistent with the axial projection of the main-chain atoms of an α-helix. A similar result had been obtained earlier (Bamford *et al.*, 1956) for an electron density synthesis from the X-ray data.

b. *Poly(γ-benzyl-L-glutamate)*

The α form of $[\text{Glu(Bzl)}]_n$ has been studied extensively using both X-ray diffraction (Bamford *et al.*, 1951a, 1956; Tomita *et al.*, 1962; A. Elliott *et al.*, 1965) and electron diffraction (Parsons and Martius, 1964; Vainshtein and Tatarinova, 1967a) and a plot of the observed X-ray pattern is shown in Fig. 9.3a. This differs from the patterns obtained from $[\text{Ala}]_n$ and $[\text{Glu(Me)}]_n$ in that sharp reflections are observed only on the equator, all other layer lines consisting of streaks. From the discussion of the effects of disorder on diffraction patterns given in Chapter 1, Section A.IV it seems likely that disorders of chain rotation and translation and possibly also chain direction are present in crystallites of the α form of $[\text{Glu(Bzl)}]_n$.

(i) *Helical Parameters.* With the exception of the meridional arcs with $Z \simeq 0.1$ and $0.2\,\text{Å}^{-1}$, the observed layer lines can be indexed on an axial repeat of structure of $c = 27\,\text{Å}$, suggesting that a distorted helix with 18 residues in five turns is present. Further evidence of regular distortion is contained in the fact that intensity appears on layer lines, for example, $l = 2$, 4, and 6, at values of R for which the predicted value for a perfect helix with $u = 18$ and $v = \pm 5$ would be negligible.

The mean unit heights $\langle h \rangle$ in normal and N-deutero-$[\text{Glu(Bzl)}]_n$ were measured by Tomita *et al.* (1962) and values of 1.494 and

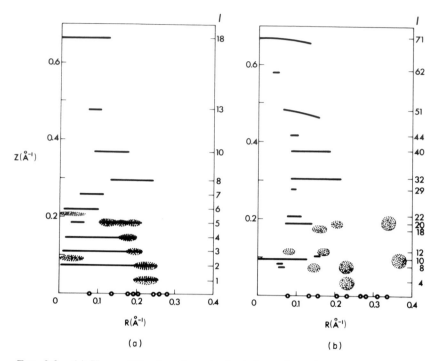

FIG. 9.3.   (a) X-ray diffraction data obtained from the α form of poly(γ-benzyl-L-glutamate). Discrete reflections are found on the equator ($l = 0$) but only streaks are observed on other layer lines. The layer lines are indexed on an axial repeat of $c = 27.0$ Å. (b) Data obtained from a mixture of equal parts of poly(γ-benzyl-L-glutamate) and poly(γ-benzyl-D-glutamate). The layer-line streaks can be indexed on an axial repeat of $c = 106$ Å. Discrete reflections are indicated by circles, sharp layer-line streaks by full lines, and diffuse streaks are dotted. (Redrawn from A. Elliott *et al.*, 1965.)

1.501 ± 0.002 Å, respectively, were obtained. This change was inter-preted as being due to an increase in the hydrogen bond length of 0.025 Å. An increase in the mean pitch $\langle P \rangle$ of 0.027 ± 0.008 Å was also observed so that the mean unit twist $\langle t \rangle = 360\langle h \rangle / \langle P \rangle$ was not greatly affected by the increase in hydrogen bond length.

(ii) *Liquid Crystals.* [Glu(Bzl)]$_n$ forms liquid crystals in various solvents (Robinson, 1966) and these have been studied by X-ray diffraction (Luzzati *et al.*, 1961, 1962; Parry and Elliott, 1965, 1967; Saludjian and Luzzati, 1967; Traub *et al.*, 1967; Squire and Elliott, 1969; Go *et al.*, 1969; Samulski and Tobolsky, 1971). The nature of the ordering is a function of solvent and of polymer concentration. A so-called "cholesteric" phase is observed in *m*-cresol at polymer concentrations of about 20–36% (w/w) which is characterized by the appearance of a diffuse ring in the X-ray diffraction pattern. On the basis of the variation of the observed spacing of this ring with concentration Luzzati *et al.* (1961, 1962) concluded that the main chain had a conformation similar to that of the 3_{10}-helix (Donohue, 1953). However, Parry and Elliott (1965) and Traub *et al.* (1967) succeeded in producing oriented specimens and the additional reflections observed in the diffraction pattern in both cases enabled an unequivocal assignment to the α-helix conformation to be made. A reappraisal of the original observations confirmed this conclusion (Saludjian and Luzzati, 1966, 1967).

At higher concentrations of polymer two further phases occur. In *m*-cresol a "hexagonal" phase appears and is characterized by sharp reflections which index on a hexagonal array (Luzzati *et al.*, 1961, 1962; Parry and Elliott, 1965). In dimethylformamide and pyridine, on the other hand, a "complex" phase is observed at high concentrations and Luzzati *et al.* (1961, 1962) concluded from their diffraction data that three-stranded ropes of α-helices distorted into a coiled-coil conformation (Chapter 1, Section A.II.e) were present. Initially this conclusion was supported by Parry and Elliott (1965) but further studies on well-oriented specimens showed that it was incorrect (Parry and Elliott, 1967).

The observed pattern, which is depicted in Fig. 9.4, shows a number of features not previously observed with synthetic polypeptides. In particular there is a meridional reflection at $Z = 1/5.06$ Å^{-1} and a strong near-equatorial layer-line streak with $Z = 1/55$ Å^{-1}. Although similar effects are produced by distortion of the α-helix into a coiled-coil conformation, other features of the pattern, such as the strong meridional reflection at $Z = 1/1.50$ Å, are not consistent with this interpretation. An alternative departure from helical symmetry was suggested to account for the observed pattern, in which the main-chain atoms had an α-helical conformation while the side chains formed ordered arrays of different helical symmetry to the main chain (Parry and Elliott, 1967; Squire and Elliott, 1969).

The aggregates in the "cholesteric" phase of liquid-crystalline solutions of [Glu(Bzl)]$_n$ can be aligned by means of a magnetic field (Sobajima, 1967; Samulski and Tobolsky, 1968) to give a "nematic" structure which is retained when the solvent is evaporated (Go *et al.*, 1969; Samulski and Tobolsky, 1971). The latter authors found that the X-ray diffraction patterns given by films cast in this manner from methylene dichloride or *cis*-1,2-dichloroethylene suggested the presence of α-helices with $u = 18$ and $v = \pm5$.

Markedly different patterns were obtained, however, from films cast from chloroform. These could be indexed on an axial repeat of $c = 10.35$ Å and it was concluded that the helices present in these films had values of $u = 7$ and $v = \pm 2$, that is, exactly 3.5 residues

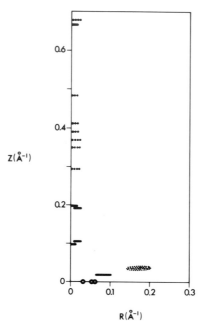

FIG. 9.4. X-ray diffraction data obtained from the "complex" phase of poly(γ-benzyl-L-glutamate) which forms at high concentrations in solution in dimethylformamide. The full lines indicate data obtained at 55 % polymer concentration and the broken lines indicate additional data obtained at higher concentrations. (Redrawn from Parry and Elliott, 1967.)

per turn. An interesting feature of these studies was the observation that the pattern reverted to the more usual one, with 3.6 residues per turn, when the films were heated.

(iii) *Epitaxial Crystallization.* Single crystals of the α form of $[\text{Glu(Bzl)}]_n$ have been prepared from dilute solution by epitaxial crystallization onto freshly cleaved alkali halide crystals (Carr *et al.,* 1968). Electron diffraction studies were used to establish the orientation of the helical axes, and the crystallite dimensions parallel to these axes were found to be substantially less than the lengths of the individual molecules. It was concluded, therefore, that crystal reentering folds were present.

c. *Racemic Poly(γ-benzylglutamate)*

Oriented fibers were prepared from a racemic solution containing equal parts of poly(γ-benzyl-L-glutamate) and poly(γ-benzyl-D-glutamate) by Tsuboi *et al.* (1961) and shown to have an unusual X-ray diffraction pattern which differed from that of fibers prepared from either of the enantiomorphs. The pattern was originally interpreted in terms of an α-helix with $u = 7$ and $v = \pm 2$, that is, exactly 3.5 residues per turn. More detailed studies (A. Elliott *et al.*, 1965) showed that the pattern was more complex than originally supposed and the observed layer lines (Fig. 9.3b) were tentatively indexed on an axial repeat of $c = 106$ Å. Patterns obtained from better-oriented specimens (Squire and Elliott, 1972) showed that the true axial period was 320 Å, but most of the reflections could be indexed on an orthogonal cell with $a = 14.76$ Å, $b = 25.40$ Å, and $c = 64.11$ Å (fiber axis). The observed distribution of intensity was shown to be consistent with an α-helix having 43 residues in 12 turns, giving $\langle h \rangle = 1.491$ Å, $\langle t \rangle = \pm 100.5°$, and $\langle P \rangle = \pm 5.34$ Å compared with the values $\langle h \rangle = 1.505$ Å, $\langle t \rangle = \pm 100°$, and $\langle P \rangle = \pm 5.42$ Å for the enantiomorphic structure.

Appreciable intensity occurs on layer lines for which a helical structure with $u = 43$ and $v = \pm 12$ would not be expected to diffract strongly and this is presumed to derive from a side-chain arrangement which has a symmetry that is different from the main chain. Mitsui *et al.* (1967) suggested a helical arrangement with three units per turn but this was discounted by Squire and Elliott (1972), who proposed a more complex model.

Racemic poly(γ-benzylglutamate) forms liquid crystals in dimethylformamide and acetophenone (Squire and Elliott, 1969, 1972) and the observed reflections can be indexed on a tetragonal cell with $a = 30.03$ Å and $c = 64.16$ Å (fiber axis). An analysis of the distribution of intensity in the diffraction pattern suggested that the molecular conformation in the liquid crystalline forms is very similar to that which occurs in the dried fibers.

d. *Other Esters of Poly(L-glutamic acid)*

A series of polymers of γ-glutamate esters prepared by Ledger and Stewart (1965, 1966) were examined by X-ray diffraction (Fraser *et al.*, 1967b) in a study of the effects on the α-helix of intermolecular and intramolecular interactions between the outer portions of the side chains. Significant differences were found between the helical parameters for the *ortho* and *meta* isomers and the *para* isomer of poly(γ-nitrobenzyl-L-glutamate) and the values, together with those for other polymers studied, are included in Table 9.8a (see p. 214). On the basis of these and other values obtained from studies of synthetic polypeptides and proteins a correlation between unit height h and unit twist t was established which is discussed further in Section D.

III. ESTERS OF POLY(L-ASPARTIC ACID)

a. *Poly(β-benzyl-L-aspartate)*

Blout and Karlson (1958) found that the constant b_0 in Eq. (8.1) for the dispersion of optical rotation for solutions of $[Asp(Bzl)]_n$ in chloroform was of opposite sign to that of other polymers of L-amino acid derivatives in the α conformation. Further studies (Karlson *et al.*, 1960; Bradbury *et al.*, 1960a) established with reasonable certainty that the helix sense in solution was left-handed, that is, opposite to that established for solid films of $[Ala]_n$ by Elliott and Malcolm (1958).

When $[Asp(Bzl)]_n$ is cast from chloroform solution at room temperature a film is obtained which yields a poorly developed X-ray diffraction pattern of the α-helix type. After heating the film *in vacuo* to 160°C, however, a different pattern with sharp crystal-like reflections is obtained (Bradbury *et al.*, 1959, 1962a). The observed reflections can be indexed on an axial repeat of structure $c = 5.30$ Å and a tetragonal unit cell (which has a square base) of side $a = 13.85$ Å.

(i) *Helix Parameters.* The absence of layer lines requiring a value of c larger than 5.30 Å indicates that the helix has an integral number of residues in one turn of a helix of pitch $P = 5.30$ Å. The density calculated for four residues in one turn agreed satisfactorily with the observed density and the main-chain conformation was assumed to be a left-handed α-helix so distorted that each turn contained four rather than around 3.6 residues per turn as in the α form of $[Ala]_n$ and $[Glu(Me)]_n$. This modification of the α-helix, termed the ω-helix, was superficially similar to a model which had been suggested earlier as a possible conformation of the polypeptide chain by Bragg *et al.* (1950).

(ii) *Refinement of Model Structure.* A model for the arrangement of the atoms in a residue was developed by Bradbury *et al.* (1962a) from a consideration of the infrared dichroism of various localized vibrational modes of the side chain combined with studies of scale atomic models of the molecular framework. The model was refined by comparison of optical diffraction patterns (Chapter 4) with the observed X-ray diffraction data. As with $[Ala]_n$, it was found necessary to assume that a random distribution of "up" and "down" chains was present (Section B.II) in order to obtain satisfactory agreement between the predicted and observed intensity distribution. The coordinates quoted for the main-chain atoms and for C^β in the refined model are given in Table 9.4.

TABLE 9.4

Atomic Coordinates of a Residue in the ω-Helical Conformation of Poly(β-benzyl-L-aspartate)[a]

Atom	x (Å)	y (Å)	z (Å)	r (Å)	ϕ (deg)
N	−0.41	1.75	−0.80	1.80	103.2
C^α	0.56	2.48	0	2.54	77.3
C^β	−0.17	3.66	0.70	3.66	92.7
C'	1.19	1.54	1.05	1.95	52.3
O	1.27	1.90	2.25	2.29	56.2

[a] Calculated from data given by Bradbury *et al.* (1962a). Helix parameters: unit height $h = 1.325$ Å, unit twist $t = -90°$, repeat of structure $c = 5.30$ Å contains $u = 4$ residues in $v = -1$ turn of pitch $P = -5.30$ Å. Bond angles and bond distances calculated from these coordinates are: $b(N, C^\alpha) = 1.45$ Å; $b(C^\alpha, C^\beta) = 1.55$ Å; $b(C^\alpha, C') = 1.54$ Å; $b(C', O) = 1.26$ Å; $b(C', N) = 1.37$ Å; $\tau(N, C^\alpha, C') = 109.9°$; $\tau(C^\alpha, C', N) = 114.2°$; $\tau(C^\alpha, C', O) = 120.1°$; $\tau(C', N, C^\alpha) = 123.1°$. Torsion angles calculated from these coordinates: $\phi = 62.8°$, $\psi = 54.9°$, $\omega = 175.4°$.

Because the refinement was limited to adjustments of the side-chain conformation and molecular rotation, the main-chain conformation is that arrived at by a consideration of the stereochemistry of a left-handed helix with four residues per turn. According to Bradbury *et al.* (1962a), the amide group in the model structure has nonplanar amide groups with $\omega = 175°$, a hydrogen bond length $b(N_i, O_{i-4}) = 2.9$ Å, and $\tau(H_i, N_i, O_{i-4}) = 25°$.

The refinement of this model for the ω form of $[\text{Asp(Bzl)}]_n$ is incomplete since there are some residual discrepancies between the predicted and observed intensity distributions. One problem which must be overcome before further refinement can be carried out is the fact that the diffraction pattern is not unique and varies slightly from specimen to specimen. It should also be noted that the assignment of a left-handed screw sense to the helix in the solid state is based on indirect evidence.

McGuire *et al.* (1971) attempted to refine the structure proposed by Bradbury *et al.* (1962a) on the basis of a minimization of the sum of the calculated intramolecular and intermolecular potential energies. Bond angles and distances were maintained at standard values used in earlier work (Scott and Scheraga, 1966; Ooi *et al.*, 1967) except for

b(C, H), which was increased to 1.07 Å for aliphatic CH bonds in the side chain, and minimization was carried out with respect to the torsion angles ϕ, ψ, ω, and χ_1 to χ_4 (Chapter 6) and the orientation of the molecule about the helix axis. It was claimed that the conformation and packing of minimum calculated potential energy were similar to those proposed by Bradbury *et al.* (1962a) on the basis of X-ray diffraction and infrared studies, but even allowing for the fact that the diagrams seem to be incorrectly drawn, the two conformations appear to be significantly different.

Several aspects of the refinement are open to criticism since no account was taken of the random chain sense believed to be present in the crystal (Bradbury *et al.*, 1962a) or of the possibility that the bond angles, in particular τ(N, C$^\alpha$, C'), have values different from the "standard" values. In view of the distortion present in the ω-helix, this possibility cannot be excluded and the results obtained by Gibson and Scheraga (1966) indicate that the energy penalty due to steric hindrance is likely to be overestimated if no flexibility in bond angles and distances is allowed. An unexplained feature of the calculations was the fact that the total potential energy decreased rapidly with decrease in h when this parameter was allowed to vary, that is, the observed axial period does not coincide with a minimum in the sum of the intramolecular and intermolecular potential energies.

(iii) *Monolayers.* Malcolm (1970) prepared molecular monolayers of [Asp(Bzl)]$_n$ at the air–water interface and examined collapsed films by electron diffraction. Air-dried specimens were poorly crystalline but the diffraction patterns showed the main features expected for an α-helical conformation. Parallel studies using infrared spectra indicated that the amide frequencies were appropriate to a right-handed helix (Section C.I) rather than a left-handed helix as observed in solution. A transformation to a left-handed helical form, again as evidenced by infrared spectra, could be induced without loss of orientation by exposing the dried film to the vapor of chloroform containing 10% (v/v) dichloroacetic acid.

A transition to an ω form, yielding a diffraction pattern in many respects similar to that obtained from bulk specimens with X rays, occurred when films were heated briefly to 160°C. In addition to the reflections observed in the X-ray diffraction pattern, a meridional reflection on the first layer line was recorded. This reflection is forbidden in a helix with $u = 4$ residues in $v = \pm 1$ turns (Chapter 1, Section A.III) and it must be concluded that the four residues are not completely equivalent.

(iv) *Liquid Crystals.* Saludjian and Luzzati (1967) reported that [Asp(Bzl)]$_n$ forms liquid crystals in *m*-cresol solutions. At about 60% (w/w) concentration of polymer a "hexagonal" phase was observed, while at higher concentrations a phase with a tetragonal cell was observed in equilibrium with the "hexagonal" phase. When the solvent was removed the tetragonal form persisted and the spacings of the observed reflections suggest that the unit cell was essentially similar to that observed by Bradbury *et al.* (1962a) in the heated films containing the ω form.

b. *Poly(β-p-chlorobenzyl-L-aspartate)*

Hashimoto and co-workers (Hashimoto and Aritomi, 1966; Hashimoto and Arakawa, 1967) found that in contrast to [Asp(Bzl)]$_n$ the polymer [Asp(pClBzl)]$_n$ formed right-handed α-helices in solution and Takeda *et al.* (1970) have obtained X-ray diffraction patterns from films of this polymer.

(i) α-*Helical Form.* After heating for two hours *in vacuo* at 150°C highly crystalline specimens were obtained which yielded a diffraction pattern rich in sharp reflections together with a few layer-line streaks. The sharp reflections can be indexed on a hexagonal unit cell of side $a - 14.9$ Å and axial repeat $c - 27.0$ Å. A strong meridional reflection was observed on $l = 18$ and a strong off-meridional reflection on $l = 5$ as predicted for a helical conformation with 18 residues in five turns (Chapter 1, Section A.III). The appearance of additional meridional reflections on $l = 6$ and 12 indicates, however, that there is some departure from perfect helical symmetry and a plausible explanation was given in terms of a repeating unit consisting of three residues.

The layer-line streaks were interpreted as indicating a random distribution of chain directions as regards the sequence $-NH-CHR-CO-$. This conclusion was supported by the better agreement obtained between the observed intensities and those calculated for models with random chain sense compared with the agreement obtained when a uniform chain sense was assumed.

The location of the chlorine atoms was determined by applying helix diffraction theory (Chapter 1, Section A.III) and models for the side-chain conformations arrived at by combining this information with observations of infrared dichroism and with stereochemical data. The helix was assumed to be right-handed. Refinement of the model on the basis of the X-ray data was attempted and reasonable agreement between calculated and observed intensities was obtained for the equator. The agreement on other layer lines, however, was generally poor and the atomic coordinates cannot be regarded as being fully refined.

(ii) *ω-Helical Form.*   Takeda *et al.* (1970) further reported that when the α-helical form of $[Asp(pClBzl)]_n$ was heated to 190°C a different X-ray pattern was obtained which could be indexed on a tetragonal cell with side $a = 23.3$ Å and an axial repeat of $c = 5.20$ Å. It was concluded that the structure contained right-handed helices with exactly four residues per turn of pitch $P = 5.20$ Å and that the unit cell contained one "up" and one "down" chain. In view of the afore-mentioned change in helix sense in $[Asp(Bzl)]_n$ observed by Malcolm (1970), the assumption that the ω-helical form of $[Asp(pClBzl)]_n$ is right-handed is open to question.

IV. OTHER POLYMERS

a. *Poly(S-benzylthio-L-cysteine)*

The polymer poly(*S*-benzylthio-L-cysteine), in which the side chain is $-CH_2-S-S-CH_2-C_6H_5$, was studied by Fraser *et al.* (1962a). No solvent could be found but oriented films were prepared by rolling and examined by X-ray diffraction. The pattern contained sharp reflections which could be indexed on a tetragonal cell with edge $a = 14.28$ Å and an axial repeat of $c = 5.55$ Å. The tetragonal symmetry, together with the absence of meridional reflections except on the fourth layer line, indicated that the chains had a helical symmetry with $u = 4$ and $v = \pm 1$. From density considerations only one chain could be accommodated in the unit cell and the conformation of the main chain was presumed to be similar to that observed in the ω form of $[Asp(Bzl)]_n$. The helix sense in this polymer is not known.

b. *Poly(L-tryptophan)*

Peggion *et al.* (1968) studied oriented films of $[Try]_n$, cast from dimethylformamide, by X-ray diffraction and found a pattern characteristic of the α-helical conformation with a sharp meridional reflection with $Z = 1/1.49$ Å^{-1} and strong near-meridional reflections on a layer line with $Z = 1/5.42$ Å^{-1}. Thus the unit height is $h = 1.49$ Å and the unit twist is $t = 360 \times 1.49/5.42 = 99.0°$.

The determination of the helix sense of $[Try]_n$ by optical methods is complicated by the presence of side-chain chromophores and Peggion *et al.* (1968) studied the circular dichroism spectra in solution of a series of copolymers of Try and Glu(Et). Since no abrupt transition occurred with increasing proportions of Try, it was reasoned that both $[Try]_n$ and $[Glu(Et)]_n$ had a right-handed α-helical conformation in solution. A similar conclusion was reached by Damle (1970) and Engel *et al.* (1971).

c. Poly(L-lysine) and Poly(L-arginine)

Studies of hydrohalides of $[Lys]_n$ by a variety of methods have established that this polymer can exist in a range of conformations depending upon the state of ionization and degree of hydration of the side-chain amine group (Blout and Lenormant, 1957; Elliott et al., 1957a; Applequist and Doty, 1962).

(i) *Films.* The structure of films of the hydrochloride of $[Lys]_n$ has been studied using X-ray diffraction by Johnson (1959) and Shmueli and Traub (1965). The pattern obtained was found to depend upon the water content of the specimen. The dried material gave a pattern characteristic of the β conformation (Chapter 10) and with increasing relative humidity, the intersheet distance gradually expanded, but above 84% relative humidity a transition to a pattern characteristic of the α-helix was observed. In this pattern sharp equatorial reflections were observed which could be indexed on a hexagonal lattice with cell edge $a = 16.8$ Å, but on other layer lines there were continuous distributions of intensity, suggesting that random axial displacements and rotations of the molecules were present (Chapter 1, Section A.IV.d). The cell dimensions at 84% relative humidity correspond to about five water molecules per residue and the cell was found to expand laterally with increasing relative humidity. Above a water content equivalent to about 15 water molecules per residue, however, the regular packing of the helices disappeared. The conformational change from a β form to an α form was found to be reversible.

The structure of films of the hydrochloride of $[Arg]_n$ has also been investigated by X-ray diffraction (Suwalsky and Traub, 1972). Specimens containing up to about five water molecules per residue gave diffraction patterns showing features characteristic of an α-helical conformation, while specimens containing 5–20 water molecules per residue gave a β pattern.

(ii) *Single Crystals.* Padden et al. (1969) succeeded in obtaining crystals from aqueous solutions of $[Lys]_n$ in the presence of divalent anions. Single crystals of the monohydrogen phosphate were obtained in the form of hexagonal lamellae about 150 Å thick (Fig. 9.5) in which the conformation was shown both by electron and X-ray diffraction to be α-helical. The electron diffraction data indicated that the chain axes were perpendicular to the plane of the lamellae and since the average length of the molecule in the α-helical conformation was predicted to be about 1100 Å, it was concluded that the molecules must be folded.

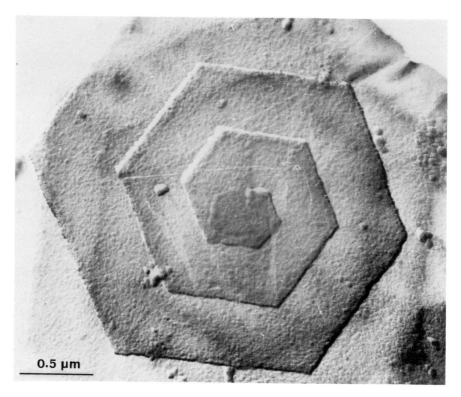

0.5 μm

FIG. 9.5. Portion of a single crystal of poly(L-lysine) monohydrogen phosphate freeze-dried from the mother liquor (Padden *et al.*, 1969).

The packing in the crystal was found to be hexagonal, with a cell edge $a = 19.55$ Å, and a lateral contraction of the cell was observed in water/ethanol mixtures as the proportion of ethanol was increased. A transformation to the β conformation occurred at high ethanol concentrations and this could be reversed by resuspending the crystals in the mother liquor. A transformation to the β conformation was also observed when crystals of the α form were dried.

d. *Sequential Polypeptides*

Diffraction patterns characteristic of the α-helical conformation have been obtained from a number of polymers containing ordered sequences of residues (sequential polypeptides). These include [Glu(Me)Glu(Me) Glu(Me)ValGlu(Me)]$_n$ (Fraser *et al.*, 1965b), [Glu(Et)Glu(Et)Glu(Et) Cys(Bzl)Glu(Et)]$_n$, [Glu(Et)Cys(Bzl)Glu(Et)]$_n$, and [Glu(Et)Cys(Bzl) Glu(Et)Glu(Et)]$_n$ (Fraser *et al.*, 1965d).

The pattern obtained from the polymer $[\text{Glu(Et)Cys(Bzl)Glu(Et)}]_n$ contained sharp reflections characteristic of a crystalline phase and these could be indexed on a hexagonal unit cell with side $a = 13.2$ Å and an axial repeat $c = 26.8$ Å (Fraser et al., 1965c). These dimensions are consistent with an α-helical conformation of the main chain in which the mean unit height is $\langle h \rangle = 26.8/18 = 1.489$ Å and mean unit twist is $\langle t \rangle = \pm 100°$. Intensity was observed on layer lines for which the contribution from a perfect helix with $u = 18$ units in $v = \pm 5$ turns would be negligible and this was correlated with the fact that the chemical repeating unit of the polymer is a group of three residues. If all groups were equivalent in the crystal, the expected helical symmetry of the chains would be $u = 6$, $v = \mp 1$, and the observed pattern is consistent with this symmetry. An unexplained feature of the observed pattern was the absence of meridional reflections with $l = 6$ and 12 which are permitted with a helix in which $u = 6$ and $v = \mp 1$.

e. *Liquid Crystals*

Luzzati and co-workers (Luzzati et al., 1961, 1962, 1966; Saludjian and Luzzati, 1966, 1967; Saludjian et al., 1963a,b) have studied the X-ray diffraction patterns of unoriented liquid crystalline forms of a number of synthetic polypeptides and the results obtained with $[\text{Glu(Bzl)}]_n$ have been discussed in Section B.II.b. The interpretation of the diffraction patterns of unoriented specimens is more difficult than for oriented specimens but the α-helical form was reported to be present in the "hexagonal" phase (Section B.II.b) of $[\text{Glu}]_n$ and $[\text{Lys(Cbz)}]_n$ in dimethylformamide, and ω-helical forms were reported to be present in $[\text{Glu}]_n$, $[\text{Lys(Cbz)}]_n$, and $[\text{Asp(Bzl)}]_n$ at high concentrations of polymer. More recently Squire and Elliott (1969) have questioned the assignment of an ω conformation on the basis of a tetragonal (square-based) cell. Working with oriented specimens, they were able to show that under certain circumstances liquid crystals with tetragonal cells are formed which contain α-helices. Thus the observation of a square-based cell does not automatically imply the presence of a helix with four residues per turn.

C. INFRARED STUDIES

The infrared spectra of synthetic polypeptides in the α form have been studied extensively and useful reviews of this topic have been given by Miyazawa (1962, 1967) and Susi (1969). Much of this work has been concerned with the detailed interpretation of infrared spectra on the basis of knowledge gained primarily from X-ray diffraction studies. In the present context the question of what useful information about the α-helix can be gained from these studies is considered with particular reference to features which have been or may be of value in studies of the conformation in fibrous proteins.

I. FREQUENCIES OF AMIDE BANDS

Elliott and Ambrose (1950) established a correlation between the frequencies of the amide vibrations and the conformation of the polypeptide chain in $[Glu(Me)]_n$ and found that the α form was characterized by frequencies of 1658 cm^{-1} for the parallel component of amide I and 1550 cm^{-1} for the perpendicular component of amide II. A small shift in the frequency of the amide A vibration was also noted between the α and β forms of this polymer. This work was extended to other enantiomorphic synthetic polypeptides (Ambrose and Elliott, 1951a; Elliott, 1953a; Bamford et al., 1956) and narrow ranges of 1652–1657 and 1545–1551 cm^{-1} were reported for the frequencies of the amide I and II bands, respectively (Elliott, 1953a). Further studies (Elliott et al., 1957a) showed that it was necessary to exercise caution in the use of the amide I frequency in the diagnosis of the α form, because the frequency for unordered material was close to that for the α form.

An explanation of the dependence of the vibrational frequencies of the amide modes on conformation was provided by the analysis given by Miyazawa (1960a), which was discussed in detail in Chapter 5. In the case of the α-helix each mode of the isolated amide group gives rise to two infrared-active modes with frequencies

$$\nu_{\parallel}(0) = \nu_0 + D_1 + D_3 \tag{9.2}$$

$$\nu_{\perp}(t) = \nu_0 + D_1 \cos t + D_3 \cos 3t \tag{9.3}$$

where ν_0 is the unperturbed frequency, D_1 is a coefficient which depends on the interaction between adjacent groups in the helix, and D_3 is a coefficient which depends on the interaction between groups coupled by hydrogen bonds. Miyazawa and Blout (1961) suggested values of $\nu_0 = 1658$ cm^{-1}, $D_1 = 10$ cm^{-1}, and $D_3 = -18$ cm^{-1} for the amide I mode of the α form of $[Glu(Bzl)]_n$, giving predicted values of $\nu_{\parallel}(0) = 1650$ cm^{-1} and $\nu_{\perp}(100°) = 1647$ cm^{-1} compared with observed values of 1650 and 1652 cm^{-1}, respectively. For the amide II mode values of $\nu_0 = 1535$ cm^{-1}, $D_1 = -21$ cm^{-1}, and $D_3 = 2$ cm^{-1} were suggested, giving predicted values of $\nu_{\parallel}(0) = 1516$ cm^{-1} and $\nu_{\perp}(100°) = 1540$ cm^{-1}

compared with the observed values of 1516 and 1546 cm^{-1}, respectively. In both cases the agreement between the calculated and observed values of $\nu_{\parallel}(0)$ is not significant since this value was used in the determination of D_1. Other amide bands exhibit frequency differences between $\nu_{\parallel}(0)$ and $\nu_{\perp}(t)$ but so far these have not been studied in detail.

Potentially the frequencies of the amide vibrations provide information, through the values of ν_0, D_1, and D_3, on a number of important molecular parameters which cannot be obtained readily by other means. The unperturbed frequency ν_0 depends upon the spatial configuration of the atoms in the repeating unit of the polymer, the electronic structure of the bonds, and the nature of the hydrogen bond, while the interaction coefficients depend upon the mutual relationships of the interacting amide groups. To date, explicit interpretation in such terms has not proved possible but differences between the amide I frequencies and amide II frequencies in the α forms of various synthetic polypeptides have been observed and these are summarized in Tables 9.5a and 9.5b.

The ranges for esters of $[\text{Glu}]_n$ are quite small even though the unit height h and unit twist t of the helix vary somewhat from polymer to polymer (see Table 9.8a). On the other hand, the ranges for esters of $[\text{Asp}]_n$ are considerably greater and this can be attributed to three main effects. First, t varies over a considerably greater range (see Table 9.8a) and the values for the perpendicular component predicted from Eq. (9.3) vary from $\nu_{\perp}(90°) = \nu_0$ to $\nu_{\perp}(100°) = \nu_0 - 0.17D_1 + 0.5D_3$ for the ω and α forms, respectively. Second, distortion of the amide-group geometry and changes in hydrogen bond strength will affect ν_0, while changes in the spatial relationships of the amide groups will affect D_1 and D_3.

For the ω form, $t = 90°$ and so from Eqs. (9.2) and (9.3) $\nu_{\parallel}(0) - \nu_{\perp}(t) = D_1 + D_3$. From the values given in Table 9.5a an estimate of $D_1 + D_3 = 7$ cm^{-1} is obtained for the amide I modes of $[\text{Asp}(\text{Bzl})]_n$ in the ω form, which may be compared with the value of -8 cm^{-1} suggested by Miyazawa and Blout (1961) for the α form. The coefficient D_3 depends upon interactions between hydrogen-bonded residues in adjacent turns and thus might be expected to be more sensitive to changes in t than D_1. The discrepancy between the observed and calculated values for $D_1 + D_3$ probably arises from this effect, particularly since D_3 has a large magnitude for amide I. In contrast, the estimate of -18 cm^{-1} obtained from Table 9.5a for $D_1 + D_3$ applicable to the amide II modes of the ω form is close to the value of -19 cm^{-1} predicted for the α-helix by Miyazawa and Blout (1961). This is perhaps to be expected since the dominant term is D_1. The predicted value of D_3 in the α-helix is small (2 cm^{-1}) and would be expected to be even less in the ω form

TABLE 9.5a

Amide I and II Frequencies of Homopolymers and Random Copolymers in the α Form

Polymer[a]	Conformation[b]	Amide I $\nu_\parallel(0)$ (cm^{-1})	Amide I $\nu_\perp(t)$ (cm^{-1})	Amide II $\nu_\parallel(0)$ (cm^{-1})	Amide II $\nu_\perp(t)$ (cm^{-1})	Ref.[c]
[Ala]$_n$	α_R	1658	—	—	1548	1
[Glu]$_n$	α_R	1650	1650	1515	1550	2
Esters of [Glu]$_n$						
[Glu(Me)]$_n$	α_R	1654	1659	1519	1550	1
[Glu(Bzl)]$_n$	α_R	1652	1655	1518	1549	3
[Glu(oNO$_2$Bzl)]$_n$	α_R	1653	1657	—	1552	4
[Glu(mNO$_2$Bzl)]$_n$	α_R	1652	1656	—	1550	4
[Glu(pNO$_2$Bzl)]$_n$	α_R	1654	1657	—	1546	4
[Glu(pIBzl)]$_n$	α_R	1655	1658	1516	1549	4
[Glu(NpMe)]$_n$	α_R	1654	1656	1518	1550	4
[Glu(Me$_3$Bzl)]$_n$	α_R	1656	1658	1517	1550	4
[Glu(Et)]$_n$	α_R	1658	—	1522	1550	1
[Glu(isoAm)]$_n$	α_R	1656	—	—	1550	1
Range		1652–1658	1655–1659	1516–1522	1546–1552	
Esters of [Asp]$_n$						
[Asp(Bzl)]$_n$	ω_L	1677	1670	1520	1538	5
[Asp(Bzl)]$_n$	α_R	1658	—	—	1552	6
[Asp(Bzl)]$_n$	α_L	1664	—	—	1561	5
[Asp(Bzl)]$_n$	α_L	1668	—	—	1560	7
[Asp(pMeBzl)]$_n$	α_R	1656	—	—	1555	7
[Asp(pClBzl)]$_n$	α_R	1657	—	—	1556	7
[Asp(PhEt)]$_n$	α_R	1657	—	—	1554	7
[Asp(nPr)]$_n$	α	1662	—	—	1554	8
[Asp(pIBzl)]$_n$	α	1660	1660	1525	1555	4
Copolymers	α_R	1657–1661	—	—	1551–1555	9
Copolymers	α_L	1664–1668	—	—	1555–1559	9
Copolymers	ω	1667–1673	—	—	1534–1538	9
Range		1656–1677	1660–1670	1520–1525	1534–1561	

[a] Abbreviations defined on p. xvii.

[b] The assignment of helix sense as left (L) or right (R) is in most cases based on indirect evidence.

[c] References: 1, Masuda et al. (1969); 2, Miyazawa and Blout (1961); 3, Tsuboi (1962); 4, Suzuki (1971); 5, Bradbury et al. (1962a); 6, Malcolm (1970); 7, Hashimoto and Arakawa (1967) (values for solution in chloroform); 8, Bradbury et al. (1960b); 9, Bradbury et al. (1968b).

TABLE 9.5b

Amide I and II Frequencies of Sequential Polymers in the α Form

Polymer[a]	Amide I		Amide II		Ref.[b]
	$\nu_\parallel(0)$ (cm^{-1})	$\nu_\perp(0)$ (cm^{-1})	$\nu_\parallel(0)$ (cm^{-1})	$\nu_\perp(0)$ (cm^{-1})	
[{Glu(Me)}$_m$-Val-Glu(Me)]$_n$, $m = 1$–3	1654–1656	1655–1657	—	1546–1549	1
[{Glu(Et)}$_m$-Cys(Bzl)-Glu(Et)]$_n$, $m = 0$–3	1656–1657	1658	—	1548–1550	2
[Glu(Et)-Cys(Bzl)-{Glu(Et)}$_2$]$_n$	1656	1657	—	~1550	2
[Cys(Bzl)-{Glu(Et)}$_2$]$_n$	1659	1661	—	1550	2
[Glu(Et)-{Cys(Bzl)-Glu(Et)}$_2$]$_n$	1657	1658	—	1549	2
[Glu(Et)-{Cys(Bzl)}$_2$-{Glu(Et)}$_2$]$_n$	1657	1658	—	1550	2
[{Glu(Et)}$_m$-Gly]$_n$, $m = 3$–4	1657	1658	—	1548	3
[{Glu(Et)}$_2$-Gly-{Glu(Et)}$_3$]$_n$	1657–1658	1658	—	1548–1557	3
[{Glu(Me)}$_m$-Ser(Ac)-Glu(Me)]$_n$, $m = 2$–3	1654–1656	1656–1657	1515–1518	1547–1548	4
Range for all polymers[c]	1650–1677	1650–1670	1515–1525	1534–1561	

[a] Indirect evidence suggests that all these polymers exist as right-handed α-helices. Symbols defined on p. xvii.

[b] References: 1, Fraser *et al.* (1965b); 2, Fraser *et al.* (1965d); 3, Fraser *et al.* (1967a); 4, Fraser *et al.* (1968a).

[c] This table and Table 9.5a.

since the angle between the planes of the bonded amide groups is greater than in the α form (Miyazawa and Blout, 1961).

The third factor which operates to increase the range of values for esters of [Asp]$_n$ is the occurrence of both right- and left-handed helical forms. For L-residues these two forms cannot be transformed one into the other by reflection or any other symmetry operation and so would be expected *a priori* to have different normal modes of vibration. Miyazawa *et al.* (1967) calculated the vibrational frequencies for the left-handed and right-handed forms of [Ala]$_n$ and reported that the amide I and II frequencies were higher for the left-handed form by 8 and 2 cm^{-1}, respectively. The senses and magnitudes of these differences are in reasonable agreement with those observed by Hashimoto and Arakawa (1967) and Bradbury *et al.* (1968b) quoted in Table 9.5a.

It is clear from the range of values reported for the α forms of various synthetic polypeptides that while correlations exist between the frequencies of the amide I and II vibrations and the α-helical conformation, these must be applied with circumspection in dealing with materials of unknown conformation. In addition to the variability of frequency, overlap occurs with the frequency ranges for unordered and for collagen-like conformations. In view of these difficulties considerable attention has been given to a search for other conformation-sensitive amide bands, and the amide V band in particular has been studied in con-siderable detail (Miyazawa *et al.*, 1962; Masuda and Miyazawa, 1967; Miyazawa *et al.*, 1967; Masuda *et al.*, 1969). Although the frequency for the α form is well separated from that of the β form, the bands associated with the α and unordered forms overlap.

The frequencies of NH and OH bond stretching vibrations are generally lowered and the absorption bands broadened and intensified when the group acts as donor in the formation of a hydrogen bond. The magnitude of the frequency change from the nonbonded to the bonded state has been correlated with the length of the hydrogen bond (Pimentel and Sederholm, 1956; Pimentel and McClellan, 1960) and this correlation is potentially of value in the study of the conforma-tions of regular arrangements such as the α-helix. It cannot be applied directly, however, since the NH stretching mode of the amide group and the overtone of the amide II vibration have comparable frequencies and similar symmetries and interact through Fermi resonance (Chapter 5). The frequencies of the resulting amide A and B components ν_A and ν_B are related to the true stretching frequency ν_{NH} and the

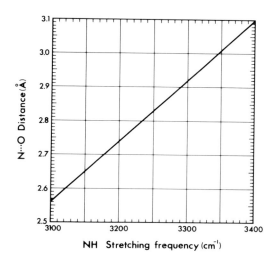

FIG. 9.6.   Relationship between $b(N_i, O_{i-4})$ in the α-helix and the NH stretching frequency assuming that ν°_{NH} in Eq. (9.5) has a value of 3457 cm^{-1}.

frequency of the amide II vibration ν_{II} by the expression (Miyazawa, 1960c)

$$\nu_{NH} = \nu_A + \nu_B - 2\nu_{II} \tag{9.4}$$

Some calculated values of ν_{NH} for different synthetic polypeptides are given in Table 9.6. It should be noted that these values refer to the $\nu_{\parallel}(0)$ mode and need to be corrected by subtracting the appropriate value of $D_1 + D_3$. The values of D_1 and D_3 for the NH stretching vibration are unknown but the available evidence suggests that they are small (<10 cm^{-1}).

In order to assess the hydrogen bond length $b(N, O)$ in the α-helix it is also necessary to know ν_{NH}° for the non-hydrogen-bonded state. The value of ν_{NH}° has been determined for a range of N-substituted amides by Jones (1966) and of the compounds investigated the most appropriate model would appear to be N-ethylpropionamide, $C_2H_5CONHC_2H_5$, for which $\nu_{NH}^\circ = 3457$ cm^{-1}. With this value the relationship given by Pimentel and Sederholm (1956)

$$\nu_{NH}^\circ - \nu_{NH} = 548[3.21 - b(N, O)] \tag{9.5}$$

yields

$$b(N, O) = 0.001825\nu_{NH} - 3.1 \tag{9.6}$$

This relationship is illustrated in Fig. 9.6.

When Eqs. (9.4) and (9.6) are applied to the observed data for [Ala]$_n$ (Table 9.6) a value of $b(N, O) = 2.85$ Å is obtained, which is close to the measured value of 2.86 Å derived from the X-ray diffraction data by Arnott and Dover (1967). The error incurred in neglecting to correct for the interaction coefficients D_1 and D_3 is 0.002 Å/cm^{-1}. The values of $b(N, O)$ in the esters of [Glu]$_n$ are generally similar to that in [Ala]$_n$, except for the p-nitrobenzyl ester, for which the calculated hydrogen bond length is appreciably longer (Table 9.6). This can be correlated with the greater value of the unit height h in this particular polymer (see Table 9.8a). The values of $b(N, O)$ in esters of [Asp]$_n$ are generally greater than in the esters of [Glu]$_n$, suggesting that the hydrogen bonding is weaker in esters of [Asp]$_n$.

II. DICHROISM OF AMIDE BANDS

Measurements of the dichroism of the amide bands of synthetic polypeptides provide information about the spatial orientation of the amide groups relative to the molecular axis. Originally the dichroic ratios observed for the amide A and amide I bands of synthetic

TABLE 9.6

Values of the Hydrogen Bond Distance $b(N, O)$ for Synthetic Polypeptides in
the α Form Calculated from the Observed Values of ν_A, ν_B, and ν_{II}

Polymer[a]	Conformation[b]	$\nu_A{}^c$ (cm^{-1})	$\nu_B{}^c$ (cm^{-1})	$\nu_{II}{}^d$ (cm^{-1})	$\nu_{NH}{}^e$ (cm^{-1})	$b(N, O)^f$ (Å)	Ref.[g]
[Ala]$_n$	α_R	3293	3060	1545	3263	2.85	1
[Glu(Bzl)]$_n$	α_R	3291	3065	1549	3258	2.85	2
[Glu(oNO$_2$Bzl)]$_n$	α_R	3291	3062	1550	3253	2.84	2
[Glu(mNO$_2$Bzl)]$_n$	α_R	3292	3065	1549	3259	2.85	2
[Glu(pNO$_2$Bzl)]$_n$	α_R	3313	3065	1548	3282	2.89	2
[Glu(pIBzl)]$_n$	α_R	3294	3063	1549	3259	2.85	2
[Glu(NpMe)]$_n$	α_R	3292	3061	1550	3253	2.84	2
[Glu(Me$_3$Bzl)]$_n$	α_R	3294	3065	1550	3259	2.85	2
[Asp(nPr)]$_n$	α	3305	3075	1554	3272	2.87	3
[Asp(Bzl)]$_n$	α_L	3308	3082	1561	3286	2.90	4
[Asp(Bzl)]$_n$	ω_L	3298	3075	1538	3301	2.92	4

[a] Abbreviations defined on p. xvii.
[b] Assignment of helix sense based on solution studies.
[c] $\nu_{\parallel}(0)$ measured from parallel spectrum.
[d] $\nu_{\perp}(t)$ measured from perpendicular spectrum.
[e] Calculated from Eq. (9.4) assuming $D_1 + D_3 = 0$.
[f] Calculated from Eq. (9.6).
[g] References: 1, Elliott (1954); 2, Fraser et al. (1967b); 3, Bradbury et al. (1960b); 4, Bradbury et al. (1962a).

polypeptides were thought to be incompatible with the α-helix (Bamford et al., 1952) but recognition of the complex nature of amide-group vibrations enabled the observed ratios to be reconciled, in a qualitative manner, with the α-helix (Fraser and Price, 1952). However, lack of precise knowledge about directions, relative to the amide group, of the transition moments associated with these and other amide group vibrations precluded a quantitative test (Elliott, 1953b). Subsequent measurements on crystals of simple model compounds provided estimates for the transition moment directions of several in-plane amide vibrations relative to the CO direction (Chapter 5, Section A.I), and the information obtained from crystalline N,N'-diacetylhexamethylenediamine (Sandeman, 1955) was used by Tsuboi (1962) to calculate the inclination of the transition moment directions to the axis for the model of the α-helix described by Bamford et al. (1956). These were compared with the values deduced from measurements of dichroism in [Glu(Bzl)]$_n$, but the agreement was generally poor (Table 9.7). It was noted by Tsuboi (1962) that some improvement could be obtained if the amide

TABLE 9.7

Comparison of Calculated and Observed Transition Moment Directions[a]
Relative to the Axis of the α-Helix

Method	Amide vibration (deg)				
	A	I	II	III	V
Calculated for an α-helix[b] with $h = 1.50$ Å	22.7	34.6	87.5	45.6	86.7
Observed in [Glu(Bzl)]$_n$ (Tsuboi, 1962)	28	39	75	40	>80
Observed in [Glu(Bzl)]$_n$ (Suzuki, 1971)	17.3	27.6	75.3	—	76.4
Calculated for an α-helix[c] with $h = 1.525$ Å	22.9	34.9	87.9	45.1	88.5
Observed in [Glu(pNO$_2$Bzl)]$_n$ (Suzuki, 1971)	19.0	31.2	—	—	80.5

[a] The sense of the transition moment direction cannot be determined from infrared measurements; of the two possibilities, the angle in the range 0–90° has been given.

[b] Calculated from the coordinates given in Table 9.1 and the transition moment directions, relative to the CO group of 8° for amide A, 20° for amide I, 73° for amide II, and −60° for amide III. The transition moment for amide V was assumed to be normal to the plane of the amide group.

[c] Calculated from the coordinates of an α-helix with the standard amide-group geometry (Table 6.1), $\tau(N, C^\alpha, C') = 109.47°$, $h = 1.525$ Å, $t = 99.6°$. The values of h and t correspond to those observed in [Glu(pNO$_2$Bzl)]$_n$ (Table 9.8a). The transition moment directions were assumed to be the same as given in footnote b.

group was rotated through an angle of 10° about the normal to the plane of the group.

Although the work of Tsuboi (1962) represented a considerable advance over earlier studies in that a quantitative assessment was made of the effects of imperfect orientation, several other factors need to be considered. The procedure for estimating the inclination of a transition moment direction to the axis of the α-helix involves the following:

(1) Determination, usually by X-ray diffraction studies, of the type of orientation texture present in the specimen. This texture determines the way in which the observed dichroism must be interpreted (Chapter 5, Section B.II).

(2) Analysis of the spectra into individual absorption bands and a baseline (Chapter 5, Section C).

(3) Correction of the observed dichroism for the effects of imperfect orientation.

If the $v_\parallel(0)$ and $v_\perp(t)$ modes are well separated then the corrected dichroic ratio R_0, in the case of fiber orientation, can be calculated from the expression

$$R_0 = ([A_0]_\pi + 2[A_0]_\sigma)/(\tfrac{1}{2}[A_t]_\pi + [A_t]_\sigma) \tag{9.7}$$

where $[A_0]_\pi$ is the integrated area beneath the $\nu_\parallel(0)$ band in the parallel (π) spectrum and $[A_0]_\sigma$ is the integrated area of this component in the perpendicular (σ) spectrum, etc. If the texture is such that the molecular axes all lie in a plane perpendicular to the incident radiation but have azimuthal dispersion, the coefficients 2 and $\frac{1}{2}$ are omitted from Eq. (9.7) (Miyazawa and Blout, 1961). This method, which relies on a result obtained by Higgs (1953), is only applicable when the $\nu_\parallel(0)$ and $\nu_\perp(t)$ components are well resolved, as is the case, for example, with the amide II components.

If the components are not well separated, the overlapped pair must be treated as a single band and a correction applied for the effects of disorientation. The orientation parameter f (Chapter 5, Section B.II) can be determined from the X-ray diffraction pattern (Chapter 5, Section B.III) or from the observed absorbances for a well-separated component such as the $\nu_\perp(t)$ component of amide II (Miyazawa and Blout, 1961; Miyazawa, 1962; Tsuboi, 1962). For fiber orientation the value of f in the case of a $\nu_\perp(t)$ mode is given by

$$f = 2([A_t]_\sigma - [A_t]_\pi)/(2[A_t]_\sigma + [A_t]_\pi) \tag{9.8}$$

and in the case of a $\nu_\parallel(0)$ mode by

$$f = ([A_0]_\pi - [A_0]_\sigma)/([A_0]_\pi + 2[A_0]_\sigma) \tag{9.9}$$

[see Eq. (5.52)]. The value of R_0 is then calculated from the expression

$$R_0 = [f(R + 2) + 2(R - 1)]/[f(R + 2) - (R - 1)] \tag{9.10}$$

where

$$R = ([A_0]_\pi + [A_t]_\pi)/([A_0]_\sigma + [A_t]_\sigma) \tag{9.11}$$

The numerator in Eq. (9.11) is simply the observed integrated intensity of the overlapped pair in the parallel spectrum and the denominator is the integrated intensity in the perpendicular spectrum.

(4) Assuming that all amide groups are equivalent, the inclination of the transition moment direction to the helix axis can then be calculated from Eq. (5.51), giving

$$\alpha = \cot^{-1}[\pm(\tfrac{1}{2}R_0)^{1/2}] \tag{9.12}$$

It should be noted that the correct choice of sign in Eq. (9.12) cannot be determined from dichroism measurements alone. Thus if α is the principal value of $\cot^{-1}[(\tfrac{1}{2}R_0)^{1/2}]$ then $180° - \alpha$ is also a solution of Eq. (9.12).

If values of α can be obtained for two amide modes, information can also be obtained about the inclination to the helix axis of the normal to the plane of the amide group. If α_1 and α_2 are values in the range 0–90° for the inclinations of two transition moments to the helix axis and ψ is the angle between them (Table 5.2), it can be shown that the inclination of the normal to the helix axis α_N is given by

$$\cos \alpha_N = \pm (\sin \alpha_2) \left[1 - \left(\frac{\cos \alpha_1 - \cos \alpha_2 \cos \psi}{\sin \alpha_2 \sin \psi} \right)^2 \right]^{1/2} \qquad (9.13)$$

The plus and minus signs give two complementary solutions which reflect an uncertainty of 180° in the directions of the transition moments, which is discussed later. In addition, two solutions are in general obtained, one of which is incorrect, depending on whether ψ is taken in the range 0–90° or 90–180°. If a value of α for a third mode can be obtained, the correct solution can be found by comparing the results obtained by using α_1 and α_2 with those obtained by using α_1 and α_3 or α_2 and α_3.

According to Beer (1956), both the orientation parameter f and the spatial orientation of the amide group can be uniquely determined directly from measurements of R for three bands together with the information given in Table 5.2. However, the fact that α_1, α_2, α_3, and ψ cannot be distinguished from their complements appears to have been overlooked and in general a fourth observation would be required to eliminate incorrect solutions. Even then there would be ambiguity, since two solutions would remain, corresponding to the correct spatial orientation and its reflection in the xy plane.

The dichroism in the spectrum of oriented films of $[\text{Glu(Bzl)}]_n$ has been reinvestigated by Suzuki (1971) and interpreted along the lines just indicated. The inclinations of the transition moments, calculated from the observed data (Table 9.7), differ from those calculated for the "standard" α-helix and it seems likely that some refinement of the atomic positions in the model is required. Measurements were also made on $[\text{Glu}(p\text{NO}_2\text{Bzl})]_n$ and the results obtained with both polymers are included in Table 9.7. In this case the observed transition moment directions for the amide A and amide I bands agree, within experimental error, with those calculated for an α-helix having the "standard" amide geometry and $h = 1.525$ Å. The agreement in the case of the amide V transition moment direction is less satisfactory.

III. SIDE-CHAIN BANDS

Studies of the dichroism of absorption bands associated with vibrational modes localized in side chains have been used in a number of instances to devise models for the conformation of the side chains in synthetic polypeptides in the α form (Tsuboi, 1962; Mitsui *et al.*, 1967; Takeda *et al.*, 1970) and the ω form (Bradbury *et al.*, 1962a). As explained earlier, the conformation cannot be determined uniquely from infrared measurements, and stereochemical and other data must be used to reduce the number of possibilities. Changes in the frequencies and dichroisms of side-chain bands have also been used to detect side-chain rearrangements (Tsuboi *et al.*, 1961; Bradbury *et al.*, 1962a, 1968b; Tsuboi, 1964; Mitsui *et al.*, 1967).

Many side-chain localized vibrations have slightly different frequencies in spectra obtained with the electric vector vibrating respectively parallel and perpendicular to the fiber axis. These frequency differences suggest that vibrations in different residues are coupled through weak interactions and that frequencies are related by equations similar in type to those used in the analysis of the amide vibrations (Miyazawa and Blout, 1961).

D. FACTORS AFFECTING HELIX SENSE AND GEOMETRY

I. HELIX SENSE

Shortly after the announcement of the discovery of the α-helix Huggins (1952) pointed out that with L-amino acid residues there would be a short contact between O_i and C_i^β in the left-handed form of the α-helix which was not present in the right-handed form. Accordingly, it was suggested that the right-handed form should be preferred for polymers of L-amino acids. This prediction has largely been borne out in practice and the only exceptions which have been observed are in $[\mathrm{Asp(Bzl)}]_n$ (Blout and Karlson, 1958; Karlson *et al.*, 1960; Bradbury *et al.*, 1960a), $[\mathrm{Asp(Me)}]_n$ (Goodman *et al.*, 1963a; Bradbury *et al.*, 1968a), and $[\mathrm{Asp}(o\mathrm{ClBzl})]_n$ and $[\mathrm{Asp}(m\mathrm{ClBzl})]_n$ (Erenrich *et al.*, 1970), which appear to exist in the left-handed form. The preference for the left-handed form is not an intrinsic property of esters of $[\mathrm{Asp}]_n$, since closely related polymers such as $[\mathrm{Asp}(p\mathrm{MeBzl})]_n$ and $[\mathrm{Asp(Et)}]_n$ form right-handed helices (Hashimoto and Aritomi, 1966; Hashimoto, 1966; Bradbury *et al.*, 1968a). In addition, a transition from a right-handed to a left-handed helical form with increase in temperature has been observed in $[\mathrm{Asp}(n\mathrm{Pr})]_n$ by Bradbury *et al.* (1968a,b), and $[\mathrm{Asp(Bzl)}]_n$ can be isolated in the right-handed helical form from molecular monolayers (Malcolm, 1968b, 1970).

Calculations of intramolecular potential energies per mole residue have been carried out for helical conformations of $[\mathrm{Ala}]_n$ (Liquori, 1966;

Ramachandran et al., 1966b; Scott and Scheraga, 1966; Ooi et al., 1967; Brant, 1968) and deep minima are observed in regions of the (ϕ, ψ) conformational maps corresponding to the left-handed and right-handed forms of the α-helix. According to Ooi et al. (1967), the minimum for the right-handed helix is 0.38 kcal mole^{-1} lower than that for the left-handed helix, the difference in energy being attributable to differences in the contributions of the nonbonded interactions. Similar calculations have been carried out for other homopolypeptides (Ooi et al., 1967; Yan et al., 1968, 1970) and in some instances the conformation of lowest energy was found to be a left-handed α-helix. It is difficult to test these predictions experimentally since the most stable helix sense will depend upon the difference in free energy between the right-handed and left-handed forms rather than the difference in intramolecular potential energy. In many cases the observed helix sense corresponds to the helix sense of the conformation of lowest calculated potential energy but this would appear to be partly fortuitous since the side-chain conformation, which plays an important part in determining the differences between the calculated potential energies, appears to be different in the actual polymer from that of the calculated minimum-energy conformation (Bradbury et al., 1972).

A feature of the energy calculations that calls for comment is the highly empirical nature of the potential functions that are used. For example, Ooi et al. (1967) calculated minimum-energy conformations for helical forms of [Asp(Me)]$_n$ and found that with the potential functions used, the minimum for the right-handed form was lower than that for the left-handed form. The experimental data discussed earlier indicate that the opposite is true for the free energies. However, it was found that the calculated depths of the minima for the intramolecular potential energy could be brought in line with the observed data by empirical adjustment of the constants in the nonbonded potential functions used in the energy calculation. These amended functions were then used in subsequent calculations.

Despite the lack of a secure theoretical basis for the formulation of potential functions, the calculations of intramolecular potential energy have been developed to the stage where some of the factors which influence helix sense can be appreciated. In [Ala]$_n$ the contribution of the main chain to the potential energy exceeds that of the side chain but as the length of the side chain increases, the contribution from the side chain becomes progressively more important. For example, in [Glu(Bzl)]$_n$ the side-chain contribution is twice that of the main chain (Yan et al., 1968) and more than half of the total energy arises from nonbonded interactions in the side chain.

In esters of [Asp]$_n$ an unfavorable interaction occurs between the dipole of the ester group and the main-chain amide group which is less for a left-handed helix than for a right-handed helix (Ooi *et al.*, 1966, 1967; Yan *et al.*, 1968). Against this must be balanced the remaining contribution from the side-chain terms which may outweigh this effect in some instances (Yan *et al.*, 1968). In esters of [Glu]$_n$ the interaction between the dipole of the ester group and the main-chain amide group is favorable in the right-handed helical conformation of lowest energy and unfavorable in the left-handed form. No evidence was found to support an earlier suggestion by Goodman *et al.* (1963b) that there is competition between the side-chain and main-chain C=O groups for hydrogen bond formation with the main-chain N—H group in the right-handed form which destabilizes it relative to the left-handed form.

The helix sense in solution of several polypeptides has been shown to be temperature-sensitive (Bradbury *et al.*, 1968a,b; Toniolo *et al.*, 1968) and Lotan *et al.* (1969) suggested that the observed inversions of helix sense with increasing temperature could be explained on the basis of an increased amplitude of intramolecular motions. This effect was simulated in calculations of potential energies by increasing the effective radius of the hydrogen atoms and it was found that the helix sense of the lowest-energy conformations of several homopolypeptides inverted with increasing temperature.

The experimental data on synthetic polypeptides which have been gathered from optical and X-ray diffraction studies indicate that the right-handed α-helix is in general the more stable conformation for L-amino acid residues. In the case of esters of [Asp]$_n$, however, an unfavorable dipole–dipole interaction is present which tends to desta-bilize the right-handed form of the α-helix relative to the left-handed form. In the protein structures so far determined by X-ray crystal-lography no evidence has been obtained for the occurrence of left-handed α-helices.

II. HELIX GEOMETRY

Values of unit height h and unit twist t in the α-helical forms of various synthetic polypeptides, as determined by X-ray diffraction studies, are collected together in Tables 9.8a and 9.8b. Several poly-peptides, including [Ala]$_n$, have values of h and t close to those originally suggested by Pauling and Corey (1951a) but appreciable departures from these values are evident, with h varying from 1.325 Å in the ω form of [Asp(Bzl)]$_n$ to 1.525 Å in the α form of [Glu(pNO$_2$Bzl)]$_n$, while $|t|$ varies from 90° in [Asp(Bzl)]$_n$ to 102.8° in nematic films of [Glu(Bzl)]$_n$.

Some correlation is evident between h and t (Fig. 9.7) and the general trend suggests that the number of residues per turn can be varied over a small range without appreciable increase in energy, provided any increase in h is accompanied by a compensating increase in t. This behavior suggests that, to a first approximation, the polypeptide chain can be thought of as a ribbon of width w wrapped around a cylinder of radius r equal to the radial coordinate of C^α (Fig. 9.8). A relationship between h and t can be derived (Fraser *et al.*, 1967b) in terms of the distance d between successive C^α atoms and the width of the ribbon as follows:

$$h^4 - h^2\{d^2 + (w/360)^2[t^2 - 4\sin^2(\tfrac{1}{2}t)]\} + (wdt/360)^2 = 0 \qquad (9.14)$$

For the "standard peptide" geometry (Table 6.1) $d = 3.80$ Å and if w is chosen so that the α-helix with $h = 1.50$ Å and $t = 100°$ is a solution of Eq. (9.14), the relationship shown in Fig. 9.7 is obtained. Although the treatment is an oversimplification of the problem, the general trend in both synthetic polypeptides and proteins, also included in Fig. 9.7, is well represented.

Another correlation which can be made is with the normal and extreme limits of allowed conformations for $[Ala]_n$ given by Ramachandran and Sasisekharan (1968). The relevant regions in the (ϕ, ψ) steric map (Chapter 6, Section B.III) corresponding to right-handed helices with $N_iH_i \cdots O_{i-4}$ hydrogen bonding have been transformed to h and t values by means of Eqs. (6.14) and (6.15) and the result is shown in Fig. 9.9. Again the general trend is well represented but a comprehensive analysis on this basis is not possible without a detailed study of the effect on the steric map of variation in $\tau(N, C^\alpha, C')$ and other parameters. The range of allowed values of h and t is very sensitive to the choice of $\tau(N, C^\alpha, C')$ (Ramakrishnan and Ramachandran, 1965).

As regards the detailed structure of the α-helix, it must be recognized that the only synthetic polypeptide for which the atomic positions and helix sense have been satisfactorily established is $[Ala]_n$. In no other case is the agreement between the calculated and observed intensity transforms sufficiently good to place any reliance on the precise value of the atomic coordinates given for model structures. On the contrary, the lack of agreement is sufficiently great in some instances to indicate either that the coordinates are significantly in error or that the packing model is incorrect.

All the synthetic polypeptides in the α form so far examined in detail by X-ray diffraction show departures from strict helical symmetry in the solid state. These departures can be attributed to the fact that

TABLE 9.8a

Observed Values of Pitch, Unit Height, and Unit Twist in the
α-Helical Conformation in Synthetic Polypeptides[a]

Symbol[b] in Figs. 9.7 and 9.9	Polymer[c]	Pitch P (Å)	Unit height h (Å)	Unit twist t (deg)
1	18-residue/5-turn α-helix[d]	5.40	1.50	100.0
[Glu(Bzl)]$_n$ and related polymers				
2	[Glu(Bzl)]$_n$	5.42	1.505	100.0
3	[Glu(Bzl)]$_n$ nematic	5.25	1.50	102.9
4	[Glu(Bzl)]$_n$ racemic[e]	5.30	1.495	101.5
5	[Glu(Bzl)]$_n$ racemic[f]	5.35	1.492	100.4
6	[Glu(oNO$_2$Bzl)]$_n$	5.34	1.495	100.8
7	[Glu(mNO$_2$Bzl)]$_n$	5.32	1.485	100.5
8	[Glu(pNO$_2$Bzl)]$_n$	5.51	1.525	99.6
9	[Glu(pIBzl)]$_n$	5.35	1.51	101.6
10	[Glu(NpMe)]$_n$	5.39	1.505	100.5
11	[Glu(Me$_3$Bzl)]$_n$	5.47	1.50	98.7
12	[Glu(pNO$_2$Bzl)-Glu(Bzl)]$_n$	5.42	1.50	99.6
Other synthetic polypeptides				
13	[Ala]$_n$	5.41	1.496	99.6
14	[Glu(Me)]$_n$	5.43	1.493	99.1
15	[Try]$_n$	5.42	1.49	99.0
16	[Lys(Cbz)]$_n$	5.38	1.495	100.0
17	[Asp(pClBzl)]$_n$, α form	5.40	1.50	100.0
18	[Asp(pClBzl)]$_n$, ω form	5.20	1.30	90.0
19	[Asp(Bzl)]$_n$, α form		1.496	
20	[Asp(Bzl)]$_n$, ω form	5.30	1.325	90.0
21	[Cys(BzlTh)]$_n$	5.55	1.39	90.0
22	[Glu(Et)-Cys(Bzl)-Glu(Et)]$_n$	5.36	1.489	100.0
23	[Leu, DL-Phe]$_n$	5.36	1.47	98.7

[a] Helix parameters derived from unit cell: 2, 5, 13, 14, 16–22. Helix parameters derived from ratio of the Z coordinates of the "turn" layer line ($= 1/P$) and the "1.5 Å reflection" ($= 1/h$): 3, 4, 6–12, 15, 23. Where the helix is left-handed, the values given for P and t are magnitudes.

[b] References: 2, 4, Elliott, A. et al. (1965); 3, Samulski and Tobolsky (1971); 5, Squire and Elliott (1969); 6–12, Fraser et al. (1967b); 13, Brown and Trotter (1956); 14, 23, Bamford et al. (1956); 15, Peggion et al. (1968); 16, Yakel (1953); 17, 18, Takeda et al. (1970); 19, 20, Bradbury et al. (1962a); 21, Fraser et al. (1962a); 22, Fraser et al. (1965c).

[c] Abbreviations defined on p. xvii.

[d] Coordinates given in Table 9.1.

[e] Mixture of equal parts of D and L polymers.

[f] Mixture of equal parts of D and L polymers, liquid crystal.

TABLE 9.8b

Observed Values of Pitch, Unit Height, and Unit Twist in the α-Helical Conformation in Myoglobin and Lysozyme

Symbol in Figs. 9.7 and 9.9	Helical segment	Number of residues	Pitch P (Å)	Unit height h (Å)	Unit twist t (deg)
Myoglobin[a]					
24	A	16	5.45	1.50	99.1
25	B	16	5.47	1.47	96.8
26	D	7	5.27	1.45	99.1
27	E_1	10	5.49	1.52	99.7
28	E_2	10	5.47	1.49	98.0
29	F	9	5.40	1.46	97.4
30	G	19	5.49	1.53	100.3
31	H	24	5.41	1.49	99.1
Lysozyme[b]					
32	A	11	5.42	1.51	100.2
33	B	11	5.44	1.485	98.3
34	C	6	5.61	1.66	106.6
35	D	9	5.41	1.455	96.7

[a] Kendrew (1962).   [b] Blake et al. (1967).

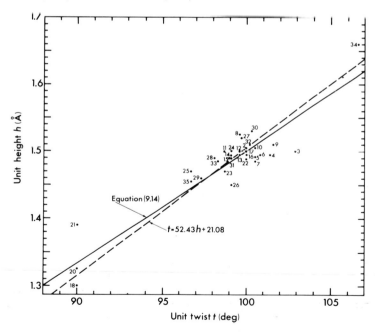

FIG. 9.7. Unit height h and unit twist t of α-helices in synthetic polypeptides and proteins. A correlation between these quantities is evident and the fitted curve $t = 52.43h + 21.08$ is compared with the curve predicted from the simplified ribbon model shown in Fig. 9.8. A key to the individual compounds is given in Tables 9.8a and 9.8b.

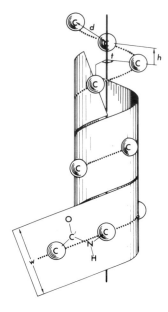

FIG. 9.8. Simplified model of the α-helix to explain the correlation between unit height h and unit twist t. The polypeptide chain is visualized as a ribbon wrapped around a cylinder so as to form a continuous tube. Any variation in h results in a change in t and these changes are related by Eq. (9.14). This relationship is illustrated in Fig. 9.7.

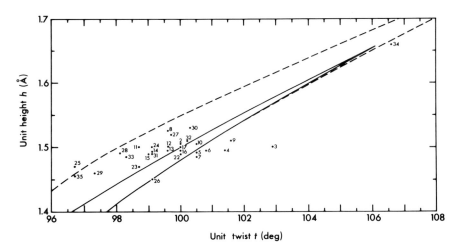

FIG. 9.9. Correlation of unit height h and unit twist t of α-helices in synthetic poly-peptides and proteins with "allowed" conformations for [Ala]$_n$ (Chapter 6). The relevant regions of the (ϕ, ψ) steric map in Fig. 6.7a for $\tau(N, C^\alpha, C') = 110°$ have been trans-formed to give "normal" (full line) and "extreme" (broken line) limits in terms of h and t. The allowed regions are sensitive to the value assumed for $\tau(N, C^\alpha, C')$.

a homopolypeptide with the screw symmetry of the α-helix cannot be packed in a crystal in such a manner that all residues have equivalent environments. The impression has been gained from the polymers studied so far that the departure from strict helical symmetry increases with distance from the helix axis and in extreme cases (Parry and Elliott, 1967; Squire and Elliott, 1969, 1972) the symmetries of the main chain and side chains are only tenuously related. Again it should be emphasized that these conclusions are based on plausible arguments rather than an objective test by comparison of observed and calculated intensity transforms.

In summary the geometry of the α-helix in $[Ala]_n$ appears to be very close in all respects to that originally suggested by Pauling and Corey (1951a) but appreciable variations in both unit height and unit twist have been observed. Similar variations have been found in globular proteins (Blake et al., 1967; Watson, 1970) and it can be concluded that the α-helix is capable of some flexibility. Appreciable departures from strict helical symmetry occur in crystals of synthetic polypeptides and regular distortion patterns of long period have been observed (A. Elliott et al., 1965; Parry and Elliott, 1967). A further result from the study of synthetic polypeptides which has considerable relevance to conformation in fibrous proteins is the observation that in single crystals the α-helix is regularly folded (Padden and Keith, 1965; Padden et al., 1969). This leads to an approximately equal number of "up" and "down" chains in the crystal and provides convincing support for the finding, from X-ray studies, that models with a single chain direction are not adequate to explain the observed intensity transforms (Elliott and Malcolm, 1958; Bradbury et al., 1962a; Arnott and Wonacott, 1966b; Takeda et al., 1970).

Chapter 10

The Beta Conformation

A. INTRODUCTION

At an early stage in structural studies of fibrous proteins it was recognized that the X-ray diffraction data obtained from silk (Meyer and Mark, 1928) and from stretched mammalian (β) keratin (Astbury and Street, 1931; Astbury and Woods, 1933) could be interpreted in terms of structures in which extended or nearly extended helical polypeptide chains with unit twist $t = 180°$ were arranged in sheets stabilized by the formation of interchain hydrogen bonds. The bonds were directed approximately perpendicular to the chain axis and in the plane of the sheet (Fig. 10.1).

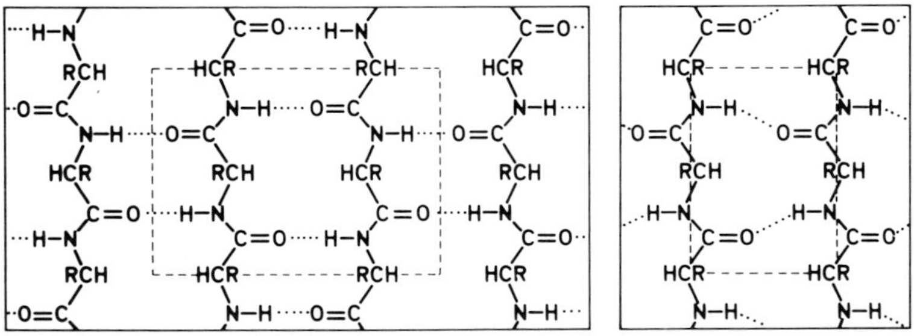

(a)   (b)

Fig. 10.1. Schematic diagram illustrating the hydrogen bonding arrangements in (a) the antiparallel-chain sheet and (b) the parallel-chain sheet model for β structures.

218

Two types of sheet can be built up in this way: In the parallel-chain sheet all the chains have the same sense with respect to the sequence —NH—CHR—CO—, while in the antiparallel-chain sheet the sequence alternates in a regular manner. Astbury and Woods (1933) obtained evidence from X-ray diffraction studies of β-keratin which suggested that the antiparallel-chain arrangement was present in this material and Huggins (1943) noted that the antiparallel arrangement was more suited to the formation of hydrogen bonds in the case of fully extended or nearly fully extended chains. The analysis of the vibrational modes of regular conformations given by Miyazawa (1960a) was applied by Miyazawa and Blout (1961) to a number of earlier studies of the infrared spectra of synthetic polypeptides and proteins and it was concluded that except in the case of β-keratin, the antiparallel arrangement was present. The chain arrangement in β-keratin was assumed to be parallel but this was later shown not to be so (Bradbury and Elliott, 1963b; Fraser et al., 1969a).

Supporting evidence for the sheetlike nature of β structures was obtained from X-ray diffraction studies of doubly oriented specimens by Astbury and Sisson (1935) in which reflections corresponding to the interchain and intersheet periodicities were identified and shown to be normal to each other and to the chain axis. In addition, studies of infrared dichroism (Ambrose and Elliott, 1951a,b; Bamford et al., 1956) confirmed that the CO and NH bonds were preferentially oriented perpendicular to the chain axis in the β conformation.

Pauling and Corey (1951c,g) investigated possible models for β structures and described two pleated-sheet conformations (Pauling and Corey, 1953a) which satisfied their stereochemical criteria (Chapter 6). In the antiparallel-chain, pleated-sheet model (Fig. 10.2c) the unit height $h = 3.50$ Å, while in the parallel-chain, pleated-sheet model $h = 3.25$ Å. The shorter repeat in the latter model was dictated by the requirement that satisfactory hydrogen bonds be established between adjacent chains in the sheet.

Experimental verification of the presence of the antiparallel-chain pleated sheet in polypeptide materials, through the achievement of quantitative agreement between the calculated and observed X-ray intensity transforms, has been obtained for $[Ala]_n$ by Arnott et al. (1967) and for β-keratin by Fraser et al. (1969a). No examples of parallel-chain pleated sheets have been authenticated in either synthetic polypeptides or fibrous proteins. Small and somewhat irregular pleated sheets have been reported to be present in the structures of several globular proteins, including lysozyme (Blake et al., 1967), carboxypeptidase A (Lipscomb et al., 1968; Quiocho and Lipscomb, 1971), lactic dehydrogenase

(Adams *et al.*, 1970), insulin (Blundell *et al.*, 1971), papain (Drenth *et al.*, 1971), ribonuclease-S (Wyckoff *et al.*, 1970), subtilisin BPN′ (Wright *et al.*, 1969), human carbonic anhydrase C (Liljas *et al.*, 1972), rubredoxin

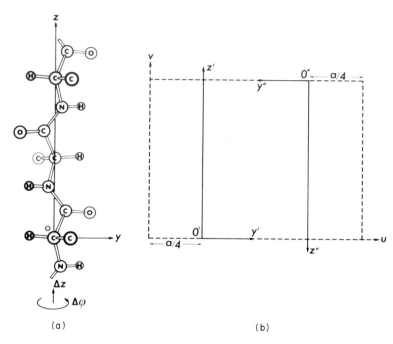

(a)   (b)

FIG. 10.2.   (a) Helical conformation of the polypeptide chain with unit twist $t = 180°$ and unit height $h = 3.5$ Å. (b) Illustration of the way in which an antiparallel-chain, pleated-sheet structure can be generated from the helical conformation in (a). The chain is displaced by a distance Δz parallel to Oz and rotated by $\Delta\phi$ around Oz in the reference frame $Oxyz$ and this frame is then set successively at $O'x'y'z'$ and $O''x''y''z''$ to give the "up" chain and "down" chain, respectively.

(Herriott *et al.*, 1970), cytochrome b_5 (Mathews *et al.*, 1972), and an extracellular nuclease (Arnone *et al.*, 1971) and in the fibrous protein feather keratin (Fraser *et al.*, 1971a).

Many studies have been described in which characteristic X-ray spacings, infrared absorption frequencies, and optical rotatory dispersion properties have been used to detect the presence of the β conformation in synthetic polypeptides but in this chapter attention will be restricted to structural studies which have provided information about the stereochemistry of the pleated-sheet structures.

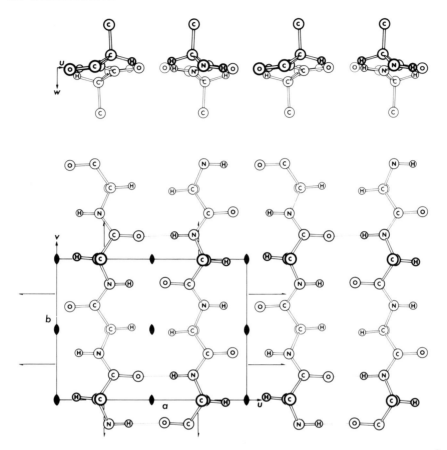

FIG. 10.2c. Model proposed by Pauling and Corey (1953a) in which $\Delta z = 0.035$ Å and $\Delta\phi = -24.8°$.

I. THE PLEATED-SHEET MODELS

In the pleated-sheet models (Pauling and Corey, 1951g, 1953a) the individual chains have helical symmetry with unit twist $t = 180°$ and so the atoms of any one residue can be generated from those of any other residue in the same chain by the operation of a twofold screw axis (Fig. 10.2a). From Fig. 6.4a it can be seen that for conformations with $t = 180°$ the torsion angles are related by the conditions $\phi \simeq -\psi$, and from the steric map in Fig. 6.6a it appears that ϕ is restricted, for L-residues, to a range of about -70 to $-180°$. Thus if values are assumed for the bond lengths and angles of the main chain and for the torsion angle ω, the conformation is uniquely defined by specifying the unit height h. Atomic coordinates for the conformations with $h = 3.25$ Å, corresponding to the parallel-chain model, and with $h = 3.50$ Å, corresponding to the antiparallel-chain model, are given in Table 10.1.

When the chains are packed side by side to form a sheet the residues will only remain equivalent if certain conditions are satisfied with respect to the symmetry of the arrangement. In the parallel-chain pleated sheet the necessary condition is that neighboring chains are related by a translational symmetry which is perpendicular to the chain axis. Thus all chains have the same orientation about their axes relative to the plane of the sheet and like atoms occur on a series of planes perpendicular to the chain axes spaced h apart. The only parameters needed to define the packing arrangement in the sheet are therefore the interchain distance and the orientation of the chain about its axis.

In the case of the antiparallel-chain pleated sheet the conditions for equivalence are first that the up-chain and the down-chain be related by a twofold rotation axis perpendicular to the plane of the sheet and second that neighboring two-chain units in the sheet be related by a translational symmetry which is perpendicular to the chain axes. It is also necessary that the magnitude of the translation be equal to twice the distance between the chain axes in the repeating unit. The residues in the sheet are then related by a series of twofold rotation axes perpendicular to the plane of the sheet and a series of twofold screw axes in the plane of the sheet (Fig. 10.2c).

In addition to the interchain distance and the orientation of the up-chain relative to the plane of the sheet, a third parameter must be specified, in the case of the antiparallel-chain model, in order to define the packing arrangement in the sheet. This relates to the relative axial positions of the up-chain and the down-chain in the repeating unit. In Table 10.1 the positions of the atoms in a residue in the up-chain are referred to a coordinate system $Oxyz$ (Fig. 10.2a) in which the C^α atom lies on the x axis. The repeating unit in the sheet for a particular antiparallel-chain model can then be obtained by generating a second residue in the coordinate system $Oxyz$ through the operation of the twofold screw axis, rotating the two residues by an angle $\Delta\phi$ and displacing them axially by a distance Δz, and then setting the coordinate system successively in the positions $O'x'y'z'$ and $O''x''y''z''$ shown in Fig. 10.2b. These are chosen so that O', in terms of the sheet coordinates u, v, and w, is at $(a/4, 0, 0)$ and O'' is at $(3a/4, b, 0)$, where $b \, (= 2h)$ is the magnitude of the lattice vector parallel to the chain axes and a is the magnitude of the lattice vector perpendicular to the chain axes. In the antiparallel-chain model proposed by Pauling and Corey (1953a)

TABLE 10.1

Atomic Coordinates of the Repeating Unit in Helices with Unit Twist $t = 180°$[a]

Atom	x (Å)	y (Å)	z (Å)	r (Å)	ϕ (deg)
(a) Unit height $h = 3.25$ Å, $\phi = -119.34°$, $\psi = 113.63°$					
N	0.245	0.336	−1.225	0.415	53.916
H	0.152	1.318	−1.388	1.327	83.434
C^α	0.985	0.000	0.000	0.985	0.000
H^α	1.053	−1.093	0.099	1.518	−46.083
C^β	2.393	0.593	−0.077	2.466	13.927
C'	0.253	0.578	1.213	0.631	66.345
O	0.160	1.803	1.378	1.811	84.926
(b) Unit height $h = 3.50$ Å, $\phi = -146.67°$, $\psi = 143.00°$					
N	0.023	0.287	−1.250	0.288	85.437
H	−0.518	1.128	−1.225	1.241	114.668
C^α	0.741	0.000	0.000	0.741	0.000
H^α	0.900	−1.085	0.090	1.409	−50.318
C^β	2.093	0.715	−0.012	2.212	18.858
C'	−0.087	0.495	1.188	0.502	99.960
O	−0.745	1.543	1.116	1.714	115.752

[a] Other residues at r, $\phi + mt$, $z + mh$, where m is an integer. Standard planar amide geometry (Table 6.1), and $\tau(N, C^\alpha, C') = 109.47°$.

the parameter values were $a = 9.5$ Å, $b = 7.0$ Å, $\Delta z = 0.035$ Å, and $\Delta \phi = -24.8°$ and the resulting structure is illustrated in Fig. 10.2c.

When pleated sheets are packed together to form three-dimensionally regular structures the symmetry elements present in the sheet may not be present in the crystal and when this is so the residues will no longer be equivalent. For example, in the structure proposed by Marsh et al. (1955a) for Tussah silk fibroin two different environments exist for residues, depending upon their position in the unit cell.

A conflict of terminology arises in the discussion of crystalline β structures since most authors follow long-established crystallographic practice in the choice of origin and the naming of the axes. The a axis is generally chosen in the manner shown in Fig. 10.2c but either b or c may be used to designate the cell edge parallel to the chain axes, depending on the space group. In many instances the situation is further confused by uncertainty regarding the correct space group, but where practicable the terminology used in the original paper will be adhered to. For ease of comparison atomic coordinates will be given in terms of the convention illustrated in Fig. 10.2c.

II. CRYSTALLITE ORIENTATION

Many synthetic polypeptides adopt the β conformation when cast from suitable solvents, particularly if the molecular weight is low, and a variety of orientation textures have been observed. In cast films the sheets of chains are often preferentially oriented parallel to the plane of the film and if the films are stroked while drying or rolled or extended after drying, the chains may also adopt a preferred orientation relative to the direction of stroking or stretching. Such films are then doubly oriented and the optical properties resemble those of a biaxial crystal (Chapter 5, Section B.II.a). Frequently the chain axes tend to be aligned with the direction of stroking or stretching but with certain polymer preparations the chain axes tend to be aligned perpendicular to this direction. The latter arrangement has also been observed in denatured globular proteins (Astbury et al., 1935, 1959; Senti et al., 1943; Ambrose and Elliott, 1951b; Burke and Rougvie, 1972) and in denatured fibrous proteins (Rudall, 1946, 1952; Astbury, 1947b; Mercer, 1949; Peacock, 1959; Filshie et al., 1964; Burke, 1969). This so-called cross-β texture is frequently observed in synthetic polypeptides of low molecular weight (Bradbury et al., 1960b) and is thought to arise through the formation of sheets which become much longer in the hydrogen bond direction than in the chain direction. The application of shear would then be expected to align the sheet so that the chain axes were perpendicular to the direction of shearing. This effect has been observed experimentally with solutions of $[Glu(Me)]_n$ in m-cresol (Elliott, 1967). An unusual type of orientation has been observed in $[Cys(Cbz)]_n$ (Elliott et al., 1964) and in $[Gly-Ala-Glu(Et)]_n$ (Rippon et al., 1971) in which the chain axes, in the cross-β texture, were oriented perpendicular to the plane of the film.

The double orientation which is developed when polypeptides in the β form are oriented by shearing is helpful in the indexing of reflections since the assignment of an index on the basis of spacing can be checked by rotating the specimen in such a manner that the reciprocal lattice point intersects the sphere of reflection. The observed intensity should then be a maximum. However, the collection of quantitative intensity data is complicated by the fact that the orientation is not good enough for single-crystal methods to be used and fiber methods are not applicable.

B. HOMOPOLYMERS

I. Poly(glycine)

Two crystalline forms of $[\text{Gly}]_n$ have been observed (Meyer and Go, 1934) and these are distinguished by the symbols I and II. X-ray diffraction data (Astbury et al., 1948b; Astbury, 1949c; Bamford et al., 1953a) and infrared data (Elliott and Malcolm, 1956a; Miyazawa and Blout, 1961; Bradbury and Elliott, 1963b) obtained from studies of form I suggest that it has a β conformation. The conformation present in form II is discussed in Chapter 11.

So far it has not proved possible to prepare well-oriented specimens of $[\text{Gly}]_n$ in the β form and the unit cell has not been completely determined. Astbury (1949c) suggested that the c-axis (chain axis) projection of the cell had the form shown in Fig. 10.3a with an interchain distance, in the plane of the sheet, of 4.77 Å and an intersheet spacing of 3.44 Å. The sheets are not packed directly one upon another but are staggered so that the projections of the cell sides are inclined at 114°. This cell, based on a single chain, is not consistent with the infrared data, which indicate that alternate chains in the sheet are antiparallel (Miyazawa and Blout, 1961; Bradbury and Elliott, 1963b). The length of the projected unit of structure in the plane of the sheet should therefore be twice the interchain distance. Figure 10.3a therefore refers to a pseudocell and a possible explanation of the discrepancy, which also exists with some other polypeptides in the β form, is discussed in Section B.II.

Insufficient X-ray data are available to enable determination of whether the unit cell is monoclinic, in which case Fig. 10.3a represents the base of the pseudocell, or whether it is triclinic, in which case Fig. 10.3a would represent an axial projection of the base of the pseudocell. Bamford et al. (1953a) observed a strong reflection of spacing 1.16 Å which they assumed to be associated with a repeat of structure parallel to the chain axis. The value of the unit height h cannot be determined directly from this observation unless both the unit cell and the index of the reflection are known but it may be argued plausibly that $h \geqslant 3.48$ Å. This may be compared with the value $h = 3.50$ Å predicted by Pauling and Corey (1953a) for the antiparallel-chain pleated sheet.

Calculations of intrachain potential energy for regular helical conformations of $[\text{Gly}]_n$ have been reported by Scott and Scheraga (1966) and deep minima occur in regions of the (ϕ, ψ) map corresponding to the left-handed and right-handed α-helical conformations. No minima

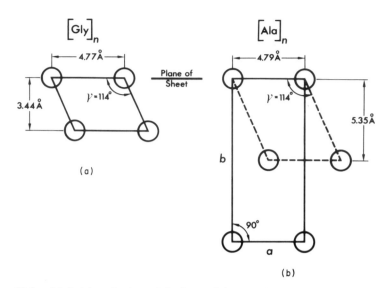

FIG. 10.3.   (a) Axial projection of the base of the pseudocell in the β form of [Gly]$_n$ deduced from the X-ray diffraction pattern (Astbury, 1949c). The chain axis is normal to the plane of the page. (b) Base of the orthogonal cell in the β form of [Ala]$_n$ derived from the X-ray diffraction pattern (Brown and Trotter, 1956; Arnott et al., 1967). The equatorial reflections can be indexed on the smaller monoclinic cell shown by the broken lines.

appear in the region corresponding to the β conformations due to the omission of interchain potential, which plays an important part in the stabilization of the pleated-sheet structure. Venkatachalam (1968a) has investigated the variation of interchain potential per residue with interchain distance within an isolated pleated sheet for [Gly]$_n$. The calculations showed minima, for both the parallel and antiparallel pleated sheets, in the vicinity of 4.75 Å (Fig. 10.4a), which may be compared with the observed value of 4.77 Å. Too much significance should not be attached to the agreement since the position of the minimum will depend upon the values assumed in the calculations for Δz and $\Delta \phi$ (Section A.I). Evidence obtained from studies of the β form of [Ala]$_n$ (Arnott et al., 1967) and β-keratin (Fraser et al., 1969a) indicate that appreciable departures occur from the values assumed for these parameters in the models described by Pauling and Corey (1953a).

Calculations of intersheet potential per residue were also carried out by Venkatachalam (1968a) for a fixed interchain distance within the sheet. The intersheet spacing was fixed at 3.45 Å and the effect of varying γ (Fig. 10.3a) explored. Minima were observed, for antiparallel-chain pleated sheets, at values of $\gamma = 49°$ and $113°$. The second of

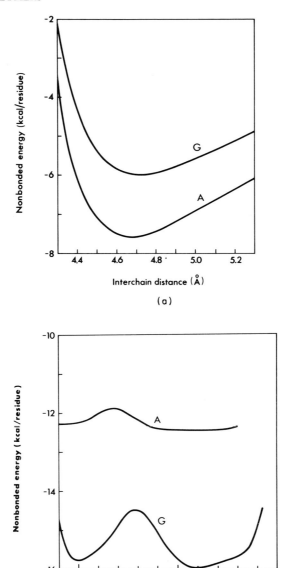

FIG. 10.4. (a) Variation of nonbonded energy with distance of separation between a pair of polypeptide chains in the antiparallel-chain pleated sheet. (b) Variation of nonbonded interaction energy between adjacent pleated sheets with the angle γ (Fig. 10.3a). Plotted from data given by Venkatachalam (1968a). G = $[Gly]_n$, A = $[Ala]_n$.

these agrees well with the value of $114°$ derived by Astbury (1949c) for the pseudocell of $[Gly]_n$. Again too much significance should not be attached to the precise agreement since the effects of varying Δz and $\Delta \phi$ were not explored. An unsatisfactory feature of the calculations for which no explanation could be offered was that the observed intersheet distance did not correspond to a minimum of potential energy.

II. POLY(L-ALANINE)

Specimens of $[Ala]_n$ containing well-oriented β-crystallites have been obtained by steam-stretching fibers spun from dichloroacetic acid solution (Brown and Trotter, 1956; Arnott et al., 1967). In addition double orientation could be induced by hot-rolling or steam-pressing of the fibers. The X-ray diffraction pattern obtained from these fibers has been studied in considerable detail and the main features are illustrated in Fig. 10.5. The sharp reflections can be indexed on an orthogonal cell containing four residues and the dimensions given by Arnott et al. (1967), rounded to the nearest 0.01 Å, are $a = 4.73$ Å, $b = 10.54$ Å, and $c = 6.89$ Å (chain axis). The area of the c-axis projection is such that two chains must pass through the cell. In addition to the sharp reflections, streaks are observed on layer lines with $l = 2$, 3, 4, and 5, indicating that some type of disorder is present. Brown and Trotter (1956) noted that the equatorial reflections could be indexed on a one-chain monoclinic cell and the relationship of this cell to the c axis projection of the cell in the β form of $[Gly]_n$ is shown in Fig. 10.3b.

Infrared spectra of thin oriented films of $[Ala]_n$ containing a high proportion of the β form were recorded by Elliott (1954) and the frequencies and dichroic characters of the two observed components of the amide I band at 1630 and 1690 cm^{-1} are consistent with their assignment to the $\nu_\perp(\pi, 0)$ and $\nu_\parallel(0, \pi)$ components expected for the antiparallel-chain pleated sheet (Chapter 5, Section A.II). As with $[Gly]_n$, this observation appears to be in conflict with the X-ray data where the unit cell is only one chain wide in the sheet direction, whereas it should be two chains wide for the antiparallel arrangement.

Marsh et al. (1955a) suggested that the β form of $[Ala]_n$ contained antiparallel-chain pleated sheets, but Brown and Trotter (1956) considered a number of packing arrangements for parallel-chain and antiparallel-chain pleated sheets and were unable to reconcile any of these with the observed intensities. The problem of the halving of the a-axis dimension from its expected value was shown by Arnott et al. (1967) to be explicable in terms of a statistical packing model in which

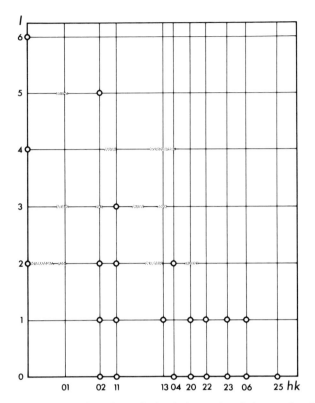

FIG. 10.5. X-ray diffraction data obtained from the β form of poly(L-alanine). Discrete reflections are indicated by circles and layer-line streaks are shaded. The discrete reflections can be indexed on an orthogonal cell with $a = 4.73$ Å, $b = 10.54$ Å, and $c = 6.89$ Å. (Redrawn from Bamford et al., 1956.)

consecutive antiparallel-chain sheets are stacked with a random displacement of $\pm\frac{1}{2}(4.73$ Å) in the a direction. Provided the number of sheets is large, the statistical arrangement can be shown to be equivalent to one in which each chain site is occupied by one-half of an up-chain and one-half of a down-chain. Arnott et al. (1967) suggested that a statistical arrangement of sheets is very likely in the case of $[\mathrm{Ala}]_n$ since the intersheet interactions would be similar for the two displacements.

A limited least squares refinement of the antiparallel-chain, pleated-sheet model suggested by Pauling and Corey (1953a) was carried out on the basis of 29 X-ray intensity data by Arnott et al. (1967) using the linked-atom method developed by Arnott and Wonacott (1966a,b). The values of the bond distances and angles were maintained at the values determined for the α-helix in $[\text{Ala}]_n$ (Table 9.2), while the torsion angles ϕ, ψ, and ω and the axial translation Δz and rotation $\Delta\phi$ of the chain were allowed to vary. The value of the normalized mean deviation R (Chapter 1, Section D.I.a) was reduced to 0.136 compared with 0.360 for the starting model. The refined value of ω was $-178.5°$, which is probably not significantly different from the planar trans configuration.

The atomic coordinates of the four residues in the repeating unit of the antiparallel-chain pleated sheet, based on the values derived by Arnott et al. (1967), are given in Table 10.2 and the refined model is illustrated in Fig. 10.6. The values of the parameters which fix the position of the chain in the cell are $\Delta z = 0.38$ Å and $\Delta\phi = -10°$, compared with $\Delta z = 0.04$ Å and $\Delta\phi = -25°$ for the original pleated-sheet model (Fig. 10.2c) due to Pauling and Corey (1953a).

Since $\Delta z = 0.38$ Å, the C^α atoms in adjacent chains are displaced

TABLE 10.2

Atomic Coordinates of the Residues in the Repeating Unit of the
Antiparallel-Chain Pleated-Sheet Structure in the Beta Form of Poly(L-alanine)[a]

Atom	"Up" chain			Hydrogen bonds	"Down" chain		
	u (Å)	v (Å)	w (Å)		u (Å)	v (Å)	w (Å)
N	2.695	−0.837	0.130		6.765	0.837	0.130
H	3.665	−0.857	−0.100		5.795	0.857	−0.100
C^α	2.225	0.383	0.790		7.235	−0.383	0.790
H^α	1.155	0.453	0.700		8.305	−0.453	0.700
C^β	2.605	0.353	2.280		6.855	−0.353	2.280
C'	2.855	1.608	0.130		6.605	−1.608	0.130
O	4.065	1.638	−0.130		5.395	−1.638	−0.130
N	2.035	2.608	−0.130		7.425	4.282	−0.130
H	1.065	2.588	0.100		8.395	4.302	0.100
C^α	2.505	3.828	−0.790		6.955	3.062	−0.790
H^α	3.575	3.898	−0.700		5.885	2.992	−0.700
C^β	2.125	3.798	−2.280		7.335	3.092	−2.280
C'	1.875	5.053	−0.130		7.585	1.837	−0.130
O	0.665	5.083	0.130		8.795	1.807	0.130

[a] Calculated from data given by Arnott et al. (1967). The bond lengths and angles have the same values as given in Table 9.3 for the α form of $[\text{Ala}]_n$. Helix parameters for the polypeptide chain are: unit height $h = 3.445$ Å, unit twist $t = 180°$, $\phi = -138.8°$, $\psi = 134.7°$, $\omega = -178.5°$. Equivalent positions: (u, v, w), $(4.73 - u, 3.445 + v, -w)$, $(9.46 - u, -v, w)$, $(4.73 + u, 3.445 - v, -w)$.

axially by 0.76 Å and the direction of the displacement is such as to bring the CO groups in neighboring chains almost to the same level. Unfortunately the refinement of Δz was complicated by a high correlation with the intersheet displacement but the sense of the shift does not seem to be in doubt.

Venkatachalam (1968a) investigated the variation of interchain potential per residue with interchain distance within a single sheet for the Pauling and Corey (1953a) models and found that a minimum occurred, for the antiparallel arrangement, at a value of around 4.7 Å (Fig. 10.4a). Calculations of intersheet potential per residue were also carried out for a fixed interchain distance in the sheet. The intersheet distance was fixed at 5.35 Å and the effect of varying γ (Fig. 10.3) was studied. An ill-defined minimum at $\gamma = 106°$ was found (Fig. 10.4b). More elaborate calculations were carried out by McGuire *et al.* (1971) using two sets of empirical potential functions, one for intrachain interactions and the other for interchain interactions. The statistical crystal model used by Arnott *et al.* (1967) to explain the X-ray data could not be treated directly but calculations were performed on supposedly equivalent model structures. The potential energies per residue for these models were shown to be lower than for the regular packing models considered by Marsh *et al.* (1955a) and Arnott *et al.*

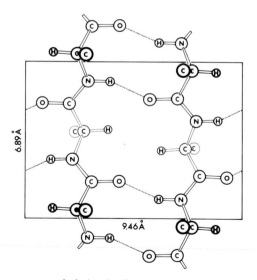

Fig. 10.6. Arrangement of chains in the antiparallel-chain, pleated-sheet structure present in the β form of [Ala]$_n$ (Arnott *et al.*, 1967). The chains are located in positions corresponding to $\Delta z = 0.38$ Å and $\Delta\phi = -10°$ (Fig. 10.2a) which differ somewhat from those suggested by Pauling and Corey (1953a) and illustrated in Fig. 10.2c.

(1967). In the minimum-energy conformations for all models the axial translation Δz and rotation $\Delta \phi$ of the chains differed from those suggested by Pauling and Corey (1953a). The changes were in the same direction but smaller in magnitude than those found experimentally by Arnott *et al.* (1967) in $[\text{Ala}]_n$.

III. POLY(γ-METHYL-L-GLUTAMATE)

X-ray diffraction patterns have been obtained from oriented specimens of $[\text{Glu(Me)}]_n$ in the β form (Bamford *et al.*, 1953a; Brown and Trotter, 1956) and the reflections were indexed on an orthogonal unit cell with $a = 4.70$ Å, $b = 21$ Å, and $c = 6.83$ Å (chain axis) containing four residues. Both the interchain distance a and the unit height h ($= c/2$) are less than in $[\text{Ala}]_n$. The arrangement of the chains was assumed to be similar to that shown in Fig. 10.3b for $[\text{Ala}]_n$ except that the intersheet distance ($= b/2$) was increased to 10.5 Å.

Infrared spectra of $[\text{Glu(Me)}]_n$ in the β form were reported by Elliott (1953c) and show clear evidence of a weak band at about 1700 cm^{-1} similar to that expected for the $\nu_\parallel(0, \pi)$ mode of the anti-parallel-chain pleated sheet (Chapter 5, Section A.II.b). This appears to be in conflict with the X-ray data since the a dimension of the unit cell implies a one-chain repeat of structure within the sheet rather than the two-chain repeat expected for an antiparallel chain arrangement. A possible explanation of this anomaly is that sheet-packing disorder similar to that found in the β form of $[\text{Ala}]_n$ is present.

Specimens of $[\text{Glu(Me)}]_n$ in the β form suitable for electron diffraction studies were prepared by Vainshtein and Tatarinova (1967a,b). The observed reflections could be indexed on a unit cell with $a = 4.725$ Å and $c = 6.83$ Å (chain axis) containing two residues and it was assumed that a parallel-chain, pleated-sheet structure was present. Three weak reflections which required a to be doubled to the value expected for an antiparallel-chain pleated sheet were attributed to occasional disorders in chain sense. The absence of reflections associated with a third cell dimension perpendicular to the plane of the sheet was assumed to be due to variations of 2–3 Å or greater in the intersheet distance and to a lack of correlation between the translation parallel to a and c of successive sheets.

Neither of these explanations is particularly plausible. A small amount of intrasheet chain disorder would not be expected to produce *sharp* reflections requiring a to be doubled and an alternative explanation which could account for the observed effect was given by Fraser *et al.* (1969a). It was shown that if a crystallite contains only a small number of antiparallel-chain pleated sheets with a packing disorder of the type found in the β form of $[\text{Ala}]_n$, then the effect is to reduce rather than to eliminate the intensity of the reflections requiring a to be doubled. Thus the observed electron diffraction data are not, as originally claimed, inconsistent with the antiparallel-chain, pleated-sheet model. The explanation given for the failure to detect a reflection corresponding to the intersheet distance is unlikely since it implies fluctuations of the order of 30% in density. The possibility does not seem to have been considered that a high degree of double orientation was present with the sheets lying parallel to the plane of the film.

An interesting feature of the electron diffraction patterns obtained from $[\text{Glu(Me)}]_n$ in the β form was the presence of weak meridional reflections on layer lines with $l = 3$ and 5. These reflections are predicted to have zero intensity for a helical conformation with $t = 180°$ and their appearance was attributed to departures from perfect helical symmetry due to disorder. This explanation is not consistent with the observed diffraction effects and from the discussion given in Chapter 1, Section A.IV.c it would seem more likely to be due to a *regular* distortion from helical symmetry similar to that observed in many synthetic polypeptides in the α form (Chapter 9).

IV. POLY(n-PROPYL-L-ASPARTATE)

Highly crystalline specimens of $[\text{Asp}(n\text{Pr})]_n$ in the cross-β form have been prepared from low-molecular-weight samples by casting films from dichloroacetic acid solution at 60°C and stroking the drying solution (Bradbury *et al.*, 1960b). A high degree of double orientation was present in the films and this enabled the X-ray diffraction pattern to be indexed with more certainty than is possible with fiber-type orientation. The unit cell proved to be monoclinic with $a = 9.57$ Å, $b = 6.79$ Å (chain axis), $c = 25.08$ Å, and $\beta = 96°$ and to contain eight residues. Four chains pass through the cell and unlike the cells derived for the β forms of $[\text{Ala}]_n$ and $[\text{Glu(Me)}]_n$, discussed earlier, the value of a is appropriate to the antiparallel-chain pleated sheet.

Measurements of infrared spectra were also reported for oriented films of $[\text{Asp}(n\text{Pr})]_n$ and the dichroisms of the amide A, I, and II bands were shown to be consistent with the presence of a cross-β conformation. A weak band was observed at 1702 cm^{-1} of opposite dichroism to the main amide I band at 1637 cm^{-1} and the subsequent analysis of the amide modes in β forms by Miyazawa and Blout (1961) make it clear that the two bands may be assigned to the $\nu_{\parallel}(0, \pi)$ and $\nu_{\perp}(\pi, 0)$ components of the amide I vibration of the antiparallel-chain pleated sheet.

Bradbury and Elliott (1963b) have suggested that a weak band observed at 1676 cm^{-1} may be assigned to the $\nu_{\parallel}(0, 0)$ mode of a parallel-chain arrangement (Chapter 5, Section A.II.b). The band was supposed to arise from the presence of disorder in the regular alternation of up-chains and down-chains within a sheet. This suggestion was correlated with the fact that certain of the reflections requiring the unit cell to be two chains wide in the a direction are more diffuse than the remainder and that the observed value of the unit height $h = 3.40$ Å is intermediate between the values for the parallel-chain and anti-parallel-chain models (Section A.I). The assignment of the band at 1676 cm^{-1} to the $\nu_{\parallel}(0, 0)$ mode of a parallel-chain arrangement relies heavily upon the assumption that the dichroism is opposite in character to the strong component of amide I at 1637 cm^{-1}. Without a proper analysis of the overlapping bands in this region of the spectrum (Chapter 5, Section C) it is doubtful whether this assumption can be justified. Should the dichroism of the weak band at 1676 cm^{-1} prove to be the same as the strong band at 1637 cm^{-1}, then the appropriate assignment would be to the $\nu_{\perp}(\pi, \pi)$ component of an antiparallel arrangement.

The unit cell of [Asp(nPr)]$_n$ in the cross-β form is two sheets thick in a direction perpendicular to the plane of the pleated sheet and the intersheet distance is consequently $\frac{1}{2}c \sin \beta = 12.47$ Å. Since 00l reflections with odd l are present in the diffraction pattern, the two sheets in the cell must have different structures.

V. POLY(S-CARBOBENZOXY-L-CYSTEINE)

Films prepared by casting [Cys(Cbz)]$_n$ from dimethylformamide or dichloroacetic acid solution have a poorly crystalline cross-β conformation and the observed X-ray reflections can be indexed on an orthogonal cell with $a = 4.76$ Å, $b = 6.95$ Å (chain axis), and $c = 32.4$ Å containing four residues (Elliott et al., 1964). No reflections were observed which required the a axis to be doubled, as expected for the antiparallel-chain pleated sheet and no component of the amide I band corresponding to the $\nu_{\parallel}(0, \pi)$ mode of an antiparallel-chain arrangement was detected in the infrared spectrum. Owing to the poor crystallinity and orientation in the specimens, neither the X-ray nor the infrared data were of sufficient quality to exclude the possibility that the chain arrangement was antiparallel or to distinguish, if they were present, between a parallel or a random arrangement.

It was found that specimens of much higher crystallinity could be obtained by moistening the polymer with dichloroacetic acid to produce a swollen gel and subjecting this to shear while drying. A considerable degree of double orientation was found to be present and the reflections in the X-ray diffraction pattern could be indexed on an orthogonal cell with $a = 4.89$ Å, $b = 6.89$ Å (chain axis), and $c = 32.8$ Å containing four residues. The differences between the unit-cell dimensions for the two methods of preparation far exceed experimental error and the two crystalline forms were distinguished by the symbols I and II. In the more crystalline preparation (form II) the a axis was again parallel to the direction of stroking, indicating a cross-β conformation, but the chain axes were found to be preferentially oriented perpendicular to the plane of the sheet. This is the reverse of the normal situation and can be correlated with the observation that the reflections associated with the repeat of structure normal to the sheet were much sharper than those associated with the a-axis repeat. Both observations are consistent with the suggestion that the crystallite dimension normal to the sheet was much greater than the dimension parallel to a.

The infrared spectra of the two crystalline forms were found to be quite different. The spectrum of form I had frequencies for the amide A, B, I, and II bands of 3286, 3070, 1634, and 1518 cm^{-1}, respectively, which are similar to those observed in other polypeptides in the β form. In form II, however, the bands were unusually sharp and the frequencies of the main components of the amide A, B, I, and II bands were 3362, 2975, 1641, and 1503 cm^{-1}, respectively. In addition a weak band was observed at 1678 cm^{-1} of opposite dichroism to that of the main component of amide I at 1641 cm^{-1}. This may be due either to the $\nu_{\parallel}(0, \pi)$ mode of an antiparallel-chain arrangement or to the $\nu_{\parallel}(0, 0)$ mode of a parallel-chain arrangement. The X-ray evidence on this point is also inconclusive since the fact that the unit cell is only one chain wide in the plane of the sheet could be due to a packing disorder similar to that observed in [Ala]$_n$ (Section B.II).

A noteworthy feature of the infrared spectrum obtained from form II is the high value of the amide A frequency compared with that found in form I and in other synthetic polypeptides in the β form. This can be correlated with the unusually large value of 4.89 Å for the interchain distance and with the habit of the crystallites which, as discussed earlier, appear to grow more rapidly parallel to the c axis rather than to the a axis.

VI. POLY(L-LYSINE) AND POLY(L-ARGININE)

Blout and Lenormant (1957) studied infrared spectra from oriented films of the hydrochloride of [Lys]$_n$ cast from aqueous solution and observed that the chain conformation underwent a spontaneous and reversible transformation from the α form to the β form when the films were dried. The transformation occurred without loss of orientation and in both conformations the chain axes were preferentially aligned parallel to the direction of stroking. Components of the amide I absorption band were observed at the frequencies expected for the $\nu_{\parallel}(0, \pi)$ and $\nu_{\perp}(\pi, 0)$ modes of an antiparallel-chain arrangement and the dichroisms were in the correct sense. Similar humidity-dependent changes in conformation were noted in all of the four possible hydro-halides of $[Lys]_n$ by Elliott *et al.* (1957a).

Shmueli and Traub (1965) investigated the crystalline structure of the hydrochloride of $[Lys]_n$ at various relative humidities by X-ray diffraction. Below 84% relative humidity the observed reflections could be indexed on an orthogonal cell which expanded laterally with increase in relative humidity. The dimensions were, depending upon the degree of hydration, $a = 4.62$–4.78 Å, $b = 15.2$–17.0 Å, and $c = 6.66$ Å (chain axis). No reflections were observed which required the a axis to be doubled as might be expected from the evidence obtained from infrared spectra for the presence of an antiparallel-chain arrangement. The diffraction data were not, however, of sufficient quality to enable one to determine whether this was due to the poor crystallinity and orienta-tion or to a disorder in sheet packing of the type observed in $[Ala]_n$ (Section B.II).

A plausible model was suggested for the structure in which the ions were arranged midway between the sheets with the negative ions belonging to one sheet opposite the positively charged side-chain groups belonging to the adjacent sheet and vice versa. The increase in the dimension of the b axis with relative humidity was attributed to a progressive hydration of the ions and the charged side-chain groups, with little change in the side-chain conformation. The breakdown of the β structure above 84% relative humidity was suggested to be due to an increased mobility of the chloride ions, leading to less effective screening of the mutual repulsion between the NH_3^+ groups. An interesting feature of this study is the small unit height of 3.33 Å, which is appreciably less than that observed in most synthetic polypeptides but similar to that observed in β-keratin (Chapter 16).

Suwalsky and Traub (1972) obtained X-ray diffraction patterns from oriented films of the hydrochloride of $[Arg]_n$ and found that specimens containing 5–20 water molecules per residue gave β patterns that could be indexed on a monoclinic unit cell with $a = 9.26$ Å, $b = 22.05$ Å, $c = 6.76$ Å (fiber axis), and $\gamma = 108.9°$. The cell dimensions did not vary appreciably with hydration, unlike the situation with the β form of the hydrochloride of $[Lys]_n$.

VII. POLY(L-GLUTAMIC ACID)

Lenormant *et al.* (1958) obtained infrared spectra from oriented specimens of the sodium salt of $[Glu]_n$ and found that reversible conformational changes occurred as the relative humidity was varied. Below a relative humidity of 70% the frequencies and dichroisms of the amide I and II components were consistent with the presence of an antiparallel-chain β conformation with the chain axes preferentially oriented parallel to the direction of stroking. At higher humidities a spontaneous transformation to an oriented α form was observed which could be reversed by drying. A similar but incomplete transformation was observed in oriented films of $[Glu]_n$.

Fibers of $[Glu]_n$ in the β form suitable for X-ray diffraction studies were prepared by Johnson (1959) and two types of pattern were obtained. In the first type six layer lines were observed and the repeat of structure parallel to the chain axes appeared to be 6.89 Å, leading to a unit height of $h = 3.44_5$ Å. Insufficient data were available, however, to allow the remainder of the unit-cell parameters to be determined. In the second type of pattern the equatorial reflections were better developed and were indexed on an orthogonal cell with $a = 4.69$ Å and $b = 12.50$ Å. The data from other layer lines were insufficient to fix the value of the chain axis repeat c.

Data were also obtained by Johnson (1959) on the sodium salt of $[Glu]_n$ and below 75% relative humidity the pattern obtained could be indexed on an orthogonal cell with $a = 4.66$ Å, $b = 12.8$–14.2 Å (depending upon the relative humidity), and $c = 6.76$ Å. No reflections were observed requiring the a axis to be doubled to the value expected for an antiparallel-chain arrangement.

Keith *et al.* (1969a,b) obtained X-ray powder data from microcrystals of the calcium salt of $[Glu]_n$ and derived values for the lengths of two of the reciprocal cell axes (a^* and c^*). No further data on the length of the third axis (b^*) or on the interaxial angles could be obtained directly from the observed diffraction rings. A unit cell was suggested on the basis of a series of assumptions regarding the internal structure of the crystal but further X-ray data would be required before this cell could be regarded as other than speculative. A detailed model was presented for the positions of the atoms in the assumed cell and its intensity transform was calculated. Since quantitative observed intensity data were not collected and only five $h0l$ reflections were resolved, the model cannot be tested on the basis of the presently available data.

Lamellar single crystals of the calcium salt of $[Glu]_n$ have been prepared (Keith *et al.*, 1969a) and on the basis of selected-area electron diffraction studies it was concluded that the chain axes were preferentially oriented perpendicular to the planes of the lamellae. Since the lamellae were estimated to be only 35–40 Å thick whereas the molecular length was about 3000 Å, crystal reentrant folds were assumed to be present at the surfaces of the lamellae. It was further suggested that for the folds to be reasonably compact the chain must loop back and forth in the same pleated sheet, thus leading naturally to an antiparallel arrangement.

VIII. Other Homopolypeptides

Amide bands assignable to the $v_\parallel(0, \pi)$ and $v_\perp(\pi, 0)$ modes of the antiparallel-chain arrangement have been observed in the infrared spectra of a number of synthetic homopolypeptides not discussed in earlier sections. These include [Ala(Me)]$_n$ (Itoh et al., 1969); [Cys(Bzl)]$_n$ (Fraser et al., 1965d); [Cys(CbzMe)]$_n$ (Ikeda et al., 1964); [Glu(Bzl)]$_n$ (Bradbury et al., 1960b; Masuda and Miyazawa, 1967); [Glu(Me)]$_n$ (Fraser et al., 1965b); [Glu(Na)]$_n$ (Fasman et al., 1970); [Lys]$_n$ (Fasman et al., 1970); [Ser]$_n$ (Koenig and Sutton, 1971); [Ser(Ac)]$_n$ (Fasman and Blout, 1960; Yahara and Imahori, 1963); [Ser(Bzl)]$_n$ (Bradbury et al., 1962b); and [Val]$_n$ (Fraser et al., 1965b; Koenig and Sutton, 1971). Similar bands appear in the spectrum of [Ala]$_6$ and thus it seems likely that an antiparallel arrangement is also present in this material (Sutton and Koenig, 1970).

C. SEQUENTIAL POLYPEPTIDES

Many insect silk fibroins contain considerable quantities of glycine, alanine, and serine and in some instances regular sequences are present in which every alternate residue is glycine (Chapter 13). Studies of sequential polypeptides containing these residues is therefore of considerable relevance to the conformation of silk fibroins and these are reviewed in this section.

I. Poly(L-alanylglycine)

Go et al. (1956) reported the synthesis of [Gly-Ala]$_n$ and found that the X-ray diffraction pattern and infrared spectrum of this material resembled those obtained from the crystalline portion of Bombyx mori silk fibroin. The related polymer [Ala-Gly]$_n$ was studied by Fraser et al. (1965a) and oriented specimens were obtained having sufficient crystallinity to enable the unit cell to be determined. The observed X-ray pattern could be indexed on an orthogonal cell with $a = 9.44$ Å, $b = 8.96$ Å, and $c = 6.94$ Å (Fraser et al., 1966) and the a axis was found to be oriented parallel to the direction of stroking. The structure was thus of the cross-β type and this was confirmed by the dichroism exhibited by the amide absorption bands in the infrared spectrum. A weak band at 1702 cm^{-1} of opposite dichroic character to the main amide I component at 1630 cm^{-1} was identified with the $\nu_\parallel(0, \pi)$ mode of an antiparallel-chain arrangement and this was consistent with the X-ray data which indicated that the unit cell was two chains wide in the a direction.

If a helix with unit twist $t = 180°$ is formed by the main-chain atoms in [Ala-Gly]$_n$, all the Gly residues project on one side of the yz plane (Fig. 10.7a) and all the Ala residues on the other. An antiparallel-chain pleated sheet with a two-chain repeat can be built up in one of two ways. In the first (Fig. 10.7b) equal numbers of Gly and Ala residues project on each side of the sheet, while in the second (Fig. 10.7c) all the Gly residues project on the same side of the sheet. The resulting sheets can be packed in many ways but for the second type of sheet there are two distinct possibilities. In the first the sheets pack with the Gly surface of one sheet in contact with the Ala surface of the adjacent sheet (Fig. 10.8, top left), while in the second the sheets pack with the Gly surfaces in contact and the Ala surfaces in contact (Fig. 10.8, bottom left). The latter arrangement was suggested by Marsh et al. (1955b) to be present in B. mori silk fibroin. The alternation in intersheet distance (Fig. 10.7c) would be expected to produce appreciable intensity in the 00l reflections with l odd compared with arrangements in which the intersheet distance was sensibly constant. Appreciable intensity was in fact observed for the 001 and 003 reflections in [Ala-Gly]$_n$ (Fraser et al., 1965a) and so it was concluded that all the Gly residues were on the same side of the sheet as in Fig. 10.7c and that the Gly surfaces of adjacent sheets were in contact as in Fig. 10.8, bottom left.

Although the disposition of the chains in the unit cell was not investigated in detail, the relative intensities of the 200 and 201 equatorial reflections were found to be consistent with a stagger of $a/4$ between adjacent sheets as suggested by Marsh et al. (1955b) for B. mori silk fibroin. The measured intensities of the 00l reflections were also used to estimate the two intersheet spacings (Fraser et al., 1965a) and in a subsequent refinement (Table 13.7b) the spacing across the Gly–Gly contact was estimated to be 3.79 Å and the spacing across the Ala–Ala contact to be 5.17 Å.

A second crystalline form of [Ala-Gly]$_n$ was prepared by Lotz and Keith (1971) in which the conformation of the polypeptide chain appears to be similar to that present in silk I (Chapter 13, Section B.II.b). The X-ray and electron diffraction patterns were indexed on an orthogonal unit cell with $a = 4.72$ Å, $b = 14.4$ Å, and $c = 9.6$ Å

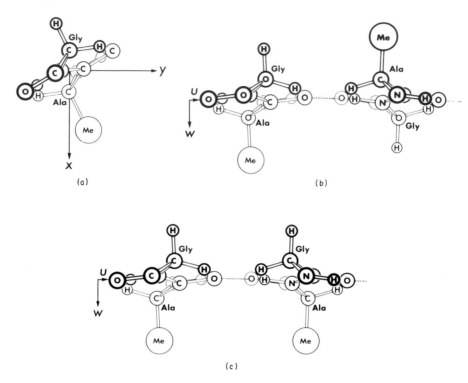

FIG. 10.7. (a) The z-axis projection of a helical polypeptide chain with unit twist $t = 180°$. In the case of the sequential polypeptide [Ala-Gly]$_n$ all the Ala residues project on one side of the yz plane and all the Gly residues project on the other. Antiparallel-chain pleated sheets may be formed (b) with equal numbers of Ala and Gly residues on each side or (c) with all the Ala residues on one side and all the Gly residues on the other.

(fiber axis). A model was proposed for the structure in which the chains were hydrogen bonded to form sheets in the ac plane, as in the pleated-sheet β models. However, the sheets in the silk I model are deeply puckered with four residues in the axial repeat of 9.6 Å. The asymmetric unit is a pair of residues and the conformation has helical symmetry with $h = 4.8$ Å and $t = 180°$. The torsion angles about the C$^\alpha$ atoms alternate between the values $\phi = -123°$, $\chi = 122°$ and $\phi = 57°$, $\chi = 72°$. These values are close to those for β conformations and for left-handed α-helices, respectively (Fig. 6.7a).

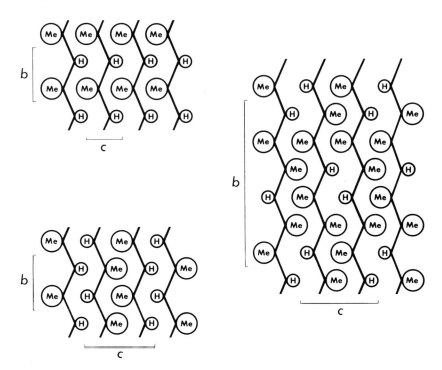

FIG. 10.8. Possible packing arrangements for pleated sheets of the type illustrated in Fig. 10.7c. In the arrangement shown at top left the unit cell is one sheet thick in the w direction, and in the arrangement shown at bottom left the cell is two sheets thick in the w direction. In the case of the polymer [Ala-Ala-Gly]$_n$ the packing arrangement shown at the right is believed to be present (Doyle *et al.*, 1970).

II. POLY(L-ALANYL-L-ALANYLGLYCINE)

Doyle *et al.* (1970) prepared the polymer [Ala-Ala-Gly]$_n$ and obtained films which exhibited a small degree of orientation. A weak band was observed in the infrared spectrum at 1700 cm^{-1} of opposite dichroic character to the main amide I component at 1630 cm^{-1} and it was concluded that an antiparallel-chain β conformation was present. The character of the dichroism in the amide bands indicated that the chain axes were preferentially oriented perpendicular to the direction of stroking as in the cross-β arrangement (Section A.II). X-ray diffraction data were also obtained from this polymer and the observed reflections were indexed on an orthogonal cell with $a = 9.44$ Å, $b = 6.98$ Å (chain axis), and $c = 9.96$ Å. No reflections were reported, however, that required $a = 9.44$ Å rather than half this value. As with [Ala-Gly]$_n$

(Section C.I), the strong intensity of the 201 reflection compared with the 200 reflection indicated that adjacent sheets were packed with a displacement of about $a/4$ as in the structure proposed for *B. mori* silk fibroin by Marsh *et al.* (1955b).

If the main-chain atoms in [Ala-Ala-Gly]$_n$ form a helix with unit twist $t = 180°$, the numbers of Ala residues will be the same on both sides of the yz plane (Fig. 10.7a) and the same will be true for Gly residues. Regardless of how the chains are arranged in the sheet, there will be equal numbers of Ala residues on each side of the sheet and so the intersheet distance will not be subject to the considerable alternation in value present in [Ala-Gly]$_n$ (Section C.I). This conclusion is in line with the fact that no strong 00l reflections with odd l were observed. Doyle *et al.* (1970) suggested that the sheets were packed in the manner indicated in Fig. 10.8(right) and if this is the case, the true b-axis repeat would be $3 \times 6.98 = 20.94$ Å and the cell used for indexing the observed reflections would be only a pseudocell. No reflections were observed corresponding to the larger cell and this is surprising in view of the considerable oscillation in electron density of period $3 \times 3.49 = 10.47$ Å parallel to b (Fig. 10.8, right). A possible explanation for this anomaly, which does not seem to have been considered, is that the chains in a sheet are staggered axially in some way so that the Gly residues do not all occur at the same axial levels.

The observed intersheet spacing of 4.98 Å is only slightly less than the values of 5.27 Å observed in [Ala]$_n$, and 5.17 Å estimated for the intersheet spacing across the Ala–Ala contacts in [Ala-Gly]$_n$ (Table 13.8). Doyle *et al.* (1970) suggest that the intersheet spacing in [Ala-Ala-Gly]$_n$ is determined primarily by Ala–Ala contacts.

III. POLY(L-ALANYLGLYCYL-L-ALANYLGLYCYL-L-SERYLGLYCINE)

The polymer [Ala-Gly-Ala-Gly-Ser-Gly]$_n$ was prepared by Fraser *et al.* (1966) as a model for the crystalline regions of *B. mori* silk fibroin and X-ray diffraction data were obtained which could be indexed on an orthogonal unit cell with $a = 9.39$ Å, $b = 6.85$ Å (chain axis), and $c = 9.05$ Å. The mean intersheet distance $c/2$ is somewhat greater than in [Ala-Gly]$_n$ and this can be attributed to the presence of an additional oxygen atom on every sixth residue. Since the repeat of structure along the chain is $6 \times 3.425 = 20.55$ Å, the observed cell only represents a pseudounit. The X-ray diffraction pattern closely resembled that obtained from the crystalline fraction of *B. mori* silk fibroin and this is discussed further in Chapter 13, Section B.II.

IV. OTHER SEQUENTIAL POLYPEPTIDES

Amide bands assignable to the $\nu_\parallel(0, \pi)$ and $\nu_\perp(\pi, 0)$ modes of the antiparallel-chain arrangement have been observed in the infrared spectra of a number of sequential polypeptides not discussed in earlier sections. These include: $[\{Glu(Me)\}_m\text{-Val-Glu(Me)}]_n$, $m = 0\text{--}3$ (Fraser et al., 1965b); $[\{Glu(Et)\}_m\text{-Cys(Bzl)-Glu(Et)}]_n$, $m = 0\text{--}3$ (Fraser et al., 1965d); $[\{Glu(Et)\}_m\text{-Gly}]_n$, $m = 1\text{--}4$ (Fraser et al., 1967a); $[\{Glu(Me)\}_m\text{--}Ser(Ac)\text{--Glu(Me)}]_n$, $m = 0\text{--}3$ (Fraser et al., 1965h, 1968a); and $[Gly\text{--Ala--Glu(Et)}]_n$ (Rippon et al., 1971).

D. SUMMARY

All the data which have been obtained from X-ray diffraction studies of synthetic polypeptides in the β form appear to be consistent with the pleated-sheet type of structure proposed by Pauling and Corey (1951g, 1953a). Only in the case of $[Ala]_n$, however, has it proved possible to test the model by the direct determination of the positions of the atoms in the unit cell (Arnott et al., 1967). The observed unit height of 3.445 Å in this material is very close to the value of 3.50 Å predicted for the antiparallel-chain, pleated-sheet model and so also is the interchain distance within the sheet, which was observed to be 4.73 Å and predicted to be 4.75 Å.

The pleated-sheet models were derived from a qualitative consideration of intrasheet potential energy and the position of the chain in the repeating unit was predicted to be such that $\Delta z = 0.04$ Å and $\Delta\phi = -25°$. In $[Ala]_n$ the determination of Δz was complicated by the high correlation, in the least squares refinement, between this parameter and intersheet parameters, but it seems clear that the position of the chain in $[Ala]_n$ is somewhat different from that predicted by Pauling and Corey (1953a). According to McGuire et al. (1971), the principal factor determining the packing arrangement is the interchain interactions and the values of Δz and $\Delta\phi$ in any particular polymer will be determined by both intrasheet and intersheet interactions. It cannot be assumed, therefore, that the values of Δz and $\Delta\phi$ determined for $[Ala]_n$ are necessarily applicable to other β structures.

In most cases, including $[Ala]_n$, no reflections are observed in the X-ray pattern that require the unit cell to be two chains wide in the plane of the sheet, as expected for the antiparallel-chain arrangement, but the infrared data indicate that the antiparallel arrangement is very common. In the case of $[Ala]_n$ it seems clear that the absence of reflections requiring the cell to be two chains wide is due to the statistical

TABLE 10.3

Observed Values of Unit Height, Interchain Distance, and Intersheet Distance in Synthetic Polypeptides in the Beta Form[a]

Symbol in Fig. 10.9	Material[b]	Unit height[c] h (Å)	Interchain distance (Å)	Intersheet[c] distance (Å)
1	$[Gly]_n$	$\geqslant 3.48$	4.77	3.44
2	$[Ala]_n$	3.44_5	4.73_4	5.26_7
3	$[Asp(nPr)]_n$	3.39_5	4.78_5	12.47
4	$[Glu(Me)]_n$	3.41_5	4.72_5	—
5	$[Glu(Me)]_n$	3.41_5	4.70	10.5
6	$[Cys(Cbz)]_n$ form I	3.47_5	4.76	16.2
7	$[Cys(Cbz)]_n$ form II	3.44_5	4.89	16.4
8	$[Lys(HCl)]_n$ 0% RH	3.33	4.62	15.2
9	$[Lys(HCl)]_n$ 84% RH	3.33	4.78	17.0
10	$[Arg(HCl)]_n$ hydrated	3.4	4.65	20.9
11	$[Glu]_n$ hydrated Na salt	3.38	4.66	12.8
12	$[Ala-Gly]_n$	3.47	4.72	4.48
13	$[Ala-Ala-Gly]_n$	3.49	4.72	4.98
14	$[Ala-Gly-Ala-Gly-Ser-Gly]_n$	3.42_5	4.69_5	4.52_5
15	$[Glu(Et)-Gly]_n$	3.47_5	4.64	8.98
16	$[Glu(Et)-Gly-Gly]_n$	3.45	4.66	7.35
17	$[Cys(Bzl)-Gly]_n$	—	4.67	9.10

[a] Derived from X-ray diffraction studies.

[b] See p. xvii for list of abbreviations. References: 1, Astbury (1949c), Bamford et al. (1953a); 2, Arnott et al. (1967); 3, Bradbury et al. (1960b); 4, Vainshtein and Tatarinova (1967a, b); 5, Brown and Trotter (1956); 6, 7, Elliott et al. (1964); 8–10, Shmueli and Traub (1965), Traub et al. (1967); 11, Johnson (1959); 12, Fraser et al. (1966); 13, Doyle et al. (1970); 14, Fraser et al. (1966); 15–17, Fraser et al. (1965h).

[c] Mean value in the case of sequential polypeptides.

nature of the sheet-packing arrangement, but in many other cases the orientation and crystallinity have not been good enough for such reflections to be observed even if they were present. Until diffraction patterns of sufficient quality have been obtained it would therefore be premature to conclude that a cell dimension one chain wide indicated a statistical sheet-packing arrangement. For example, reflections requiring the unit cell to be two chains wide in the plane of the sheet have been observed in $[Asp(nPr)]_n$ (Bradbury et al., 1960b) and $[Glu(Me)]_n$ (Vainshtein and Tatarinova, 1967a,b) and these could well have been missed had the crystallinity and orientation been lower. In $[Asp(nPr)]_n$ certain of the reflections requiring the cell to be two chains wide are rather diffuse and may indicate a degree of randomness in

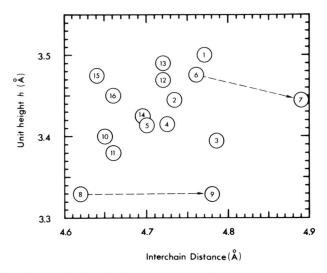

FIG. 10.9. Unit height h and distance between chains in the pleated sheet observed in synthetic polypeptides. A key to the individual polymers is given in Table 10.3 and the arrows indicate different crystalline modifications of the same polypeptide. No significant correlation exists between h and interchain distance.

chain sense, but apart from this there appears to be no evidence which unequivocally supports the presence of the parallel-chain pleated sheet in any of the synthetic polypeptides so far examined.

In the parallel-chain arrangement the hydrogen bonds cannot be formed satisfactorily unless the unit height is appreciably less than the value of 3.50 Å suggested for the antiparallel arrangement. On the basis of qualitative considerations of intrasheet potential energy, Pauling and Corey (1953a) suggested that a value of 3.25 Å was to be expected for the parallel-chain arrangement and intimated that the difference between the two values might be used to diagnose the chain arrangement. The values observed in a number of synthetic polypeptides are collected in Table 10.3 and these can be seen to range from 3.3 Å for $[Lys(HCl)]_n$ to 3.5 Å for $[Gly]_n$. In almost all cases, however, including $[Lys(HCl)]_n$, the infrared data indicate that an antiparallel-chain arrangement is present and it must be concluded that side chains play an important part in determining the unit height in the antiparallel arrangement and that the value of the unit height cannot be used to diagnose the chain arrangement.

As with α conformations (Chapter 9), the unit height in the β conformation appears to be affected both by the nature of the side chains and by the packing arrangement in the crystal. For example the unit height and the interchain and intersheet distances all differ between forms I and II of $[Cys(Cbz)]_n$ and the distance between chains in the sheet changes with hydration of the side chain in $[Lys(HCl)]_n$ (Fig. 10.9). The unit height and interchain distance do not show any significant correlation.

In a number of cases the X-ray diffraction data from homopolypeptides indicate that the unit cell contains two sheets which are crystallographically different, as, for example, in $[Asp(nPr)]_n$. This departure from equivalence is presumably dictated by the requirements of obtaining optimum side-chain packing between the sheets. In sequential polypeptides that contain an even number of residues in the repeating unit there exists the possibility that the arrangement of chains in the sheet, and of sheets in the crystal, is such as to produce two distinct types of sheet–sheet contacts which alternate regularly throughout the crystal (Fig. 10.8, bottom left). This arrangement, which is believed to be present in B. mori silk fibroin (Marsh et al., 1955b), also appears to be present in $[Ala–Gly]_n$ (Fraser et al., 1965a), $[Ala-Gly-Ala-Gly-Ser-Gly]_n$ (Fraser et al., 1966), and $[Cys(Bzl)–Gly]_n$ (Fraser et al., 1965h).

Sequential polypeptides are in many ways better models than homopolypeptides for the β conformation in proteins due to the variety of side-chain compositions in the natural material. A striking feature of the data obtained so far is that the X-ray diffraction data can be indexed on a pseudocell in which the cell sides in the plane of the sheet have lengths appropriate to a homopolypeptide rather than a sequential polypeptide. This can be interpreted as indicating that the equivalence of the atoms in the main-chain framework of the pleated sheet is little affected by the type of side chain. On the other hand, the mean value of the unit height and the interchain distance in the sheet are obviously affected by the nature of the side chains (Table 10.3).

Chapter 11

Poly(glycine) II, Poly(L-Proline) II, and Related Conformations

A. INTRODUCTION

Collagens contain appreciable amounts of glycine and proline and present-day knowledge of the chain conformation in these proteins is based largely on information which has been obtained from the study of $[Gly]_n$, $[Pro]_n$, and sequential polypeptides containing Gly, Pro, and other residues. The residues Pro and Hyp cannot be accommodated in either an α-helix or in a β conformation without disrupting the continuity of the helix (Pauling and Corey, 1951f) but in certain circumstances a third class of regular helical conformation may be formed which will accommodate the imino acid residues Pro and Hyp. Although the principal features of several of these conformations are known, it has not proved possible, on the basis of the X-ray diffraction data presently available, to refine the atomic coordinates in the manner which proved possible for $[Ala]_n$ in the α form (Chapter 9, Section B.I) and in the β form (Chapter 10, Section B.II).

I. REFINEMENT ON THE BASIS OF STEREOCHEMICAL CRITERIA

Due largely to the paucity of diffraction data on collagen and related glycine-containing and proline-containing polypeptides, extensive refinement of models has been carried out on the basis of stereochemical criteria. These procedures provide a valuable guide to the selection of possible conformations but their use for detailed refinement is questionable since the criteria used are essentially empirical and often arbitrary. In a number of instances the stereochemical criteria cannot all be met simultaneously and it is clear that a balance exists among the conflicting requirements. No entirely satisfactory method is available at present for calculating the free energies of the various possibilities and the final result of a refinement based on arbitrary stereochemical criteria is greatly influenced by the choice of criteria. Inevitably the results obtained are then open to the criticism of subjectivity. The concept of minimum allowable interatomic contact distances (Chapter 6, Section B) has proved useful in the initial selection of stereochemically feasible models but the extent to which short contacts classified as "extreme" (Table 6.2) can be tolerated in a model before its credibility becomes suspect is necessarily speculative. A similar difficulty exists with regard to hydrogen bonds in model structures, which are often classified as "satisfactory" even when the parameters are on the extreme limit of the observed range. In view of the data collected by Ramakrishnan and Prasad (1971), it is questionable whether $N-H \cdots O = C$ bonds for which $b(N, O)$ lies outside the range 2.75–3.05 Å or $\tau(H, N, O)$ exceeds 20° can be regarded as entirely "satisfactory." A critical discussion of the factors involved in hydrogen bond formation has been given by Donohue (1968), who concluded that, "A structure in which *all* of the hydrogen bonds are strained is highly unlikely." (From "Structural Chemistry and Molecular Biology," edited by Alexander Rich and Norman Davidson. N. H. Freeman and Co. Copyright © 1968.)

A further type of interaction about which no objective assessment appears to be possible at present is that of the so-called $CH \cdots O$ hydrogen bond. Ramachandran and co-workers (Sasisekharan, 1959a; Ramachandran and Sasisekharan, 1965; Ramachandran *et al.*, 1966a, 1967; Ramachandran, 1967; Ramachandran and Chandrasekharan, 1968) have interpreted the occurrence of certain unusually short $CH \cdots O$ distances in their model structures as being energetically favorable, due to the formation of a hydrogen bond. Sutor (1963) collected information on $C(H) \cdots O$ distances in molecular crystals and found that in a number of instances distances occurred which were less than the sum of the van der Waals radii of a methyl group and an oxygen atom. These occurred

> mainly in heterocyclic molecules and sometimes in compounds containing atoms of varying electronegativity with, frequently, a conjugated double-bond system or a series of double bonds. In all cases ... the carbon atom is directly attached to the heteroatom or more electronegative atom or group, or is part of a conjugated double bond system containing a heteroatom. Activation of the CH group is highly probable and thus the right conditions exist

for the formation of hydrogen bonds, but it remains to be shown whether or not hydrogen atoms are involved [p. 1106].

Donohue (1968) has also reviewed the occurrence of short CH···O contacts in molecular crystals and after a careful consideration of the available data concluded that there was no evidence for hydrogen bond formation.

Allerhand and Schleyer (1963) used infrared spectroscopy to investigate the possibility that CH groups act as proton donors in hydrogen bond formation. Compounds of the type $X—CH_2—Y$, where X, Y = Cl, Br, I, CN, showed evidence of hydrogen bonding to proton acceptors but they concluded that, "The evidence indicates that a single electron-withdrawing group attached to an sp^3-hybridized carbon is not sufficient to make the hydrogens attached to such a carbon proton donors. Methyl groups in any molecule do not hydrogen bond, nor do C—H groups in alkyl chains, nor in aldehydes, formates and formamides."[†] Thus there does not seem to be any evidence from the study of the crystal structures or spectra of small molecules to suggest that $C—H···O=C$ hydrogen bonds are likely to be present in polypeptides.

II. NOMENCLATURE

Models for the chain conformation in form II of $[Gly]_n$ (Section B) were suggested by Rich and Crick (1955) and for form II of $[Pro]_n$ (Section C) by Cowan and McGavin (1955) and these will be referred to as the PG II and PP II conformations, respectively. In both models the residues have helical symmetry with $h \simeq 3.1$ Å, with $t = \pm 120°$ in the case of $[Gly]_n$ and $t = -120°$ in the case of $[Pro]_n$. In certain sequential polypeptides containing Pro and Gly, conformations are found which may be regarded as hybrids of the left-handed PG II and the PP II conformations and these will be referred to as PG II–PP II hybrids.

In collagen and in certain sequential polypeptides containing glycine and imino acids the structural unit appears to be a three-strand rope in which the individual strands have a conformation which is basically that of the PG II–PP II type distorted into a right-handed coiled-coil (Chapter 1, Section A.II.e). Various models have been suggested for this three-strand rope structure but they are basically of three types, which are distinguishable on the basis of the topological arrangement of the interstrand hydrogen bonds. The genealogy of the three types is depicted in Fig. 11.1 and the differences in topology can be appreciated from a consideration of Fig. 11.3.

† Reprinted from *J. Amer. Chem. Soc.* **85,** 1715 (1963). Copyright 1963 by the American Chemical Society. Reprinted by permission of the copyright owner.

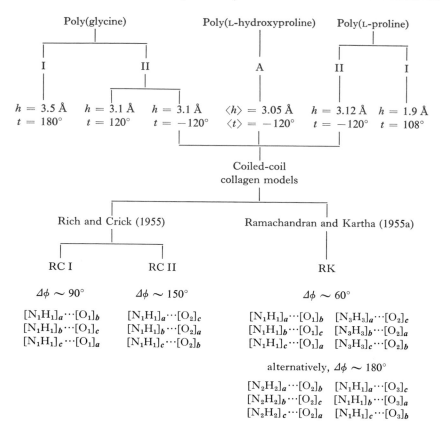

FIG 11.1. Genealogy of the three-strand rope models for collagen and collagenlike synthetic polypeptides. Each strand in the rope has a coiled-coil conformation derived from a simple helical conformation with $t = -120°$ and $h \simeq 3.1$ Å. Three types of model can be distinguished on the basis of the topological arrangement of hydrogen bonds between the three strands. These are illustrated in Fig. 11.3.

For a given choice of bond angles and distances atomic coordinates can be generated for a regular helical conformation with $h = 3.1$ Å and $t = -120°$ and the orientation of the chain fixed so that a C^α atom lies on the x axis of a coordinate system $Oxyz$ (Fig. 11.2a). Three such chains can then be packed together to form a three-chain unit with one chain lying on the u axis of the coordinate system uvw (Fig. 11.2b). In the actual rope structures the axes of the helices are then wound around the w axis so that they become coiled-coils but the coiling takes place in such a way that the relative positions of the atoms at any axial level are not greatly affected and the topological arrangement

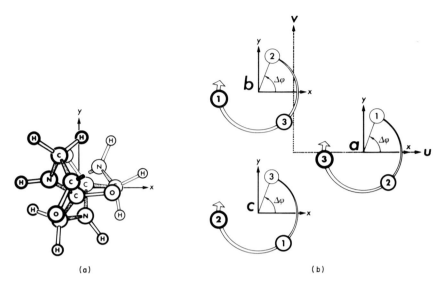

FIG. 11.2.   (a) Regular helical conformation of the polypeptide chain with $t = -120°$ and $h = 3.1$ Å. (b) Three-chain unit from which the coiled-coil models for collagen can be generated by the introduction of a right-handed twist about the w axis. The different models proposed can be approximated by an appropriate choice of $\Delta\phi$. The numbering of residues in different chains is such that residue 1 in chain a, residue 2 in chain b, and residue 3 in chain c are all at the same level.

of hydrogen bonds is unchanged. The three different types of arrangement can therefore be considered in terms of a three-chain unit without rope twist.

The models proposed by Rich and Crick (1955) can be approximated by rotating the individual chains in their local coordinate systems by an angle of $\Delta\phi \sim 90°$ for their model I and by $\Delta\phi \sim 150°$ for their model II. The residues are numbered 1, 2, 3, 1, 2, 3,... proceeding in the direction defined by $-NH-CHR-CO-$ and the chains are distinguished by the letters a, b, and c (Fig. 11.2b). The numbering of residues in different chains is such that residue 1 in chain a, residue 2 in chain b, and residue 3 in chain c occur at similar levels. It is convenient to consider the residue that provides the donor NH group as residue 1 and with this convention the hydrogen bonding in model I is of the type $[N_1H_1]_a\cdots[O_1]_b$ and in model II is of the type $[N_1H_1]_a\cdots[O_2]_c$ (Figs. 11.1 and 11.3). These will be referred to as the RC I and RC II types, respectively.

In the arrangement proposed by Ramachandran and Kartha (1955a), which will be referred to as the RK type, the relative positions of the atoms at any level can be approximated by rotating the individual chains in their local coordinate systems by $\Delta\phi \sim 60°$ (Fig. 11.2b). Two sets of systematic hydrogen bonds are then envisaged, the first set being of the type $[N_1H_1]_a\cdots[O_1]_b$ as in the RC I topology and the second of the type $[N_3H_3]_a\cdots[O_2]_c$ (Figs. 11.1 and 11.3).

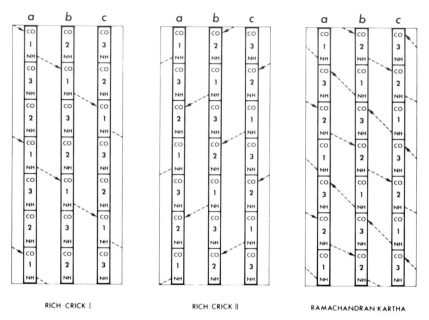

RICH CRICK I   RICH CRICK II   RAMACHANDRAN KARTHA

FIG. 11.3. Topological arrangement of hydrogen bonds in the collagen models proposed by Rich and Crick (1955) and Ramachandran and Kartha (1955a). In the RC I and RC II arrangements one hydrogen bond is formed per tripeptide unit, while in the RK arrangement two bonds are formed.

An alternative and equivalent method of reproducing the RK type of hydrogen bond topology is to take $\Delta\phi \sim (60° + 120°)$, that is, about 180°, and to renumber the residues so that the residue contributing the donor NH group for the second set of hydrogen bonds is numbered residue 1 (Fig. 11.1). In this case the relationship to the RC I type of arrangement is not so obvious and the two sets of bonds are now of the types $[N_2H_2]_a\cdots[O_2]_b$ and $[N_1H_1]_a\cdots[O_3]_c$. This choice is the one used by Ramachandran and co-workers and has the advantage that position 1 would normally be occupied by a Gly residue as in the RC I and RC II arrangements. Attention should also be drawn to the fact that Ramachandran and co-workers use the concept of a peptide unit as the repeating structure in a polypeptide chain (Chapter 6, Section A.I) so that the oxygen atom of residue 1 is termed O_1 but the nitrogen atom of residue 1 is termed N_3. Also it should be noted that while residues are numbered 1, 2, 3 in the same (positive) sense as defined earlier, the three chains in a rope are designated *A, B, C* in the *reverse* (negative) cyclic order by Ramachandran and co-workers.

The term "triple helical structure" is much used in literature on the conformation of collagen and on collagenlike conformations without any clear indication of whether rope twist is implied. For example, it has frequently been claimed that solution studies demonstrate that "a collagenlike triple helical structure is present" despite the fact that solution methods are not generally sensitive to the presence or absence of rope twist. Since the term is ambiguous, it will be avoided wherever possible. In the solid state it has generally been assumed that a meridional X-ray reflection of spacing 2.75–2.95 Å indicated the presence of rope twist while a spacing in the range 3.05–3.25 indicated that rope twist was not present. In the very few cases where structure analysis has proved to be possible the correlation has been verified.

B. POLY(GLYCINE) II

As mentioned in Chapter 10, Section B.I, two structural forms of $[Gly]_n$ have been observed (Meyer and Go, 1934). Form I has a β structure in which the chains have a helical conformation with $h \simeq 3.5$ Å and $t = 180°$ and are packed together in sheets stabilized by hydrogen bonds. In form II, which was obtained by precipitating the polymer from aqueous lithium bromide solution, the chains also have a helical conformation but $h \simeq 3.1$ Å, $t = \pm 120°$, and the packing is hexagonal, with a three-dimensional array of hydrogen bonds.

I. Diffraction Data

Bamford *et al.* (1955) obtained a powder X-ray diffraction pattern from form II of $[Gly]_n$ and noted the presence of an intense ring of spacing 4.15 Å together with a weaker ring of spacing 3.8 Å. A small amount of orientation was induced in a rolled specimen and the 4.15-Å reflection appeared to be equatorial. No further details of spacings or relative intensities were given but it was suggested that the chains might be packed in a hexagonal array. Padden and Keith (1965) obtained electron diffraction data from single crystals of form II of $[Gly]_n$ and reported reflections of spacing 4.16 Å (very strong), 3.12 Å (one weak, one strong), 2.45 Å (weak), and 1.61 Å (weak). These reflections were shown to index satisfactorily on a hexagonal unit cell with $a = 4.8$ Å and $c = 9.3$ Å.

II. MODEL STRUCTURES

Crick and Rich (1955) proposed a model for the structure for form II of $[Gly]_n$ which is illustrated in Fig. 11.4. All the chains are similarly

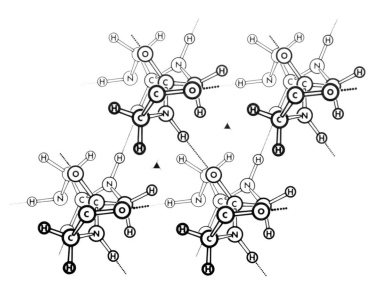

FIG. 11.4. Arrangement of chains in the crystal structure proposed for form II of $[Gly]_n$ by Crick and Rich (1955).

directed with respect to the sequence $-NH-CH_2-CO-$ and each has a threefold screw axis. The glycyl residue does not contain an asymmetric carbon atom and crystallites could be formed equally well from chains with $t = 120°$ or with $t = -120°$. In the case illustrated in Fig. 11.4 $t = -120°$ and so the helices are left-handed. The chains are packed in hexagonal array and each is hydrogen-bonded to its six nearest neighbors. The unit cell was assumed to have a dimension $a = 4.8$ Å perpendicular to the chain axes and the axial repeat of structure was $c = 9.3$ Å, so that the unit height is $h = 3.1$ Å.

Atomic coordinates are given in Table 11.1 for a regular helical polypeptide chain with $h = 3.1$ Å and $t = -120°$ in which the bond angles and distances have the "standard" values given in Table 6.1, $\tau(C^\alpha) = 109.47°$, and $\omega = 180°$. In the structure proposed by Crick and Rich (1955) the orientation of the chain (Fig. 11.2b) is such that the C^α atoms, and the NH and CO groups, all lie close to the planes containing the axes of nearest-neighbor chains (Fig. 11.4). The hydrogen

TABLE 11.1

Atomic Coordinates of a Regular Helical Conformation[a] for Poly(glycine)[b]

Atom	x (Å)	y (Å)	z (Å)	r (Å)	ϕ (deg)
N	0.724	−0.703	−1.170	1.009	44.160
H	1.131	−1.602	−1.331	1.961	54.780
C^α	1.269	0	0	1.269	0
$H^{\alpha 1}$	2.222	−0.464	0.294	2.270	11.783
$H^{\alpha 2}$	1.423	0.961	−0.230	1.717	−34.031
C′	0.276	−0.094	1.160	0.292	18.811
O	−0.419	−1.108	1.325	1.184	110.712

[a] $\phi = -76.89°$, $\psi = 145.32°$.
[b] Unit height of 3.1 Å and a unit twist of −120°, standard planar amide geometry (Table 6.1) and $\tau(N, C^\alpha, C') = 109.47°$.

bond length $b(N, O)$ calculated for the conformation given in Table 11.1 is 2.74 Å and $\tau(H, N, O) = 3.6°$.

Ramachandran *et al.* (1966a) suggested that the structure proposed by Crick and Rich (1955) could be refined by the introduction of a short CH⋯O contact, on the assumption that this would lead to stabilization through hydrogen bond formation. However, as discussed in Section A, studies of simple compounds do not appear to provide any justification for this assumption.

Although the chains in Fig. 11.4 arc all similarly directed with respect to the sequence $-NH-CH_2-CO-$ (up-chains), Crick and Rich (1955) noted that oppositely directed, or down-chains, could be accommodated in the proposed model and so either random or regular arrangements of up-chains and down-chains are possible. Meggy and Sikorski (1956) reported that a low-molecular-weight sample of $[Gly]_n$ formed single crystals of form II and Padden and Keith (1965), working with higher-molecular-weight samples, concluded that chain folding must be present in the single crystals which they observed. Krimm (1966) correlated the finding that up-chains and down-chains were present in the same crystallite, with the observation (Elliott and Malcolm, 1956a; Krimm, 1966) that form II of $[Gly]_n$ is readily converted by mild mechanical treatment to form I, in which the chain arrangement is known to be antiparallel (Chapter 10, Section B.I).

Ramachandran *et al.* (1967) carried out a detailed study of the interchain interactions in a crystallite of $[Gly]_n$ containing a mixture of up-chains and down-chains and confirmed the conclusion, reached earlier by Crick and Rich (1955), that satisfactory NH$\cdots$O hydrogen bonds could be formed between oppositely directed chains. A refinement was carried out in order to introduce short CH$\cdots$O contacts, again in the belief that this would lead to stabilization through hydrogen bonding, but as mentioned earlier in Section A, there does not appear to be any basis for this assumption. The possibility that regular arrangements of up-chains and down-chains occur in the crystallites was also discussed.

III. SPECTROSCOPIC STUDIES

Elliott and Malcolm (1956a) obtained infrared spectra from forms I and II of $[Gly]_n$ and noted that the frequencies of almost all the absorption bands were different for the two forms. An unusual and as yet unexplained feature of the spectrum of form II is the very considerable intensity of the amide B band (Chapter 5, Section A.I) which is usually very weak. Further studies of the infrared spectra of form II have been reported by Suzuki et al. (1966), Krimm et al. (1967), and Krimm and Kuroiwa (1968) and Raman spectra have been reported by Smith et al. (1969) and Small et al. (1970).

Krimm et al. (1967) assigned bands at 2850 and 2923 cm^{-1} in the spectrum of form I of $[Gly]_n$ to the symmetric and antisymmetric stretching modes, respectively, of the CH_2 group. In form II additional bands were observed at 2800 cm^{-1} and 2980 cm^{-1} and these were assigned to CH_2 groups in which one of the H atoms was involved in a CH$\cdots$O hydrogen bond. A band observed at 1418 cm^{-1} was assumed to be due to the bending mode of the bonded CH_2 groups while a weak shoulder at 1432 cm^{-1} was associated with the nonbonded groups.

It was concluded that the evidence supported the existence of CH$\cdots$O hydrogen bonds, but there are a number of difficulties in accepting this conclusion. First, the bands assigned to the "bonded" CH_2 stretching modes are weaker than those assigned to the "unbonded" modes, whereas the reverse order of intensity is observed in the bands assigned to the deformation modes. In fact the 1432-cm^{-1} band, which should be the stronger component, was not observed by Suzuki et al. (1966). Second, it was found that the frequencies of the "bonded" CH_2 bands could only be accounted for if it was assumed that one CH bond was weakened by the same amount as the other was strengthened, but when the temperature was lowered a different assumption had to be made to explain the observed frequency shifts (Krimm and Kuroiwa, 1968). The suggested assignment of the CH_2 stretching modes also appears to be inconsistent with the observed Raman spectrum (Small *et al.*, 1970). There are several unusual features of the infrared and Raman spectra of $[Gly]_n$ in both forms I and II in the range 2600–3100 cm^{-1} (Tsuboi, 1964; Suzuki *et al.*, 1966; Smith *et al.*, 1969; Small *et al.*, 1970) and until these are more fully understood the assignments proposed by Krimm *et al.* (1967) cannot be properly assessed. In the interim it would seem prudent to regard the spectroscopic demonstration of CH$\cdots$O hydrogen bonds in form II of $[Gly]_n$ as "not proven."

If the result obtained in Eq. (9.4) is applied to the spectrum of $[Gly]_n$ in form II using the values $\nu_A = 3303$ cm^{-1}, $\nu_B = 3086$ cm^{-1}, and $\nu_{II} = 1554$ cm^{-1} (Suzuki *et al.*, 1966), a value of $\nu_{NH} = 3281$ cm^{-1} is obtained, assuming that the interaction coefficients are negligible. From Fig. 9.6 this is seen to correspond to a hydrogen bond distance of $b(N, O) = 2.89$ Å, which is somewhat greater than the value of 2.76 Å originally suggested by Crick and Rich (1955).

C. POLY(L-PROLINE)

Two distinct forms of $[Pro]_n$ have been found which may be distinguished on the basis of the specific optical rotation in solution (Kurtz *et al.*, 1956, 1958). When the freshly synthesized polymer is dissolved in acetic acid it exhibits a small positive rotation and this has been designated form I (Blout and Fasman, 1958; Harrington and Sela, 1958). On standing, the rotation increases to a large negative value at a rate which depends upon the temperature (Steinberg *et al.*, 1960) to give form II. It was suggested (Kurtz *et al.*, 1958; Harrington and Sela, 1958) that the configuration about the C'N bond was cis ($\omega = 0°$) in form I and trans ($\omega = 180°$) in form II. The mutarotation was found to be reversible (Steinberg *et al.*, 1958, 1960) and Downie and Randall (1959) showed that the activation energy for the transition between the two forms was 22.9 kcal mole^{-1}, as expected for a cis–trans change. The conformational transition has been studied in considerable detail (Katchalski *et al.*, 1963; Carver and Blout, 1967; Veis *et al.*, 1967b; Mandelkern, 1967; Engel, 1967; Torchia and Bovey, 1971; Winklmair *et al.*, 1971). X-ray diffraction data on form I have been reported by Sasisekharan (1960) and Traub and Shmueli (1963a,b) and the latter authors proposed a helical model for this conformation in which the unit height is $h = 1.90$ Å, the unit twist is $t = -108°$, and the amide group is in the cis configuration, that is to say, the torsion angle is $\omega = 0°$.

Cowan and McGavin (1955) obtained X-ray diffraction data from oriented films of form II of $[Pro]_n$ and found that the observed reflections could be indexed on a hexagonal cell with $a = 6.62$ Å and $c = 9.36$ Å (chain axis). No meridional reflections were observed on layer lines with $l = 1, 2, 4,$ or 5 and this was assumed to be due to the presence of a threefold screw axis parallel to c. The observed density was found to be consistent with the presence of three residues per unit cell and it was concluded that the chain conformation was helical with unit height $h = 3.12$ Å and $t = \pm 120°$. Various models with these helical parameters were considered on the basis of the stereochemical criteria given by Pauling and Corey (Chapter 6). Both left-handed and right-handed helices with the cis configuration about the C'N bond ($\omega = 0°$) were found to be unsatisfactory, as were right-handed helices with the trans configuration about the C'N bond ($\omega = 180°$). In the model eventually arrived at (Fig. 11.5) the helix was left-handed ($t = -120°$)

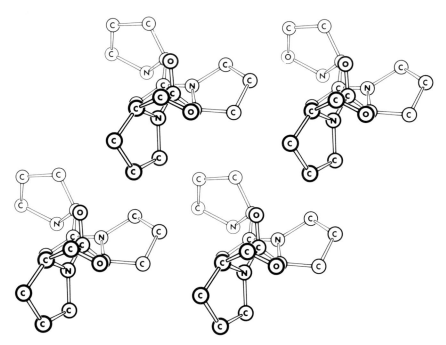

Fig. 11.5. Arrangement of chains in the crystal structure proposed for form II of [Pro]$_n$ by Sasisekharan (1959a).

and the configuration about the C'N bond was trans. Atomic coordinates for this model are given in Table 11.2 and it can be seen that the main-chain atoms occupy positions very similar to those in the PG II conformation (Table 11.1).

Sasisekharan (1959a) remeasured the diffraction pattern of form II of [Pro]$_n$ and suggested that the a dimension of the unit cell was 6.68 Å rather than 6.62 Å. Qualitative intensity data were collected and used as a basis for finding the orientation of the chain in the unit cell. Coordinates were generated for a chain in a reference frame $Oxyz$ similar to that defined in Section A and are given in Table 11.2. These

TABLE 11.2

Chain Conformation in Helical Models[a] for Form II of Poly(L-proline)

Atom	x (Å)	y (Å)	z (Å)	r (Å)	ϕ (deg)
		(a) Cowan and McGavin (1955)			
N	0.73	0.70	−1.17	1.01	44.0
C^α	1.30	0	0	1.30	0
C^β	2.57	0.76	0.32	2.68	16.5
C^γ	2.32	2.16	−0.26	3.17	43.0
C^δ	1.33	2.05	−1.37	2.44	57.0
C′	0.31	0.07	1.14	0.32	12.5
O	−0.38	1.07	1.28	1.14	109.5
		(b) Sasisekharan (1959a)			
N	0.72	0.71	−1.18	1.01	44.5
C^α	1.25	0	0	1.25	0
C^β	2.57	0.66	0.32	2.65	14.5
C^γ	2.48	2.01	−0.34	3.19	39.0
C^δ	1.40	2.04	−1.42	2.47	55.5
C′	0.25	0.09	1.16	0.27	20.5
O	−0.45	1.09	1.33	1.18	112.5

[a] Unit height $h = 3.12$ Å, unit twist $t = -120°$.

differ somewhat from the coordinates given by Cowan and McGavin (1955). A series of orientations with respect to the crystal axes was examined but none was found to give completely satisfactory agreement with the observed intensities, nor was any found which did not involve short interatomic contacts. In a later study Arnott and Dover (1968) obtained values of $a = 6.62$ Å and $c = 9.31$ Å and carried out a refinement of the model on the basis of powder diffraction data. A crystal model in which all chains had the same direction was found to give better agreement than one with a random distribution of chain directions.

Films cast from aged solutions of [Pro]$_n$ in formic acid were shown by Andries and Walton (1969) to contain spherulites and the conformation was found, by electron diffraction studies, to be that of form II. The spherulites contained lamellae about 150 Å in thickness and the diffraction data indicated that the chain axes were normal to the plane of the lamellae. Since the molecular length in the case of one sample of polymer was about 815 Å, it was concluded that the chains in the lamellae must be folded. No account was taken in earlier studies (Cowan and McGavin, 1955; Sasisekharan, 1959a) of the possible occurrence of both up-chains and down-chains in the same crystallite.

Studies of interatomic contact distances and intrachain potential energy have shown that the ranges of ϕ and ψ for which regular helical structures of $[Pro]_n$ can be formed are very limited (Steinberg *et al.*, 1960; De Santis *et al.*, 1965; Leach *et al.*, 1966a; Schimmel and Flory, 1967; Gō and Scheraga, 1970). The torsion angle ϕ about the $N-C^\alpha$ bond is limited by the near-rigid geometry of the pyrrolidine ring to a value of about $-70°$ (Leung and Marsh, 1958; Ueki *et al.*, 1969; Balasubramanian *et al.*, 1971; Matsuzaki and Iitaka, 1971; Nishikawa and Ooi, 1971) and in form II the only remaining variable, assuming $\omega = 180°$ and fixed bond angles and distances, is the torsion angle ψ about the $C^\alpha-C'$ bond.

Calculations of intrachain potential energy for various values of ψ for $[Pro]_2$ or $[Pro]_n$ have been reported by De Santis *et al.* (1965), Schimmel and Flory (1967), and Gō and Scheraga (1970). The results differ, due to different assumptions made about the chain geometry and about the nature of the nonbonded potential functions, but there is general agreement that outside the range $100° < \psi < 190°$ significant destabilization of regular helical conformations is likely through steric interference. The values of ψ in the conformations described by Cowan and McGavin (1955), Sasisekharan (1959a), and Arnott and Dover (1968) are, respectively, 145.1°, 145.9°, and 84.9°.

A knowledge of the chain conformation in form II of $[Pro]_n$ is important because of its bearing on the structure of collagen. The presently available powder diffraction data are well accounted for by the model proposed by Arnott and Dover (1968) but fiber diffraction data are desirable for a more complete refinement. In view of the evidence for chain folding, it will also be necessary to elucidate the nature of the distribution of up-chains and down-chains in the crystallites.

D. POLY(L-HYDROXYPROLINE)

Sasisekharan (1959b) studied the conformation of $[Hyp]_n$ and found two distinct forms, termed A and B, which could be distinguished on the basis of their X-ray pattern and optical activity. The two forms were found to be interconvertible by treatment with appropriate solvents and the X-ray diffraction pattern of form A, which was highly crystalline, was studied in detail. The reflections from a powder specimen were indexed on a hexagonal cell with $a = 12.3$ Å and $c = 9.15$ Å (chain axis) and the indexing was confirmed by measurements on a specimen showing a small degree of orientation.

The a axis in form A of $[Hyp]_n$ is appreciably greater than in $[Pro]_n$ and a consideration of the observed density indicated that three chains pass through the unit cell. It was concluded from a detailed study of the diffraction pattern that the three residues included in the 9.15 Å length of an individual chain were not equivalent, that is, the chains did not possess helical symmetry. The three chains in the cell were, however, related by a threefold screw axis.

A model for the structure of form A of $[Hyp]_n$ was devised by fixing the bond angles and distances at predetermined values and examining interatomic contact distances and the possibility of hydrogen bond formation for different chain conformations and orientations. Although the individual Hyp residues in a single chain are not equivalent, the arrangement approximates to a left-handed helix with unit height $h = 3.05$ Å and unit twist $t = -120°$. It was claimed that "there are practically no short contacts of any sort" but details were not given. Short interchain OH$\cdots$O contacts of 2.50, 2.53, and 2.56 Å were attributed to hydrogen bond formation but these distances are somewhat smaller than is usual for OH$\cdots$O hydrogen bonds (Donohue, 1968).

The intensity transform for the model was calculated and compared with a qualitative assessment of the observed intensity data. There were several discrepancies and the model appears to require further refinement and testing on the basis of quantitative intensity data. As with form II of $[Pro]_n$, this will probably be complicated by the need to establish whether the crystallites contain a mixture of up-chains and down-chains and, if so, to devise a means of allowing for the effect of this on the calculated intensity transform.

E. POLYTRIPEPTIDES

Two features of the composition of collagen which appear to play an important part in determining the conformation of the polypeptide chain are the presence of the imino acid residues Pro and Hyp and the occurrence of Gly in every third position in the sequence (Chapter 14). In this section studies of polytripeptides containing Gly together with Pro or Hyp residues are described. It is convenient to regard Gly as the leading residue of the triplet and in the present context the polymers will be described in this way even though the actual polymer was synthesized from a different tripeptide. For example, [Pro-Ala-Gly]$_n$ would be referred to as [Gly-Pro-Ala]$_n$ and [Ala-Pro-Gly]$_n$ would be referred to as [Gly-Ala-Pro]$_n$. In this way the structural relationships between the various polymers and collagen can more readily be appreciated.

I. POLY(GLYCYLGLYCYL-L-PROLINE)

X-ray diffraction data were obtained from powder specimens of [Gly-Gly-Pro]$_n$ (Traub and Yonath, 1966; Traub, 1969) and the observed rings were found to index satisfactorily on an orthogonal unit cell with $a = 12.2$ Å, $b = 4.9$ Å, and $c = 9.3$ Å (chain axis) containing

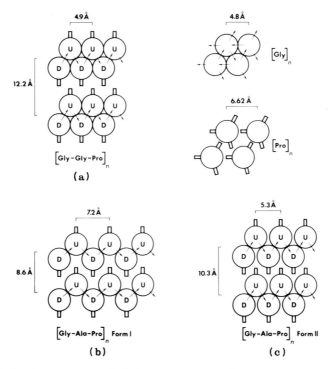

FIG. 11.6. Arrangement of chains in the crystal structures proposed (a) for [Gly-Gly-Pro]$_n$ by Traub (1969), (b) for form I of [Gly-Ala-Pro]$_n$ by Segal and Traub (1969), and (c) for form II of [Gly-Ala-Pro]$_n$ by Doyle et al. (1971). The disposition of interchain hydrogen bonds is indicated in a schematic way by arrows and the disposition of prolyl residues by bars.

two tripeptide units and two water molecules. A model structure was developed based on the assumption that the chain conformation was a hybrid between the PG II and PP II conformations (Section A). The main-chain atoms in the hybrid conformation form a left-handed helix but the Pro residues occur directly above one another, and so the complete chain is highly asymmetric. A schematic illustration of the proposed packing arrangement is given in Fig. 11.6a and it can be seen that the chains are packed together in sheets stabilized by the formation of interchain hydrogen bonds. Each sheet contains equal numbers of up-chains and down-chains arranged in a regular manner. The hydrogen bonds between similarly directed chains have $b(N_1, O_3) = 2.85$ Å and those between oppositely directed chains have $b(N_2, O_2) = 2.75$ Å. It was found that a water molecule could be located so as to form a hydrogen bond with the otherwise nonbonded C_1O_1 groups.

Support for the general correctness of the proposed arrangement was obtained from the observation that a solvated phase could be obtained with formic acid in which the unit cell was monoclinic ($\beta = 94°$) and the b and c axes had similar dimensions to the hydrated form but a was increased from 12.2 to 13.5 Å. This was presumed to indicate that the double-layered sheets became slightly staggered and that the distance between them increased as a result of the incorporation of formic acid in the vicinity of the otherwise nonbonded C_1O_1 group.

The intensity transform predicted from the model was calculated and shown to be in general agreement with a qualitative assessment of the observed intensities. However, insufficient data were collected to provide a searching test of the model.

II. POLY(GLYCYL-L-ALANYL-L-PROLINE)

The structure suggested by Traub (1969) for $[Gly\text{-}Gly\text{-}Pro]_n$ is stereochemically impossible if either of the Gly residues is replaced by a residue with a β carbon atom. The effect of introducing an Ala residue into the second position was investigated by Segal and Traub (1969) and oriented films of $[Gly\text{-}Ala\text{-}Pro]_n$ obtained by air-drying aqueous suspensions were examined by X-ray diffraction. The reflections were indexed on an orthogonal cell with $a = 8.6$ Å, $b = 7.2$ Å, and $c = 9.4$ Å (chain axis) containing two tripeptide units and 1.9–2.4 water molecules. A survey of possible structures was carried out based on the assumption that the chain conformation was a hybrid between the PG II and PP II conformations (Section A) and the model arrived at is shown schematically in Fig. 11.6b. Up-chains and down-chains alternate regularly in the sheets and only one hydrogen bond is formed per tripeptide unit between the N_1H_1 group of the Gly residue and the O_2 atom of the Ala residue in an oppositely directed chain. The hydrogen bond length $b(N_1, O_2) = 2.9$ Å and $\tau(H_1, N_1, O_2) = 12°$. It was also suggested that a water molecule might form hydrogen bonds simultaneously with the O_1 atom of a Gly residue and the N_2 atom of an Ala residue in similarly directed chains.

The intensity transform of the proposed model was calculated and compared with a qualitative assessment of the observed intensities. Some discrepancies are apparent and it is clear that further refinement of the model is required.

Segal and Traub (1969) found that a second crystalline form of [Gly-Ala-Pro]$_n$ could be prepared by treating the first form with acetone or dioxane after drying *in vacuo*. Schwartz *et al.* (1970) obtained electron diffraction patterns from this form and indexed the observed reflections on a hexagonal unit cell with $a = 12.5$ Å and $c = 9.4$ Å (chain axis). It was suggested on the basis of the c-axis repeat that the chain conformation might be of the PG II–PP II type. Doyle *et al.* (1971) also studied this crystalline form and suggested that the cell derived by Schwartz *et al.* was incorrect. The alternative cell was orthogonal with $a = 10.30$ Å, $b = 5.27$ Å, and $c = 9.40$ Å (chain axis) containing two tripeptide units. It was suggested that the conformation was of the PG II–PP II type and that the chains were packed in sheets in the manner shown in Fig. 11.6c.

A third crystalline form of [Gly-Ala-Pro]$_n$ was prepared by Doyle *et al.* (1971) from a solution of the polymer in trifluoroethanol. The X-ray diffraction pattern showed a number of similarities both in spacing and intensity with the pattern obtained from collagen including a meridional reflection at 2.85 Å. It was concluded that collagenlike three-strand ropes were present in this crystalline form.

III. POLY(GLYCYL-L-PROLYL-L-ALANINE)

The polymer [Gly-Pro-Ala]$_n$ has been shown to exist in two structural forms which may be distinguished on the basis of their X-ray diffraction patterns (Traub and Yonath, 1966, 1967). In one form, prepared from aqueous solution, the pattern can be indexed on a hexagonal cell with $a = 11.4$ Å and $c = 9.3$ Å (chain axis). Possible structures which were consistent with the unit-cell dimensions were examined systematically and it was concluded that the chain conformation was basically a PG II–PP II hybrid. Three chains pass through the cell and these were supposed to be bonded together through the formation of a system of hydrogen bonds which were topologically equivalent either to those in the RC I or to those in the RC II three-strand rope model (Section A). The RC II type of bonding was preferred but the alternative possibility could not be excluded.

A second type of diffraction pattern, obtained from thoroughly dried polymer, was more diffuse but exhibited a close resemblance to the pattern obtained from collagen. It was concluded that in the dry state, [Gly-Pro-Ala]$_n$ has a three-strand rope structure, similar to that believed to be present in collagen, that uncoils upon the addition of water. Related structural modifications had been reported earlier for the polymer [Gly-Pro-Hyp]$_n$ by Andreeva and co-workers (Andreeva *et al.*, 1963; Andreeva and Millionova, 1964).

IV. POLY(GLYCYL-L-PROLYL-L-PROLINE)

The polymer [Gly-Pro-Pro]$_n$ yields an X-ray diffraction pattern which closely resembles that obtained from collagen and considerable progress has been made in the elucidation of its molecular structure (Shibnev *et al.*, 1965; Traub and Yonath, 1966; Traub *et al.*, 1969; Yonath and Traub, 1969). The pattern is somewhat sharper than that obtained from unstretched collagen and can be indexed on a hexagonal unit cell with an a-axis dimension which varies from 12.5 to 13.6 Å depending upon the degree of hydration. When equilibrated at 52% relative humidity the polymer was found to contain three water molecules per tripeptide unit and after prolonged drying this number was reduced to one (Traub and Yonath, 1966). Very little change was observed in the c-axis dimension, which was determined to be 28.7 ± 0.4 Å in the dried material (Yonath and Traub, 1969).

The pattern was interpreted as indicating the presence of a three-strand rope structure in which the strands had a coiled-coil conformation derived from a PG II–PP II hybrid. The individual strands were supposed to follow a right-handed helix of pitch $P = 3 \times 28.7 = 86.1$ Å (Fig. 11.7) and with a unit height for the Gly-Pro-Pro triplet of $h = 8.61$ Å and a unit twist $t = 36°$.

If the three strands were similarly directed with respect to the sequence $-N-C^\alpha-C'-$ and staggered in the manner shown in Fig. 11.7, the rope would have the symmetry of a left-handed helix with $c = 28.7$ Å, $P = c/3 = 9.6$ Å, a height for the [Gly]$_a$, [Pro]$_b$, [Pro]$_c$ unit of $h = 2.87$ Å, and a unit twist $t = -108°$. If, on the other hand, the rope contained a mixture of up-chains and down-chains, the helical symmetry would be the same as that of an individual chain with $c = P = 86.1$ Å, $h = 8.61$ Å, and $t = 36°$. No evidence for the larger value of c was found and it was therefore assumed that all the

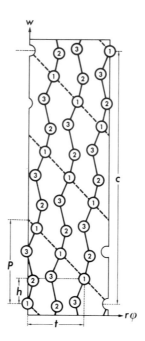

FIG. 11.7. Schematic radial projection illustrating the disposition of residues in the three-strand-rope model for [Gly-Pro-Pro]$_n$ proposed by Yonath and Traub (1969). Position 1 is occupied by Gly and positions 2 and 3 by Pro. The basic helix (Chapter 1, Section A.II.c) is indicated by a broken line and is left-handed with ten tripeptide units in three turns.

chains were similarly directed and staggered in the manner shown in Fig. 11.7.

Detailed interpretation of the diffraction pattern obtained from [Gly-Pro-Pro]$_n$ was complicated by the presence of screw disorder in the packing of the three-strand ropes in the crystallites (Chapter 1, Section A.IV.d). Sharp reflections were observed on $l = 0$ and 3, and from Eq. (1.56) it follows that the pitch of the screw disorder $P_s = nc/3$, where n is the order of the Bessel function involved. From Eq. (1.23) this can be calculated to be $n = -1$ and so $P_s = -9.6$ Å. This is the same as the pitch of the basic helix (Chapter 1, Section A.II.c). The intensities of the spots and streaks were reduced to a common scale by the methods outlined in Chapter 1, Section C.II.

A model for the crystal structure of [Gly-Pro-Pro]$_n$ was developed by a systematic exploration of possible conformations and the elimination of those that were stereochemically unacceptable. Further selection was carried out on the basis of the possibility of hydrogen bond formation and a comparison of the observed and calculated intensities. The structure that best satisfied all three criteria was further refined to give the model illustrated in Fig. 11.8. Atomic coordinates for this model are given in Table 11.3. The torsion angle ψ_2 about the $C_2{}^\alpha C_2'$ bond between the two prolyl residues has a value of 127°, which is in the expected range (Section C). The hydrogen bonding is topologically similar to that of the RC II model (Section A) but the atomic positions are appreciably different and the hydrogen bond is longer, with $b(N_1, O_2) = 2.96$ Å and $\tau(H_1, N_1, O_2) = 9°$. The diffraction patterns from which the intensity data were collected were obtained from specimens with about two water molecules per tripeptide unit and various possible positions for water in the structure were considered. In the suggested arrangement (Table 11.3) one water molecule makes hydrogen bonds to O_1 and N_1 and the other to O_2 and O_3 and they are also hydrogen-bonded to each other. These five hydrogen bond lengths are all between 2.64 and 2.90 Å.

The torsion angles ϕ_2 and ϕ_3 about the NC^α bonds in the pyrrolidine rings have values of -76 and $-45°$. As discussed earlier, very little variation in this angle is possible and these values are close to the permissible limits (Yonath and Traub, 1969; Hopfinger and Walton, 1970b).

A comparison was made between the observed intensity data and the intensity transforms calculated for the model structure, for the original RC I and RC II collagen models, and for an alternative model with $N_1H_1\cdots O_3$ hydrogen bonds. The agreement with the very limited intensity data collected was generally better for the model depicted in Fig. 11.8. However, the comparison was not entirely valid as the effect on the transform of incorporating water molecules into the RC I and RC II models does not seem to have been explored. The model with $N_1H_1\cdots O_3$ hydrogen bonds was found to be unlikely since $\tau(H_1, N_1, O_3)$ was about 40° and the C_2O_2 group was directed toward the helix axis where it was not available for hydrogen bonding to water.

Extensive studies of the solution properties of [Gly-Pro-Pro]$_n$ have been reported (Engel et al., 1966; Sakakibara et al., 1968; Kobayashi et al., 1970) and the evidence obtained indicates that ordered helical aggregates occur in solution under certain conditions. These aggregates, which can be destroyed by heating, have generally been assumed to be three-strand ropes similar to those found in solid films although the methods available do not enable this to be established directly.

Olsen et al. (1971) observed that [Pro-Pro-Gly]$_{10}$ and [Pro-Pro-Gly]$_{20}$, when dried from acetic acid solution, formed microcrystals in which one dimension was relatively constant, at about 90 Å and 180 Å, respectively. Since these are close to the values expected for the length of a molecule having the conformation described by Yonath and Traub (1969), it was concluded that the crystallites were formed by the lateral aggregation of three-strand rope structures of this type. Support for this conclusion was obtained from the appearance, in the case of [Pro-Pro-Gly]$_{20}$, of rodlike particles of about 15×180 Å which were presumed to be individual three-strand ropes. Positive staining of the amino or the carboxyl end groups showed that both types of end group occurred at the surfaces separated by a constant distance. On this basis it was assumed that the three-strand ropes were formed from similarly directed chains packed in an antiparallel arrangement but the alternative possibility that the individual ropes contained a mixture of oppositely directed chains was not specifically discussed.

Berg *et al.* (1970) attempted to distinguish between these two

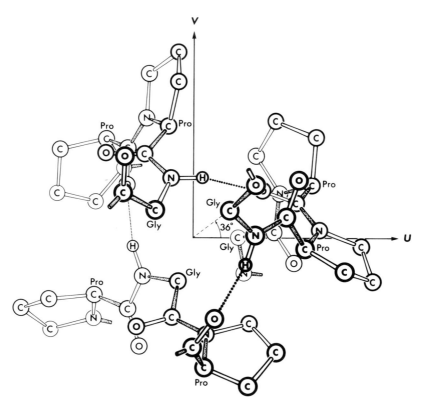

Fig. 11.8.   Three-strand rope model for [Gly-Pro-Pro]$_n$ proposed by Yonath and Traub (1969). The hydrogen bonding topology is of the RC II type (Fig. 11.3).

TABLE 11.3

Atomic Coordinates for the Chain Conformation in a Model for
Poly(glycyl-L-prolyl-L-proline)[a]

Residue	Atom	u (Å)	v (Å)	w (Å)	r (Å)	ϕ (deg)
Gly	N_1	1.65	−1.11	−0.96	1.99	−33.8
	H_1	0.86	−1.70	−1.08	1.91	−63.1
	C_1^α	1.46	0.00	0.00	1.46	0.0
	$H_1^{\alpha 1}$	0.58	−0.10	0.46	0.59	−10.0
	$H_1^{\alpha 2}$	1.33	0.85	−0.49	1.59	32.7
	C_1'	2.59	0.16	1.03	2.59	3.5
	O_1	3.23	−0.80	1.44	3.33	−14.0
Pro	N_2	2.76	1.38	1.56	3.09	26.6
	C_2^α	3.79	1.64	2.60	4.13	23.4
	H_2^α	4.65	1.20	2.34	4.80	14.5
	C_2^β	3.92	3.15	2.63	5.03	38.8
	$H_2^{\beta 1}$	4.32	3.74	3.38	5.72	39.5
	$H_2^{\beta 2}$	4.57	3.27	1.88	5.64	35.1
	C_2^γ	2.63	3.66	2.01	4.51	54.4
	$H_2^{\gamma 1}$	1.99	3.93	2.74	4.40	63.2
	$H_2^{\gamma 2}$	2.82	4.48	1.34	5.29	57.8
	C_2^δ	2.01	2.57	1.14	3.26	51.9
	$H_2^{\delta 1}$	2.19	2.75	0.21	3.52	51.4
	$H_2^{\delta 2}$	1.04	2.46	1.47	2.67	67.1
	C_2'	3.31	1.08	3.94	3.48	18.2
	O_2	2.21	1.47	4.36	2.66	33.6
	N_3	4.03	0.23	4.66	4.04	3.3
	C_3^α	3.58	−0.35	5.94	3.60	−5.6
	H_3^α	2.95	−1.11	5.74	3.15	−20.6
	C_3^β	4.83	−1.08	6.37	4.95	−12.6
	$H_3^{\beta 1}$	5.33	−0.53	7.14	5.36	−5.7
	$H_3^{\beta 2}$	4.59	−1.94	6.69	4.98	−22.9
	C_3^γ	5.72	−1.37	5.17	5.88	−13.5
	$H_3^{\gamma 1}$	6.68	−1.29	5.46	6.80	−10.9
	$H_3^{\gamma 2}$	5.48	−2.27	4.76	5.93	−22.5
	C_3^δ	5.35	−0.28	4.16	5.36	−3.0
	$H_3^{\delta 1}$	5.27	−0.67	4.20	5.31	7.2
	$H_3^{\delta 2}$	6.04	0.46	3.24	6.06	4.4
	C_3'	2.97	0.63	6.92	3.04	12.0
	O_3	3.37	1.80	7.02	3.82	28.2
$(H_2O)_1$	O	3.58	−1.46	9.09	3.87	−22.2
$(H_2O)_2$	O	2.86	−3.92	7.98	4.85	−53.9

[a] Data given by Yonath and Traub (1969). $\phi_1 = -51°$, $\psi_1 = 153°$; $\phi_2 = -76°$, $\psi_2 = 127°$; $\phi_3 = -45°$, $\psi_3 = 148°$; atoms of the next triplet in this chain are at $(r, \phi + 36°, w + 8.61$ Å$)$. Atoms in the corresponding triplets in the other two chains are at $(r, \phi - 108°, w + 2.87$ Å$)$, and $(r, \phi - 216°, w + 5.74$ Å$)$.

possibilities by measuring the pK value for the terminal amino group and by studying the effect on the melting curve of protonation of the end groups. The pK for the amino group in [Pro-Pro-Gly]$_{10}$ was found to be 0.59 units lower than for the related nonhelical polymer [Pro-Pro-Gly]$_5$ and this was attributed to a close proximity of the amino end groups in a three-strand rope with similarly directed chains. One difficulty in accepting this explanation of the effect as conclusive relates to the fact that in the three-strand-rope model of Yonath and Traub (1969) the amino end groups in the three chains would occur at different axial levels due to the stagger between similar residues (Fig. 11.7). Neither is it certain that the two amino end groups at the same end of a rope with one inverted chain would not produce the same effect.

The evidence obtained from electron microscope studies of [Pro-Pro-Gly]$_{20}$ would seem to indicate that the three strands in the rodlike assemblies are individual molecules but it has been suggested that some of the solution properties of [Gly-Pro-Pro]$_n$ are indicative of the presence of three-strand ropes formed through the folding of a single molecule (Engel et al., 1966; Engel, 1967). If this were so, one of the strands would be oppositely directed to the other two. Rich and Crick (1955, 1961) had earlier drawn attention to the fact that in the RC I and RC II models one of the chains may be inverted and this possibility was further explored, in the light of the findings on [Gly-Pro-Pro]$_n$, by Ramachandran et al. (1968). A model was proposed in which the direction of one chain (c) in an RC I-like structure was inverted and hydrogen bonding formed according to the scheme which Rich and Crick (1961) designated (I, I, II). This involves two $N_1H_1\cdots O_1$ bonds and one $N_1H_1\cdots O_2$ bond per three triplets (Fig. 11.9).[†]

Single crystals of [Pro-Pro-Gly]$_{10}$ have been prepared (Sakakibara et al., 1972) and the X-ray diffraction pattern can be indexed on an orthorhombic unit cell with dimensions $a = 26.9$ Å, $b = 26.4$ Å, and $c = 102.6$ Å, and the space group is either $P2_12_12_1$ or $P2_12_12$. The length of the c axis is close to the length expected for a molecule having the conformation described by Yonath and Traub (1969), and on the basis of the observed water content and density of the crystal it was concluded that the cell contained 12 chains.

Andries and Walton (1970) attempted to test the possibility that chain

---

[†] Ramachandran et al. (1968) state that their model was derived from a three-chain unit with $N_1H_1\cdots O_2$-type bonding (RC II topology) by inverting one chain. The diagram and the later description in their paper suggest, however, that the model was in fact derived from a three-chain unit with $N_1H_1\cdots O_1$-type bonding (RC I topology) as described here.

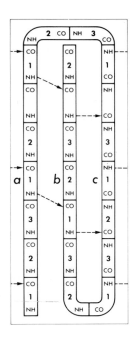

FIG. 11.9. Topological arrangement of hydrogen bonds in a three-strand rope model for [Gly-Pro-Pro]$_n$ formed from a single molecule (Ramachandran *et al.*, 1968).

folding occurs, by examining single crystals of [Gly-Pro-Pro]$_n$ by electron microscopy and electron diffraction. Lamellar crystals were observed and the unusual feature noted that the thickness was a function of the molecular weight of the polymer. It was not found possible to establish the direction of the chain axes in the crystals but they were assumed to be normal to the surface. The thickness of the lamellae was found to be of the order expected for a single molecule folded into a three-strand rope and it was concluded that structures of the type proposed by Ramachandran *et al.* (1968) were present. A number of other explanations are possible and it would seem to be necessary to establish the orientation of the ropes relative to the surface of the lamellae before this observation can be properly interpreted.

V. OTHER POLYTRIPEPTIDES

Andreeva and co-workers have carried out extensive investigations of polytripeptides containing the residues Gly, Ala, Pro, and Hyp and this work has been reviewed by Andreeva *et al.* (1967). The earliest polymer studied in detail was [Gly-Pro-Hyp]$_n$ and it was found that the X-ray diffraction pattern obtained from solid films of this material depended upon the molecular weight. Low-molecular-weight samples gave a pattern which suggested that three-strand units were present in which each strand had a PG II–PP II type of conformation, whereas high-molecular-weight samples gave a pattern which suggested that three-strand ropes were present and that the individual chains had a coiled-coil conformation (Andreeva *et al.*, 1963; Millionova *et al.*, 1963; Rogulenkova *et al.*, 1964). Coiled-coil structures, as evidenced by the occurrence of a spacing of about 2.85 Å in the X-ray diffraction pattern, were also found to be present in [Gly-Pro-Pro]$_n$, [Gly-Hyp-Pro]$_n$, and [Gly-Hyp-Hyp]$_n$ (Table 11.4).

The polytripeptide [Gly-Gly-Ala]$_n$ has been studied by Brack and Spach (1968), Andries *et al.* (1971), and Lotz and Keith (1971) and two forms have been obtained, depending on the method of precipitation. In one form the polypeptide chains appear to have a β conformation and in the other a PG II type of conformation. So far, details of the chain arrangements in the two forms have not been elucidated but Lotz and Keith (1971) have suggested a plausible model for the PG II type of conformation. Insufficient diffraction data are available, however, to provide a proper test of the model.

F. POLYHEXAPEPTIDES

The structures of a series of polyhexapeptides have been examined (Traub *et al.*, 1969; Segal, 1969; Segal *et al.*, 1969) with particular reference to the possibility that a three-strand rope of the RK type might be present (Section A). In the polymer [Gly-Pro-Ala-Gly-Pro-Pro]$_n$, position 2 is occupied in both triplets by an imino acid and one interstrand hydrogen bond per triplet would be formed in either the RK, RC I, or RC II structure (Fig. 11.3). In the polymers [Gly-Ala-Pro-Gly-Pro-Pro]$_n$, [Gly-Ala-Pro-Gly-Pro-Ala]$_n$, and [Gly-Ala-Ala-Gly-Pro-Pro], position 2 in the first triplet is occupied by an amino acid and an additional interstrand hydrogen bond would be formed per hexapeptide unit in the RK type of structure (Fig. 11.3).

All the polyhexapeptides gave X-ray diffraction patterns which resembled that obtained from collagen, and the pattern obtained from [Gly-Pro-Ala-Gly-Pro-Pro]$_n$ closely resembled that obtained from [Gly-Pro-Pro]$_n$ (Section E.IV). As with the polytripeptide, the appearance of spots on the equatorial and third layer lines and streaks on other layer lines was indicative of screw disorder (Chapter 1, Section A.IV.d). The patterns obtained from all four polyhexapeptides were interpreted

TABLE 11.4

Conformations of Poly(glycine) and Proline-Containing Synthetic Polypeptides

Sequence[a]	Mean unit height (Å)	Conformation[b]	Ref.[c]
[Gly]$_n$	$\geqslant 3.48$	Antiparallel β	1
	3.1	PG II	2
[Pro]$_n$	1.90	PP I	3
	3.12	PP II	4
[Gly-Pro-Gly]$_n$	3.1	PG II–PP II	5
[Gly-Ala-Pro]$_n$			
From aqueous solution	3.13	PG II–PP II	6
From trifluoroethanol	2.85	Three-strand rope	7
[Gly-Pro-Ala]$_n$			
Hydrated	3.10	PG II–PP II	8
Dry	2.88	Three-strand rope	8
[Gly-Pro-Hyp]$_n$			
Low MW	3.25	PG II–PP II	9
High MW	2.82	Three-strand rope	9
[Gly-Hyp-Pro]$_n$	2.82	Three-strand rope	9
[Gly-Ala-Hyp]$_n$	2.92	Three-strand rope	9
[Gly-Hyp-Hyp]$_n$	2.75	Three-strand rope	9
[Gly-Pro-Phe]$_n$	2.9	Three-strand rope	10
[Gly-Pro-Lys(HCl)]$_n$	nr[d]	Three-strand rope	11
[Gly-Pro-Pro]$_n$	2.87	RC II three-strand rope	12
[Gly-Pro-Ser]$_n$	2.88	Three-strand rope	13
[Gly-Pro-Ala-Gly-Pro-Pro]$_n$	2.87	RC II three-strand rope	11
[Gly-Ala-Pro-Gly-Pro-Pro]$_n$	2.95	RC II three-strand rope	11
[Gly-Ala-Ala-Gly-Pro-Pro]$_n$	2.95	RC II three-strand rope	11
[Gly-Ala-Pro-Gly-Pro-Ala]$_n$	2.95	RC II three-strand rope	11

[a] For a list of abbreviations used see p. xvii. The repeating sequence as given does not always coincide with the sequence of the peptide used in the synthesis of the polymer.

[b] Symbols used: PG II, poly(glycine) II conformation; PP I, poly(L-proline) I conformation; PP II, poly(L-proline) II conformation; PG II–PP II, hybrid conformations related to PG II and PP II; three-strand rope, conformation derived from three PG II–PP II chains by coiled-coil distortion; RC II three-strand rope, rope with hydrogen bond pattern as in structure II of Rich and Crick (1955).

[c] 1, Astbury (1949c), Bamford et al. (1953a); 2, Bamford et al. (1955); 3, Traub and Shmueli (1963a); 4, Cowan and McGavin (1955); 5, Traub and Yonath (1966), Traub (1969); 6, Segal and Traub (1969); 7, Doyle et al. (1971); 8, Traub and Yonath (1967); 9, Andreeva et al. (1963, 1967); 10, Scatturin et al. (1967); 11, Segal et al. (1969); 12, Yonath and Traub (1969); 13, Brown et al. (1972).

[d] Not reported.

in terms of a three-strand rope model and possible structures were analyzed in a systematic way. The structures were evaluated on the basis of interatomic contacts and on the calculated intensities for equatorial reflections. No acceptable structures were found which involved $N_1H_1 \cdots O_3$ hydrogen bonds of the type envisaged in the RK model (Section A), but satisfactory structures were found which involved $N_1H_1 \cdots O_2$ hydrogen bonds of the type envisaged in the RC II model. It was concluded that all four polyhexapeptides had structures of the RC II type similar to that given in Table 11.3 for [Gly-Pro-Pro]$_n$, although it was noted that in the three polymers where Ala followed Gly the mean axial translation per residue was slightly longer than in [Gly-Pro-Ala-Gly-Pro-Pro]$_n$ (Table 11.4). Independent evidence for the formation of three-strand ropes with only one hydrogen bond per triplet was also obtained from solution studies (Segal, 1969).

Although the results obtained in this investigation appear to exclude the possibility that a three-strand rope of the RK type is present in these polyhexapeptides, the X-ray data used were very limited and further exploration on the basis of more comprehensive intensity data is clearly desirable.

G. SUMMARY

Considerable progress has been made toward the elucidation of the structures of form II of [Gly]$_n$ and [Pro]$_n$ but in neither case are the atomic positions known with any certainty. Objective refinement on the basis of accurate X-ray intensity data as was carried out for the α and β forms of [Ala]$_n$ has not so far been accomplished. While the application of stereochemical criteria has played a valuable role in the development of models for these structures, so many arbitrary assumptions and empirical rules are involved that the detailed refinements which have been carried out on this basis are of doubtful validity.

The results obtained with polytripeptides have shown that collagenlike three-strand ropes can be formed with the repeating sequences Gly-X-I, Gly-I-X, or Gly-I-I, where I is an imino acid residue and X is a residue other than Gly, Pro, or Hyp. The polymer formed from the sequence Gly-Ala-Ala crystallizes in a β conformation (Chapter 10, Section C.II) but when this tripeptide unit is combined with Gly-Pro-Pro in the polyhexapeptide [Gly-Ala-Ala-Gly-Pro-Pro]$_n$ a collagenlike three-strand rope is formed. The finding that certain polymers can form three-strand units either with or without rope twist is particularly important. The factors which determine the relative stabilities of the two forms are not readily apparent but they must be comparatively subtle since a transition from the PG II–PP II to the coiled-coil conformation is observed in [Gly-Pro-Hyp]$_n$ with increase in molecular weight and in [Gly-Pro-Ala]$_n$ with reduction in water content. It has been suggested (Ramachandran et al., 1962; Ramachandran, 1963b) that the coiled-coil conformation is stabilized by the formation of two systematic sets of hydrogen bonds as in the RK type of arrangement (Section A). However, this explanation is not applicable to polymers of the type [Gly-I-I]$_n$ or [Gly-I-X]$_n$, in both of which coiled-coil conformations have been observed.

In the case of [Gly-Pro-Pro]$_n$ sufficient X-ray data were available for a test of the various types of three-strand rope models discussed in Section A to be carried out. Both in this polymer and in the polyhexapeptides the evidence appears to favor structures with the RC II type of hydrogen bonding rather than the RC I or the RK types. In the RK type of bonding two intrastrand hydrogen bonds are formed per tripeptide and [Gly-Ala-Pro]$_n$ in particular might have been expected to form such a structure but this was not found to be so. Of the hydrogen bond arrangements which have been suggested to occur in collagenlike three-strand rope structures, the only one which has so far been demonstrated to occur in sequential polypeptides is the RC II type.

Chapter 12

The Influence of Composition
on Conformation

A. INTRODUCTION

The conformation adopted by a polypeptide chain is influenced by both its composition and its environment. Studies of synthetic polypeptides in solution have thrown considerable light on the factors that determine whether or not a regular helical conformation is adopted (Bamford *et al.*, 1956; Fasman, 1967b; Mandelkern, 1967) but in the solid state the situation is more complex since the mechanical, chemical, or thermal history of the specimen and the presence of solvent molecules can influence conformation. Nevertheless, some important generalizations have emerged from studies of synthetic polypeptides and globular protein crystals regarding the influence of composition on conformation and these are discussed in the present chapter. A detailed discussion of the influence of sequence on conformation in fibrous proteins and the effects of such factors as the distribution of polar and nonpolar residues will be reserved for later chapters.

The profound influence of Pro residues on the ability of a polypeptide chain to adopt the α-helix conformation was recognized at an early stage (Pauling and Corey, 1951f; Szent-Györgyi and Cohen, 1957) but the first attempt at a systematic classification of residues according to their propensity for different conformations was that of Blout *et al.* (1960). On the basis of infrared studies of films of homopolypeptides it was suggested that two classes of amino acids could be distinguished, those that formed α-helical structures when polymerized into homo-

276

polypeptides, and those that formed either random or β structures when polymerized into homopolypeptides. It was noted that the amino acids classified as non-α-helix-forming were of two types; in the first a dialkyl substitution occurred at C^β and in the second an O or an S atom occupied the γ position in the side chain. The amino acids classified in this way are listed in Table 12.1 (Blout, 1962). It was recognized

TABLE 12.1

Classification of Residues According to Propensity for
α-Helix Formation in Homopolypeptides[a]

α-Helix-forming		Non-α-helix-forming	
Residue	Side chain	Residue	Side chain
Ala	$-CH_3$	β-Dialkyl-substituted	
Asp esters	$-CH_2-COO-Alk^b$	Val	$-CH(CH_3)_2$
Glu	$-(CH_2)_2-COOH$	Ile	$-CH(CH_3)-C_2H_5$
Glu esters	$-(CH_2)_2-COO-Alk^b$	Hetero-γ-atom	
Leu	$-CH_2-CH(CH_3)_2$	Ser(Ac)	$-CH_2-O-CO-CH_3$
Lys	$-(CH_2)_4-NH_2$	Ser	$-CH_2-OH$
Met	$-(CH_2)_2-S-CH_3$	Cys(Me)	$-CH_2-S-CH_3$
Phe	$-CH_2-C_6H_5$	Thr(Ac)	$-CH(CH_3)-COO-CH_3$
Tyr	$-CH_2-C_6H_4OH$	Thr	$-CH(CH_3)-OH$

[a] Blout et al. (1960); Bloom et al. (1962); Blout (1962).
[b] Alkyl group.

that the conformation adopted by a homopolypeptide depends, in some cases, upon the molecular weight, and the classification was based on results obtained with polymers in which the degree of polymerization was estimated to exceed 100. Degrees of polymerization appreciably less than this appear to favor the β conformation in the solid state (Bamford et al., 1951b; Blout and Asadourian, 1956; Bradbury et al., 1960b; Tsuboi and Wada, 1961; Goodman et al., 1971). In solution the effects of degree of polymerization on conformation vary with solvent composition and have been studied in considerable detail (see, for example, Fasman, 1967b).

Blout et al. (1960) suggested that correlations between composition and conformation noted for homopolypeptides might also be operative in proteins and since that time a number of studies have been made with a view to establishing correlations between secondary structure in proteins and various features of composition and sequence. These correlations have been based on the use of the optical rotatory dispersion parameter b_0 (Chapter 8, Section B) as an index of α-helix content (Davies, 1964; Havsteen, 1966; Goldsack, 1969), analyses of the distribution of residues in globular proteins of known three-dimensional structure (Perutz *et al.*, 1965; Guzzo, 1965; Prothero, 1966; Cook, 1967; Periti *et al.*, 1967; Schiffer and Edmundson, 1967; Low *et al.*, 1968; Ptitsyn, 1969; Pain and Robson, 1970; Finkelstein and Ptitsyn, 1971; Robson and Pain, 1971), considerations of stereochemistry (Leach *et al.*, 1966a,b; Némethy *et al.*, 1966; Liquori, 1969; Ramachandran and Sasisekharan, 1968), or potential energy calculations (Kotelchuck and Scheraga, 1968, 1969; Scheraga, 1971). The ultimate aim of such studies has generally been to devise rules which hopefully will enable the occurrence of secondary structure in proteins to be predicted from a knowledge of the composition or sequence. Although some progress has been made, no reliable set of rules has been devised thus far and this approach has had little impact on the formulation of model structures for fibrous proteins. Some of the factual matter gathered in these studies, however, is pertinent to the studies of synthetic polypeptides discussed in this chapter and has been included where appropriate. The statistical analysis of the distribution of residues among the α-helical, β, and nonregular regions of globular proteins of known structure (Ptitsyn, 1969; Finkelstein and Ptitsyn, 1971) has been particularly informative and the correlations established in the latter study are reproduced in Table 12.2.

TABLE 12.2

Statistically Significant Differences in Amino Acid Content of
Regions with Regular Secondary Structure in Globular Proteins[a]

α-Helix		β Structure		Nonregular	
Significantly above average	Significantly below average	Significantly above average	Significantly below average	Significantly above average	Significantly below average
Ala	Gly	Ile	Nil	Gly	Ala
Glu	Tyr	Leu		Ser	Leu
Leu	Asn	Val		Asn	Glu
His	Ser			Pro	Val
	Pro			Arg	
	Cys				

[a] Finkelstein and Ptitsyn (1971).

B. CONFORMATIONAL EFFECTS OF SOME
INDIVIDUAL RESIDUES

I. VALINE

The effect of mixing "α-helix-forming" and "non-α-helix-forming" residues in a copolymer was investigated by Bloom *et al.* (1962) and it was found that in the case of (Met, Val) copolymers the helix content decreased with increasing Val content both in solution and in solid films. This approach is limited, however, by the fact that the copolymers are unlikely to be random and that the composition and sequence of the individual molecules are unknown, and the influence of Val residues on α-helix formation was investigated in a more detailed way (Fraser *et al.*, 1965b) by studying a series of sequential polypeptides containing Val and Glu(Me) residues. In solid films the homopolypeptide [Glu(Me)]$_n$ forms α-helices and it was found that the extent of helix formation was not materially different in the polymer [{Glu(Me)}$_3$-Val-Glu(Me)]$_n$, which contains a 20% molar proportion of Val residues. However, higher concentrations of Val residues reduced the helix content of the films and the polymer [Val-Glu(Me)]$_n$ was found to be almost completely in the β conformation in the solid state. The solution properties, in dichloroacetic acid/ethylenedichloride mixtures, generally paralleled the solid-state properties with respect to the extent of α-helix formation. However, the stability of the helix, as judged by the solvent composition at which a random coil → helix transition occurred, showed some dependence on sequence and the polymer [{Glu(Me)}$_3$-{Val}$_2$-Glu(Me)]$_n$ formed helices which were more stable to disruption by dichloroacetic acid than those formed by [Glu(Me)]$_n$.

The results obtained in this investigation were interpreted in terms of the environment of the Val residues on the surface on an α-helix (Fig. 12.1). In the case of [{Glu(Me)}$_3$-Val-Glu(Me)]$_n$, which has properties very similar to [Glu(Me)]$_n$, each Val residue is entirely surrounded by Glu(Me) residues, but in all other cases one or two of the nearest-neighbor residues is a Val residue. It was suggested that the reduction in the stability of the α-helix observed in the polymers [{Glu(Me)}$_m$-Val-Glu(Me)]$_n$, $m = 0, 1, 2$, might be due to unfavorable Val–Val side-chain interactions of the type $i \cdots i + 3$ and $i \cdots i + 4$ (Fig. 12.1). Interactions of the type $i \cdots i + 1$, on the other hand, were considered, on the basis of the result obtained with [{Glu(Me)}$_3$-{Val}$_2$-Glu(Me)]$_n$, to stabilize the α-helical conformation in the presence of polar solvents. It would be desirable to test the general validity of

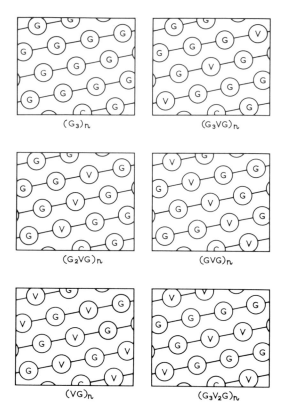

FIG. 12.1.   Schematic radial projections showing the side-chain arrangement on the surface of an α-helix in a series of sequential polypeptides of γ-methyl-L-glutamate (G) and valine (V).

these conclusions by investigating the properties of similar sequential polypeptides with other α-helix-forming residues in place of Glu(Me).

It is apparent from these studies that appreciable proportions of Val residues can be accommodated in an α-helix but that the effect on stability is sequence-dependent. All possible types of Val···Val side-chain interactions in the α-helix were present in the series of polymers studied and thus none appears to preclude α-helix formation. This has been convincingly demonstrated by Epand and Scheraga (1968), who prepared a block copolymer from Val and D,L-lysine hydrochloride and showed that the Val block was partly helical in solution. In the analysis of residue distribution in globular proteins (Table 12.2) it was found that the content of Val residues in α-helical and non-α-helical regions was not significantly different. Taken together, the observations on synthetic polypeptides and globular proteins suggest that in the concentrations normally encountered in fibrous proteins, Val residues are unlikely to have a profound effect on α-helix formation.

The possibility that steric interference occurs in the α-helical conformation of $[Val]_n$ has been investigated by the methods outlined in Chapter 6 (Davies, 1965; Leach *et al.*, 1966a,b; Némethy *et al.*, 1966; Fraser *et al.*, 1968a). Rotation of the side chain about the $C^\alpha C^\beta$ bond is restricted by steric interference between the side-chain and the main-chain atoms and neither of the $C^\beta C^\gamma$ branches can occupy the minimum-energy orientation corresponding to $\chi_1 = 60°$ [Chapter 6, Section C.I.b(iv)]. If the orientation about $C^\alpha C^\beta$ is chosen so that the two branches occupy positions corresponding to the minima in torsional energy at $\chi_1 = -60°$ and $180°$, steric interference occurs between the methyl group on branch 2 of the side chain of residue i and the oxygen atom O_{i-4} of the turn below. However, this can be relieved by a small rotation about the $C^\alpha C^\beta$ bond away from torsional energy minima.

It is difficult to assess how important steric interference will be in practice since the conclusions just reached involve a great many assumptions about bond angles, torsion angles, and allowable interatomic contact distances. If any of the assumed values are appreciably in error, the conclusions may well be invalid. Ooi *et al.* (1967) carried out a more comprehensive study of the problem by calculating the intramolecular potential energy of regular helical conformations of $[Val]_n$. It was found that the right-handed α-helix was the conformation of lowest energy but it is difficult to relate this finding to the observed data on solid films since intermolecular interactions, which are an important source of stabilization in β structures, were not considered. In globular proteins the β segments of chains contain above-average amounts of Val residues (Table 12.2).

II. CYSTEINE

The residue Cys(Me) was classified by Blout (1962) as non-α-helix-forming and this classification also appears to be appropriate to Cys(Cbz) (Elliott *et al.*, 1964), Cys(CbzMe) (Harrap and Stapleton, 1963; Ikeda *et al.*, 1964) and Cys(Bzl) (Blout *et al.*, 1960; Fraser *et al.*, 1965d). However, the homopolypeptide $[Cys(BzlTh)]_n$ was found to have an ω-helical structure in solid films (Fraser *et al.*, 1962a).

The effect of combining Cys(Me), which had been classified as non-α-helix-forming (Table 12.1), and Met, which had been classified as α-helix-forming, in a series of copolymers was investigated by Bloom *et al.* (1962). Below 50% molar concentration of Cys(Me), helix formation in solution was not greatly affected, but in solid films the α-helix content steadily decreased and the β content steadily increased with increasing proportion of Cys(Me). In a similar study Ascoli and De Cupis (1964) prepared a 1:1 copolymer of Lys(Cbz), which had earlier been shown to be helical in solutions of appropriate composition (Fasman *et al.*, 1961), and Cys(Cbz). The maximum helix content attained in solutions of the copolymer was about half that observed in $[Lys(Cbz)]_n$. As mentioned previously, results obtained with co-polymers of this type are difficult to interpret since the distribution of residues is unlikely to be random and different molecules may have markedly different compositions and sequences.

A series of polypeptides containing ordered sequences of Cys(Bzl) and Glu(Et) was studied by Fraser *et al.* (1965d) and the proportion of α-helix in solid films was found to decrease steadily with increasing proportion of Cys(Bzl). In no case was α-helix formation completely inhibited, although the α-helix content of solid films of [Cys(Bzl)-

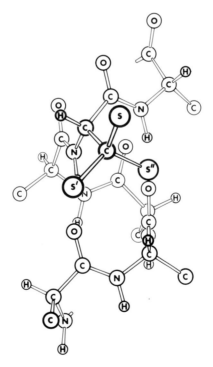

FIG. 12.2. Steric environment of the S^γ atom of a cysteine residue in the α-helical conformation. The three positions illustrated correspond to $\chi_1 = -60°$ (S′), $\chi_1 = 60°$ (S″), and $\chi_1 = 180°$ (S).

Glu(Et)]$_n$ was low. Positional effects on the surface of the α-helix were found to be relatively minor and the destabilizing effect of the Cys(Bzl) residues was thus attributed to unfavorable interactions between the side chain and the main chain. This possibility had also been envisaged in earlier studies (Bloom et al., 1962; Fraser et al., 1962a).

A search for unfavorable steric interactions when a Cys derivative is incorporated into a right-handed α-helix was carried out using the methods described in Chapter 6. The torsional potential for rotation about the $C^\alpha C^\beta$ bond has minima for $\chi_1 = -60°$, $60°$, and $180°$ and the corresponding positions of the S atom are illustrated in Fig. 12.2. The orientations $\chi_1 = \pm 60°$ were found to involve short interatomic contacts, but orientations in the vicinity of $\chi_1 = 180°$ appeared to be free of short interatomic contacts. It was noted, however, that in a number of simple compounds the orientation about the $C^\alpha C^\beta$ bond corresponded to a value of $\chi_1 = 60°$ (Lakshminarayanan et al., 1967) which is precluded by short contacts in the α-helical conformation. In view of this, it was suggested that the destabilizing effect might be due to the unfavorable rotational conformation about the $C^\alpha C^\beta$ bond imposed by the α-helical conformation. Birshtein and Ptitsyn (1967) pointed out that the $C^\beta S$ bond was polar and that interaction between the $C^\beta S$ dipole and the main-chain amide dipoles might be important in determining conformation; the energy per residue was calculated to be 0.1, -0.5, and 0.5 kcal for $\chi_1 = -60°$, $60°$, and $180°$, respectively. As mentioned earlier, values of $\chi_1 = -60°$ and $60°$ are precluded by short interatomic contacts and the unfavorable dipole–dipole interaction in the vicinity of $\chi_1 = 180°$ thus provides a plausible explanation for the destabilizing influence of Cys residues.

In the statistical analysis of residue distribution in globular proteins carried out by Finkelstein and Ptitsyn (1971) it was found that the Cys content of the α-helical regions was significantly below average. Taken together the results obtained with synthetic polypeptides and globular proteins suggest that Cys residues probably play an important role in destabilizing α-helix formation in Cys-rich fibrous proteins. When disulfide linkages are formed between Cys residues an additional effect on α-helix formation is introduced since intrahelical disulfide linkages are not sterically feasible (Lindley and Rollett, 1955).

III. SERINE

The residues Ser and Ser(Ac) were classified by Blout (1962) as non-α-helix-forming (Table 12.1) and this classification also appears to be appropriate to Ser(Bzl) (Bradbury et al., 1962b; Bohak and Katchalski, 1963). Sarathy and Ramachandran (1968) have suggested that [Ser]$_n$

might form left-handed α-helices stabilized by both $N_iH_i\cdots O_{i-4}$ and $O_i{}^\gamma H_i{}^\gamma\cdots O_{i-1}$ hydrogen bonds but so far this has not been observed to be so. Poly-O-acetyl-L-homoserine, $[NH\cdot CH(CH_2\cdot CH_2\cdot OH)\cdot CO]_n$, in which the hydroxyl group is one position further removed from the main chain than in $[Ser]_n$, has been observed to form α-helices (Fasman, 1967b).

Unordered copolymers of Glu(Bzl) and Ser(Ac) were prepared by Kulkarni et $al.$ (1967) and it was found that in copolymers containing less than 18% Ser(Ac) the conformation was essentially α-helical, whereas copolymers with higher contents of Ser(Ac) contained both α and β phases. Copolymers of Glu and Ser were also studied and it was found that the helix content attainable decreased steadily with increasing content of Ser. Imahori and Inouye (1967) investigated the conformation of unordered copolymers of Glu(Bzl) and Ser(Bzl) in solution and found that in chloroform containing a small amount of dichloroacetic acid the helix content decreased steadily with increasing proportion of Ser(Bzl). In polymers with more than 50 mole % of Ser(Bzl) an appreciable fraction of the polymer had a β conformation.

A series of sequential polypeptides with the repeating sequence $[\{Glu(Me)\}_m\text{-}Ser(Ac)]_n$, $m = 1\text{-}4$, was investigated by Fraser et $al.$ (1968a) and it was found that the polymers with $m = 1$ and 2 were nonhelical both in solution and in solid films. X-ray diffraction and infrared studies showed that the films had a β structure. Films of the polymers with $m = 3$ and 4 contained both α and β phases but the helix content was lower than in films of $[Glu(Me)]_n$ of similar molecular weight. Freshly prepared solutions of the polymers with $m = 3$ and 4 contained an appreciable proportion of helical material but, on standing, a gel was formed and the helix content dropped to $<10\%$.

These data indicate that the introduction of Ser(Ac) residues into a Glu(Me) polymer causes a dramatic reduction of the stability of the α-helical conformation relative to a random structure or to a β structure. The possibility that this effect was due to steric interference was investigated but it was found that no short interatomic contacts were present for $\chi_1 = -60°$ or $180°$. An alternative possibility was suggested to be competition between $O_i{}^\gamma$ and O_{i-3} for hydrogen bond formation with $N_{i+1}H_{i+1}$ since the former bond can only be formed if the α-helix is disrupted. According to Birshtein and Ptitsyn (1967), the dipole–dipole interactions between the $C^\beta O^\gamma$ bond and the main-chain amide groups are 0.1 and 0.5 kcal mole^{-1} for $\chi_1 = -60°$ and $180°$, respectively, in the α-helix compared with 0.1, 0, and -0.1 kcal mole^{-1} for $\chi_1 = -60°$, $60°$, and $180°$, respectively in the antiparallel-chain pleated sheet. These differences may be sufficient to explain the observed effects.

An additional factor which is operative with Ser residues is the ability of the $O^\gamma H^\gamma$ group to form hydrogen bonds with the C'O groups of the main chain. In globins such bonds appear to be formed in helical regions between $O_i{}^\gamma H_i{}^\gamma$ and O_{i-4} and are also involved in the termination of helical segments (Kendrew, 1962; Perutz et al., 1965). It is interesting that in myoglobin the torsion angle χ_1 about the $C^\alpha C^\beta$ bond has a value of about 60° for one Ser residue in an α-helical segment (Watson, 1970). This value of χ_1 normally leads to steric interference between the side-chain atoms of residue i and the O_{i-4} atom. In this particular instance an $O_i{}^\gamma H_i{}^\gamma \cdots O_{i-4}$ hydrogen bond is formed and the helix is distorted. In lysozyme (Phillips, 1967) all the Ser residues except two occur either in nonhelical regions or at the end of an α-helical segment. This reinforces the view that Ser residues tend to destabilize the α-helix. Finkelstein and Ptitsyn (1971) found that the proportion of Ser residues in the α-helical segments was significantly below average in globular proteins and that Ser residues were concentrated in regions devoid of regular secondary structure.

IV. Glycine

As discussed in Chapters 10 and 11, $[\text{Gly}]_n$ has been prepared in the β form and in the PG II form but so far no α-helical form has been observed. Gly was not included in the classification of amino acids given by Blout et al. (1960) but the effect of Gly residues on the stability of the α-helix has been investigated experimentally by the preparation of copolymers of Gly with residues classified as α-helix-forming (Block and Kay, 1967; Fraser et al., 1967a; Iio and Takahashi, 1970).

A series of sequential polypeptides having the repeating sequence $\{\text{Glu(Et)}\}_m\text{-Gly}$, $m = 1-5$, was studied by Fraser et al. (1967a) and the helix content in solution was found to decrease markedly with Gly content. The α-helix contents of solid films were generally lower, and above 25% molar concentration of Gly the films were entirely in the β form. The helix content of the polymer with 83% molar concentration of Glu(Et), $m = 5$, was found to have more than twice the α-helix content of the polymer with 80% molar concentration of Glu(Et), $m = 4$, and it was noted that a complete turn of Glu(Et) residues could be formed in the case of $m = 5$ but not in the case of $m = 4$. Although the helix content of the films obtained from the polymer with $m = 5$ was estimated to be 63%, the X-ray diffraction pattern indicated that, as with the other members of the series, only the β phase was crystalline. It was concluded from this study that Gly residues cause a reduction in the stability of the α-helix relative to a random conformation in solution and relative to a β conformation in the solid. The Gly residue differs from other residues in that it has no side chain, and Bamford *et al.* (1956) had earlier recognized that this might be the origin of the β-favoring influence of this residue. In the absence of side-chain atoms the adverse effect on helix stability cannot be attributed to steric interference, and it was suggested (Fraser *et al.*, 1967a) that the explanation might be found in the additional flexibility conferred on the main chain by the absence of side-chain atoms. It has already been noted that the range of (ϕ, ψ) combinations for Gly is appreciably greater than for other residues (Chapter 6, Section B.III) and Némethy *et al.* (1966) suggested that the entropy of unfolding per residue from a restricted structure such as the α-helix to a random conformation is appreciably greater than for other residues.

Iio and Takahashi (1970) investigated α-helix formation in the sequential polypeptides $[Ala_3\text{-}Gly]_n$, $[Ala_2\text{-}Gly]_n$, and $[Ala\text{-}Gly_2]_n$. Infrared spectra obtained from solid films showed a steady decrease in α-helix content with increase in molar proportion of Gly and all the films contained a β phase. All the polymers appeared to be partly helical in solution but the stability of the helix to disruption by dichloroacetic acid was not simply dependent on molar concentration of Gly, since the polymer $[Ala_3\text{-}Gly]_n$ was less helical than the polymer $[Ala_2\text{-}Gly]_n$. It was suggested that this effect might be related to the way in which the Ala and Gly residues are distributed on the surface of an α-helix in the two cases.

The general conclusion to be drawn from the studies of solid films is that Gly residues tend to destabilize the α-helix and favor the formation of β structures. In the analysis of globular proteins reported by Ptitsyn (1969) it was found that the proportion of Gly residues in the α-helical segments was significantly below average and that Gly residues were concentrated in regions devoid of regular secondary structure.

V. PROLINE

The effects of Pro residues on the conformation of polytripeptides were considered in Chapter 11 and in the present section the effects of less regular distributions will be considered. Pro residues cannot be incorporated either into a continuous α-helix or a β structure (Pauling and Corey, 1951f; Lindley, 1955; De Santis *et al.*, 1965; Leach *et al.*, 1966a) but it is sterically possible for Pro to occur at sites 1, 2, 3, or $n + 1$ of an α-helix containing n residues (Lindley, 1955; Perutz *et al.*, 1965). In this description residues are regarded as part of the helix if either or both their NH or CO groups are hydrogen-bonded within the helix and the numbering proceeds in the usual direction —NH—CHR—CO→. Thus Pro residues effectively limit the possible lengths of the α-helical segments in a polypeptide chain.

The role of Pro in the determination of chain conformation in solution was studied by Szent-Györgyi and Cohen (1957) using rotatory dispersion and specific rotation as an index of α-helix content and although the latter method is no longer thought to be reliable, the qualitative aspects of the investigation are probably still valid. It was concluded that (i) if the Pro content is less than 3% by weight and distributed statistically, the helix content may exceed 50%, (ii) about 8% by weight of Pro leads to a random coil, and (iii) very high contents of Pro may favor a PP II type of structure (Chapter 11, Section A). Since the residue weight of Pro is close to the mean residue weight in most proteins, the contents quoted also apply approximately to the molar concentrations of Pro.

In accord with expectation, Pro residues in globins are confined to the ends of helical segments or to nonhelical regions (Perutz et al., 1965). In both papain (Drenth et al., 1971) and ribonuclease (Kartha et al., 1967), Pro residues occur at bends in sections of antiparallel-chain β structure.

VI. ACIDIC AND BASIC RESIDUES

Extensive studies of homopolypeptides of amino acids with acidic and basic side chains have been carried out in aqueous solution and it has been found that conformational transitions can be induced by variation of the state of ionization of the side chain. In $[Glu]_n$ a helix to random-coil transition is observed as the pH is increased and is associated with the ionization of the side-chain carboxyl group (Doty et al., 1957). A similar effect is observed with $[Lys]_n$ as the pH is decreased, again associated with the development of charge on the side chain (Applequist and Doty, 1962). In both cases the transition can be attributed to repulsion between like charges, and the effects of temperature, solvent composition, and other factors on the transition have been investigated in considerable detail (Fasman, 1967b). Related conformational changes between α-helical and β structures in solid films have already been discussed in Chapters 9 and 10.

Ptitsyn (1969) did not find any significant concentration of acidic (Asp + Glu) residues or basic (Arg + His + Lys) residues in the helical or nonhelical regions of globular proteins but noted that significantly greater than average proportions of acidic residues were found at the imino ends of helical segments and significantly greater than average proportions of basic residues were found at the carbonyl ends of helical segments.

C. CLASSIFICATION OF RESIDUES

The results obtained in studies of copolymers of residues classified as α-helix-forming and non-α-helix-forming (Table 12.1) largely substantiate the expectation that the helical content would decrease as the proportion of non-α-helix residues was progressively increased. The results obtained with Val residues (Section B.I) suggest that the destabilizing effect is also dependent on the sequential distribution in this case and that under certain circumstances the stability of the α-helix might even be increased by the presence of Val residues.

The Gly residue was not included in the original classification and has been considered by a number of authors as neither favoring nor interfering with α-helix formation (Havsteen, 1966; Kotelchuck and Scheraga, 1968, 1969). The results obtained with copolymers of Gly and residues classified as α-helix-forming clearly demonstrate, however, that this residue has a profound tendency to destabilize the α-helix, and this is confirmed by the analysis of globular proteins (Table 12.2).

Reservations must be made about the extension to fibrous proteins of a classification based on the study of synthetic polypeptides since the residues studied have, in the main, been derivatives of the naturally occurring residues and since organic, rather than aqueous, solvents have generally been employed. In addition, the studies reported in Chapters 9–11 emphasize the importance of the effect of intermolecular interactions on conformation and the O-benzyl derivative of a serine residue, for example, will be a poor model for a serine residue in this regard.

Several tests of the classification given in Table 12.1 have been carried out on the basis of data obtained from globular proteins and Ptitsyn (1969) found that the group (Cys, Ile, Ser, Thr, Val) was concentrated in the nonhelical regions to an extent which was significant at the 5% level. The two types of residue classified as non-α-helix were tested separately and it was found that the β-dialkyl-substituted type (Val, Ile) was not significantly concentrated in the nonhelical regions, while the hetero-γ-atom type (Ser + Thr + Cys) was significantly concentrated in the nonhelical regions at the 0.5% level. Numerous classifications, some qualitative and some quantitative, of the tendency for individual amino acids to favor helical or nonhelical conformations have been given and some representative examples are given in Table 12.3. There appears from this table and from Table 12.1

TABLE 12.3

Classification of Residues According to Propensity for α-Helix Formation

Residue	Reference[a]					
	1	2	3	4	5	6
Ala	+	+	+	0	+0.9	+1.3
Arg		+	+	0	+0.2	+0.4
Asn	−	−	−	0	−0.4	−0.2
Asp		−	−	−	−0.2	+0.1
Cys	−	+	+	0	+0.3	−0.5
Gln		+	+	0	+0.7	+0.8
Glu	+	+	+	+	+1.2	+1.4
Gly	−	0	0	−	−0.5	+0.2
His	+	+	−	0	+0.8	+0.8
Ile		+	+	+	+0.7	+0.6
Leu	+	+	+	+	+1.1	+1.2
Lys		−	−	0	−0.3	+0.7
Met	+	+	+	+	+1.0	+0.7
Phe		+	+	+	+0.3	+0.1
Pro	−	−	−	−	—[b]	−3.2
Ser	−	−	−	−	−0.7	−0.1
Thr		+	−	0	−0.1	+0.2
Trp		−	+	+	+1.0	+0.9
Tyr	−	−	−	+	−0.2	−0.3
Val		+	+	0	+0.4	+0.7

[a] 1. Finkelstein and Ptitsyn (1971); +, significantly concentrated in helical regions; −, significantly concentrated in nonhelical regions. 2. Kotelchuck and Scheraga (1968); +, residues designated as "helix-making"; −, residues designated as "helix-breaking"; 0, residues designated as "helix-indifferent." Assignments based on energy calculations. 3. Kotelchuck and Scheraga (1969); symbols as in 2. Assignments arbitrarily adjusted to improve accuracy of prediction of helical segments in known structures. 4. Lewis *et al.* (1970); symbols as in 2 and 3. Assignments based on studies of synthetic copolymers in the case of Gly, Ala, and Leu and the remainder on a series of *ad hoc* assumptions. 5. Pain and Robson (1970); figures quoted are the calculated "helix-forming powers" × 10. 6. Robson and Pain (1971); figures quoted are the calculated "helix-forming information" rounded to one decimal place.

[b] Treated as a special case (Pain and Robson, 1970).

to be fairly general agreement that Ala, Glu, Leu, and Met favor α-helix formation while Pro, Asn, and Ser do not. Also, on the basis of the studies on synthetic copolymers and the analyses of Finkelstein and Ptitsyn (1971) and Pain and Robson (1970), Gly should be included with Pro, Asn, and Ser as residues which do not favor α-helix formation.

The suggestion by Blout *et al.* (1960) that residues could be classified according to their propensity for α-helix conformation appears to be soundly based and can be attributed to the fact that local interactions between side-chain and main-chain atoms appear to constitute an important conformation-determining factor (Schimmel and Flory, 1968; Finkelstein and Ptitsyn, 1971; Scheraga, 1971). Many other factors are involved, however, and such factors as the spatial distribution of polar and nonpolar groups (Perutz *et al.*, 1965; Schiffer and Edmundson, 1967; Dunnill, 1968), other sequence-dependent effects (Periti *et al.*, 1967; Low *et al.*, 1968; Kotelchuck and Scheraga, 1969; Kotelchuck *et al.*, 1969; Welscher, 1969; Lewis *et al.*, 1970; Robson and Pain, 1971; Scheraga, 1971), and interchain interactions (Schiffer and Edmundson, 1967; Atassi and Singhal, 1970; Hermans and Puett, 1971) need to be taken into consideration. In addition to the empirical or semiempirical approach to the prediction of conformation, there has also been a considerable interest in obtaining a deeper understanding of the factors involved through both theoretical and experimental studies of the problem. Some of these have already been discussed in earlier sections and valuable reviews of the extensive work on this topic have been given in monographs edited by Fasman (1967a) and Timasheff and Fasman (1969).

CONFORMATION IN FIBROUS PROTEINS

Chapter 13

Silks

A. INTRODUCTION

Arthropods produce a great variety of silks which are used in the fabrication of structures external to the body of the animal. Characteristically the silk is produced in specialized glands and stored in a fluid state in the lumen of the gland. As the fluid passes out of the body of the animal, a rapid transformation to the solid state takes place and the silk becomes water-insoluble.

The major component in most silks is a fibrous protein which in the final insoluble state has a high content of regular secondary structure. Wide variations in composition and structure have been found among different silks and their study offers a unique possibility for obtaining an understanding of the effect of composition on conformation and of the principles underlying the assembly of fibrous proteins into ordered aggregates. In this section a brief outline will be given of the range of compositions and conformations which have been encountered in silks and detailed investigations of these conformations will be discussed in following sections.

Comprehensive reviews of chemical and physical studies of silks have been given by Lucas *et al.* (1958), Rudall (1962), Seifter and Gallop (1966), Lucas and Rudall (1968a), and Rudall and Kenchington (1971).

I. CLASSIFICATION OF SILKS

Silks are often classified on the basis of the predominant form of regular secondary structure in the final product as determined by

X-ray diffraction studies (Fig. 13.1). Thus it is convenient to distinguish beta silks, alpha silks, and collagenlike silks and a list of silks which have been classified in this way is given in Table 13.1. Warwicker (1960a) classified beta silks into five groups on the basis of features in the diffraction pattern, as discussed in Section B.I, and later workers (Lucas and Rudall, 1968a; Rudall and Kenchington, 1971) have

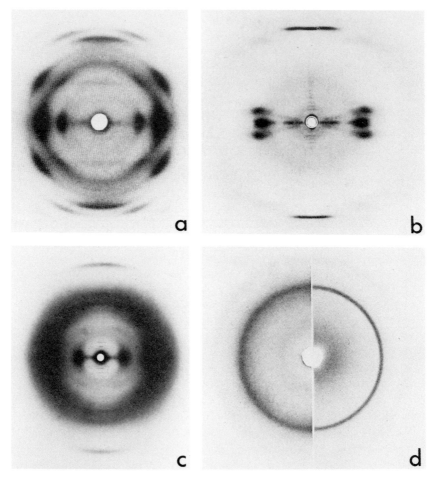

FIG. 13.1. X-ray diffraction patterns obtained from (a) *Digelansinus diversipes* silk (β); (b) *Apis mellifera* silk (α); (c) *Nematus ribesii* silk (collagenlike); (d) denatured *Phymatocera aterrima* silk (left) and polyglycine II (right). Reproduced from F. Lucas and K. M. Rudall (1968a), Extracellular fibrous proteins: the silks, *in* "Extracellular and supporting structures, Comprehensive Biochemistry," Vol. 26B, (M. Florkin and E. Stotz, eds.) Elsevier, Amsterdam, 1968.

suggested that the number of groups should be increased as indicated in Table 13.2 to embrace a further type of beta silk (group 6), alpha silks (group 7), and collagenlike silks (group 0).

TABLE 13.1

Classification of Arthropod Silks on the Basis of Their X-Ray Diffraction Patterns

Class, family or subfamily, genus, and species	Ref.[a]	X-ray classifi- cation[b]	Content of selected residues[e] (residues/100 residues)					
			Gly	Ala	Ser	Glu	Pro	Ref.[a]
Arachnida								
Argiopidae								
Araneus diadematus								
(cocoon)	1	$\beta(5)$	8	24	30	9	1	6
(drag line)	3	$\beta(4)$	37	23	6	9	11	6
Nephila madagascariensis								
(drag line)	1	$\beta(3b)$	41	32	4	12	—	7
Nephila senegalensis								
(cocoon)	1	$\beta(5)$	12	29	22	10	—	7
Theraphosidae								
Avicularia avicularia	1	$\beta(4)$	—	—	—	—	—	—
Scodra griseipes								
(nest web)	1	$\beta(4)$	—	—	—	—	—	—
Tapinaouchenius plumipes								
(nest web)	1	$\beta(4)$	—	—	—	—	—	—
Insecta								
Agrotidae								
Bena prasinana	1	$\beta(1)$	43	25	14	3	0	7
Allantinae								
Allantus sp.	2	PG II(0)	—	—	—	—	—	—
Emphytus sp.	2	PG II(0)	—	—	—	—	—	—
Empria sp.	2	PG II(0)	—	—	—	—	—	—
Monostegia sp.	2	β	—	—	—	—	—	—
Apidae								
Apis mellifera	3	$\alpha(7)$	—	—	—	—	—	—
Bombus lucorum	3	$\alpha(7)$	—	—	—	—	—	—
Arctiidae								
Arctia caja	1	$\beta(4)$	22	23	11	6	3	7
Argidae								
Arge ustulata	2	$\beta(6) + \alpha$	2	41	8	33[d]	2	3
Digelansinus diversipes	2, 3	$\beta(6)$	2	38	9	36[d]	—	3
Pachylota audouinii	2, 3	$\beta(6)$	2	36	12	36[d]	—	3

TABLE 13.1 (*continued*)

Class, family or subfamily, genus, and species	Ref.[a]	X-ray classification[b]	Content of selected residues[c] (residues/100 residues)					
			Gly	Ala	Ser	Glu	Pro	Ref.[a]
Blennocampinae								
Ardis sp.	2	PG II(0)	—	—	—	—	—	—
Blennocampa sp.	2	PG II(0)	—	—	—	—	—	—
Erythraspides sp.	2	PG II(0)	—	—	—	—	—	—
Monophadnoides sp.	2	PG II(0)	—	—	—	—	—	—
Phymatocera aterrima	2	PG II(0)	66	2	3	1	3	3
Tethida sp.	2	PG II+C	—	—	—	—	—	—
Tomostethus sp.	2	PG II+C	—	—	—	—	—	—
Bombycidae								
Bombyx huttoni	1	β(1)	—	—	—	—	—	—
Bombyx mandarina	1	β(1)	48	30	10	1	0	4
Bombyx meridionalis	1	β(1)	45	30	11	1	—	7
Bombyx mori	1	β(1)	45	29	12	1	0	3, 8, 9
Rondotia menciana	1	β(3b)	—	—	—	—	—	—
Braconidae								
Apanteles bignelli	1	β(4)	—	—	—	—	—	—
Macrocentrus resinelli	2	β	—	—	—	—	—	—
Macrocentrus thoracicus	2, 3	β	—	—	—	—	—	—
Chrysopidae								
Chrysopa flava	2, 3	×β	25	21	43	1	—	10
Italochrysa stigmata	3	×β	—	—	—	—	—	—
Cimbicidae								
Cimbex fermorata	3	β(3)	—	—	—	—	—	—
Trichiosama sp.	3	β(3)	—	—	—	—	—	—
Cosmopterigidae								
Mompha ochracella	1	β(3b)	—	—	—	—	—	—
Diprionidae								
Diprion pini	3	β(3)	—	—	—	—	—	—
Gilpinia sp.	3	β(3)	—	—	—	—	—	—
Neodiprion pratti	3	β(3)	—	—	—	—	—	—
Dolerinae								
Dolerus sp.	2	β	—	—	—	—	—	—
Galleriidae								
Achroia grisella	1	β(3b)	—	—	—	—	—	—
Galleria mellonella	1	β(3b)	32	25	17	1	5	7

TABLE 13.1 (*continued*)

Class, family or subfamily, genus, and species	Ref.[a]	X-ray classification[b]	Content of selected residues[c] (residues/100 residues)					
			Gly	Ala	Ser	Glu	Pro	Ref.[a]
Heterarthrinae								
Caliroa sp.	2	PG II(0)	—	—	—	—	—	—
Endelomyia sp.	2	PG II(0)	—	—	—	—	—	—
Fenusa sp.	2	PG II(0)	—	—	—	—	—	—
Hydrophilidae								
Hydrophilus piceus	4	×β	—	—	—	—	—	—
Ichneumonidae								
Dusona sp.	3	β	—	—	—	—	—	—
Phytodietus sp.	3	β	—	—	—	—	—	—
Lasiocampidae								
Braura truncata	1	β(3a)	35	29	9	1	—	7
Catalebeda violescens	1	β(3a)	—	—	—	—	—	—
Dendrolimus pini	1	β(3b)	—	—	—	—	—	—
Eriogaster lanestris	1	β(3b)	—	—	—	—	—	—
Gastropacha quercifolia	1	β(3b)	—	—	—	—	—	—
Grammodora nigrolineata	1	β(3b)	—	—	—	—	—	—
Lasiocampa quercus	1	β(4)	19	23	17	7	—	7
Malacosoma neustria	1	β(3b)	29	30	18	2	1	5
Pachymeta flavia	1	β(3b)	31	34	11	2	—	7
Pachypasa otus	1	β(3b)	35	27	9	1	—	7
Philudoria potatoria	1	β(3b)	—	—	—	—	—	—
Trabala vishnu	1	β(3b)	—	—	—	—	—	—
Lymantriidae								
Euproctis chrysorrhoea	1	β(3b)	—	—	—	—	—	—
Lymantria dispar	1	β(3b)	—	—	—	—	—	—
Mantidae								
Mantis (ootheca)	2, 12	α	6	15	5	21	2	2
Megalodontoidea								
Macroxyela sp.	2	β	—	—	—	—	—	—
Neurotoma sp.	2	β	—	—	—	—	—	—
Mycetophilidae								
Arachnocampa luminosa	4	×β	—	—	—	—	—	—
Nematinae								
Cladius sp.	2	β	—	—	—	—	—	—
Hemichroa sp.	2	C	—	—	—	—	—	—
Nematus ribesii	2	C	28	14	11	10	10	3
Pachynematus sp.	2	C	—	—	—	—	—	—

TABLE 13.1 (*continued*)

Class, family or subfamily, genus, and species	Ref.[a]	X-ray classification[b]	Content of selected residues[c] (residues/100 residues)					
			Gly	Ala	Ser	Glu	Pro	Ref.[a]
Pikonema sp.	2	C	—	—	—	—	—	—
Pristiphora sp.	2	C	—	—	—	—	—	—
Trichiocampus sp.	2	β	—	—	—	—	—	—
Nymphidae								
Nymphes mymeloides	3	×β	—	—	—	—	—	—
Papilionidae								
Papilio machaon	1	β(3b)	—	—	—	—	—	—
Pergidae								
Perga affinis	3	β(3)	—	—	—	—	—	—
Pseudoperga sp.	3	β(3)	—	—	—	—	—	—
Pieridae								
Pieris brassicae	14	β(3b)	—	—	—	—	—	—
Plutellidae								
Argyresthia sp.	1	β(3b)	—	—	—	—	—	—
Psychidae								
Canephora asiatica	5	β(2b)	34	37	11	3	2	5
Clania sp.	1	β(2b)	37	34	10	3	1	7
Pulicidae								
Xenopsylla cheopis	2	α	—	—	—	—	—	—
Saturniidae								
Actias selene	1	β(3a)	—	—	—	—	—	—
Antheraea mylitta	1	β(3a)	27	41	11	1	0	3
Antheraea pernyi	1	β(3a)	27	44	12	1	0	11
Antheraea roylei	1	β(3a)	—	—	—	—	—	—
Antheraea yamamai	1	β(3a)	28	44	10	1	0	3
Attacus atlas	1	β(3a)	31	46	7	1	0	7
Automeris viridescens	1	β(3b)	—	—	—	—	—	—
Caligula eucalypti	1	β(3b)	—	—	—	—	—	—
Caligula grotei	1	β(3b)	—	—	—	—	—	—
Caligula japonica	13	β(3)	—	—	—	—	—	—
Callosamia promethea	1	β(3a)	34	46	8	6[e]	—	7
Cricula andrei	1	β(3b)	28	34	15	9[e]	—	7
Dictyoploca japonica	1	β(3b)	23	35	13	2	1	3
Dictyoploca simla	1	β(3b)	—	—	—	—	—	—
Graellsia isabellae	1	β(3b)	—	—	—	—	—	—
Loepa katinka	1	β(3a)	27	41	9	2	—	7
Philosamia cynthia	1	β(3a)	32	46	7	1	0	3
Rhodinia fugax	1	β(3b)	—	—	—	—	—	—

TABLE 13.1 (*continued*)

Class, family or subfamily, genus, and species	Ref.[a]	X-ray classification[b]	Content of selected residues[e] (residues/100 residues)					
			Gly	Ala	Ser	Glu	Pro	Ref.[a]
Rothschildia forbesi	1	β(3a)	—	—	—	—	—	—
Rothschildia jacobeae	1	β(3a)	—	—	—	—	—	—
Rothschildia orizaba	1	β(3a)	—	—	—	—	—	—
Samia cecropia	1	β(3a)	—	—	—	—	—	—
Samia rubra	1	β(3a)	—	—	—	—	—	—
Saturnia pavonia	1	β(3a)	—	—	—	—	—	—
Tela polyphemus	1	β(3a)	—	—	—	—	—	—
Tropaea luna	1	β(3b)	—	—	—	—	—	—
Selandriinae								
Strongylogaster sp.	2	β	—	—	—	—	—	—
Tenthredininae								
Aglaostigma sp.	2	β + C	—	—	—	—	—	—
Macrophya sp.	2	β	—	—	—	—	—	—
Thaumetopoeidae								
Anaphe infracta	1	β(2a)	31	57	3	1	1	3
Anaphe moloneyi	1, 15	β(2a)	42	53	0	0	0	3
Anaphe reticulata	1	β(2a)	—	—	—	—	—	—
Anaphe venata	1	β(2a)	42	54	0	0	0	3
Hypsoides unicolor	1	β(2a)	—	—	—	—	—	—
Thaumetopoea pityocampa	1	β(4)	24	14	22	5	—	7
Vespidae								
Vespa cabro	4	α	—	—	—	—	—	—
Vespa sylvestris	4	α	—	—	—	—	—	—
Yponomeutidae								
Yponomeuta evonymella	1	β(3b)	—	—	—	—	—	—

[a] 1, Warwicker (1960a); 2, Rudall and Kenchington (1971); 3, Lucas and Rudall (1968a); 4, Rudall (1962); 5, Kirimura (1962); 6, Lucas (1964); 7, Lucas *et al.* (1960); 8, Lucas (1966); 9, Lucas *et al.* (1958); 10, Lucas *et al.* (1957); 11, Schroeder and Kay (1955); 12, Rudall (1956); 13, Warwicker (1960b); 14, Hunt (1971); 15, Brown and Trotter (1956).

[b] The patterns are classified as α, β, cross-β ($\times \beta$), polyglycine II (PG II), and collagen-like (C). A number in parentheses following the classification indicates the grouping (Table 13.2).

[c] Rounded to nearest whole number.

[d] Largely as glutamine.

[e] Includes aspartic acid.

TABLE 13.2

Classification of Silks on the Basis of Conformation[a]

Group	Example	Conformation[b]	Special features of silks
—	*Phymatocera aterrima*	PG II	Contains about two-thirds Gly
0	*Nematus ribesii*	Collagenlike	Contains about one-third Gly and one-tenth Pro
1	*Bombyx mori*	β, $c = 9.3$ Å	Contains [Gly-(Ala, Ser)]$_n$ sequences
2a	*Anaphe moloneyi*	β, $c = 10.0$ Å	High proportion of (Gly + Ala + Ser)
2b	*Clania* sp.	β, $c = 10.0$ Å	—
3a	*Antheraea mylitta*	β, $c = 10.6$ Å	Contains [Ala]$_n$ sequences
3b	*Dictyoploca japonica*	β, $c = 10.6$ Å	—
4	*Thaumetopoea pityocampa*	β, $c = 15.0$ Å ⎫	Appreciable proportions of
5	*Nephila senegalensis*	β, $c = 15.7$ Å ⎬	bulkier side chains
6	*Digelansinus diversipes*	β, $c = 13.8$ Å	Rich in Ala and Gln, low in Gly
7	*Apis mellifera*	α	Low in Gly

[a] Warwicker (1960a), Lucas and Rudall (1968a), Rudall and Kenchington (1971).

[b] In the case of β structures, c is the unit-cell dimension normal to the plane of the pleated sheet.

II. CHEMICAL COMPOSITION

In many silks the fibrous component, known as fibroin, occurs in combination with a second material termed sericin and methods for the isolation of fibroin free of sericin have been studied in some detail in the case of *Bombyx mori* silk (Dunn *et al.*, 1944; Waldschmidt-Leitz and Zeiss, 1955) and some tussah silks (Lucas *et al.*, 1955; Schroeder and Kay, 1955). Similar methods have been used with other silks without such detailed study and it is likely that some of the published analyses may not be representative of pure fibroin. In certain silks, such as those produced by spiders, no sericin is present.

Extensive data have been collected on the amino acid compositions of silks and silk fibroins and these have been summarized by Lucas and Rudall (1968a). Wide variations in composition are observed from silk to silk but a general feature is the preponderance of a few types of amino acid residues. Where available the contents of selected residues have been included in Table 13.1 and the complete compositions of typical members of the conformation-based classifications are given in Table 13.3.

Some correlations between composition and conformation have emerged but a quantitative analysis is precluded due to the prevalence

TABLE 13.3

Amino Acid Compositions of Arthropod Silk Proteins
(residues/100 residues)

	Bombyx mori[a] fibroin	Antheraea mylitta[a] fibroin	Digelansinus diversipes[a] silk	Chrysopa flava[b] silk	Mantis[c] ribbons	Phymatocera aterrima[a] silk	Nematus ribesii[a,d] silk
Conformation	β	β	β	×β	α	PG II	collagen
Alanine	29.3	41.4	38.2	21.2	14.7	1.5	13.8
Arginine	0.5	3.9	0.5	0.5	5.0	2.9	4.0
Aspartic acid	1.3	6.3	2.2	6.0	9.5	7.6	5.8
Glutamic acid	1.0	0.9	36.3	0.8	21.4	0.7	9.8
Glycine	44.5	26.6	2.2	24.6	5.5	66.2	27.5
Half cystine	0.2	—	—	—	—	—	—
Histidine	0.2	1.0	0.6	—	2.7	1.1	0.5
Hydroxylysine	—	—	—	—	—	—	1.8
Isoleucine	0.7	0.3	1.9	—	2.4	2.2	2.0
Leucine	0.5	0.3	4.1	—	6.5	3.4	2.2
Lysine	0.3	0.2	0.5	—	10.1	2.1	1.7
Methionine	0.1	—	—	—	—	—	—
Phenylalanine	0.6	0.3	—	—	1.9	0.1	0.4
Proline	0.3	0.4	—	—	1.9	0.8	9.7
Serine	12.1	11.1	9.1	42.7	5.4	3.4	11.0
Threonine	0.9	0.4	0.5	3.1	—	0.3	3.5
Tryptophan	0.2	1.2	—	0.3	3.1	—	—
Tyrosine	5.2	4.7	2.1	0.8	3.8	5.7	2.5
Valine	2.2	0.9	1.9	—	3.2	2.0	3.8
Ammonia	—	0.8	35	—	—	3.1	—

[a] Lucas and Rudall (1968a).
[b] Lucas et al. (1957).
[c] Rudall and Kenchington (1971).
[d] Purified extract of prepupal glands has about 34 Gly, 12 Ala, and 10 Pro residues/100 (Rudall and Kenchington, 1971).

of a nonuniform distribution of residues between the crystalline and amorphous regions of the fiber. Lucas et al. (1960) analyzed fibroins from a wide range of beta silks and found considerable variations in composition although the content of amino acids with small side chains (Gly, Ala, Ser) was always high. Alpha silks contain very little glycine and appreciably greater quantities of charged side chains than do beta silks. The presence of the PG II conformation (Chapter 11, Section B.II) in silk from *Phymatocera aterrima* can be correlated with the very high glycine content (Table 13.3), and the presence of a collagenlike conformation in silk from *Nematus ribesii* can be correlated with the presence of about one-third glycine residues and one-tenth proline residues (Rudall and Kenchington, 1971).

B. BETA SILKS

I. CLASSIFICATION OF BETA SILKS

A number of comparative studies of the X-ray diffraction patterns of beta silks have been reported (Brill, 1923; Kratky and Kuriyama, 1931; Trogus and Hess, 1933; Warwicker, 1956, 1960a) and the last-named author suggested that they could be classified into five groups. All the patterns analyzed by Warwicker (1960a) could be indexed on orthogonal cells with $a = 9.44$ Å, $b = 6.95$ Å (fiber axis), and c in the range 9.3–15.7 Å, the magnitude of the c axis dimension forming the basis for classification (Table 13.2). Groups 2 and 3 were further subdivided on the basis of intensity differences between the equatorial reflections corresponding to 002 and (201 + 102). A further group of beta silks with c values intermediate between those of groups 3 and 4 were reported by Lucas and Rudall (1968a) and these were placed in a separate group (Table 13.2). The group numbers of the various beta silks so far classified are included in Table 13.1.

A study of fibroins produced by moth larvae suggested that the X-ray classification might be related to the biological classification of the insect producing it (Warwicker, 1956), but a later study (Warwicker, 1960a) showed that the correspondence was not complete and that anomalies occurred. Lucas *et al.* (1960) analyzed the fibroins which had been classified by Warwicker (1960a) but could find no precise relationships between composition and either X-ray or biological classification. In some instances the groupings have subsequently been shown to have a structural basis, as discussed in later sections.

II. BOMBYX MORI FIBROIN

The cocoon silk produced by the larva of the moth *Bombyx mori* has been used as a textile fiber for over 3000 years and the chemical and physical structure of the fibroin has been investigated in great detail. Useful reviews of early studies have been given by Zahn (1952), Kendrew (1954), Bamford *et al.* (1956), and Lucas *et al.* (1958).

a. *Chemical and Physicochemical Studies*

Fibroin and sericin are stored in a liquid form in the lumen of the silk gland and during the spinning process the fibroin is converted to a filamentous form and the sericin is deposited as an external coating. Both materials are then water-insoluble and attempts to separate the fibroin without degradation and to prepare it in a soluble form suitable for chemical and physicochemical studies have only met with limited success. Cupriethylene diamine has been used to dissolve cocoon fibroin (Coleman and Howitt, 1947) but this causes appreciable degradation and recent work has been carried out mainly on fibroin solubilized with concentrated lithium bromide or isocyanate solutions (von Weimarn, 1932; Signer and Strässle, 1947; Ambrose *et al.*, 1951; Waldschmidt-Leitz and Zeiss, 1955). The salt can subsequently be removed by dialysis, leaving an aqueous solution of fibroin which is metastable and gels on prolonged standing. An alternative means of obtaining aqueous solutions of fibroin is to disperse the fluid contents of the silk glands (Foà, 1912; Mercer, 1954; Iizuka, 1968; Tashiro and Otsuki, 1970a). Mechanical shearing of the contents of silk gland causes the fibroin to precipitate (Foà, 1912; Ramsden, 1938; Meyer and Jeannerat, 1939; Iizuka, 1966) and aqueous dispersions of cocoon fibroin or of gland fibroin exhibit similar properties.

Estimates of the molecular weight of fibroin based on measurements of solubilized cocoon fibroin have been reviewed by Lucas *et al.* (1958), Hyde and Wippler (1962), and Tashiro and Otsuki (1970a,b). A wide range of values has been reported and this can be attributed, in part, to different extents of degradation during the removal of sericin or during solubilization. These problems can be avoided by using aqueous dispersions of gland fibroin; Iizuka (1963) obtained an estimate of 435,000 for the molecular weight by light scattering and Lucas (1966) obtained a value of 400,000 by ultracentrifuge measurements. Tashiro and Otsuki (1970a) identified a particle of similar weight as the fibroin component of the gland contents and obtained a value of 370,000 for the molecular weight by sedimentation analysis. The particle weight was not influenced by the presence of denaturants but was reduced to about 170,000 in the presence of reducing agents. The original particle weight was restored on removal of the reducing agent and it was concluded that the fibroin molecule consists of two subunits with molecular weight about 170,000 linked by a disulfide bridge.

Although the sedimentation data suggest that the fibroin molecule comprises two subunits, the chain structure of the molecule is uncertain. Four types of N-terminal end groups have been detected in fibroin preparations and on this basis Shaw (1964) suggested that the molecule had a four-chain structure. A similar conclusion was reached by Lucas (1966) on the basis of the measured cystine content, which corresponded to four cystine residues per 400,000 molecular weight. Robson *et al.*

(1970) identified the cysteine-containing sequence and suggested that fibroin consists of a single type of chain of weight 103,000 daltons containing a single intrachain disulfide link.

Measurements of the infrared spectra of aqueous dispersions of fibroin (Fig. 13.2) show that the native β conformation is not retained in aqueous solution (Iizuka and Yang, 1966; Iizuka, 1968). Parallel

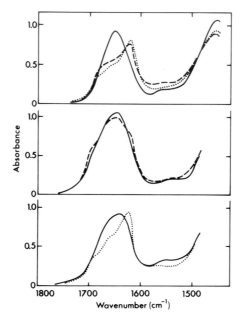

FIG 13.2. Infrared spectra of silk fibroins in water (solid line), 50% v/v water–methanol (dotted line), and 50% v/v water–dioxane (broken line). Top spectra, *Bombyx mori* fibroin; middle spectra, *Anaphe moloneyi* fibroin; bottom spectra, *Antheraea pernyi* fibroin (Iizuka, 1968).

studies of optical rotatory dispersion showed that the magnitude of the parameter b_0 was about zero so that the solubilization process does not involve a transformation from a β conformation to an α-helical conformation. Upon the addition of methanol or dioxane in excess of 30% (v/v), the infrared spectrum was found to change, with the appearance of maxima at 1620 and 1690 cm^{-1} (Fig. 13.2) indicating a partial transition to an antiparallel-chain, pleated-sheet conformation (Chapter 5, Section A.II.c).

Sequence studies of *B. mori* fibroin, which have been reviewed by Lucas *et al.* (1958) and Lucas and Rudall (1968a), suggest that segments of different character are present in the sequence. In the first type there appears to be a regular repetition of the sequence Gly-Ala-Gly-Ala-Gly-Ser; in the second type, Gly, Ala, Val, and Tyr occur, with Gly frequently occupying alternate positions in the sequence; and in the third type, polar and large neutral residues predominate. Treatment of fibroin solutions with chymotrypsin leads to the formation of a granular precipitate, representing about 60% of the original material, consisting almost entirely of the first type of segment. The precipitate has been termed the Cp fraction and shown to have the sequence

$$\text{Gly-Ala-Gly-Ala-Gly-[Ser-Gly-(Ala-Gly)}_n]_8\text{-Gly-Ala-Ala-Gly-Tyr}$$

where n has a mean value of two and is usually two (Lucas et al., 1957, 1958). Warwicker (1954) obtained a powder-type X-ray diffraction pattern from the Cp fraction and found that the spacings of the rings were similar to those observed in powdered cocoon fibroin, suggesting that the crystalline regions contain the repeating sequence Gly-Ala-Gly-Ala-Gly-Ser. Further support for this view was obtained from studies of a model hexapeptide (Schnabel and Zahn, 1958) and the model sequential polypeptide $[\text{Ala-Gly-Ala-Gly-Ser-Gly}]_n$ (Fraser et al., 1966).

b. *Silk I*

During the spinning of silk the liquid contents of the silk glands are converted to a fibrous, water-insoluble form and it seems likely that the process is accompanied by a conformational transition. A high proportion of the spun silk has a β conformation but the nature of the conformation of the fibroin molecule in the silk gland is unknown. Shimizu (1941) found that if the contents of the silk gland were dried without mechanical disturbance, the fibroin crystallized in a form which gave a powder diffraction pattern (Table 13.4) that was unrelated to the pattern obtained from cocoon fibroin. Kratky et al. (1950) discovered this modification independently and termed it silk I. So far it has not been established whether the silk I form of fibroin is involved in the spinning of silk *in vivo* but a transition from the silk I form to a disoriented β form can be induced by stretching or rolling swollen films.

Silk I has also been prepared from aqueous dispersions of cocoon fibroin (Bamford et al., 1956; Kratky, 1956) and the Cp fraction (Konishi et al., 1967; Konishi and Kurokawa, 1968; Hayakawa et al., 1970). As with the silk gland contents, the diffraction pattern can be transformed to a disoriented β pattern by stretching or rolling. This property makes it difficult to prepare specimens of silk I with fiber-type

TABLE 13.4

Spacings of Powder Rings (Å) in the X-ray Pattern of the Silk I
Modification of *B. mori* Fibroin

	Gland silk			
	Form I[b]			Cocoon silk fibroin
				Cp fraction
β Form[a]	1 ·	2	3	form I[c]
9.7	7.55	7.30	7.25	7.20
4.69	5.64	5.66	5.56	5.64
4.30	4.56	4.50	4.50	4.59
3.67	—	4.16	—	4.26
2.78	4.00	3.92	3.90	3.93
2.27	3.67	3.62	3.60	3.64
2.07	3.20	3.18	3.16	3.24
—	2.77	2.74	2.77	2.81
—	2.47	2.40	2.44	2.44
—	2.25	2.26	2.25	2.27
—	—	2.01	—	2.04

[a] Shimizu (1941).
[b] 1, Shimizu (1941); 2, Kratky (1956); 3, Hirabayashi *et al.* (1967); see also Mercer (1954).
[c] Konishi and Kurokawa (1968).

orientation suitable for structural studies. Konishi *et al.* (1967) and Konishi and Kurokawa (1968) obtained specimens of the Cp fraction with preferred uniplanar orientation and on the basis of X-ray and electron diffraction studies suggested that the unit cell was orthogonal, with $a = 4.59$ Å, $b = 9.08$ Å (fiber axis), and $c = 7.20$ Å, and contained a single chain. It was also suggested that the b dimension corresponded to the projected length of four residues.

Measurements of the infrared spectra of films of silk I (Ambrose *et al.*, 1951; Lenormant, 1956; Iizuka and Yang, 1966; Iizuka, 1968; Konishi and Kurokawa, 1968; Hayakawa *et al.*, 1970) have shown that the frequencies of the amide bands differ from those associated with the β conformation (Chapter 5, Section A.II.c). On the other hand the optical rotatory dispersion parameter b_0 has a value of about zero (Elliott *et al.*, 1957b), which would seem to rule out the possibility of an α-helical conformation. Hirabayashi *et al.* (1967, 1968) succeeded in preparing oriented specimens which exhibited a small amount of infrared dichroism and the character of the dichroism suggested that the hydrogen bonds were preferentially oriented perpendicular to the fiber axis. This observation is consistent with studies of a related crystalline form of [Ala-Gly]$_n$ (Chapter 10, Section C.I) which suggest that the chains form sheetlike structures stabilized by laterally directed interchain hydrogen bonds. The X-ray data for the synthetic polymer can be indexed on a cell similar to that suggested by Konishi and Kurokawa (1968) provided that the c axis is doubled (Lotz and Keith, 1971). In the model for silk I suggested by the latter authors the chain directions within a sheet are parallel and it is not immediately clear how the transformation to an antiparallel-chain pleated sheet takes place.

c. *Unit Cell*

Diffraction patterns were recorded from silk fibers at a very early stage in the development of the X-ray method (Nishikawa and Ono, 1913; Herzog and Jancke, 1920, 1921). The discrete reflections (Fig. 13.3b)

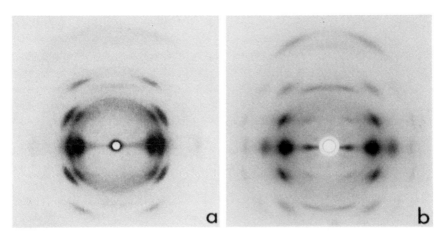

FIG. 13.3. X-ray diffraction patterns obtained from (a) *Antheraea mylitta* fibroin; (b) *Bombyx mori* fibroin (Lucas and Rudall, 1968b).

originate from the crystalline regions of the fiber and Brill (1923) showed that a pseudorepeat of structure of around 7 Å exists parallel to the fiber axis, leading to the appearance of layer lines. Kratky and Kuriyama (1931) estimated the value of the repeat to be 6.95 ± 0.25 Å and later studies gave values of 6.94 Å (Bamford *et al.*, 1953a), 6.97 ± 0.03 Å (Marsh *et al.*, 1955b), and 6.95 ± 0.05 Å (Warwicker, 1960a,b).

The elucidation of the remaining cell parameters proved more difficult. Following the discovery that doubly oriented specimens (Chapter 5, Section B.II.a) could be prepared either from silk gut

(Herzog and Jancke, 1929) or from silkworm gland (Kratky, 1929) by rolling, Kratky and Kuriyama (1931) identified prominent reciprocal lattice vectors perpendicular to the fiber axis and used these to construct a series of possible monoclinic cells. Little further progress was made until it became apparent, through the studies of Pauling and Corey (1953a), that the crystal structures of the beta silks were based on the antiparallel-chain pleated sheet (Chapter 10, Section A.I). The additional requirements imposed by this feature led to the suggestion that the unit cell was orthogonal with dimensions estimated to be $a = 9.44$ Å, $b = 6.95$ Å, and $c = 9.29$ Å by Warwicker (1954) and $a = 9.40$, $b = 6.97$, and $c = 9.20$ Å by Marsh *et al.* (1955b). This orthogonal cell is closely related to one of the cells proposed by Kratky and Kuriyama (Warwicker, 1954) and the a and c axes are oriented respectively parallel and perpendicular to the direction of rolling in doubly oriented specimens. A comparison of the observed and calculated spacings of the equatorial reflections is given in Table 13.5 and the

TABLE 13.5

Equatorial Reflections from *B. mori* Fibroin

Observed spacing[a] (Å)			Character[b]	Preferred[b] orientation	Index[c]	Calculated spacing[c] (Å)
1	2	3				
9.18	9.29	9.70	Broad	‖	001	9.20
4.59	4.68	4.70	Broad	‖	002	4.60
4.33	4.30	4.25	Medium sharp	~70°	201, 20$\bar{1}$	4.19
3.07	3.06	3.05	Broad	‖	003	3.07
2.38	2.38	2.35	Sharp	⊥	400	2.35
—	—	2.10	—	—	401, 40$\bar{2}$	2.09
—	—	1.80	Broad	‖	005	1.84
—	1.56	1.56	Sharp	⊥	600	1.57
					601, 60$\bar{1}$	1.54
—	—	1.20	Sharp	⊥	800	1.18

[a] 1, Kratky and Kuriyama (1931); 2, Warwicker (1954); 3, Marsh *et al.* (1955b).

[b] Kratky and Kuriyama (1931); Marsh *et al.* (1955b). The preferred orientation of the reflecting planes in doubly oriented specimens is referred to the plane of rolling.

[c] Marsh *et al.* (1955b), based on an orthogonal cell with $a = 9.40$ Å, $b = 6.97$ Å, and $c = 9.20$ Å.

agreement can be seen to be satisfactory except in the case of the 001 reflection. Possible reasons for this discrepancy were considered by Marsh *et al.* (1955b) and Rudall (1962) but no entirely satisfactory explanation was found.

d. *Crystal Structure*

Meyer and Mark (1928), from a consideration of the X-ray and chemical data then available, put forward the suggestion that the crystalline regions of the fiber contained chains oriented parallel to the fiber axis with long runs of alternating Gly and Ala residues. The repeat of about 7 Å parallel to the fiber axis was associated with the length of the two residue units and the possibility was considered that there might be occasional replacement of Gly or Ala by other types of residue. The importance of interactions between the amide CO and NH groups in neighboring chains was also recognized. These concepts were further refined and developed by Brill (1943) and Huggins (1943) but little progress toward the elucidation of the crystal structure was possible until the stereochemistry of the polypeptide chain was better understood.

Following the description by Pauling and Corey (1953a) of the pleated-sheet conformations, Warwicker (1954) and Marsh *et al.* (1955b) described model structures in which antiparallel-chain pleated sheets of the type shown in Fig. 10.7c were packed together with the Gly surfaces in contact and the Ala surfaces in contact (Fig. 10.8b). The intersheet distance thus alternated between values of around 3.5 Å appropriate to the contact between the Gly surfaces and around 5.7 Å between the Ala surfaces, giving a repeat of structure parallel to *c* of 9.2 Å.

In the model proposed by Warwicker (1954) the space group was orthorhombic and the chains in consecutive sheets were directly in line with each other when viewed from a direction perpendicular to the plane of the sheet, i.e., down the *c* axis. Marsh *et al.* (1955b) pointed out that the strong equatorial reflection of spacing 4.25 Å can be indexed unambiguously as 201 on the basis of data obtained from doubly oriented specimens, and yet the value of the intensity calculated for this orthorhombic structure was very low.

In the structure proposed by Marsh *et al.* (1955b) the space group is monoclinic and the sheets are packed so that the chains in consecutive sheets are staggered when viewed down the *c* axis (Fig. 13.4). Atomic coordinates for the model are given in Table 13.6 and a comparison of the observed and calculated intensities for the equatorial reflections is given in Table 13.7a.

The calculated intensity transform agrees with the observed data to the extent that strong diffraction is predicted in the correct regions of reciprocal space, but the relative intensities predicted from the model are not in agreement with the observed intensities. Marsh *et al.* (1955b)

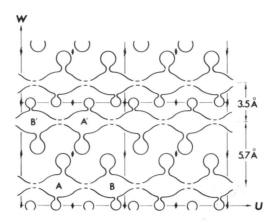

FIG. 13.4. Projection, down the fiber axis, of the model proposed by Marsh *et al.* (1955b) for the pseudo repeat of structure in the crystalline regions of *Bombyx mori* fibroin.

TABLE 13.6

Atomic Coordinates in the Model Proposed for the Crystallites of *B. mori* Silk Fibroin by Marsh *et al.* (1955b)[a]

Atom		u (Å)	v (Å)	w (Å)	Atom		u (Å)	v (Å)	w (Å)
Chain A	N_1	3.80	0.52	1.50	Chain B	N_1	8.50	6.45	1.98
	$C_1{}^\alpha$	3.20	1.76	1.04		$C_1{}^\alpha$	7.90	5.21	2.44
	$C_1{}'$	3.97	2.97	1.59		$C_1{}^\beta$	7.90	5.21	3.96
	O_1	5.20	2.97	1.59		$C_1{}'$	8.67	4.00	1.89
	N_2	3.25	4.00	1.98		O_1	0.50	4.00	1.89
	$C_2{}^\alpha$	3.85	5.24	2.44		N_2	7.95	2.97	1.50
	$C_2{}^\beta$	3.85	5.24	3.96		$C_2{}^\alpha$	8.55	1.73	1.04
	$C_2{}'$	3.08	6.45	1.89		$C_2{}'$	7.78	0.52	1.59
	O_2	1.85	6.45	1.89		O_2	6.55	0.52	1.59

[a] Monoclinic unit cell, $a = 9.40$ Å, $b = 6.97$ Å, $c = 9.20$ Å, and $\beta = 90°$. Space group $P2_1$, equivalent positions u, v, w (chains A and B) and $-u, v + \frac{1}{2}b, -w$ (chains A′ and B′).

commented that some improvement might be effected by the application of an anisotropic temperature factor, but it is clear from the data presented in Table 13.7a that the anomaly between the intensities of the 002 and 201 reflections could never be removed by this means alone. In the observed pattern these reflections overlap and Marsh *et al.* (1955b) separated them by an empirical procedure which yielded the

TABLE 13.7a

Comparison of Observed Equatorial Intensities (I_o) in the X-Ray Diffraction Pattern of *Bombyx mori* Silk Fibroin with Intensities (I_c) Calculated[a] for the Model Proposed by Marsh *et al.* (1955b) and a Refined Model

Index hl	Original model I_c[b]	I_o[c]	Refined model I_c[b,d]	I_o[c,e]	Index hl	Original model I_c[b]	I_o[c]	Refined model I_c[b,d]	I_o[c,e]
00	—	—	—	—	33	0	—	0	—
10	0	—	0	—	42	2	—	1	—
01	56	20	12	12	24	4	<1	1	<1
11	0	—	0	—	50	0	—	0	—
20	0	—	0	—	43	3	—	1	—
02	100	100	100	100	34	0	—	0	—
21	90	200	67	82	05	10	—	2	—
12	1	—	0	—	51	1	—	0	—
22	8	—	2	—	15	1	—	0	—
30	0	—	0	—	52	1	—	0	—
03	75	40	25	24	25	0	—	0	—
31	1	—	1	—	44	0	—	0	—
13	1	—	1	—	53	0	—	0	—
32	1	—	0	—	35	0	—	0	—
23	0	—	1	—	60	0	—	0	—
40	11	4	6	3	61	6	4	2	2
04	0	—	3	—	06	1	—	0	—
41	0	—	0	—	16	1	—	0	—
14	1	—	0	—	62	2	—	0	—

[a] Calculated using atomic scattering factors given by Ibers (1962).

[b] The values quoted are for LPF^2, where LP is the combined Lorentz–polarization factor and F is the structure factor. The values have been scaled to make $I_c = 100$ for the 002 reflection.

[c] Data of Marsh *et al.* (1955b) for integrated intensities scaled to give $I_o = 100$ for the 002 reflection.

[d] Intersheet spacing across the glycyl surfaces increased from 3.48 Å to 3.87 Å and an anisotropic temperature factor applied with $B_a = 5$ Å^2 and $B_c = 10$ Å^2.

[e] The relative contributions of the 002 and 201 reflections have been adjusted on the basis of the integrated areas determined from a least squares analysis of the composite reflection (Fraser and Suzuki, 1966).

integrated intensities given in Table 13.7a. In a later study (Dobb *et al.*, 1967; Fraser and Suzuki, 1966) the contributions were estimated by more refined procedures and appreciably different values obtained for the integrated intensities (Table 13.7a). The agreement between the calculated and observed intensities was thereby improved but need for refinement remained.

TABLE 13.7b

Comparison of the Observed Intensities (I_o) of the 00l Reflections in [Ala-Gly]$_n$
with Intensities (I_c) Calculated[a] for the Model Proposed by Marsh *et al.* (1955b)
and a Refined Model

| Index | Original model | Refined model | |
hkl	I_c [b]	I_c [c]	I_o [d]
001	40	10	10
002	100	100	100
003	65	24	24
004	1	4	3

[a] Calculated using atomic scattering factors given by Ibers (1962),

[b] The values quoted are for LPF^2, where F is the structure factor, and are scaled to make $F^2 = 100$ for the 002 reflection.

[c] The values quoted are for $LPF^2 \exp(-\frac{1}{2}B_c D^2)$, where D is equal to the reciprocal of the observed spacing and $B_c = 9$ Å^2. The value of c was taken as 8.96 Å (Fraser *et al.*, 1966) and the intersheet spacing across the glycyl surfaces was increased from 3.48 to 3.79 Å.

[d] Integrated observed intensities.

The sequential polypeptide [Ala-Gly]$_n$ has been shown to have a crystal structure in the β form which is very similar to that of *B. mori* fibroin (Go *et al.*, 1956; Fraser *et al.*, 1965a) and intensities for the 00l reflections were calculated, by the latter authors, on the assumption that the model of Marsh *et al.* (1955b) was also applicable to the sequential polypeptide. The relative intensities of the 00l series were found to be very sensitive to the choice of the distances between the pleated sheets and satisfactory agreement with the observed data could only be obtained if the intersheet distance across the Gly surfaces was increased. A reexamination of this problem (Table 13.7b) showed that values of 3.79 and 5.17 Å for the intersheet spacings gave the best fit. The revised intersheet distance of 3.79 Å across the Gly surfaces is appreciably greater than that in the model (3.48 Å) and the revised value across the Ala surfaces (5.17 Å) is equal, within experimental error, to the value of 5.27 Å observed in the β form of [Ala]$_n$ by Arnott *et al.* (1967).

A similar refinement has been carried out on the basis of the observed data for silk fibroin (Table 13.7a) and once again an increase in the intersheet distance across the Gly surfaces was required in order to obtain satisfactory agreement. The revised intersheet distance is 3.87 Å, which is not significantly different from that estimated for [Ala-Gly]$_n$.

TABLE 13.8

Comparison of Structural Parameters in the Beta Conformation in
Silks and in Model Polymers

Material	Intersheet distance (Å)	Mean axial rise per residue[a] h (Å)	Interchain distance[a] (Å)	Ref.[b]
Homopolymers				
[Gly]$_n$	3.44	$\geqslant 3.48$	4.77	1, 2
[Ala]$_n$	5.27	3.45	4.73	3
[Ser]$_n$	5.4–5.5	—	—	4
Sequential polymers				
[Ala-Gly]$_n$	3.79, 5.17	3.47	4.72	5, 6
[Ala-Gly-Ala-Gly-Ser-Gly]$_n$	3.79,[c] 5.26	3.43	4.70	5
Beta silks				
B. mori fibroin	3.87, 5.33	3.49	4.70	6, 7
Cp fraction	3.87,[c] 5.26	3.44	4.69	5
Anaphe moleneyi fibroin	4.90	3.47	4.72	8
	5.00	3.48	4.72	9
Antheraea pernyi fibroin	5.30	3.48	4.72	10
Antheraea mylitta fibroin	5.30	3.48	4.72	9
Nephila madagascarensis silk	5.36	3.46	4.84	8
Chrysopa flava egg stalk	5.45	3.45	4.74	4
Lasiocampa quercus fibroin	7.50	3.48	4.72	9
Nephila senegalensis silk	7.85	3.48	4.72	9

[a] Rounded to two decimal places.

[b] 1, Astbury (1949c); 2, Bamford *et al.* (1953a); 3, Arnott *et al.* (1967); 4, Geddes *et al.* (1968); 5, Fraser *et al.* (1966); 6, Table 13.7; 7, Marsh *et al.* (1955b); 8, Brown and Trotter (1956); 9, Warwicker (1960b); 10, Marsh *et al.* (1955a).

[c] Assumed.

The intersheet spacings deduced for *B. mori* fibroin and for related synthetic polypeptides are compared in Table 13.8. It should be noted that the value given for [Gly]$_n$ can only be regarded as being very approximate. The extent and precision of the intensity data for the $h0l$ reflections (Table 13.7a) are insufficient to permit further refinement of the *b*-axis projection of the model.

The intensity data obtained by Marsh *et al.* (1955b) for reflections with $k \neq 0$ was only semiquantitative and agreement between the calculated and observed intensities was generally poor. It seems likely that the relative axial positions and rotations of the chain are incorrect and that refinement of the parameters Δz and $\Delta\phi$ (Chapter 10, Section A.I) is required. A further possibility which might need to be considered is that some randomness in sheet packing is present as in the crystal structures of $[\text{Ala}]_n$ (Chapter 10, Section B.II) and β-keratin (Chapter 16, Section D.III).

The model proposed by Marsh *et al.* (1955b) is based on a repeating Ala-Gly sequence, but as explained earlier, one-third of the alanine residues are replaced by serine residues in *B. mori* fibroin. The model therefore refers to a pseudo repeat of structure. Weak meridional reflections of spacing 20.9, 10.36, and 6.95 Å are observed in silk fibroin (Fig. 13.5) both in the native material (Marsh *et al.*, 1955b) and after

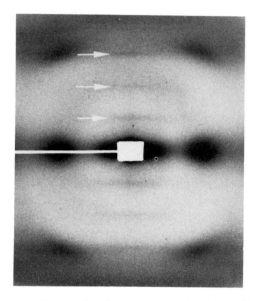

FIG. 13.5. Low-angle X-ray diffraction pattern obtained from *Bombyx mori* fibroin showing meridional streaks (arrowed) on the first three layer lines of a periodicity of about 21 Å.

nitration (Zahn *et al.*, 1951) or treatment with alkali (Kratky and Schauenstein, 1951). The reflections of spacing 20.9 and 10.36 Å show appreciable spreading along the layer line and it seems likely that the 20.9-Å periodicity is attributable to a periodic distortion of the pseudo-structure associated with the occurrence of serine residues at regular intervals in the sequence. Since nitration leads to the formation of 3-nitrotyrosine and 3,5-dinitrotyrosine, Zahn (1952) suggested that the enhancement of these reflections in nitrated silk indicated that tyrosine residues were regularly disposed on the surface of the crystallites or between them. Friedrich-Freksa *et al.* (1944), on the other hand, reported evidence of a 70-Å periodicity associated with tyrosine in iodinated silk. Further studies would be necessary before the regular disposition of tyrosine could be regarded as definitely established. Artificially produced silk gut exhibits a diffuse low-angle meridional reflection of spacing 80 Å (Fraser and MacRae, unpublished data) which is about one-half of the length of the sequence [Ser-Gly-Ala-Gly-Ala-Gly]$_8$ present in the Cp fraction.

e. *The Cp Fraction*

The precipitate obtained when solubilized silk is treated with chymotrypsin was examined by X-ray diffraction (Warwicker, 1954) and shown to give a powder pattern very similar to that obtained from powdered silk fibroin. Certain differences were noted, however, in the intensities and spacings of the rings. In a later study Fraser *et al.* (1966) obtained patterns from partially oriented specimens of this material, thus enabling the reflections to be indexed in an unambiguous manner and the unit cell was determined to be orthogonal, with $a = 9.38$ Å, $b = 6.87$ Å, and $c = 9.13$ Å. These dimensions are very close to values obtained for the sequential polypeptide [Ala-Gly-Ala-Gly-Ser-Gly]$_n$ (Chapter 10, Section C.III) and support the finding by Lucas *et al.* (1957, 1958) that the Cp fraction contains a repetition of this sequence. The similarity between the patterns of the Cp fraction and the model polymer is very striking (Fig. 13.6). The *c*-axis repeat of 9.05 Å in [Ala-Gly-Ala-Gly-Ser-Gly]$_n$ is slightly greater than in

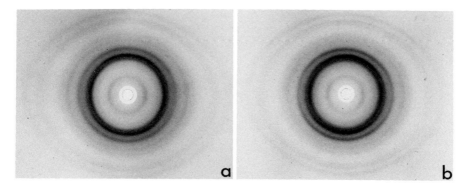

FIG. 13.6. Comparison of the X-ray diffraction patterns obtained from oriented films of (a) [Ala-Gly-Ala-Gly-Ser-Gly]$_n$; (b) the Cp fraction of *Bombyx mori* fibroin. The direction of preferred orientation is parallel to the length of the page and the X-ray beam was passed parallel to the plane of the film (Fraser *et al.*, 1966).

[Ala-Gly]$_n$ (8.96 Å), as would be expected from the presence of an additional oxygen atom on every sixth residue. The small difference between the c-axis repeat in [Ala-Gly-Ala-Gly-Ser-Gly]$_n$ (9.05 Å) and in the Cp fraction (9.13 Å) seems to be beyond experimental error and may be due to irregularities in the repeating sequence in the Cp fraction. These could either be present in the native fibroin molecule or could be introduced during solubilization by racemization of a proportion of the serine residues.

The differences in spacing between the powder rings of the Cp fraction and powdered silk observed by Warwicker (1954) can be explained on the basis of the differences in the cell dimensions of the two materials. The a dimensions are the same to within experimental error but both the b and c dimensions are greater in the native material. If the intersheet distance across the glycine surfaces of the sheet is assumed to be constant, it can be seen from Table 13.8 that there is a steady progression in the remaining intersheet distance from [Ala-Gly]$_n$ to [Ala-Gly-Ala-Gly-Ser-Gly]$_n$ to the Cp fraction to native silk fibroin.

Infrared spectra obtained from the Cp fraction show amide frequencies characteristic of the β conformation (Konishi and Kurokawa, 1968; Hayakawa et al., 1970).

f. Crystalline Texture

The reflections in the X-ray diffraction pattern of cocoon fibroin or of artificially produced silk gut are appreciably sharper in the Z direction of reciprocal space than in the R direction (Chapter 1, Section A.I.g) indicating that the crystallites are elongated in the direction of the fiber axis (Chapter 1, Section A.I.e). The lateral breadths of the reflections are different (Table 13.9) and studies of doubly oriented specimens have shown that the more diffuse reflections originate from planes perpendicular to the c axis, that is, parallel to the plane of the pleated sheet (Kratky, 1929; Marsh et al., 1955b; Dobb et al., 1967). From the discussion given in Chapter 1, Section A the diffuseness of the 00l reflections could be due either to lattice distortions (Chapter 1, Section A.IV.b) or to a small number of repeats of structure parallel to c (Chapter 1, Section A.I.e). In the former case the half-width of the reflection should increase with l but quantitative data obtained by Dobb et al. (1967) suggest that this is not so and thus the half-width can be correlated with the crystallite dimension measured in the direction of the normal to the planes (see, for example, Klug and Alexander, 1954).

From the measured half-widths of the equatorial reflections (Table 13.9) the lateral dimensions of the crystallites were estimated to be approximately 59 Å parallel to the a axis and 22 Å parallel to the

TABLE 13.9

The Lateral Dimensions of the Crystallites in *B. mori* Fibroin[a]

Reflection			Crystallite dimension (Å)	
Index	Spacing (Å)	Half-width (Å⁻¹)	Parallel to *a* axis	Parallel to *c* axis
000	—	—	62[b]	—
001	9.70	0.0425	—	24[c]
002	4.70	0.048	—	21[c]
201	4.25	0.024	—	19[d]
003	3.05	0.043	—	23[c]
400	2.35	0.018	56[c]	—
		Mean value	59	22

[a] Estimated from the equatorial X-ray diffraction data by Dobb *et al.* (1967).

[b] Estimate obtained by assuming that the innermost low-angle maximum at 43.5 Å is the first subsidiary maximum of a crystallite of rectangular cross section.

[c] Estimate obtained by taking reciprocal of half-width.

[d] Estimate obtained by taking (sin 27°)/half-width since the normal to the 201 planes is inclined at 27° to the *a* axis.

c axis. About five pleated sheets each with 12 chains could be accommodated in a crystallite with these lateral dimensions. The preferred orientation about the *b* axis which may be induced by rolling (Kratky, 1929; Marsh *et al.*, 1955b) is consistent with such a form for the crystallites since the face with the longer lateral dimension (*ab*) lies in the plane of rolling.

The *c* dimension of the pseudocell proposed for the crystallites by Marsh *et al.* (1955b) is insufficient to accommodate residues larger than serine and alanine alternating with glycine, and thus about 40%, at least, of the fibroin molecule must be external to the crystalline regions since the Cp fraction constitutes about 60% of the fibroin (Lucas *et al.*, 1958). The picture therefore emerges of lamellar crystallites aligned parallel to the fiber axis embedded in a less crystalline, or possibly amorphous, matrix made up of those portions of the fibroin molecule that do not contain the repeating sequence Ser-Gly-Ala-Gly-Ala-Gly. Support for the concept of a crystalline–amorphous texture of this type has been obtained from studies of infrared spectra, discussed in Section B.II.g, and from studies of the effect of hydration on the X-ray diffraction pattern. Warwicker (1960b) found that the spacings of the discrete reflections in the high-angle pattern were not affected by the uptake of moisture by the fiber and it must therefore be accommodated in a noncrystalline phase.

Electron microscope studies of *B. mori* fibroin have been reviewed by Zahn (1952) and Sikorski (1963). Fragmented fibroin shows a fibrillar structure (Hegetschweiler, 1949; Zahn, 1949; Mercer, 1952; Marsh *et al.*, 1955b) but no clearly defined filaments of uniform size have been detected. A reported observation of ribbonlike filaments (Dobb *et al.*, 1967) obtained by ultrasonic fragmentation was later shown to be suspect because of the difficulty of obtaining preparations free of traces of cellulosic contaminants, which give abundant yields of such filaments when irradiated with ultrasonics (Millward, 1969).

Marsh *et al.* (1955b) observed diffuse equatorial reflections of spacing 30–40 and 15 Å associated with planes parallel to the plane of rolling. It was suggested that these reflections were due to a systematic variation of intersheet distance in imperfect crystals of large lateral extent parallel to the c axis. An alternative possibility which would be in better accord with the evidence, discussed earlier, for the presence of amorphous material is that the two diffuse reflections are subsidiary maxima associated with the lamellar shape of the crystallites. Dobb *et al.* (1967) estimated the spacing of the innermost maximum to be 43.5 Å and obtained an estimate for the extent of the crystallite parallel to the a axis which was in good agreement with the value obtained from the half-width of the 400 reflection (Table 13.9).

g. *Infrared Dichroism*

Very little work has been done on the infrared spectrum of natural silk fibers and most studies of dichroism have been carried out on sections of artificially produced silk gut, or suture, which is prepared commercially by hand-drawing from silk glands. The X-ray diffraction pattern of silk suture is very similar to that of the cocoon silk and so conclusions drawn from studies of suture are likely to be applicable, at least with regard to the crystalline regions, to the cocoon fibroin. Studies of dichroism in the amide bands have been reported by a number of investigators (Goldstein and Halford, 1949; Ambrose and Elliott, 1951b; Elliott, 1954; Sutherland, 1955; Elliott and Malcolm, 1956a; Hecht and Wood, 1956; Beer *et al.*, 1959; Suzuki, 1967) and the spectra obtained by Suzuki (1967) are reproduced in Fig. 13.7. The perpendicular dichroism of the amide A band and the principal component of the amide I band, together with the parallel dichroism of the principal component of the amide II band, are consistent with a β conformation in which the chain axes are preferentially oriented parallel to the fiber axis.

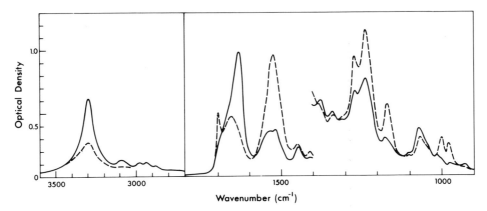

FIG. 13.7. Infrared spectrum of a thin section cut from a suture prepared from a *Bombyx mori* silk gland. Full line, electric vector vibrating perpendicular to the fiber axis; broken line, electric vector vibrating parallel to the fiber axis (Suzuki, 1967).

The amide I band has a weak parallel component at 1699 cm^{-1} (Fig. 13.7) which can be identified with the $\nu_\parallel(0, \pi)$ mode of the antiparallel-chain pleated sheet and a strong perpendicular component at 1630 cm^{-1} which is attributable to the $\nu_\perp(\pi, 0)$ mode (Chapter 5, Section A.II). In addition a broad underlying band with a maximum at 1656 cm^{-1} is present. The three components of the amide I vibration overlap considerably and the individual bands were separated by an iterative nonlinear least squares procedure (Fraser and Suzuki, 1966) as shown in Fig. 13.8. The $\nu_\perp(\pi, \pi)$ mode (Chapter 5, Section A.II) is likely to be weak (Miyazawa, 1960a) and was not detected.

The band parameters of the amide I components are summarized in Table 13.10. There is a marked difference between the half band-widths of the $\nu_\parallel(0, \pi)$ and $\nu_\perp(\pi, 0)$ modes which can be correlated with the different crystallite dimension parallel to the *b* and *a* axes, respectively (Suzuki, 1967). The inclination of the transition moment to the chain axis for the unperturbed amide I vibration of the crystalline regions was calculated from the results given in Table 13.10 and found to be consistent with the model proposed by Marsh *et al.* (1955b) provided the transition moment direction was assumed to be inclined at an angle of 19° to the CO direction. This is close to the value estimated from studies of simple compounds and β-keratin (Table 5.2).

The broad underlying band (Fig. 13.7) has a frequency close to that expected for a polypeptide chain without regular secondary structure (Miyazawa and Blout, 1961) but exhibits a small amount of perpendicular dichroism. The nature of the material responsible for this band was

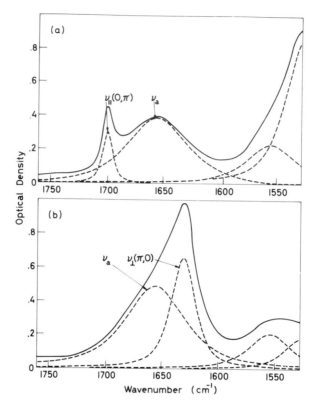

FIG. 13.8. Separation of the components of the amide I band in the infrared spectrum shown in Fig. 13.7. (a) Spectrum obtained with the electric vector vibrating parallel to the fiber axis; (b) spectrum obtained with the electric vector vibrating perpendicular to the fiber axis (Suzuki, 1967).

TABLE 13.10

Band Parameters of Amide I Components in the Infrared Spectrum of
Bombyx mori Fibroin[a]

Component[b]	Frequency of maximum (cm^{-1})	Half bandwidth (cm^{-1})	Peak absorbance	
			Parallel	Perpendicular
$\nu_\perp(\pi, 0)$	1630	25.9	0.000	0.655
$\nu_\parallel(0, \pi)$	1699	11.3	0.300	0.005
ν_r	1656	66.5	0.380	0.490

[a] Suzuki (1967).
[b] $\nu_\perp(\pi, 0)$ and $\nu_\parallel(0, \pi)$ are vibrational modes of the antiparallel-chain pleated sheet (Chapter 10, Section A.II.b), ν_r is associated with noncrystalline material.

further investigated by remeasuring the spectrum after the specimen had been exposed to deuterium oxide vapor for seven days. The $\nu_\perp(\pi, 0)$ and $\nu_\parallel(0, \pi)$ components were not affected but the broad band was displaced by about 10 cm^{-1} to a lower frequency. This displacement indicated that exchange of hydrogen for deuterium occurred in the amide groups of the material responsible for the broad band but did not penetrate the crystalline regions. This is in accord with the finding of Warwicker (1960b), mentioned earlier, that the spacings of the equatorial X-ray reflections are unchanged when the fiber is hydrated. If the material associated with the broad amide I band is regarded as amorphous, the proportion of amide groups in the crystalline material can be estimated from the integrated areas of the various components. A value of 30% is obtained in this way.

The presence of water-inaccessible amide groups in silk fibroin can be demonstrated directly by observation of the effect on the amide A band of exposing the specimen to deuterium oxide (Beer et al., 1959), and in a quantitative study (Suzuki, 1967) it was found that 32% of the amide groups were inaccessible to deuterium oxide. This corresponds closely to the estimate of the proportion of amide groups in crystalline regions but is only half the estimated content of Cp fraction (Lucas et al., 1958). The origin of this difference is not clear but it is possible that crystallization is incomplete in the artificially produced silk gut used for the infrared studies.

A number of studies of the combination bands which occur in the range 3800–5400 cm^{-1} have been reported (Bath and Ellis, 1941; Ambrose and Elliott, 1951b; Elliott et al., 1954; Fraser, 1955, 1956c; Hecht and Wood, 1956; Fraser and MacRae, 1958c) and spectra obtained by the last-named authors are reproduced in Fig. 13.9. The prominent peak at 4860 cm^{-1} is associated with a combination of the amide II and NH stretching modes (Bath and Ellis, 1941; Glatt and Ellis, 1948) and the peaks at 4530 and 4600 cm^{-1} with a combination of the amide III and NH stretching modes (Fraser, 1956a,b; Fraser and MacRae, 1958c) in the β-crystallites and in the noncrystalline regions, respectively (Ambrose and Elliott, 1951b; Elliott et al., 1954). Treatment of oven-dried cocoon fibers with deuterium oxide resulted in an appreciable increase in the dichroic ratios of the bands at 4530 and 4860 cm^{-1} (Fig. 13.9) due to preferential exchange of hydrogen for deuterium in the noncrystalline regions.

It is difficult to make quantitative measurements of intensity in the combination band region of the spectrum (Fraser, 1956c) but the dichroic ratio of the band at 4860 cm^{-1} in dried cocoon fibroin (Figure 13.9) appears to be around 4. Even though the transition

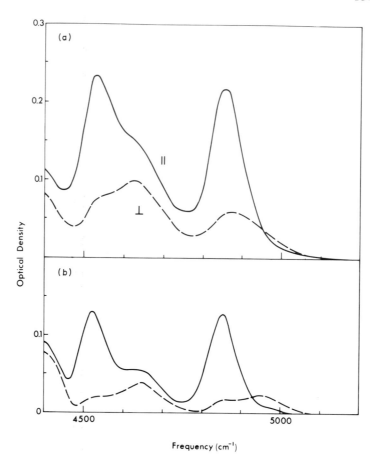

FIG. 13.9. Spectra obtained from *Bombyx mori* silk fibroin showing the amide combination bands in (a) dried fibers and (b) fibers dried after wetting with deuterium oxide. The full curves were obtained with the electric vector vibrating parallel to the fiber axis and the broken curves with the electric vector vibrating perpendicular to the fiber axis (Fraser and MacRae, 1958c).

moment direction is unknown, this enables a lower limit of about 50% to be set for the fraction of oriented material (Fraser, 1958b). This is appreciably greater than the β-crystallite content in silk gut estimated from the amide I components and may reflect a higher degree of crystallinity in the biologically produced fiber. Iizuka (1963) measured the crystallinity of cocoon fibroins of different diameters using an X-ray method and found that the estimated value varied between 40% and 46% with an average of 42%. It was also found that crystallinity tended to decrease with increasing fiber diameter. If this relationship may be extrapolated to include silk gut, which has an appreciably greater diameter than cocoon silk, this would account for the lower value of the crystallinity in the artificially produced fiber.

III. TUSSAH SILK FIBROINS

Tussah or wild silk is the product of larvae of moths belonging principally to the genus *Antheraea* and the X-ray diffraction patterns given by the fibroins belong to group 3a (Warwicker, 1960a). Amino acid analyses of *Antheraea* fibroins (Schroeder and Kay, 1955; Lucas and Rudall, 1968a) show a uniformly high content of alanine and a typical analysis is given in Table 13.2. Studies of partially hydrolyzed fibroin (Kay *et al.*, 1956; Shaw and Smith, 1961; Warwicker, 1961) suggest that the crystalline regions contain sequences with long repeats of alanyl residues.

The infrared spectrum of cocoon fibers of *A. mylitta* has been recorded in the combination band region (Elliott *et al.*, 1954) and bands with parallel dichroism were observed at about 4530 and 4650 cm^{-1}. The former frequency is characteristic of the β conformation (Elliott *et al.*, 1954) and the parallel dichroism can therefore be taken as an indication that the chains have a preferred orientation parallel to the fiber axis.

a. *X-Ray Diffraction Studies*

Early investigators (Kratky and Kuriyama, 1931; Trogus and Hess, 1933) noted that the X-ray diffraction patterns obtained from Tussah silks had many features in common with the pattern obtained from *B. mori* fibroin but significant differences were also observed (Fig. 13.3). Subsequent studies have established that both structures are based on the antiparallel-chain pleated sheet and that the differences between the diffraction patterns are related to different modes of sheet packing.

Bamford *et al.* (1953a) observed a meridional reflection of spacing 1.153 Å in the pattern from Tussah silk which they interpreted as the 060 reflection of an orthogonal or monoclinic cell giving a value of $b = 6.92$ Å for the identity distance parallel to the fiber axis. It was also found (Bamford *et al.*, 1954) that the pattern bore a striking resemblance to that given by $[Ala]_n$ in the β form, the only major difference being a lower intensity, in the silk, of the equatorial reflection with spacing 7.4 Å. In a more detailed study Brown and Trotter (1956) indexed the pattern on an orthogonal cell with $a = 4.80$ Å, $b = 6.91$ Å (fiber axis), and $c = 10.66$ Å which was essentially similar to that derived for the β form of $[Ala]_n$.

Marsh *et al.* (1955a, 1956) obtained a value of 6.95 ± 0.01 Å for the axial repeat of structure in fibroin from *A. pernyi* from measurements of the spacing of a reflection indexed as 060 and also measured the spacings and relative intensities of reflections on layer lines with $k = 0$–3. The pattern was indexed on an orthogonal unit cell with $a = 9.44$ Å, $b = 6.95$ Å (fiber axis), and $c = 10.6$ Å. The a-axis and b-axis identity distances are close to the values of 9.5 and 7.0 Å, respectively, suggested by Pauling and Corey (1953a) for the antiparallel-chain pleated sheet (Chapter 10, Section A.I) and on this basis a detailed model was derived for the crystalline regions of Tussah silk fibroin. The b-axis projection of the model is illustrated in Fig. 13.10 and the

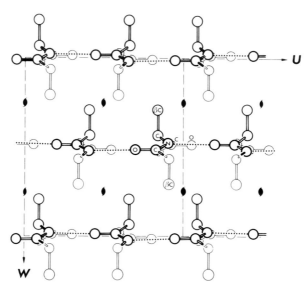

FIG. 13.10. Projection, down the fiber axis, of the model proposed by Marsh *et al.* (1955a) for the pseudo repeat of structure in the crystalline regions of *Antheraea pernyi* fibroin.

atomic positions are listed in Table 13.11. Adjacent sheets are separated by a constant distance of 5.30 Å, which is very similar to the intersheet distance in $[Ala]_n$ and also to the distance between sheets with Ala surfaces in contact in $[Ala-Gly]_n$ (Table 13.8). The space group of the structure is $P2_12_12_1$ and the cell contains four asymmetric units each comprising two residues. The calculated intensity transform of the structure was found to be in general agreement with the observed intensity data but since this was only qualitative it is not possible to assess the extent to which refinement of the model is required.

TABLE 13.11

Atomic Coordinates of the Repeating Unit in a Model for Tussah Silk Fibroin[a]

Atom	u (Å)	v (Å)	w (Å)	Atom	u (Å)	v (Å)	w (Å)
N_1	3.27	2.25	5.53	N_2	3.81	5.73	5.07
C_1^{α}	3.87	3.49	6.00	C_2^{α}	3.21	6.96	4.60
C_1^{β}	3.87	3.49	7.52	C_2^{β}	3.21	6.96	3.08
C_1'	3.10	4.70	5.45	C_2'	3.98	8.17	5.15
O_1	1.86	4.70	5.45	O_2	5.22	8.17	5.15

[a] Proposed by Marsh et al. (1955a). Orthorhombic unit cell, $a = 9.44$ Å, $b = 6.95$ Å, and $c = 10.60$ Å. Space group $P2_12_12_1$, equivalent positions u, v, w; $4.72 - u, -v, 5.3 + w$; $-u, 3.475 + v, 5.3 - w$; $4.72 + u, 3.475 - v, -w$.

The 00l reflections were found by Marsh et al. (1955a, 1956) to be diffuse, whereas the h00 and h0l reflections were found to be relatively sharp. It was suggested that the diffuseness of the 00l series was due to disorder in sheet packing, but the possibility that the breadth is due to a smaller crystallite dimension parallel to the c axis must also be considered. In the proposed model the a-axis translation between adjacent sheets is alternately $a/4$ and $-a/4$ (Fig. 13.10), whereas in [Ala]$_n$ this translation appears to be randomly $a/4$ or $-a/4$. In the case of an infinite crystallite dimension parallel to the c axis the latter situation leads to an effective halving of the a-axis identity distance and reflections with odd h are no longer observed (Chapter 10, Section B). When the crystallite only contains a limited number of sheets, however, the reflections with odd h are attenuated rather than eliminated (Chapter 16, Section D.III) and the data presently available on Tussah silk do not enable this possibility to be eliminated.

b. *Studies of Solubilized Fibroin*

Iizuka (1968) studied the optical rotatory dispersion of aqueous solutions of the gland silk of *A. pernyi* and obtained a value of -110 deg cm^2 dmole^{-1} for the value of the parameter b_0 (Chapter 8, Section B). This was taken as an indication that about 15% of the material had an α-helical conformation and it was found that in the presence of 8 M urea the helix content was negligible. Infrared spectra obtained from a solution of gland silk in deuterium oxide (Fig. 13.2) exhibited a broad amide I band with a maximum at about 1650 cm^{-1} but on the addition of methanol or dioxane in excess of 30% (v/v) an additional maximum was observed at 1620 cm^{-1} together with a shoulder near 1690 cm^{-1}.

This was interpreted as indicating a partial transformation to an antiparallel-chain β conformation.

Films cast from aqueous solutions of *A. mylitta* fibroin (Elliott and Malcolm, 1956b) and *A. pernyi* fibroin (Hirabayashi *et al.*, 1970) exhibit X-ray diffraction patterns of the powder type with a strong ring at a spacing of 7.4 Å. This corresponds to the $10\bar{1}0$ reflection of a hexagonal cell with $a = 8.55$ Å such as is found in the α form of $[Ala]_n$ (Chapter 9, Section B.I) and it seems likely that a proportion of the fibroin in these films has an α-helical conformation. In both cases this was confirmed by measurements of the infrared spectrum. Elliott and Malcolm (1956b) showed that the spectrum in the range 850–1350 cm^{-1} closely resembled that of the α form of $[Ala]_n$, while Hirabayashi *et al.* (1970) observed an amide V component with a frequency of around 620 cm^{-1}, believed to be characteristic of the α-helix. In stretched films (Hirabayashi *et al.*, 1970) or in films cast from 50% methanol (Iizuka, 1968) the β conformation was found to predominate.

IV. CROSS-BETA SILKS

The occurrence of the cross-β texture in certain preparations of synthetic polypeptides was discussed in Chapter 10, Section A.II and attributed to the formation of pleated sheets which were more extensive in the hydrogen bond direction than in the chain direction. Subsequent shearing then tended to align the sheets so that the chain axes were preferentially oriented perpendicular to the direction of shear. As might be expected, the cross-β texture is common in synthetic polypeptides with a low degree of polymerization since the development of the pleated sheet in a direction parallel to the chain axes is necessarily limited (Fig. 13.11, left). The cross-β texture has been observed in a variety of denatured globular proteins (Astbury *et al.*, 1935, 1959; Senti *et al.*, 1943; Ambrose and Elliott, 1951b; Burke and Rougvie, 1972) and denatured fibrous proteins (Rudall, 1946, 1952; Astbury, 1947b; Mercer, 1949; Peacock, 1959) but in these cases the length of the polypeptide chain greatly exceeds the dimension of the pleated sheet parallel to the b axis and it seems likely that the chains are folded, possibly as indicated in Fig. 13.11(right) (Rudall, 1946; Astbury *et al.*, 1959).

The examples of cross-β textures mentioned thus far were all produced by artificial means, but Parker and Rudall (1957) found that this texture also occurs naturally in the egg stalk of a green lace-wing fly, *Chrysopa flava*, and subsequently other examples have been found among the insect silks.

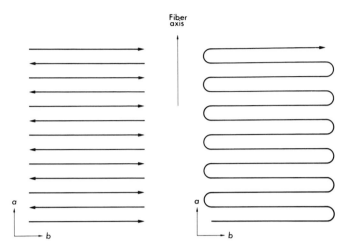

FIG. 13.11. Modes of formation of the cross-β texture. In synthetic polypeptides with a low degree of polymerization (left) sheets are formed that become elongated parallel to the *a* axis. A similar texture can result from multiple folding of a single polypeptide chain (right).

a. *Chrysopa flava Egg Stalk*

The conformation and crystalline texture in *Chrysopa* egg stalk have been investigated in some detail (Parker and Rudall, 1957; Rudall, 1962; Geddes *et al.*, 1968). The X-ray diffraction pattern (Fig. 13.12a) consists

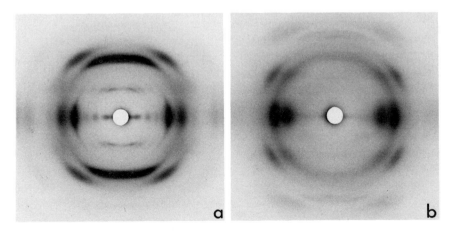

FIG. 13.12. X-ray diffraction patterns given by *Chrysopa flava* egg stalk; (a) untreated; (b) stretched fivefold (Geddes *et al.*, 1968).

of a series of layer lines corresponding to an axial repeat or pseudorepeat of 9.48 Å, which is close to the identity distance of the antiparallel-chain pleated sheet in the hydrogen bond direction. In addition, there is a strong equatorial reflection of spacing 5.45 Å which, on the basis of the composition of the egg stalk (Table 13.3) and X-ray studies of other silks (Table 13.8), can be attributed to an intersheet spacing. These observations suggested that the chain axes are preferentially aligned perpendicular to the fiber axis.

Infrared spectra obtained from *Chrysopa* egg stalks (Parker and Rudall, 1957; Parker, 1969) exhibit a nondichroic amide I component at 1650 cm^{-1}, presumably attributable to noncrystalline material, and a component of frequency 1625 cm^{-1} with parallel dichroism. This is close to the frequency associated with the $\nu_{\perp}(\pi, 0)$ mode of the anti-parallel-chain pleated sheet (Chapter 5, Section A.II) and the parallel dichroism is consistent with a cross-β texture. Complementary studies of combination bands confirmed the inferences drawn from the fundamentals (Parker and Rudall, 1957).

Freshly produced egg stalks can be stretched in water to about five or six times their original length and the high-angle X-ray diffraction pattern of the stretched material (Fig. 13.12b) closely resembles that of Tussah silk fibroin (Fig. 13.3a), in which the chain axes are preferentially aligned parallel to the fiber axis. In addition to the reflections associated with the pleated-sheet structure, a series of meridional reflections which were orders of 55.2 Å were also observed in the pattern of stretched egg stalk (Geddes *et al.*, 1968). This distance corresponds to a repeating unit 16 residues long and it was suggested that in the native material the polypeptide chains were folded as shown in Fig. 13.13d so that the sheets were eight residues wide in the direction of the chain axis. Thus the 16 residues would occupy an axial distance of 9.48 Å in the native egg stalk and 55.2 Å in the stretched material. The ratio of these two distances is of the order of six, which was the maximum extension most often observed.

In addition to reflections that can be attributed to the pseudorepeat of structure in the pleated sheet, weak equatorial maxima are observed at medium angles in wet specimens (Fig. 13.14a). Geddes *et al.* (1968) showed that these maxima could be accounted for by supposing that the crystalline portion of the egg stalk consisted of ribbonlike filaments about 25 Å thick (Fig. 13.15a). In dried specimens additional maxima were noted (Fig. 13.14e) and it was shown that these were probably due to the superposition of an interference function (Chapter 1, Section A.I.e) due to intercrystallite effects. The positions and relative intensities of the observed maxima were found to be explicable on the

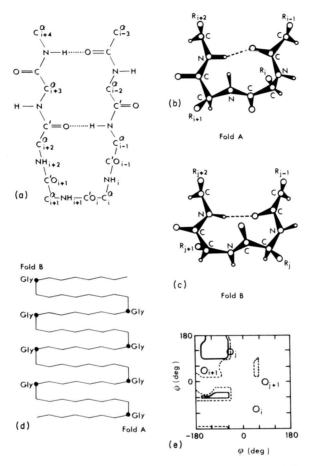

FIG. 13.13. (a) Schematic projection perpendicular to the pleated sheet of a simple reversal of chain direction. (b, c) Views, inclined at about 40° to the plane of the pleated sheet, of two folds that allow a hydrogen bond to be formed between residues $i-1$ and $i+2$ (or $j-1$ and $j+2$). (d) Schematic view of folded chain in which folds A and B occur alternately, giving a 16-residue periodicity. (e) Steric map for the Ala residue showing the (ϕ, ψ) values for the bend residues. If plotted on a steric map for Gly the (ϕ, ψ) values for residue i lies in an "allowed" region (Geddes *et al.*, 1968).

basis of a model in which the crystallites were packed with the planes of the ribbons parallel to one another. The mean distance between opposing faces was supposed to vary in a random manner about some mean value and it was found that a mean fluctuation of 4 Å about a value of 15 Å gave reasonable agreement between the observed and calculated intensity transforms (Fig. 13.14d).

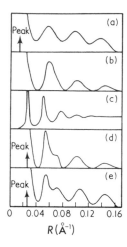

FIG. 13.14.   (a) Equatorial distribution of intensity in the
X-ray diffraction pattern of hydrated *Chrysopa flava* egg stalk.
(b) Distribution calculated for a slab of uniform electron
density 24.4 Å wide. (c) Interference function for the disordered linear lattice described in the text. (d) Product of (b)
and (c). (e) Equatorial distribution of intensity in the pattern
obtained from dried egg stalk (Geddes *et al.*, (1968).

Additional support for the ribbonlike model of the crystallites in
Chrysopa egg stalk was obtained from the observation that reflections
with $k \neq 0$ were diffuse while those with $k = 0$ were relatively sharp.
This implies that the crystallite has appreciably greater dimensions
parallel to a and c than parallel to b (Fig. 13.15a). Flower (1968) also

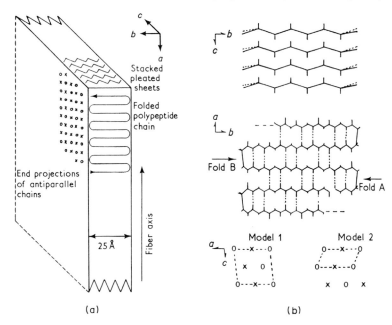

FIG. 13.15.   (a) Model for the ribbonlike filaments that occur in the *Chrysopa flava*
egg stalk; (b) details of chain conformation and packing (Geddes *et al.*, 1968).

obtained evidence for the presence of ribbonlike filaments in specimens prepared by drying down aqueous dispersions of the contents of the silk gland. Electron micrographs of negatively stained preparations (Fig. 13.16) show filaments of variable profile consistent with the

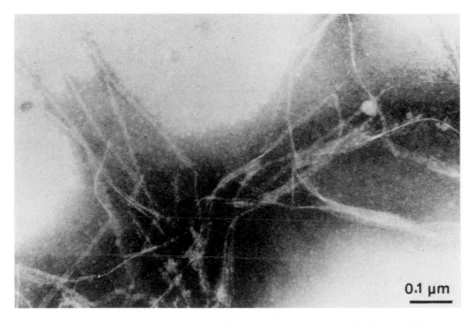

FIG. 13.16.   Electron micrograph of ribbons dispersed from the silk gland of *Chrysopa flava*. In some places the ribbons are viewed edge on (Flower, 1968).

appearance of twisted ribbons. The width at the narrowest points is consistent with a ribbonlike filament about 25 Å in thickness viewed edge on (Lucas and Rudall, 1968a).

Scale atomic models were used to investigate ways in which a polypeptide chain could be folded at regular intervals to give a structure of the type envisaged in Fig. 13.15a and two model bends were found which enabled a hydrogen bond to be formed between the NH group of residue $i - 1$ and the CO group of residue $i + 2$ (Fig. 13.13a). Atomic coordinates for these two model bends are given in Table 13.12 and the chain conformations are illustrated in Fig. 13.13b, c. The two conformations are topologically similar to those suggested by Venkatachalam (1968b) for 180° bends but additional restrictions are imposed in the present case since the residues adjacent to the bend form part of an antiparallel-chain pleated sheet. The dihedral angles ϕ and ψ for residues $i + 1$ in model A and j in model B lie within the extreme limits for an alanyl residue (Figs. 6.6 and 13.13e) but the values for residues i in model A and $j + 1$ in model B are outside these limits. In the case of residue i in model A the values fall in the "allowed" region for glycyl residues but the values for residue $j + 1$ in model B do not (Fig. 6.5b). However, values similar to those for residue $j + 1$ have been observed for glycyl residues in crystalline proteins

TABLE 13.12

Atomic Coordinates of the Bend Regions in a Model for *Chrysopa* Egg Stalk[a]

	Bend A[b]				Bend B[c]		
Atom	u (Å)	v (Å)	w (Å)	Atom	u (Å)	v (Å)	w (Å)
C'_{i-1}	0.91	-1.21	0.01	C'_{j-1}	0.87	-1.15	0.55
O_{i-1}	2.12	-1.08	-0.27	O_{j-1}	2.00	-0.92	0.98
N_i	0.33	-2.33	0.32	N_j	0.33	-2.35	0.52
C_i^α	1.00	-3.63	0.36	C_j^α	1.03	-3.55	1.03
C_i'	2.21	-3.56	1.31	C_j'	2.46	-3.71	0.53
O_i	2.06	-3.60	2.54	O_j	2.63	-4.09	-0.60
N_{i+1}	3.39	-3.48	0.71	N_{j+1}	3.39	-3.44	1.35
C_{i+1}^α	4.66	-3.45	1.44	C_{j+1}^α	4.82	-3.62	1.06
C_{i+1}'	5.48	-2.27	0.95	C_{j+1}'	5.52	-2.35	0.62
O_{i+1}	6.72	-2.29	0.94	O_{j+1}	6.76	-2.37	0.42
N_{i+2}	4.77	-1.24	0.52	N_{j+2}	4.77	-1.28	0.43
C_{i+2}^α	5.38	0	0	C_{j+2}^α	5.38	0	0

[a] Proposed by Geddes *et al.* (1968).
[b] See Fig. 13.13. Right-hand orthogonal system of coordinates with origin at C_{i-1}^α, u axis along $C_{i-1}^\alpha \rightarrow C_{i+2}^\alpha$, v axis along $C_{i-1}^\alpha \rightarrow C_{i-3}^\alpha$.
[c] See Fig. 13.13. Similar coordinate system to bend A with suffix i replaced by j.

(Ramachandran and Sasisekharan, 1968; Pohl, 1971; Quiocho and Lipscomb, 1971). It was suggested that bends of type A and B occurred alternately as shown in Fig. 13.13d and if this were so, the sequence in the crystalline regions would contain glycyl residues in positions $i + 16m$ and $i + 7 + 16m$, where $m = 0, 1, 2,\ldots$. This would provide a possible explanation for the periodicity of 55.2 Å parallel to the fiber axis in the stretched egg stalk.

In order to obtain a complete model for the crystallite it is necessary to define the interrelationship between adjacent sheets. On the basis of the observed spacings of the high-angle reflections two possible arrangements were derived and these are depicted in Fig. 13.15b.

Neither cell, however, leads to an entirely satisfactory explanation of the observed pattern of reflections. Intensity transforms were calculated for ribbonlike crystallites with cells of type 1 and type 2 (Fig. 13.15b) but apart from the equator, the agreement was poor (Fig. 13.17). Some improvement in overall fit was achieved by the inclusion of additional scattering matter at the bends but the relative intensities of the two main equatorial reflections in the calculated intensity transform were no longer in agreement with the observed ratio (Fig. 13.17a). It seems

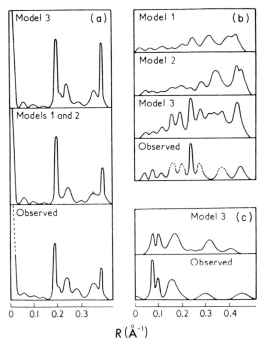

FIG. 13.17. Comparison of observed intensity transform of *Chrysopa flava* egg stalk with that calculated for the two models depicted in Fig. 13.15b and for a third model with additional scattering material at the bends of the chain. (a) Equator, (b) first layer line, (c) second layer line. (Geddes *et al.*, 1968).

likely that the model developed by Geddes *et al.* (1968) is correct with regard to the broad features of the crystalline texture and structure, but further refinement of the detailed arrangements of residues appears to be required. The intercrystalline regions are presumably occupied by polypeptide chains without a regular secondary structure and in future calculations it would be necessary to allow for the effects of this material on the intensity transform (Chapter 1, Section D.III).

b. *Other Cross-Beta Silks*

Further examples of naturally occurring cross-β textures in insect silks have been recorded (Rudall, 1962; Lucas and Rudall, 1968a). These include the egg stalks from other species of *Chrysopa* and from *Italochrysa stigmata*, the cocoon fibers of the water beetle *Hydrophilus piceus*, and threads secreted by the glow-worm *Arachnocampa luminosa*. According to Rudall (1962), the X-ray diffraction patterns of the *Hydrophilus* and *Arachnocampa* silks differ from those of the *Chrysopa* egg stalks and are more closely related to the cross-β patterns obtained from denatured mammalian keratins (Rudall, 1946, 1952; Peacock, 1959).

V. OTHER BETA SILKS

X-ray diffraction patterns have been obtained from a variety of beta silks (Table 13.1) but apart from the materials discussed in Sections B.II–IV, no detailed investigations of conformation have been reported. Information on the structure of these less-well-characterized silks is summarized in the present section on the basis of the group classification of Table 13.2. Possible correlations among taxonomic classification, X-ray grouping, and composition have been discussed by Lucas *et al.* (1960), Warwicker (1960a), Rudall (1962), Lucas and Rudall (1968a), and Rudall and Kenchington (1971). As mentioned earlier, the overall amino acid composition is not necessarily representative of the crystalline regions and thus correlations between X-ray grouping and overall composition are subject to this reservation. Preliminary studies of certain beta silks in which the c-axis repeat appears to be outside the range of the present scheme have been reported by Rudall and Kenchington (1971).

a. *Group 1*

The diffraction patterns and compositions of the group 1 fibroins are all very similar (Table 13.1) and it seems likely that the crystalline regions have structures very similar to that proposed for *Bombyx mori* fibroin (Section B.II.d). In all the group 1 fibroins so far studied the ratio of Gly:Ala:Ser has been found to be approximately 3:2:1, suggesting that the repeating sequence Ser-Gly-Ala-Gly-Ala-Gly, found in the Cp fraction of *B. mori* fibroin, may also be present in the other group 1 fibroins.

The distinguishing feature of group 1 is the regular alternation of intersheet spacing between a value of around 3.8 Å and a value of around 5.4 Å and this implies a strict alternation of Gly with Ala or Ser in the sections of chain in the crystalline regions. The alternation of intersheet spacing results in a nonzero intensity for the 00l reflections with odd l (Table 13.5). This contrasts with the situation in the Tussah silk fibroins where the intersheet spacing is constant and the intensity of the $00l$ reflections with odd l is zero (Marsh et al., 1955a).

b. *Group 2*

In the group 2 fibroins the c-axis repeat of 10 Å (Table 13.2) is intermediate between the values of 8.96 Å observed in $[\text{Ala-Gly}]_n$ (Fraser et al., 1966) and 10.535 Å observed in $[\text{Ala}]_n$ (Arnott et al., 1967). Taken in conjunction with the fact that Ala and Gly account for a high proportion of the residues in group 2 fibroins (Table 13.1), it seems likely that the crystalline regions contain more Ala than Gly and that the surfaces of the pleated sheets contain mixed populations of Ala and Gly residues. The c-axis repeat of 9.96 Å in $[\text{Ala-Ala-Gly}]_n$ (Doyle et al., 1970) is close to the value observed in group 2 fibroins and this suggests that the content of Gly residues in the crystallites may be of the order of one-third. Reflections of the $00l$ series with odd l appear to be absent from the diffraction pattern (Warwicker, 1960a) and so the intersheet spacing must be relatively constant. Rudall and Kenchington (1971) have suggested that insertion or deletion of amino acids in sequences of the type found in the crystalline regions of group 1 fibroins might account for the evolution of group 2 fibroins.

Of the group 2 fibroins, that of *Anaphe moloneyi* has been the most extensively investigated. Brown and Trotter (1956) obtained values of 3.47 Å for the axial length per residue and 4.9 Å for the mean intersheet spacing from X-ray diffraction studies. In a later study Warwicker (1960a,b) obtained corresponding values of 3.475 and 5.0 Å. Elliott et al. (1954) measured the infrared spectrum of *A. moloneyi* fibroin in the combination band region and the spectra obtained were essentially similar to those obtained from *Bombyx mori* fibroin (Fig. 13.7). Iizuka (1968) studied the optical rotatory dispersion of aqueous solutions of gland silk from *A. moloneyi* but there was no evidence for α-helix formation. Infrared spectra obtained from these solutions (Fig. 13.2) showed little evidence of pleated-sheet formation but bands characteristic of the antiparallel-chain pleated sheet were present in a 50% (v/v) water–dioxane mixture. Elliott and Malcolm (1956a,b) obtained water-soluble preparations of fibroin by dissolving fibers in trifluoroacetic acid, diluting the solution with water and freeze-drying. Infrared spectra and X-ray diffraction patterns were obtained from films of this material and depending upon solvent and treatment, pleated-sheet and non-pleated-sheet forms were observed. Water-soluble preparations of *A. infracta* (Elliott and Malcolm, 1956b) and *A. reticulatae* (Iizuka, 1968) have also been studied.

c. *Group 3*

The fibroins classified as belonging to group 3 (Table 13.1) all contain a high proportion of Ala residues and it seems likely, from the general similarity between the diffraction patterns of group 3 fibroins and [Ala]$_n$ in the β form, that the portions of the chain in the crystalline regions contain long runs of Ala residues. X-ray diffraction studies of certain group 3 fibroins were discussed in Section B.III and other studies have been reported by Brown and Trotter (1956), who obtained an estimate of 3.455 Å for the axial length per residue and 5.36 Å for the intersheet spacing in silk from the spider *Nephila madagascariensis*, and by Warwicker (1960a), who obtained corresponding values of 3.475 and 5.3 Å for fibroin from *Dictyoploca japonica*. The infrared spectrum of the *Nephila* silk in the combination band region was found to be very similar to that of Tussah silk (Bamford *et al.*, 1956).

Iizuka (1968) investigated the optical rotatory dispersion of aqueous dispersions of the gland silk from *Philosamia cynthia ricini* and obtained evidence for an α-helix content of about 24% which was destroyed in 8 M urea. Hunt (1971) studied the silk produced by larvae of the butterfly *Pieris brassicae* and found that the c-axis repeat was about 10.6 Å; however, the composition more closely resembled that of a group 4 fibroin. In parallel studies electron micrographs were obtained from silk which had been negatively stained after ultrasonic degradation. Ribbonlike filaments 250–500 Å in width and 60–100 Å thick were observed.

d. *Groups 4 and 5*

Silks classified as belonging to groups 4 and 5 generally contain significant amounts of the bulkier side chains and it seems likely that a proportion of such residues are present in the crystalline regions (Warwicker, 1960a) thus accounting for the increased c-axis dimension (Table 13.2). The mean intersheet spacing is 7.5 Å for group 4 and 7.85 Å for group 5 and it is instructive to compare these with the list given in Table 10.3 of intersheet spacings observed in synthetic polypeptides. Some idea of the space-filling properties of different residues can be gained from the fact that the intersheet spacing increases from 4.48 Å in [Ala-Gly]$_n$ to 8.98 Å in [Glu(Et)-Gly]$_n$.

e. *Group 6*

Warwicker (1960a) classified the diffraction patterns obtained from beta silks into five groups on the basis of the c-axis repeat but later Lucas and Rudall (1968a) found that the patterns obtained from the silk of sawflies belonging to the family *Argidae* could not be classified in this way and suggested that they be designated group 6. The Argid silks have a *c*-axis repeat of 13.8 Å and on the basis of the appearance of a weak 001 reflection and other features of the diffraction pattern it was suggested that there was an alternation in intersheet spacing.

The composition of the group 6 silks is unusual (Table 13.1), the major constituents being Ala and Gln in approximately equal proportions. It was suggested (Lucas and Rudall, 1968b; Rudall and Kenchington, 1971) that long sections of the chain may have the sequence [Ala-Gln]$_n$ and that the crystallites have a structure similar to that of *B. mori* fibroin except that the Gly residues are replaced by Gln residues.

C. ALPHA SILKS

A number of insect silks have been found to give an X-ray diffraction pattern which resembles, in certain respects, the patterns obtained from α-keratins and other members of the k-m-e-f group (Chapter 15, Section E.II.e). The compositions of these silks differ from those of most other silks in that the content of Gly is much lower and the content of charged residues is higher (Rudall and Kenchington, 1971) and this can be correlated with the known effects of these residues on α-helix formation (Chapter 12). Details of silks that have been found to give an α pattern are given in Table 13.1, but only in two instances have detailed studies of the conformation been carried out.

I. *Apis mellifera* Silk

Rudall (1962, 1965) showed that the silks of Hymenopteran aculeate larvae (bees, wasps, ants) give diffraction patterns of the α type with meridional arcs of spacings about 5.1 and 1.5 Å and Atkins (1967) has reported a detailed study of the pattern obtained from silk produced by the larvae of the honeybee *Apis mellifera* (Fig. 13.1b). The meridional arc has a spacing of 5.06 ± 0.05 Å and is accompanied by a pair of near-meridional arcs which appear to be on the same layer line. Strong diffraction is also present on the equator and on a pair of layer lines with a spacing of 35 ± 1 Å. The maximum on the equator occurs at an *R* coordinate of $1/9.4$ Å^{-1} and the maximum on the near-equatorial layer line occurs at an *R* coordinate of $1/8.4$ Å^{-1}.

The observed distribution of intensity on the equatorial and near-equatorial layer lines was compared with the intensity transforms calculated for three-strand and four-strand coiled-coil ropes (Fig. 13.18).

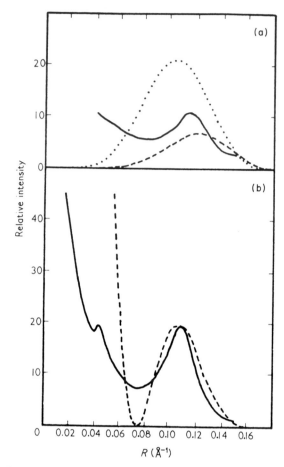

FIG. 13.18.   Comparison of observed intensity transform of *Apis mellifera* silk (full line) with that calculated for three-strand (dotted line) and four-strand (broken line) rope models with $r_0 = 5.2$ Å. (a) Near-equatorial layer line; (b) equator (Atkins, 1967).

The positions and relative intensities of the observed maxima were similar to those calculated for a four-strand rope with a major-helix radius of 5.2 Å. The pitch of the coiled-coils making up the rope is equal to the product of the number of strands and the spacing of the near-equatorial layer line (Chapter 15, Section E.II.f), which gives a value of $4 \times 35 = 140$ Å. The value for the unit twist of the undistorted helix calculated from Eq. (15.11) is $t = 99.1°$.

Although the four-strand-rope model appears at first sight to provide a satisfactory explanation of the equatorial and near-equatorial diffraction, there are a number of alternative possibilities which would need to be excluded. These include models in which a pair of two-strand ropes are twisted around each other and models in which two-strand ropes are packed together in a geometrically regular fashion so that the distribution of intensity on near-equatorial layer lines is subject to an interrope interference function. In the latter case the spacing of the near-equatorial layer lines will be determined by the packing of the ropes rather than the pitch of the coiled-coil. Measurements of the low-angle meridional pattern visible in Fig. 13.1b indicate that the axial period is probably 280 Å (Rudall, 1965) and the observed near-equatorial layer line probably therefore corresponds to $l = 8$. The spacing would then be $280/8 = 35$ Å, as observed.

Flower and Kenchington (1967) found that the silk was stored in the lumen of silk-producing glands in the form of banded tactoids (Fig. 13.19a) which could be dispersed to give suspensions of fine

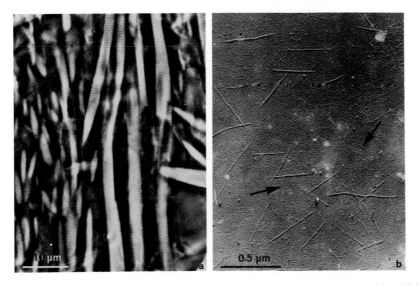

FIG. 13.19. (a) Optical micrograph of a longitudinal section of silk tactoids within the gland lumen of *Apis mellifera*. (b) Electron micrograph of a preparation of tactoids suspended in 0.1 M KCl for 1 hr. In addition to the rods about 100 Å in diameter, there are finer filaments present (arrowed) (Flower and Kenchington, 1967).

filaments (Fig. 13.19b). Similar studies were carried out on the silk-producing glands of *Bombus* (bumblebee) larvae and in this case fibrous aggregates were observed which consisted of bundles of filaments exhibiting a 140-Å longitudinal periodicity. This again emphasizes the fact that the observed layer-line spacing will be determined by long-range order, rather than by the pitch of the coiled-coil.

II. Mantid Oothecal Protein

During the deposition of the egg case of the praying mantis, globules of fibrous protein produced in the main colleterial gland are transformed into thin ribbons about 15 μm long, 1–2 μm wide, and 200–300 Å thick (Rudall, 1956, 1962; Kenchington and Flower, 1969). The ribbons exhibit a regular banding pattern when stained with phosphotungstic acid (Fig. 13.20), the bands being inclined at about 20° to the edge

0.1 µm

Fig. 13.20.   Electron micrograph of part of a ribbon from the mantis ootheca (Rudall, 1956).

of the ribbon and spaced about 120 Å apart. The ribbons yield an X-ray diffraction pattern (Fig. 13.21) which indicates a very high degree of long-range order, and the presence of strong diffraction in the 5-Å region of the meridian and the 10-Å region of the equator suggests that the ribbons are made up of a near-crystalline array of molecules having a coiled-coil, α-helical conformation. Various features of the pattern have been discussed by Rudall (1956) in relation to possible arrangements of two-strand coiled-coil ropes but no systematic study of the pattern has so far been reported. The pattern is dominated by a set of row lines associated with a lateral periodicity of about 17 Å (Fig. 13.21) and this was thought to be associated with the diameter

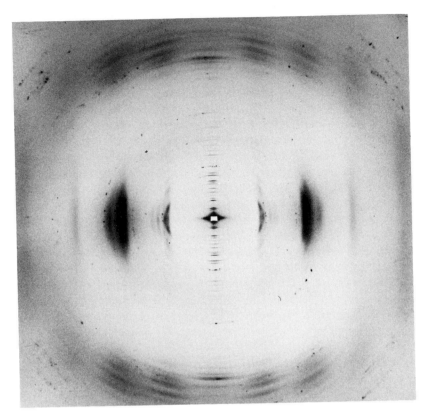

FIG. 13.21. X-ray diffraction pattern obtained from a mantis ootheca.

of a two-strand coiled-coil rope. Rudall (1962) noted that many reflections in the low-angle meridional pattern could be indexed on a repeat of 90 Å but measurements of the spacings of the two innermost reflections visible in Fig. 13.21 indicate that the axial period is in excess of 480 Å.

D. OTHER SILKS

In addition to the alpha silks and beta silks discussed in the preceding sections, silks have been found which give diffraction patterns resembling in some respects those obtained from form II of [Gly]$_n$ and from collagen. The diffraction pattern given by another type of silk, obtained from *Aspidomorpha puncticosta*, has not so far been related to any of the recognized regular chain conformations (Rudall, 1962; Atkins *et al.*,

1966; Lucas and Rudall, 1968a). The pattern shows evidence of precise long-range order and electron microscope studies indicate that the material consists of a close-packed assembly of filaments about 75 Å in diameter. The equatorial low-angle X-ray diffraction pattern was shown to be consistent with this type of structure.

I. GLYCINE-RICH SILKS

The silk obtained from the Solomon Seal sawfly *Phymatocera aterrima* has a very high content of Gly residues (Table 13.3) and it was found (Lucas and Rudall, 1968a; Rudall, 1968a) that when fibers drawn from the gland silk were steamed, an X-ray diffraction pattern similar to that obtained from form II of $[Gly]_n$ was obtained (Fig. 13.1). Further studies (Lucas and Rudall, 1968b; Rudall and Kenchington, 1971) showed that the PG II conformation (Chapter 11, Section B.II) also occurred naturally in fibers produced by the prepupa. Further examples of silks containing the PG II conformation are listed in Table 13.1. Extended crystallites of chains having the PG II conformation can only be formed from lengths of chain having the sequence $[Gly]_n$ and it must be concluded that uninterrupted runs of Gly residues occur in the sequences of these fibroins.

II. COLLAGENLIKE SILKS

Rudall (1962) obtained X-ray diffraction patterns from silk produced by the larvae of the gooseberry sawfly (*Nematus ribesii*) and noted a resemblance to the pattern given by collagen. Improved patterns were obtained by drawing fibers from the gland silk (Fig. 13.1c) and in addition to the high-angle pattern a low-angle pattern indicative of long-range axial order was observed. The spacing of the prominent high-angle meridional arc was given as 2.86 Å, which suggests that a collagenlike three-strand rope structure is present (Chapter 11, Section A.II) and if this is the case, there must be sections of the chain in which Gly residues occur in every third position. The amino acid composition of a purified preparation has been determined (Rudall and Kenchington, 1971) and Gly residues were found to constitute one-third of the total while Pro residues accounted for about one-tenth. When precipitated from solution with acetic acid vapor the purified material forms tactoids (Fig. 13.22) with a banded structure of period about 550 Å.

Rudall (1968a) investigated the effect of hydration on the pattern obtained from *Nematus* silk and observed equatorial reflections of spacing 15 and 26 Å in the hydrated material. These were indexed

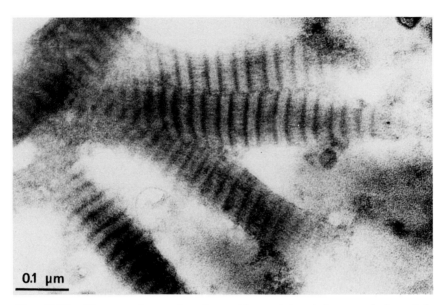

FIG. 13.22. Electron micrographs of tactoids precipitated from a solution of the collagenlike protein produced by *Nematus ribesii* and stained with phosphotungstic acid. The period of the banding pattern is about 550 Å (Rudall and Kenchington, 1971).

as the 10 and 11 reflections of a hexagonal cell with a lateral dimension of 30 Å. A low-angle meridional pattern was also found to be present in *Nematus* silk and could be indexed on an axial repeat of 108–110 Å, that is, one-fifth of the banding period mentioned earlier.

Chapter 14

Collagens

A. INTRODUCTION

Collagen proteins are widely distributed throughout the animal kingdom and in combination with elastin, mucopolysaccharides, and mineral
salts form the connective tissue responsible for the structural integrity
of the animal body. The polypeptide chains in collagens have a distinctive
conformation which is intimately related to their structural function and
in this chapter an account will be given of the information presently
available on the nature of this conformation. It is generally accepted
that the polypeptide chains form three-strand-rope structures of the
type discussed in Chapter 11, Section A.II but there has been considerable controversy about the precise nature of the conformation. The
reason for the lack of a definitive model is to be found in the paucity of
diffraction and spectroscopic data suitable for the objective assessment
of the various possibilities and for the subsequent refinement of the
most satisfactory model. Much use has therefore been made of stereochemical criteria and extremely detailed models have been developed
on this basis. In the present state of knowledge concerning interatomic
interactions it is doubtful whether stereochemical criteria alone can
provide a suitable basis for detailed refinement. The solution to the
dilemma clearly lies in the collection of more comprehensive diffraction
and spectroscopic data. As with other fibrous proteins, studies of model
polymers have played an important part in the development of ideas on
the polypeptide chain conformation in collagen and this work has
already been reviewed in Chapter 11.

The chemical nature of collagen is outlined briefly in Section B, physical studies of collagen structure are reviewed in Section C, and in Section D the results obtained are discussed in relation to models for the chain conformation in collagen. The nature of the collagen molecule and the packing arrangement in the fibril are discussed in Section E.

B. CHEMICAL NATURE OF COLLAGEN

Comprehensive reviews of the chemical and physicochemical properties of collagen are available (Harrington and von Hippel, 1961; Seifter and Gallop, 1966; Piez, 1967; von Hippel, 1967; Bailey, 1968; Traub and Piez, 1971) and in the present section a brief outline will be given of certain features of the chemistry and solution properties of collagen which are relevant to studies of the native conformation.

I. CHAIN STRUCTURE

Most studies of the chain structure of collagens have been carried out on the soluble material, sometimes termed tropocollagen, which may be extracted in variable amounts from collagen-containing tissues by the use of neutral salt solutions, acid buffers, or dilute acetic acid. The collagen so obtained is generally contaminated with traces of other materials and requires further purification. Physicochemical studies of these soluble collagens, which have been reviewed by von Hippel (1967), indicate that the vertebrate collagen molecule has a molecular weight in the vicinity of 300,000 and exists in solution as a rodlike particle approximately 2800 Å in length and 15 Å in diameter. Rodlike particles of similar dimensions have been observed directly by electron microscopy in dried preparations (Hall, 1956).

The rodlike nature of the molecule may be destroyed by heating or by the addition of hydrogen-bond-breaking reagents to the solution. Under appropriate conditions this transformation can be induced without any apparent rupture of covalent linkages and the product is generally referred to as gelatin. Extensive studies of the denaturation of soluble collagen have been carried out (Piez, 1967; Veis, 1967; Bailey, 1968; Traub and Piez, 1971) and these have established that the collagen molecule consists of three polypeptide chains (α chains) which may be covalently cross-linked to an extent which depends upon the origin and age of the collagen. The α chains have weights about one-third that of the collagen molecule and in vertebrate collagens two types of α chain are generally found to be present. These are designated α1 and α2 and quantitative studies indicate that they are present in the ratio of 2:1, and it has been shown (Nold *et al.*, 1970) that each collagen molecule contains two α1 chains and one α2 chain. The three N-termini all occur at the same end of the molecule (Section E.II.a).

Miller and Matukas (1969) have shown that chick cartilage contains two types of collagen; one has the normal complement of chains, whereas the other contains only one type of chain termed [α1(II)] which is similar, but not identical to, the α1 chain of the more usual type [α1(I)]. The chain compositions have been shown to be [α1(I)]$_2$ α2 and [α1(II)]$_3$, respectively (Trelstad *et al.*, 1970). Electrophoretic analyses of soluble fractions obtained from lower vertebrates (Kulonen and Pikkarainen, 1970) and from invertebrates (Pikkarainen *et al.*, 1968; Nordwig and Hayduk, 1969) suggest that collagens with three identical α chains may be present in these animals but confirmation of this by other techniques is desirable (Traub and Piez, 1971).

McBride and Harrington (1967a,b) have studied an unusual collagen, isolated from *Ascaris* cuticle, which appears to be stabilized by disulfide linkages. The molecular weight was estimated to be 900,000 but this was reduced to 62,000 when the disulfide linkages were reduced and carboxymethylated. It was suggested that in the native structure the 62,000-molecular-weight chains are folded in a manner similar to that indicated in Fig. 11.9 and that the folded chains are linked end-to-end by disulfide bridges.

II. PRIMARY STRUCTURE

The amino acid compositions of a wide variety of collagens have been determined (Gross, 1963; Eastoe, 1967; Cain, 1970) and the most striking finding is the relative constancy of the glycine content, which in vertebrate collagens is always close to one-third of the total number of residues. Collagens are also rich in the imino acids proline and hydroxyproline but the contents and relative proportions of these residues vary with the source of the protein. Mammalian collagens all have very similar compositions, but appreciable variations in composition are found in collagens isolated from lower vertebrates and from invertebrates. The compositions of the individual α1 and α2 chains have been determined in a number of instances (Piez *et al.*, 1963; Bornstein and Piez, 1964; Lewis and Piez, 1964; Piez, 1965) and the two chains appear to be very similar with regard to overall composition. The α2 chains generally contain a slightly higher proportion of residues with hydrophobic side chains and are slightly more basic in character than the α1 chains. The compositions of the α1 and α2 chains from rat-tail tendon are given in Table 14.1 and it can be seen that the imino acid content of the α1 chain

TABLE 14.1

Amino Acid Compositions of Collagen from Rat-Tail Tendon and
Its Component α Chains[a]

Amino acid	Whole collagen	α1 Chain	α2 Chain
Alanine	10.7	11.0	10.3
Arginine	5.0	4.9	5.1
Aspartic acid[b]	4.5	4.7	4.4
Glutamic acid[b]	7.1	7.4	6.8
Glycine	33.1	32.9	33.5
Histidine	0.4	0.2	0.7
Hydroxylysine	0.7	0.5	1.0
Hydroxyproline[c]	9.4	9.7	8.6
Isoleucine	1.0	0.6	1.5
Leucine	2.4	1.8	3.1
Lysine	2.7	3.0	2.1
Methionine	0.8	0.9	0.7
Phenylalanine	1.2	1.2	1.1
Proline	12.2	12.9	11.5
Serine	4.3	4.1	4.3
Threonine	2.0	2.0	2.0
Tyrosine	0.4	0.4	0.4
Valine	2.3	1.9	3.0
Gly	33.1	32.9	33.5
Pro + Hyp	21.6	22.6	20.1
Other	45.3	44.5	46.4

[a] Given in residues/100 residues. Data of Piez et al. (1963) rounded to nearest 0.1 residue.

[b] Includes amide.

[c] Includes both 3-hydroxyproline and 4-hydroxyproline.

is slightly greater than that of the α2 chain. This appears to be generally true in mammalian collagens.

Some progress has been made in the determination of the sequence of amino acids in the α chains of collagens from higher vertebrates and this has been summarized by Traub and Piez (1971). These collagens contain 5–9 methionyl residues per chain and cyanogen bromide has been used to cleave the chain into a series of well-defined fragments (Bornstein and Piez, 1966; Epstein et al., 1971). By a combination of chemical and electron microscope studies and the use of pulse-labeling techniques, the order in which the fragments occur in the native structure has also been established.

Data on the sequence of the first 139 residues at the N-terminal end of the α1 chain in rat-tendon collagen, corresponding to the first five cyanogen bromide peptides, has been assembled by Traub and Piez (1971) and is reproduced in Fig. 14.1a. After the first 16 residues, which correspond to the so-called telopeptide region (Section E), the sequence consists of a regular succession of glycine-led triplets. In relation to the overall composition of the α1 chain the region 15–64 contains an appreciably greater than average number of imino acid residues and the region 64–139 contains an appreciably greater than average number of charged residues. The sequence of the first 112 residues in the C-terminal fragment of the α1 chain from calf-skin collagen has also been determined

pGlu-Met-Ser-Tyr-	38 Gly-Phe-Gln-	74 Gly-Lys-Pro-	110 Gly-Leu-Asp-
5 Gly-Tyr-Asp-	41 Gly-Pro-Hyp-	77 Gly-Arg-Hyp-	113 Gly-Ala-Lys-
8 Glu-Lys-Ser-	44 Gly-Glu-Hyp-	80 Gly-Gln-Arg-	116 Gly-Asn-Thr-
11 Ala-Gly-Val-	47 Gly-Glu-Hyp-	83 Gly-Pro-Hyp-	119 Gly-Pro-Ala-
14 Ser-Val-Pro-	50 Gly-Ala-Ser-	86 Gly-Pro-Gln	122 Gly-Pro-Lys-
17 Gly-Pro-Met-	53 Gly-Pro-Met	89 Gly-Ala-Arg-	125 Gly-Glu-Hyp-
20 Gly-Pro-Ser-	56 Gly-Pro-Arg-	92 Gly-Leu-Hyp-	128 Gly-Ser-Hyp-
23 Gly-Pro-Arg-	59 Gly-Pro-Hyp-	95 Gly-Thr-Ala-	131 Gly-Glx-Asx-
26 Gly-Leu-Hyp-	62 Gly-Pro-Hyp-	98 Gly-Leu-Hyp-	134 Gly-Ala-Hyp-
29 Gly-Pro-Hyp-	65 Gly-Lys-Asn-	101 Gly-Met-Hyl-	137 Gly-Gln-Met-
32 Gly-Ala-Hyp-	68 Gly-Asp-Asp-	104 Gly-His-Arg-	
35 Gly-Pro-Gln-	71 Gly-Glu-Ala-	107 Gly-Phe-Ser-	

(a)

1 Gly-Pro-Hyp-	31 Gly-Ala-Lys-	61 Gly-Lys-Ser-	91 Gly-Pro-Arg-
4 Gly-Leu-Ala-	34 Gly-Asp-Arg-	64 Gly-Asp-Arg-	94 Gly-Asx-Hyl-
7 Gly-Pro-Hyp-	37 Gly-Glu-Thr-	67 Gly-Glu-Thr-	97 Gly-Glx-Thr-
10 Gly-Leu-Ser-	40 Gly-Pro-Ala-	70 Gly-Pro-Ala-	100 Gly-Glx-Glx-
13 Gly-Arg-Glu-	(Gly, Pro, Pro,	73 Gly-Pro-Ile-	103 Gly-Asx-Arg-
16 Gly-Ala-Hyp-	Gly, Ala, Hyp,	76 Gly-Pro-Val-	106 Gly-Ile-Hyl-
19 Gly-Ala-Glu-	Gly, Ala, Hyp,	79 Gly-Pro-Ala-	109 Gly-His-Arg-
22 Gly-Ser-Hyp-	Gly, Pro, Ala)-	82 Gly-Ala-Arg-	112 Gly-
25 Gly-Arg-Asp-	55 Gly-Pro-Val-	85 Gly-Pro-Ala-	
28 Gly-Ser-Hyp-	58 Gly-Pro-Ala-	88 Gly-Pro-Gln-	

(b)

FIG. 14.1. (a) Amino acid sequence of 139 residues at the N-terminal end of the α1 chain of rat tendon (taken from Traub and Piez, 1971). (b) Amino acid sequence of the first 112 residues in the C-terminal cyanogen bromide fragment of the α1 chain from calf-skin collagen (von der Mark et al., 1970). pGlu = pyroglutamyl, Glx = Glu or Gln, Asx = Asp or Asn.

(von der Mark et al., 1970) and is given in Fig. 14.1b. Evidence for partial gene duplication is evident in the repetition of the section 34–43 at the position 64–73.

Less information is available on the sequence in α2 chains but it has been shown that, as with the α1 chains, the first 10–15 residues do not have glycine residues in every third position. A segment of 30 residues remote from the ends of the α2 chain has been shown to consist of a series of glycine-led triplets (Highberger et al., 1971).

The N-terminal sequences, in the telopeptide region, have been determined for the α1 chains from several species including rat skin and chick skin and a high degree of homology is evident (Traub and Piez, 1971). The same is true for the N-terminal sequences of the α2 chains from these two materials but there is little in common between the N-terminal sequences of the α1 and α2 chains. Studies of molecules regenerated solely from α1 chains or from α2 chains, discussed in Section E, suggest, however, that the two chains are homologous (Kühn, 1969).

The available sequence data has been analyzed by Traub and Piez (1971) to determine whether imino acid residues show any preferential distribution with respect to positions 2 and 3 of the glycine-led triplets. The proportions found for the four possible types of triplet are given in Table 14.2 together with the proportions which would be expected

TABLE 14.2

Distribution of Imino Acid Residues in Glycine-Led Triplets

Triplet type[a]	Proportion assuming random distribution[b]	Observed proportion[c]
Gly-X-X	$(1 - \frac{3}{2}p)^2 = 0.46$	0.44
Gly-X-I	$\frac{3}{2}p(1 - \frac{3}{2}p) = 0.22$	0.20
Gly-I-X	$\frac{3}{2}p(1 - \frac{3}{2}p) = 0.22$	0.27
Gly-I-I	$(\frac{3}{2}p)^2 = 0.10$	0.09

[a] X = amino acid residue; I — imino acid residue.
[b] p = fraction of total residues which are proline or hydroxyproline. The proportions given are calculated on the basis of $p = 0.216$ (Table 14.1); no allowance has been made for the presence of telopeptide regions.
[c] Data of Traub and Piez (1971).

for a random distribution of imino acids between positions 2 and 3. The observed proportions do not appear to differ significantly from those calculated for a random distribution. The individual distributions of proline and hydroxyproline in the sequences given in Fig. 14.1 follow the trends suggested by earlier studies (Greenberg et al., 1964) whereby no hydroxyproline occurs in position 2 and proline rarely occurs in position 3. It is also noteworthy that no triplet contains two glycine residues.

III. STEREOCHEMICAL CONSIDERATIONS

It was recognized at an early stage (Astbury, 1933, 1940) that the unique conformation of collagen probably stemmed from a repetition in the sequence of glycine-led triplets and from the high content of imino acid residues. The presence of these residues restricts the rotational freedom about the main-chain bonds and they cannot be incorporated in continuous α-helices or in β conformations (Pauling and Corey, 1951f). The important part played by imino acid residues in the stabilization of the native collagen conformation is illustrated by their influence on its thermal stability (Burge and Hynes, 1959; Piez and Gross, 1960; Josse and Harrington, 1964; Harrington and Rao, 1967; Rigby, 1968a,b). Intramolecular hydrogen bonds have also been recognized as an important conformation-determining factor but recently it has been suggested that their role may have been overemphasized (Cooper, 1971). Attention has been directed to the possibility that hydrophobic bonding within the interior of the collagen molecule might make a significant contribution to the stability (Cassel, 1966; Harrap, 1969).

The influence of particular imino acid residues on the stereochemistry of the polypeptide chain has been investigated in some detail for the idealized case in which all the amide groups have the trans configuration with $\omega = 180°$ (Chapter 6, Section A.I). The results can be conveniently discussed in terms of the torsional angles ϕ and ψ (Chapter 6, Section A.I) as follows:

$$-\text{CONH}\overset{\phi_1}{-}\text{CH}_2\overset{\psi_1}{-}\text{CONH}\overset{\phi_2}{-}\text{CHR}\overset{\psi_2}{-}\text{CONH}\overset{\phi_3}{-}\text{CHR}\overset{\psi_3}{-}\text{CONH}-\text{CH}_2-\text{CONH}-$$

In the case of a polypeptide chain consisting only of amino acid residues the *local* conformation-determining interactions, which limit the possible combinations of ϕ_i and ψ_i, are essentially those between atoms in the grouping

$$-\text{C}_{i-1}\text{O}_{i-1}-\text{N}_i\text{H}_i-\text{C}_i\text{H}_i\text{R}_i-\text{C}_i\text{O}_i-\text{N}_{i+1}\text{H}_{i+1}-$$

(Schimmel and Flory, 1968) and so the nature of the preceding and succeeding residues has little effect on the allowed combinations of ϕ_i and ψ_i. The steric maps for glycine and alanine residues are shown in Figs. 6.5b and 6.6a, respectively, and it will be seen that the rotational freedom about C_1^α in the glycine-led triplet is much greater than that about C_2^α or C_3^α, which rarely, if ever, belong to glycine residues.

When an amino acid residue is succeeded by an imino acid residue the H_{i+1} atom will be replaced by a methylene group and the area of the steric map corresponding to possible combinations of ϕ_i and ψ_i will be somewhat reduced (Schimmel and Flory, 1968; Damiani *et al.*, 1970).

The rotational freedom about the C^α atom of an imino acid residue is very restricted (Chapter 11, Section C). The angle ϕ is limited to values around $-70°$ by the near-rigid geometry of the pyrrolidine ring and ψ is limited to two ranges centered around values of $-50°$ and $150°$ (Schimmel and Flory, 1968; Gō and Scheraga, 1970; De Santis and Liquori, 1971). When both positions 2 and 3 are occupied by imino acid residues the restriction on ψ_2 is even greater, since the range centered around $-50°$ is excluded by steric interference between the two imino acid residues (Schimmel and Flory, 1967; Gō and Scheraga, 1970). Thus triplets of the type Gly-I-I, which account for about one triplet in ten in vertebrate collagens (Table 14.2), place severe restrictions on possible forms of regular helical structures in which all glycine-led triplets are structurally equivalent. The possible effect of Gly-I-I triplets on the thermal stability of the collagen molecule has been considered by Josse and Harrington (1964).

C. PHYSICAL STUDIES OF CHAIN CONFORMATION

I. HIGH-ANGLE X-RAY PATTERN

The high-angle X ray diffraction pattern of vertebrate collagen has been studied for more than half a century and comprehensive lists of references to early studies have been given by Astbury (1933, 1940), Clark and Schaad (1936), and Bear (1951, 1952). A typical pattern, obtained from a rat-tail tendon dried under slight tension, is illustrated in Fig. 14.2a and essentially similar patterns have been obtained from stretched films of gelatin (Gerngross and Katz, 1926) and from soluble collagens (Wyckoff and Corey, 1936). The main features of the pattern are a strong meridional arc of spacing about 2.9 Å, discrete equatorial reflections of spacing about 12 and 6 Å, and a diffuse equatorial blob of spacing about 4.5 Å. Except in the neighborhood of the meridian, the layer lines are not well developed but it was recognized at an early stage that the sharpness of the pattern was influenced both by hydration and stretching. High-angle diffraction patterns have also been obtained from a variety of invertebrate collagens (Marks *et al.*, 1949) and the main features of the patterns were found to be essentially similar to those obtained from vertebrate collagens.

a. *Axial Pseudoperiod*

Herzog and Jancke (1926) suggested that the diffraction pattern could be indexed on an axial repeat of structure of 29.06 Å but later workers favored values of 9.3–9.8 Å (Herrmann *et al.*, 1930; Trillat, 1930;

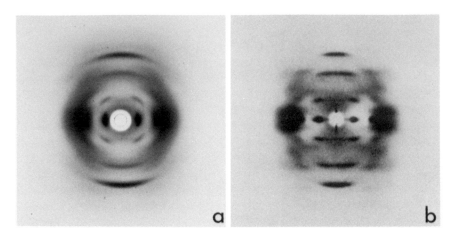

FIG. 14.2.   High-angle X-ray diffraction patterns obtained from dried rat-tail tendon
(a) unstretched and (b) stretched by 8%. From Cowan *et al.* (1955a).

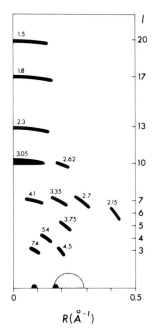

FIG. 14.3.   Distribution of diffracted intensity in the
X-ray diffraction pattern of dried and stretched kangaroo-
tail tendon (after Ramachandran and Ambady, 1954).
The reflections fall on layer lines corresponding to an
axial periodicity of 29.4 Å.

Küntzel and Prakke, 1933; Koisi, 1941) or around 8.5 Å (Meyer, 1929;
Katz, 1934). Neither of these shorter periods, however, provided a
satisfactory basis for indexing all the arcs observed even at moderate

resolution. Bear (1952), following a suggestion by Herzog and Gonnell (1931), attempted to index the data then available on kangaroo-tail tendon in terms of an axial period of 20 Å, but discrepancies were found which appeared to exceed the errors of measurement. Additional meridional and near-meridional reflections were reported by Cohen and Bear (1953) and these were also indexed on an axial period of 20 Å.

The problems encountered in indexing the diffraction pattern of collagen stem from the rather diffuse and ill-defined nature of the intensity maxima, which has generally been attributed to poor orientation and crystallinity. Astbury (1940) observed that the pattern could be sharpened by stretching the fiber and Cowan *et al.* (1953, 1955a) obtained a pattern of greatly improved quality by stretching tendon about 10% (Fig. 14.2b). On the basis of this improved pattern it was suggested that the axial period in the stretched fiber was either 31 Å or 21 Å. Ramachandran and Ambady (1954) collected similar data from stretched kangaroo-tail tendon (Fig. 14.3) and found that all the observed arcs could be satisfactorily indexed on an axial period of 29.4 Å. The period in unstretched collagen was estimated to be 28.6 Å. Cowan *et al.* (1955a) indexed the pattern obtained from a rat-tail tendon which had been stretched 8% on an axial period of 30 Å, but some discrepancies were noted.

So far, the high-angle pattern of collagen has been discussed as if collagen were a polymer with a simple repeating unit but it is clear from the sequence data given in Section B.II that this is not so. Nevertheless, the occurrence of a discrete high-angle pattern shows that there is a pseudo repeat of structure which is much shorter than the length of a collagen molecule. It is therefore more correct to speak of an "axial pseudo period" of 28.6 Å and to reserve the term "axial period" for the true repeat, which is intimately connected with the molecular length (Section E.I.a). In a fiber with perfect axial order the high-angle reflections would be resolvable into closely spaced layer lines, which would index on the true axial repeat, but in most cases this fine structure is either not present, due to disorder in the lattice (Chapter 1, Section A.IV), or else is obscured by the effects of disorientation. Some evidence of carryover of the low-angle pattern into the high-angle pattern has been observed, however, in particularly well-ordered specimens (North *et al.*, 1954; Miller and Wray, 1971) and is illustrated in Fig. 14.4.

Although the observed arcs in the high-angle pattern of stretched collagen index reasonably satisfactorily on an axial pseudoperiod of about 29 Å, the true pseudoperiod may be very much longer. This is related to the fact that the pseudostructure shows little evidence of three-dimensional crystallinity and is not therefore limited to twofold,

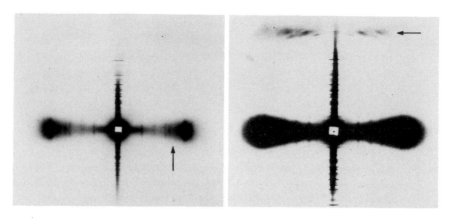

FIG. 14.4. X-ray diffraction patterns of rat-tail tendon maintained in the wet state, showing sampling of the high-angle pattern (arrow) by interference functions associated with large-scale order (Miller and Wray, 1971).

threefold, fourfold, or sixfold screw axes of symmetry. This point is discussed further in Section C.I.c.

b. *Equatorial Pattern*

The position of the prominent equatorial arc near $R = 0.09$ Å^{-1} in Fig. 14.3 depends upon the water content of the specimen and early studies showed that the R coordinate was a continuous function of humidity, varying from a value corresponding to an interplanar spacing of 10.4 Å in the dry state to an upper limit of 15–16 Å in fully hydrated collagen (Küntzel and Prakke, 1933) and about 17 Å in gelatin (Gerngross *et al.*, 1930, 1931; Herrmann *et al.*, 1930). A detailed study of the effects of moisture content on the diffraction pattern of kangaroo-tail tendon was reported by Rougvie and Bear (1953) and the results obtained are depicted in Fig. 14.5. The minimum spacing observed for the dried material was 10.6 Å and the maximum value for tendon immersed in water was about 14.5 Å. It was also noted that the dispersion of the arc was considerably reduced in the fully hydrated specimen. In high-resolution patterns from fully hydrated rat-tail tendon (Fig. 14.4) this reflection is resolved into several discrete components (North *et al.*, 1954; Miller and Wray, 1971) emphasizing that the high-angle pattern can only be regarded as originating from a pseudo repeat of structure. The remainder of the high-angle equatorial pattern has not been studied in such great detail but Cowan *et al.* (1955a) noted that behavior of an arc near 0.18 Å^{-1} on hydration was consistent with its indexing as the second order of the arc near 0.09 Å^{-1}. Changes in the diffuse scattering

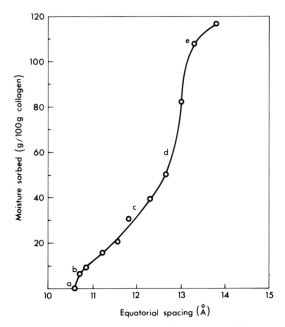

FIG. 14.5. The effect of moisture content on the spacing of the prominent high-angle equatorial reflection in the X-ray diffraction pattern of kangaroo-tail tendon (after Rougvie and Bear, 1953). The letters correspond to those in Fig. 14.6.

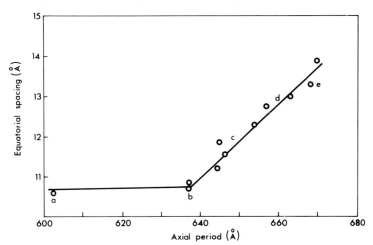

FIG. 14.6. Relationship between the spacing of the prominent equatorial reflection in the high-angle X-ray diffraction pattern of kangaroo-tail tendon and the axial period with variation in water content. The letters correspond to those in Figure 14.5, e corresponds to hydrated tendon, b to tendon dried at room temperature, and a to tendon dried *in vacuo* at an elevated temperature.

have been discussed by Bear (1952), and the relationship between changes in the high-angle equatorial pattern (Fig. 14.6) and the axial period is discussed in Section E.III.

c. *Direct Interpretation*

In early studies it was generally assumed that the high-angle arcs could ultimately be related to the reflections from a three-dimensionally crystalline structure, but with the improvements obtained by stretching (Cowan *et al.*, 1953) it became apparent that the pattern was more akin to the transform of a rodlike structure with helical symmetry. Except on the equator, the pattern in wet tendon appears to consist of maxima in continuous distributions of intensity along the layer lines. This type of pattern has also been observed with certain synthetic polypeptides, for example, $[Glu(Bzl)]_n$ (Chapter 9, Section B.III.b), and is attributable to a type of disorder in which rodlike structures are packed together with their axes parallel, but are subject to random longitudinal displacements and random rotations about their axes (Chapter 1, Section A.IV.d). As water is extracted from the specimen, the distribution of intensity on the third layer line become concentrated into narrow ranges of R, suggesting that some correlation of rotation and displacement is present between neighboring chains in the partially collapsed structure (Bear, 1955a). Rich and Crick (1961) and Yonath and Traub (1969) have suggested that the combination of spots and streaks is indicative of screw disorder (Chapter 1, Section A.IV.d) which is fairly common with helical structures containing pronounced grooves. Similar effects were observed by Yonath and Traub (1969) in the diffraction pattern of $[Gly-Pro-Pro]_n$.

The interpretation of the high-angle pattern in terms of the helix diffraction theory outlined in Chapter 1, Section A.III has been attempted by several authors. The meridional reflection of spacing about 2.9 Å in unstretched tendon was in every case assumed to arise from a zeroth-order Bessel function and so the unit height h was taken to have this value. Cohen and Bear (1953) suggested that the axial repeat c was 20 Å and that the helix contained seven units in two turns, while Cowan *et al.* (1953) considered several possibilities with $c = 21$ and 31 Å for stretched tendon. Ramachandran and Ambady (1954) showed that the data given in Fig. 14.3 could be explained very well in terms of a helix with a repeat distance, for stretched tendon, of $c = 29.4$ Å containing ten units in three turns and subsequently this has come to be generally accepted (Bear, 1955a; Cowan *et al.*, 1955b; Rich and Crick, 1961). The possibility that an $\mathcal{N}$-fold parallel rotation axis might be present (Chapter 1,

Section A.II.d) was considered by Rich and Crick (1961) but was shown to be unlikely. The helical pseudostructure in collagen appears therefore to belong to line group s, and in this event the selection rule for the appearance of Bessel functions has the form given in Chapter 1, Section A.III.b, namely

$$l = um + vn \tag{14.1}$$

where l is the layer-line index, $u = 10$, $|v| = 3$, n is the order of the Bessel function, and m is an integer. The values of n appropriate to $v = -3$ are indicated in Fig. 14.3 and, as noted by Ramachandran and Ambady (1954), meridional or near-meridional reflections are found on layer lines with contributions from low-order Bessel functions, thus supporting the assignments of u, v, and c. The question of whether the basic helix is right-handed ($v = 3$) or left-handed ($v = -3$) cannot be decided from the X-ray data alone.

Although the helix associated with the high-angle pattern of tendon collagen appears to contain ten units in three turns, there is no *a priori* reason why such a simple relationship should exist between u and v. Methods of determining precise helical parameters were investigated by Lakshmanan *et al.* (1962), who concluded that the height h was 2.95 Å and the magnitude of the unit twist $|t|$ was 109.8° in native collagen. This corresponds to a value of $|u/v| = 3.28$ compared with 3.33 for the simple relationship $u = 10$, $|v| = 3$. The value for h in stretched tendon was found to be 3.05 Å and the value of $|u/v|$ to be 3.27. The method used to determine $|u/v|$ depends critically upon the value taken for h, and this was determined by a novel method in which the inside edge of the appropriate arc was measured in the pattern from a tilted specimen (Chapter 1, Section B.IV.a). For a specimen with infinitely sharp layer lines and an infinitely narrow X-ray beam this procedure would be correct, but the natural breadth of the layer lines (Davies, 1965) and instrumental broadening will both tend to make the value of h so obtained too large. Further studies, in which a correction was applied for instrumental effects, have been summarized by Ramachandran (1967), who gives values for unstretched collagen of $h = 2.91 \pm 0.1$ Å and $|u/v| = 3.25 \pm 0.06$, which leads to a value of $t = 110 \pm 2°$. For most purposes the assumption that $|u/v| = 10/3$ will not lead to sensible error. No significant difference was found between the values of h determined for dry and hydrated tendon.

As mentioned earlier, the appearance of both spots and layer-line streaks in the diffraction pattern of certain specimens is indicative of the presence of screw disorder in the molecular packing arrangement and in the case of a helix with $u = 10$ and $|v| = 3$ the occurrence of spots on layer lines with $l = 0$ and 3 suggests, as explained in Chapter 11, Section E.IV, that the pitch of the screw disorder has a magnitude of 9.7 Å.

Very little further information can be obtained directly from the high-angle pattern of collagen. The distribution of intensity along the layer lines is determined by the intensity transform of the pseudo repeat of structure but no means exists for determining the atomic positions directly from the cylindrically averaged intensity transform. However, a comparison of the observed intensity transform with that calculated for a model of the pseudo repeat of structure provides a searching test of its correctness and also a basis for detailed refinement. Measurements of the intensity transform have been reported by Bradbury *et al.* (1958), Rich and Crick (1961), and Yonath and Traub (1969) and their comparison with model structures is discussed in Section D.IV.

II. INFRARED SPECTRUM

The infrared spectrum of air-dried mouse-tail tendon was recorded by Fraser (1950a,b, 1951) and the frequencies of the amide A, B, I, II, and III bands were found to be, respectively, 3325, 3085, 1655, 1550, and 1245 cm^{-1}. Ambrose and Elliott (1951b) showed that stretched gelatin gave a similar spectrum and Bradbury *et al.* (1958) examined the spectra of films from soluble rat-skin collagen in considerable detail. More recently Susi *et al.* (1971) studied the effects of humidity on the spectra of films of fragmented bovine tendon. A number of other investigations dealing with particular aspects have been reported (Randall *et al.*, 1952, 1953; Fitton Jackson *et al.*, 1953; Seeds, 1953; Badger and Pullin, 1954; Sutherland *et al.*, 1954; Beer *et al.*, 1959; Watson and Silvester, 1959; Furedi and Walton, 1968; Huc and Sanejouand, 1968) and spectra have also been determined in the overtone region (Ellis and Bath, 1938; Ambrose and Elliott, 1951b; Hecht and Wood, 1956; Fraser and MacRae, 1958c, 1959b). These will be discussed, where appropriate, in the following sections.

a. *Amide A Frequency*

It was noted (Fraser, 1950b, 1951) that the frequency of the amide A band in tendon collagen was appreciably higher than in α-keratin or β-keratin and further experiments were undertaken to ascertain whether the higher frequency was characteristic of the native collagen structure. It was found that the frequency in thermally contracted fibers, which no longer gave the collagen high-angle X-ray diffraction pattern, dropped to 3300 cm⁻¹, which is close to the value observed in keratins. When the contracted fibers were dissolved by heating in water to 90°C, films cast from the solution at room temperature gave the higher frequency, but the lower frequency was obtained when the same films were heated to 100°C in water and then dried rapidly. Films cast from acid-soluble collagen were also found to give the higher frequency. All these data point to the higher amide A frequency being associated with the chain conformation present in native collagen. It was suggested that the higher frequency might be due either to weaker hydrogen bonding or to a distortion of the amide group.

Ambrose and Elliott (1951b) found that the higher amide A frequency, which they measured as 3330 cm⁻¹, was also present in oriented films of gelatin. Robinson and Bott (1951) noted, however, that in films obtained by evaporation of hot gelatin solutions the frequency was reduced to 3310 cm⁻¹. Bradbury *et al.* (1958) measured the spectra of films of acid-soluble rat-skin collagen and found that after deuteration of the NH groups the ND stretching frequency was also correspondingly higher than normal. From this it was concluded that the higher than normal amide A frequency in native collagen was not due to an accidental degeneracy of vibrational energy levels. Susi *et al.* (1971) obtained spectra from films of fragmented bovine tendon collagen at various relative humidities (Fig. 14.7) and found that the frequency of the amide A vibration decreased with increase in relative humidity, falling to a value of 3320 cm⁻¹ at 75% relative humidity. The downshift upon hydration was attributed to an increase in hydrogen bond strength.

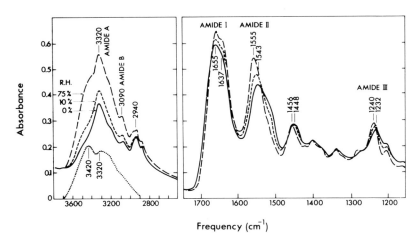

FIG. 14.7. Infrared spectra obtained from films of fragmented bovine tendon at various relative humidities (after Susi *et al.*, 1971).

Several authors have suggested that the higher-than-normal amide A frequency might be associated with weaker-than-normal hydrogen bonding (Fraser, 1951; Seeds, 1953; Ramachandran *et al.*, 1962; Ramachandran, 1963b) but allowance must be made for the fact that the amide A band is not due to a pure NH stretching vibration but is a component of the Fermi resonance between this vibration and the first overtone of the amide II vibration (Badger and Pullin, 1954; Miyazawa, 1960b,c). The direct application of empirical NH stretching frequency–$b(N, O)$ correlations (Ramachandran *et al.*, 1962; Ramachandran, 1963a,b, 1967) is therefore not justified (A. Elliott, 1963).

Additional displacement of the observed frequency may result from coupling of the vibrations in different amide groups (Chapter 5, Section A.II). No information has been obtained so far on the magnitude of the frequency shifts of the amide A or amide II bands due to coupling but if these are assumed to be small, the value of the NH stretching frequency can be calculated by the method outlined in Chapter 9, Section C.I. Using the values given earlier for air-dried mouse-tail tendon, a value of $b(N, O) = 2.94$ Å is obtained, while the values $\nu_A = 3320$, $\nu_B = 3090$, and $\nu_{II} = 1545$ cm^{-1} recorded by Susi *et al.* (1971) for fragmented bovine tendon at 75% relative humidity lead to a value of $b(N, O) = 2.92$ Å.

These estimates can be compared with the value of $b(N_1, O_2) = 2.96$ Å in the model for [Gly-Pro-Pro]$_n$ suggested by Yonath and Traub (1969) on the basis of X-ray diffraction and stereochemical data. The available information on hydrogen bonds of the type NH$\cdots$O$=$C, taken from the survey by Ramakrishnan and Prasad (1971), is shown in Fig. 14.8 and it can be seen that the estimate of $b(N, O)$ in collagen derived from the infrared data is in the modal range and while the hydrogen bond length appears to be greater than is general in the α-helix (Table 9.6), it could not thus be classed as abnormally long. The alternative suggestion (Fraser, 1951) that the higher amide A frequency in collagen might be due to a distortion of the amide-group geometry has not been explored further but in view of the recent suggestion (Winkler and Dunitz, 1971) that the requirement of precise planarity might be rather readily relaxed in actual structures, this must be considered as a possibility in collagen.

An amide A frequency of about 3330 cm^{-1} has come to be regarded as diagnostic of a collagenlike structure (Bradbury *et al.*, 1958; Ramachandran, 1963b) and some support for this view has been obtained from denaturation studies (Fraser, 1951; Robinson and Bott, 1951; Bradbury *et al.*, 1958) and from studies of developing collagen (Seeds, 1953). In the three-strand-rope models there are three structurally different types of NH group and the proportions attached to C_1^α, C_2^α, and C_3^α in mammalian tendon collagen are shown in Section C.II.b to be, respectively, 0.42, 0.29, and 0.29. It is not immediately obvious which

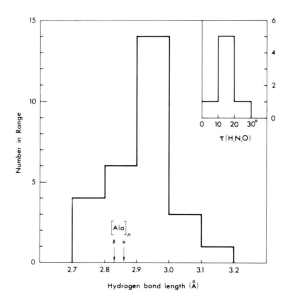

FIG. 14.8. Summary of data from crystallographic studies on hydrogen bond length in bonds of the type NH···O=C (Ramakrishnan and Prasad, 1971).

type or types are likely to be associated with the characteristic frequency but evidence which supports the assignment to the glycyl NH groups attached to C_1^α has been obtained from studies of synthetic polypeptides. Amide A frequencies in the range 3315–3360 cm^{-1} have been observed in a series of polytripeptides (Andreeva *et al.*, 1970) and four polyhexapeptides (Traub and Piez, 1971) having collagenlike structures. In particular the polymer [Gly-Pro-Pro]$_n$, which has no NH groups other than the glycyl set, has an amide A frequency of 3345 cm^{-1}. As discussed in Chapter 11, Section E.IV, this polymer forms a three-strand rope with the RC II type of hydrogen bond topology and in this instance the higher frequency may be attributed to the NH groups involved in interchain bonding.

According to the results quoted by Andreeva *et al.* (1970), polymers of the type [Gly-I-I]$_n$ have amide A frequencies in the range 3345–3360 cm^{-1}, while polymers of the type [Gly-I-X]$_n$ or [Gly-X-I]$_n$ have frequencies in the range 3315–3320 cm^{-1}, where I is an imino acid residue and X is an amino acid residue. The value observed in collagen varies slightly with humidity (Susi *et al.*, 1971) but appears to be in the range 3320–3330 cm^{-1}. This could be related to the relative infrequency in collagen of triplets containing two imino acid residues (Table 14.2).

b. *Vibrational Modes of the Three-Strand-Rope Models*

The changes in the frequencies and fine structures of the amide bands of collagen which take place when the native structure is denatured (Fraser, 1951; Robinson and Bott, 1951; Bradbury *et al.*, 1958) will be due, at least in part, to a rearrangement of the hydrogen bonds. A second effect which may contribute to these changes is the breakdown in the regular pattern of interactions between amide groups. In the α-helix and in β structures certain of the amide frequencies are appreciably displaced through such interactions and similar effects are likely to be present in three-strand-rope structures of the type considered in Chapter 11, Section A.II. The analysis of the vibrational modes of such structures is complicated, in the case of collagen, by the fact that the tripeptide units have a variable composition with imino acids randomly distributed between positions 2 and 3 (Section B.II). With other regular main-chain conformations the variable side-chain composition is no impediment to an analysis of the vibrational modes since the amide-group vibrations are localized in the main chain and the groups are similar. In the case of collagen, however, a proportion of the amide groups are tertiary rather than secondary and the localized vibrational modes are quite different (Bellamy, 1958, 1968).

The problem is most simply approached through a consideration of the vibrational modes of a polytripeptide and it is convenient to consider the asymmetric unit as a set of three amide groups $-C_1^\alpha CONH\ C_2^\alpha CONH\ C_3^\alpha CONH-$ in the same chain. The amide group following C_3^α will always be a secondary amide since C_1^α belongs to a Gly residue, but the NH groups preceding C_2^α and C_3^α will be substituted if these atoms belong to imino acid residues. The asymmetric units are related by helical symmetry (Fig. 11.7) with unit twist $t \simeq -108°$ and from the treatment given by Higgs (1953), which is discussed in Chapter 5, Section A.II, each vibrational mode of the asymmetric unit would be expected to be split into two components in the helical structure. In the first, the atomic motions in all the tripeptide units would be in phase, while in the second the phase difference between the motions in successive units in the helix would be $-108°$.

The extent to which the frequencies of the two components differed would depend upon the nature of the coupling between asymmetric units and the observable frequencies would be related by Eq. (5.2), which would take the forms

$$\nu_\parallel(0) = \nu_0 + \sum_j D_j \tag{14.2}$$

and

$$\nu_\perp(-108°) = \nu_0 + \sum_j D_j \cos(-j \cdot 108°) \tag{14.3}$$

where the symbols have the same significance as in Chapter 5. The coefficient D_3, for example, would represent coupling with the next tripeptide unit in the same chain (Fig. 11.3) and the coupling through interchain hydrogen bonding would be represented by the coefficient D_2 in the case of the RC I hydrogen bond topology and D_1 in the case of the RC II or RK topology. The value of the coefficient D_1 would be expected to be different for the RC II and RK models. In the latter model, coupling occurs through two types of hydrogen bond and one set forms a continuous chain of coupled amide groups (Fig. 11.3). In principle it should therefore be possible to distinguish between these two models on the basis of the fine structure of the amide bands.

The nature of the fine structure will depend critically upon the composition of the asymmetric unit. The three amide groups are not structurally equivalent and so multiple frequencies may be present even in an isolated asymmetric unit. Each localized vibration of the asymmetric unit will be split into two components by interactions in the helical array. When imino acid residues are present intrachain coupling will be reduced since the vibrational modes of secondary and tertiary amides are different. In the absence of an NH group, the amide group preceding an imino acid residue will not contribute to the amide A, B, II, III, or V bands but will contribute a component in the amide I region from the CO stretching mode of the tertiary amide group. The frequency of this mode will depend upon the environment of the CO group. In [Pro]$_n$ the CO stretching frequency was found to be about 1650 cm^{-1} in both form I and form II (Blout and Fasman, 1958) and using higher resolving power de Lozé and Josien (1969) estimated values of 1651 cm^{-1} for form I and 1643 cm^{-1} for form II in dried films. Some lowering of the CO stretching frequency in the amide group preceding an imino acid residue is to be expected if a hydrogen bond of the type $C{=}O\cdots H{-}O{-}H$ is formed to a water molecule.

Very little progress has been made in the analysis of the fine structure of the amide bands observed in the spectrum of collagen. The amide I band has at least two components with frequencies of 1637 and 1655 cm^{-1} in tendon which is in equilibrium with an atmosphere of 75% relative humidity (Fig. 14.7). To the extent to which collagen approximates a polytripeptide, these could feasibly be the parallel and perpendicular components arising from the quasihelical symmetry, but the CO groups are oriented in such a manner that the parallel component is expected to be weak. An alternative, and more likely, explanation is that the two peaks correspond to the perpendicular components of two separate vibrations of the asymmetric unit. The relative proportions of CO groups of different types, assuming a random distribution of imino acid residues between positions 2 and 3 in an RC II type of structure, would be as shown in Table 14.3a. It will be seen that there are five potentially different frequencies in the glycine-led triplets but insufficient data are available at present to be able to assign the observed fine structure in the amide I region to particular sets of CO groups.

TABLE 14.3

Distribution of Groups in a Three-Strand Rope with the
RC II Hydrogen Bond Topology

(a)   CO Groups

Type of CO group	Proportion[a]
Groups involved in interchain hydrogen bonds	
Secondary amide following C_2^α	$0.33 - 0.5p = 0.22$
Tertiary amide following C_2^α	$0.5p = 0.11$
Groups not involved in interchain hydrogen bonds	
Secondary amide following C_1^α	$0.33 - 0.5p = 0.22$
Secondary amide following C_3^α	0.33
Tertiary amide following C_1^α	$0.5p = 0.11$

(b)   Secondary amide groups

Type of secondary amide group	Proportion[a]
Following C_1^α and not involved in interchain hydrogen bonds	$(2 - 3p)/[6(1 - p)] = 0.29$
Following C_2^α with CO acting as acceptor for interchain hydrogen bonds	$(2 - 3p)/[6(1 - p)] = 0.29$
Following C_3^α with NH acting as donor for interchain hydrogen bonds	$1/[3(1 - p)] = 0.42$

[a] Calculated assuming a random distribution of imino acid residues between positions 2 and 3; p is the fraction of total residues which are proline or hydroxyproline. The proportions given are for a value of $p = 0.216$ appropriate to mammalian collagen (Table 14.1).

The amide A, II, III, and V bands only arise from secondary amide groups but again a distinction must be drawn among the three types of environment. The proportions, again assuming a random distribution of imino acid residues between positions 2 and 3 in an RC II type of structure, will be as shown in Table 14.3b. Bradbury et al. (1958) investigated the rate of exchange of hydrogen for deuterium in films of soluble rat-skin collagen and distinguished three types of secondary amide group. The first, in which exchange occurred very rapidly, were assumed to be present in partly degraded material; the second, in which exchange was completed within an hour, were assumed to be bonded to water molecules; and the third, which exchanged very slowly, were assumed to be involved in the formation of interchain $N-H \cdots O = C$ hydrogen bonds. The last two groups were estimated to be present in the proportion 0.56:0.44, respectively, which agrees very closely with the proportion 0.58:0.42 given in Table 14.3b for an RC II type of hydrogen bond topology.

As with the amide I band, effects due to the quasihelical symmetry may be present, but so far it has not proved possible to distinguish these from fine structure due to the three types of amide-group environment. High-resolution studies, using polarized radiation, of oriented specimens at various stages of hydration are required before detailed interpretation can be contemplated.

c. *Effects of Hydration*

The effects of hydration on the spectrum of films of soluble rat-skin collagen were studied by Bradbury et al. (1958), who noted changes in the band contours and frequencies of the amide II and III vibrations. In the spectra obtained by Susi et al. (1971) from films of fragmented bovine tendon at various relative humidities the maximum of the amide A absorption band decreased in frequency with increase in relative humidity while the amide II, III, and V maxima increased in frequency (Fig. 14.7). The changes were found to be progressive over the range 0–75% relative humidity and were attributed to the gradual attachment of water molecules accompanied by an increase in hydrogen bond strength. It was claimed that a number of the bands were intensified upon hydration and this was attributed to the formation of stronger hydrogen bonds, but inspection of the spectrum suggests that the integrated intensity is not in fact very different and it seems more likely that the increase in peak height on hydration is due to a reduction in half-bandwidth or to overlap by water bands. Evidence obtained from X-ray studies (Section C.I) suggests that when collagen is dried the structure is partially collapsed and the broadening of the amide bands which is observed on drying is consistent with this notion.

A band of frequency 1448 cm⁻¹ in the dried material was observed to increase in frequency to 1456 cm⁻¹ in the hydrated material and this was attributed to an increase in CH···O hydrogen bonding. In the absence of supporting evidence this conclusion must be treated with considerable reservation since similar frequency shifts have been observed, for example, in structural transformations in $[Asp(Bzl)]_n$ (Bradbury et al., 1962a) and $[Cys(Cbz)]_n$ (Elliott et al., 1964) where no CH···O bonds are likely to be present. As mentioned earlier, water plays a major role in the stabilization of the native collagen structure and its removal leads to a structural transformation. It does not seem to be necessary, on the basis of the data available at present, to invoke CH···O hydrogen bonds to explain the observed frequency shift of the CH_2 deformation band.

Humidity-dependent effects have also been noted in spectra obtained

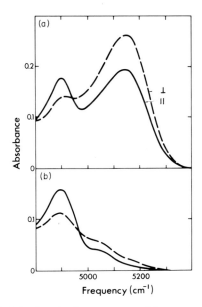

FIG. 14.9. Infrared dichroism in the spectrum of kangaroo-tail tendon equilibriated with an atmosphere of relative humidity about 50% (upper spectrum) and after drying at 100°C *in vacuo* (lower spectrum). The prominent dichroic peak at 5150 cm⁻¹ in the upper spectrum is due to a combination band of sorbed water. The full curve corresponds to the spectrum measured with the electric vector vibrating parallel to the length of the tendon and the broken curve to the spectrum measured with the electric vector vibrating perpendicular to the length of the tendon.

in the overtone region (Fraser and MacRae, 1958c) and information on the structural role of water was obtained from studies of kangaroo-tail tendon using polarized radiation (Fraser and MacRae, 1959b). The prominent band at 5150 cm^{-1} (Fig. 14.9a) is due to sorbed water (Ellis and Bath, 1938) and is virtually eliminated by drying *in vacuo* at 100°C for 3 hr (Fig. 14.9b). The band exhibits appreciable perpendicular dichroism in the partially hydrated specimen, indicating that a proportion of the sorbed water molecules are partially immobilized and disposed in a manner such that the line joining the two hydrogen atoms has a preferred orientation perpendicular to the rope axis. It seems likely that these preferentially oriented water molecules are involved in interchain hydrogen bonding within a three-strand rope.

d. *Dichroism*

A number of studies have been made of infrared dichroism both in tendon and in oriented films of gelatin and soluble collagen and the results are summarized in Table 14.4a. The amide A and amide I bands exhibit perpendicular dichroism, while the amide II band exhibits parallel dichroism. Qualitatively at least, this suggests that the NH and CO bonds are preferentially oriented perpendicular to the fiber axis.

Attempts have been made to interpret the observed dichroic ratios in a quantitative manner in order to provide a test of proposed models. Bradbury *et al.* (1958) estimated that the inclinations to the fiber axis of the transition moments of the amide bands were as listed in Table 14.4b. In order to obtain these values it was necessary to estimate the orientation parameter (Chapter 5, Section B.III.b) but details of the calculations were not given. In the terminology of Chapter 5 these values are estimates of α, the angle the transition moment direction makes to the rope axis, derived from Eqs. (5.45) and (5.46) by assuming a value for $\langle \cos^2\theta_c \rangle$, where θ_c is the inclination of the rope axis to the fiber axis. The estimate so obtained represents a type of average defined by

$$\cos^2\alpha = \left(\sum_i M_i^2 \cos^2\alpha_i \right) \Big/ \sum_i M_i^2 \qquad (14.4)$$

where M_i is the magnitude of the transition moment of an individual amide group, α_i is the inclination of the transition moment to the rope axis, and the summation extends over the entire rope. In regular structures such as the α-helix, where the asymmetric unit is a single secondary amide group, Eq. (14.4) reduces to $\cos^2\alpha = \cos^2\alpha_i$ and so a direct estimate of α_i is obtained. In collagen, however, the situation is more complex since three values of α_i are involved in the case of the amide A,

TABLE 14.4

(a)   Measurements of Infrared Dichroism in the Amide Bands of
Collagens and of Gelatin

Material	Dichroic ratio[a]					Ref.[b]
	A	I	II	III	V	
Mouse-tail tendon	0.58	0.45	1.31	1.6	—	1
Stretched gelatin	0.75	0.67	1.20	—	—	2
Kangaroo-tail tendon	0.62	0.48	1.25	—	—	3
Soluble-collagen film	0.53	0.44	1.42	(2.0)[c]	—	4
partially deuterated	0.20	0.46	—	—	—	4
Rat-tail tendon	0.70	—	1.20	—	1.00	5

(b)   Estimates of Transition Moment Direction

Structure	Transition moment direction[d] (deg)				Ref.
	A	I	II	V	
Soluble collagen film	80–85	68–73	43–45	—	4
RC II Model	85	65	45	—	4
RC II Model	83	—	50	49	5
RK Model	77	—	44	52	5

[a] Where values of the dichroic ratio were not given these have been calculated from the observed spectra. No allowance for background or overlap was made and peak heights rather than integrated intensities were used.

[b] 1, Fraser (1950a, 1951); 2, Ambrose and Elliott (1951b); 3, Badger and Pullin (1954); 4, Bradbury et al. (1958); 5, Beer et al. (1959).

[c] The value given in Ref. 4 for the ratio $D_\perp/D_\parallel$ is 5.0, but this is presumably a typographical error. If it is assumed that 0.5 was meant, the value given in parentheses would be appropriate.

[d] Inclination to fiber axis.

II, and III bands and five in the case of the band in the amide I region, due to the contributions from tertiary amide groups. The analysis of amide group distribution given in Section C.II.b leads to the result

$$\cos^2\alpha = [(2 - 3p)(\cos^2\alpha_1 + \cos^2\alpha_2)/6(1 - p)] + [(\cos^2\alpha_3)/3(1 - p)] \quad (14.5)$$

for the amide A, II, and III bands on the assumption that the magnitudes of the transition moments are equal, where p is the proportion of imino acid residues. The suffix 1 refers to a secondary amide group following

C_1^{α}, etc. For mammalian tendon collagen, with $p = 0.216$, Eq. (14.5) reduces to

$$\cos^2\alpha = 0.29(\cos^2\alpha_1 + \cos^2\alpha_2) + 0.42\cos^2\alpha_3 \tag{14.6}$$

Interpretation of the significance of α in the case of the band in the amide I region is complicated by the fact that the magnitude of the transition moment M cannot be assumed to be the same for the CO modes of secondary and tertiary amides.

Beer *et al.* (1959) used an alternative approach in which the orientation parameter was calculated for each amide band on the basis of various assumed structures, as described in Chapter 5, Section B.III.b. The sets of orientation parameters so obtained were then tested for internal consistency.

The results obtained in these studies can only be regarded as semiquantitative since all the amide bands are overlapped to some extent by other bands and corrections for this are difficult to estimate. Interpretation of the amide I region is particularly difficult since the local transition moment direction of the CO stretching mode of the tertiary amide groups associated with the imino acid residues is unknown and, as mentioned earlier, the magnitude of the transition moment, which determines the intensity, cannot be assumed to be the same as for secondary amides.

D. MODELS OF CHAIN CONFORMATION

The gradual evolution of ideas on the nature of the pseudostructure responsible for the high-angle X-ray pattern of collagen has been very elegantly described by Dickerson (1964), who also gives details and illustrations of early models. With characteristic insight into the problem, Astbury (1933) considered the possibility that the structure contained a Gly-Pro-Hyp repeating unit in which the average axial height per residue was 2.8 Å. Early attempts to devise detailed models were all based, by analogy with silk fibroin and β-keratin, on sheet structures (Astbury, 1938; Astbury and Bell, 1940; Huggins, 1943; Zahn, 1948; Ambrose and Elliott, 1951b; Randall *et al.*, 1952, 1953). Following the realization in the early 1950's of the possibility that helical structures with noncrystallographic screw axes might be present in naturally occurring high polymers, a number of helical models were proposed (Pauling and Corey, 1951f; Bear, 1952; Crick, 1954; Huggins, 1954, 1957; Ramachandran and Kartha, 1954; Millionova and Andreeva, 1958, 1959). In each case, however, there was serious disagreement between predictions from the model and at least one feature of the X-ray diffraction pattern, infrared spectrum, or the stereochemical criteria discussed in Chapter 6, Section B.

I. THE RAMACHANDRAN–KARTHA MODEL

The first three-strand-rope model for collagen was that suggested by Ramachandran and Kartha (1955a). This was derived from an earlier model, consisting of three left-handed helical chains joined by inter-chain hydrogen bonds, by the introduction of a right-handed rope twist so that the individual chains had a coiled-coil conformation (Chapter 1, Section A.II.e). The asymmetric unit of structure was a group of three residues with Gly always occupying position 1 (Chapter 11, Section A.II), and two interchain hydrogen bonds were formed per asymmetric unit

TABLE 14.5

Atomic Coordinates of the Repeating Unit in the Helical Collagen Model[a]
Proposed by Ramachandran and Co-workers[b]

Residue	Atom	u (Å)	v (Å)	w (Å)	r (Å)	ϕ (deg)
Gly	N_1	1.75	−0.94	−0.94	1.99	−28.3
	H_1	1.38	−1.86	−0.75	2.31	−53.4
	C_1^α	1.15	0.00	0.00	1.15	0
	$H_1^{\alpha 1}$	0.87	0.80	−0.54	1.18	42.5
	$H_1^{\alpha 2}$	0.37	−0.53	0.39	0.65	−55.0
	C_1'	2.21	0.39	1.05	2.24	10.0
	O_1	3.23	−0.30	1.20	3.24	−5.3
2	N_2	1.96	1.57	1.60	2.51	38.7
	H_2	1.04	1.98	1.40	2.24	62.3
	C_2^α	2.80	2.10	2.69	3.50	36.8
	H_2^α	3.76	2.24	2.43	4.38	30.8
	C_2^β	2.33	3.51	3.13	4.21	56.4
	C_2'	2.64	1.14	3.88	2.88	23.4
	O_2	1.65	0.39	3.96	1.70	13.4
3	N_3	3.67	1.06	4.70	3.82	16.1
	H_3	4.45	1.66	4.76	4.75	20.5
	C_3^α	3.53	0.21	5.91	3.54	3.4
	H_3^α	3.31	−0.72	5.61	3.39	−12.2
	C_3^β	4.90	0.22	6.65	4.90	2.6
	C_3'	2.41	0.83	6.77	2.55	18.9
	O_3	1.92	1.93	6.41	2.72	45.2

[a] Atoms of the next triplet in this chain are at $(r, \phi + 30°, w + 8.73$ Å$)$. Atoms in the corresponding triplets in the other chains are at $(r, \phi − 110°, w + 2.91$ Å$)$ and $(r, \phi − 220°, w + 5.82$ Å$)$.

[b] Ramachandran (1967).

as illustrated in Fig. 11.4. The helical symmetry satisfied the requirements of the high-angle pattern (Section C.I.b). Rich and Crick (1955) pointed out that the structure, as described, was unlikely to be correct, first, because some very short interatomic contacts were involved, and second, because imino acids could not be incorporated in position 2 although there was good evidence (Section B.II) that this position was occupied by imino acids in the protein. A further criticism concerned the larger-than-normal values of $\tau(H, N, O)$ for the hydrogen bonds.

In the light of these criticisms a modified version of the original model was put forward (Ramachandran and Sasisekharan, 1961) which was stereochemically somewhat more satisfactory, but again was not suited to the incorporation of an imino acid residue in position 2. Later a further modification was described in which a short CH···O contact was supposed to lead to additional stabilization by the formation of a hydrogen bond. Atomic coordinates for the asymmetric unit of this model are given in Table 14.5 and an axial projection of the structure is depicted in Fig. 14.10. The helical parameters are $h = 2.91$ Å,

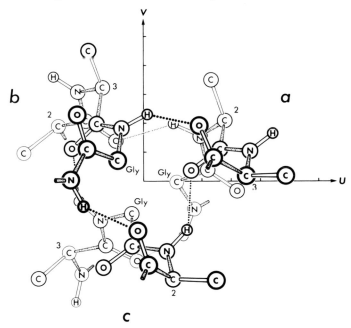

FIG. 14.10.   Axial projection of the model proposed by Ramachandran and co-workers for the structure of collagen. The asymmetric unit is a group of three residues with Gly always occupying position 1. Two interchain hydrogen bonds are formed per asymmetric unit but Pro cannot be accommodated in position 2. The scale markings represent angstroms and the axes correspond to those of Table 14.5.

$t = -110°$, and $u/v = -3.27$, and Ramachandran (1967) has listed the values of the bond angles and the interatomic distances in the model. The hydrogen bond parameters are $b(N_1, O_3) = 3.06$ Å and $\tau(H_1, N_1, O_3) = 30°$, and $b(N_3, O_2) = 2.95$ Å and $\tau(H_3, N_3, O_3) = 27°$.

In order to overcome the problem of accommodating an imino acid residue in position 2 it was suggested (Ramachandran et al., 1962; Ramachandran, 1963a,b, 1967) that the chains would adopt one of two alternative conformations depending on whether imino acid residues occurred in this position in one or more than one of the chains at any particular level along the three-strand rope. Coordinates for these two alternative models have been given by Ramachandran (1967). The model

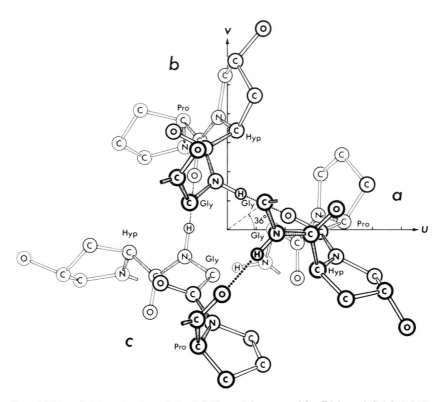

FIG. 14.11. Axial projection of the RC II model proposed by Rich and Crick (1955) for the structure of collagen, illustrated for the sequence Gly-Pro-Hyp. The asymmetric unit is a group of three residues with Gly always occupying position 1. One interchain hydrogen bond is formed per asymmetric unit. The scale markings represent angstroms and the axes correspond to those of Table 14.6.

which will accommodate the sequence Gly-I-I, where I is an imino acid residue, has the RC II type of hydrogen bond topology.

Thus in its final form the model proposed for the chain structure of collagen by Ramachandran and co-workers differs from that considered by other workers in that the hydrogen bond topology is not constant along the length of the rope but oscillates between alternative arrangements depending upon the local distribution of imino acid residues. In the models proposed by Rich and Crick, which are discussed later, imino acid residues can be accommodated in either positions 2 or 3 without interrupting the regularity of the structure. The extent to which the model proposed by Ramachandran and co-workers is consistent with the observed physical data is discussed in Section E.

II. THE RICH–CRICK MODELS

As mentioned earlier, Rich and Crick (1955) expressed the view that the model for the chain conformation of collagen proposed by Rama-

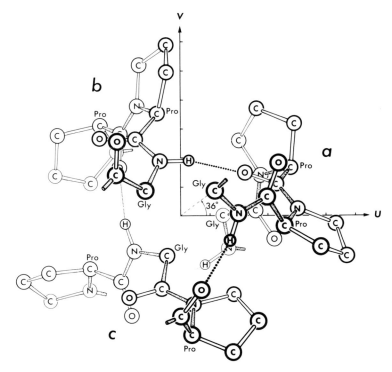

FIG. 14.12.   Axial projection of the model proposed by Yonath and Traub (1969) for the structure of collagen, illustrated for the sequence Gly-Pro-Pro. This model may be regarded as a refinement of the RC II model illustrated in Fig. 14.11.

chandran and Kartha (1955a) was not stereochemically feasible and an alternative proposal, based on their studies of the structure of form II of [Gly]$_n$ (Chapter 11, Section B.II), was put forward. The model, which may be derived by taking a group of three chains taken from a crystal of form II of [Gly]$_n$ and imparting a rope twist, has already been described in Chapter 11, Section A.II. Of the two possible arrangements of hydrogen bonds, the RC II topology has generally been preferred (Ramachandran, 1956; Rich and Crick, 1958; Burge *et al.*, 1958). This model is illustrated in Fig. 14.11 and atomic coordinates for the asymmetric unit are given in Table 14.6. Stereochemical features of the RC I

TABLE 14.6

Atomic Coordinates of the Repeating Unit in the Helical Collagen II Model[a]
Proposed by Rich and Crick (1955, 1961)

Residue	Atom	u (Å)	v (Å)	w (Å)	r (Å)	ϕ (deg)
Gly	N_1	1.32	−1.11	−0.90	1.73	−40.01
	H_1	0.35	−1.28	−1.06	1.33	−74.57
	C_1^{α}	1.61	0	0	1.61	0
	C_1'	2.45	−0.45	1.19	2.50	−10.38
	O_1	2.44	−1.64	1.59	2.93	−33.85
Pro	N_2	3.18	0.50	1.74	3.22	9.02
	C_2^{α}	4.06	0.27	2.89	4.07	3.84
	C_2^{β}	4.86	1.55	3.11	5.10	17.69
	C_2^{γ}	3.99	2.60	2.45	4.76	33.13
	C_2^{δ}	3.25	1.93	1.31	3.78	30.72
	C_2'	3.23	−0.05	4.14	3.23	−0.85
	O_2	2.13	0.50	4.33	2.18	13.11
Other	H_2	3.18	1.45	1.40	3.49	24.47
Hyp	N_3	3.79	−0.91	4.96	3.90	−13.55
	C_3^{α}	3.16	−1.36	6.21	3.44	−23.28
	C_3^{β}	4.12	−2.35	6.88	4.74	−29.69
	C_3^{γ}	5.46	−2.08	6.18	5.84	−20.88
	O_3^{δ}	6.22	−3.33	6.09	7.06	−28.14
	C_3^{δ}	5.11	−1.58	4.77	5.35	−17.16
	C_3'	2.92	−0.15	7.12	2.92	−2.97
	O_3	3.80	0.71	7.28	3.87	10.60
Other	H_3	4.68	−1.33	4.78	4.86	−15.87

[a] Atoms of the next triplet in this chain are at $(r, \phi + 36°, w + 8.58 \text{ Å})$. Atoms in the corresponding triplets in the other chains are at $(r, \phi − 108°, w + 2.86 \text{ Å})$ and $(r, \phi − 216°, w + 5.72 \text{ Å})$.

and RC II models have been discussed in considerable detail by Rich and Crick (1961) and the major difference is that either amino acid or imino acid residues can be accommodated in positions 2 and 3 in the RC II model, whereas in the RC I model imino acids can only be accommodated in position 3 if the structure is deformed. In both models position 1 can only be occupied by Gly and a single interchain hydrogen bond is formed per asymmetric unit of three residues. The hydrogen bond parameters in the RC II model are $b(N_1, O_2) = 2.8$ Å and $\tau(H_1, N_1, O_2) = 1.5°$.

Cowan et al. (1955b), on the basis of their earlier studies on $[\text{Pro}]_n$ (Chapter 11, Section C), independently suggested models similar to the RC I and RC II structures and Bear (1956) described a method of systematic model building which led to the same two possibilities. Burge et al. (1958) published an alternative set of coordinates for the RC II model but Rich and Crick (1961) expressed a preference for the values given in Table 14.6.

The sequential polymer $[\text{Gly-Pro-Pro}]_n$ was shown by Yonath and Traub (1969) to form an RC II type of structure (Chapter 11, Section E.IV) and it was suggested that the chain conformation determined for this polymer might be representative of the conformation throughout the whole of the collagen molecule. In support of this proposal it was also noted (Traub et al., 1969; Segal et al., 1969) that four polyhexapeptides, including $[\text{Gly-Ala-Pro-Gly-Pro-Ala}]_n$, had similar structures. The $[\text{Gly-Pro-Pro}]_n$ model (Table 11.3) can be regarded as a refinement of the original model proposed by Rich and Crick (1961) and it can be seen from Figs. 14.11 and 14.12 that the main difference is in the relative angular positions of the Gly residue and the imino acid residues.

III. FIXED WATER MOLECULES

The part played by water in the maintenance of the native structure of the collagen molecule has been widely recognized (Bear, 1952; Harrington and von Hippel, 1961; Tait and Franks, 1971; Traub and Piez, 1971) and several specific proposals have been made regarding the possible locations of water molecules occupying fixed positions relative to the pseudo repeat of structure. Evidence obtained from studies of infrared dichroism (Section C.II) strongly supports the presence of water molecules in which both OH bonds have fixed directions and the HH line is preferentially oriented perpendicular to the molecular axis. This implies that the water molecules form bridges through hydrogen bonds with CO groups in different chains. Burge et al. (1958) considered ways in which water molecules might be systematically bound to pairs of CO groups not otherwise implicated in interchain hydrogen bonding in the RC II model. Only one type of interchain water bridge was found to be possible linking $[O_1]_a$ to $[O_3]_c$ (Fig. 14.11), but this was not considered to be entirely satisfactory on stereochemical grounds. Instead, positions were specified for water molecules which might be singly bonded to NH or CO groups not involved in interchain bonding. These were included in calculations of the Fourier transform (Section D.IV). Rich and Crick (1961) also considered the possibility that water molecules were singly bonded to the NH groups when they occurred in positions 2 and 3. Another model involving water molecules singly bonded to the exposed main-chain NH and CO groups has been discussed by Berendsen (1968).

The formation of a single bond between a water molecule and the amide group of the polypeptide chain would only lead, in general, to a partial immobilization of the water molecule, but water molecules that were doubly or triply bonded to main-chain atoms would occupy more precisely defined positions. Suggestions for the location of such multiply bonded water molecules have been made, as mentioned earlier, by Burge et al. (1958) and also by Esipova et al. (1958), Ramachandran and Chandrasekharan (1968), and Yonath and Traub (1969).

Although there is good evidence for the presence of structural water molecules in collagen, the actual positions assigned by various investigators can only be regarded as speculative. Potentially, the location of the oxygen atoms of the immobilized water molecules could be determined, as part of the refinement of a model structure, on the basis of the X-ray intensities, and the orientations by a combination of stereochemical considerations and direct interpretation of the observed infrared dichroism, following the principles outlined in Chapter 5, Section B.III. Of the suggestions made so far for interchain bridges, the $[O_1]_a$ to $[O_3]_c$ type envisaged by Burge et al. (1958) and Ramachandran and Chandrasekharan (1968) would seem to provide a straightforward explanation for the infrared data (Fraser and MacRae, 1959b).

IV. COMPARISON WITH OBSERVED DATA

a. *Optical Transforms*

Rich and Crick (1955) reported that optical transforms of the RC I and RC II models gave a rough agreement with the X-ray diffraction pattern obtained from collagen and Bear (1956) published an optical transform of the RC II model which showed that the general distribution of intensity was similar to that in the X-ray pattern obtained from moist tendon (Fig. 14.13). In constructing the mask, the atoms were represented by annular holes to simulate the atomic scattering factors and all the positions 2 and 3 were occupied by imino acids. Only one

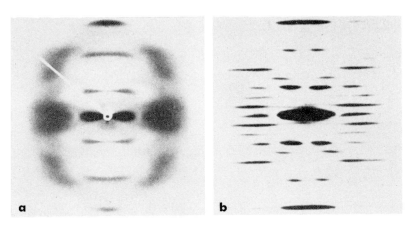

FIG. 14.13.   (a) High-angle X-ray diffraction pattern obtained from moist kangaroo-tail tendon. (b) Optical transform of the RC II structure (Bear, 1956).

view was used to obtain the optical transform and so it only represents an isolated section through the intensity transform of the structure. As explained in Chapter 4, Section B, the patterns obtained from several views must be combined in order to obtain an adequate representation of the intensity transform. If the values $R = 0.2 \text{ Å}^{-1}$, $r = 6 \text{ Å}$, and $u - 10$ are inserted in Eq. (4.2), a value of $n_B = 2$ is obtained for the number of Bessel functions contributing to each layer line and so a minimum of two projections uniformly spaced in the range $\phi = 2\pi/u = 36°$ must be used. This procedure was used by Rich and Crick (1961) to compare the optical transforms of the RC I and RC II models. No imino acid residues were included, but β-carbon atoms were incorporated for positions 2 and 3. It was concluded that the pattern obtained from the RC II model was in better agreement with the observed X-ray pattern. Burge et al. (1958) also showed that the optical transform of their version of the RC II model was in qualitative agreement with the pattern from dry, stretched rat-tail tendon.

    The problems encountered in carrying out precise tests of model structures by the use of optical transforms have been discussed in Chapter 4, Section B and a further difficulty was commented on by Bear (1956): "Experience with optical transforms of other models devised for collagen suggests, however, that the optical examinations provide what may be termed a 'necessary' rather than a 'sufficient' test, because of lack of detail in the X-ray diffraction and incomplete knowledge of side-chain distribution."

b. *Calculated Transforms*

A quantitative comparison between the intensity transform calculated for a model structure and the observed intensity transform provides a searching test of correctness but such a comparison, in the case of collagen models, is limited both by the paucity of the data so far collected and by problems associated with calculation of the transforms of pseudo-structures (Chapter 1, Section D.III).

Quantitative intensity data for layer lines with $l = 0, 3, 7$, and 10 in the diffraction pattern of wet, slightly stretched rat-tail tendon were collected by Bradbury *et al.* (1958) but Rich and Crick (1961), recognizing the importance of obtaining the maximum amount of data, estimated the distribution of intensity on all the layer lines up to $l = 12$ in hydrated, stretched rat-tail tendon. The data obtained (Fig. 14.14a) can only be regarded as semiquantitative since no corrections were applied for overlap between layer lines, polarization, geometric factors, or the fact that the transform appears to be sampled on $l = 3$ (Chapter 1,

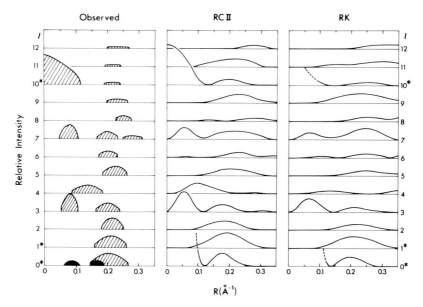

FIG. 14.14. Semiquantitative intensity data for the high-angle X-ray diffraction pattern of hydrated, stretched rat-tail tendon compared with the calculated intensity transforms of the RC II and RK models. The ordinates have been halved on the layer lines marked with an asterisk. After Rich and Crick (1961) and Ramachandran (1967).

Section A.I). Nevertheless, it represents the only attempt so far reported to map the intensity transform of collagen. Limited data of higher accuracy for portions of layer lines with $l = 0, 3, 7$, and 10, obtained from unstretched, partially hydrated sheep mucosa collagen, were reported by Yonath and Traub (1969) and are shown in Fig. 14.15c.

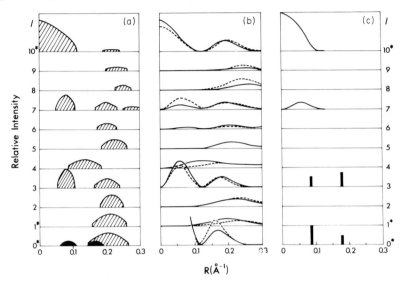

FIG. 14.15. (a) Semiquantitative intensity data for the high-angle X-ray diffraction pattern of hydrated, stretched rat-tail tendon; (b) the calculated intensity transform (full curve, dry; broken curve, hydrated) for the modified RC II model proposed by Yonath and Traub (1969); (c) limited quantitative intensity data from unstretched, partially hydrated sheep mucosa collagen. The ordinates have been halved on layer lines marked with an asterisk. After Rich and Crick (1961) and Yonath and Traub (1969).

It was assumed that the spots on $l = 0$ and $l = 3$ could be treated as lattice reflections and these were placed on the same scale as the streaks by the methods outlined in Chapter 1, Section C.II.

The calculation of the cylindrically averaged intensity transform of collagen models is complicated by the problem of side-chain distribution. The appearance of a discrete, high-angle X-ray pattern is due to the fact that a pseudorepeat of structure is present in the three-strand rope and contributions from side chains must be treated in different ways depending on whether the atoms occupy a constant position in relation to the pseudorepeating unit or occur in variable positions. The same considerations apply to water molecules in hydrated specimens. In most fibrous proteins the β-carbon atom is assumed to be the only side-chain atom which needs to be included in the pseudorepeat of structure, but in the case of collagen the contribution of the C^γ and C^δ atoms of the imino acid side chains may also have to be considered.

Rich and Crick (1961) carried out two types of calculation in order to test their models. In the first they combined the intensity transforms of models with imino acid residues in positions 2 and 3, and β-carbon atoms in positions 2 and 3, in the ratio 1:2. This corresponds to an arrangement where the imino acids are clumped together in all three chains. In the second calculation the contributions from C^γ, C^δ, and O^δ atoms were reduced to one-third of their normal value. This corresponds to the idealized case of a random distribution of imino acids between positions 2 and 3 and would, from the limited sequence data available (Section B.II) appear to be the more appropriate choice. A similar method was used by Bradbury *et al.* (1958).

Contributions from water molecules that occupy fixed sites in the repeating unit must also be included and allowance made for the fact that the occupancy of these fixed sites may depend upon the distribution of imino acids. In all the calculations which have been reported so far the contributions to the intensity transform of side chains and water molecules not occupying fixed positions relative to the pseudorepeat of structure have been ignored. Methods of approximating the contribution from such a "background" of atoms have been discussed in Chapter 1, Section D.III and the most appropriate way to deal with the problem in collagen would seem to be to calculate a Fourier transform

$$F = F_1 - \rho v F_2 + F_3 + \tfrac{3}{2}p(F_4 + F_5) \tag{14.7}$$

where F_1 is the Fourier transform of the main-chain atoms in the asymmetric unit, ρ is the mean electron density of the "background" of side chains plus water, F_2 is the normalized transform of the space occupied by the main-chain atoms, v is the volume of this space, F_3 is the contribution above background of the β-carbon atoms in positions 2 and 3, p is the fraction of imino acid residues, F_4 is the contribution above background of the C^γ and C^δ atoms in position 2, and F_5 is the contribution above background of the C^γ, C^δ, and O^δ atoms in position 3, with the O^δ contribution suitably weighted. Further terms would need to be included to cover the contribution of fixed water molecules, appropriately weighted, again above background.

The volume of the rope occupied by the main chains would be well approximated by a cylinder, and so

$$F_2 = J_1(2\pi Rr)/\pi Rr, \qquad l = 0 \tag{14.8}$$

$$F_2 = 0, \qquad\qquad\quad l \neq 0 \tag{14.9}$$

where r has a value of 4–5 Å. Such a term would definitely need to be included for a meaningful comparison of observed and calculated transforms to be made for the equator. The evaluation of the remaining terms would be straightforward except for the fact that it is the scattering above background which must be included. In the calculations which have been reported normal atomic scattering factors have been used for side-chain atoms and so their contribution to the Fourier transform has been overestimated.

Bradbury *et al.* (1958) calculated the magnitude of the Fourier transform of an RC II type of structure with the repeating sequence Gly-Pro-Hyp and the scattering factors of the C^γ, C^δ, and O^δ atoms were reduced to one-third of their normal value. In addition, the transform was also calculated for a similar structure in which water molecules were singly bonded to all the NH and CO groups not involved in inter-chain hydrogen bonding. The results were in reasonable agreement with the observed data on the layer lines with $l = 0$ and 10 but gross discrepancies were apparent on the layer lines with $l = 3$ and 7.

A more comprehensive test of the RC I and RC II models was reported by Rich and Crick (1961) and the intensity transforms of the two structures were found to be rather similar, but some preference was expressed for the RC II model. Various methods of allowing for the contributions of the imino acid residues were tried and the effects of incorporating water molecules in fixed positions was examined. The intensity transform of the RC II model appropriate to the case of imino acids clumped together in all three chains was given and is reproduced in Fig. 14.14b. The overall agreement with the observed data is good but a detailed comparison is not justified, because of the semiquantitative nature of the data.

The intensity transform of the RK model (Table 14.5) has been calculated and is reproduced in Fig. 14.14c (Ramachandran, 1967). Contributions from the main-chain atoms and from β-carbon atoms in positions 2 and 3 were included. The intensity transform is not greatly different from that of the RC II model and it is clear that until more accurate intensity data are available and calculations are performed in which *all* the scattering matter is taken into account, a choice between the two alternatives will not be possible on the basis of the high-angle diffraction pattern. The calculation of an intensity transform representative of the model proposed by Ramachandran and co-workers (Ramachandran, 1963a,b, 1967) presents special difficulties since the conformation is supposed to fluctuate between three alternative forms depending upon the local distribution of amino acids.

As discussed earlier, the polymer [Gly-Pro-Pro]$_n$ appears to adopt an

RC II type of structure and Yonath and Traub (1969) have compared the intensity transform for the model developed by them for this polymer with the observed data from collagen (Fig. 14.15). The transform was also calculated with two fixed water molecules per asymmetric unit (Table 11.3) and the effect on the intensity distribution on certain layer lines is appreciable. This emphasizes the importance of including this contribution, although, as explained earlier, the effects are likely to be exaggerated unless allowance is made for background. In addition, the contributions from the C^γ and C^δ atoms in positions 2 and 3 will also be exaggerated both through this effect and due to neglect of the factor $\frac{3}{2}p$ in Eq. (14.7). Thus it is difficult to make a quantitative assessment of the extent to which this refinement is an improvement over the original set of coordinates given by Rich and Crick (1961). The semiquantitative data obtained by the latter authors are included in Fig. 14.15 for comparison with the intensity transform calculated for $[Gly-Pro-Pro]_n$.

c. *Infrared Dichroism*

As explained in Section C.II.d, the calculation of the dichroic ratio R_0 predicted by a particular model for the structure of collagen must take account of the magnitudes and inclinations to the molecular axis of the transition moments associated with all the amide groups in the structure that contribute to the band. Values for $\alpha = \cot^{-1}[(\frac{1}{2}R_0)^{1/2}]$ have been given by Bradbury *et al.* (1958) for the RC II model and by Beer *et al.* (1959) for both the RC II and RK models and these are given in Table 14.4b. In neither case was the basis for calculation stated and it is not clear whether appropriate allowance was made for the partial occupancy of positions 2 and 3 by imino acid residues or how the value for amide I, which contains contributions from both secondary and tertiary amides, was arrived at. In view of the uncertainties both in the measurement of dichroic ratio and in the calculation of predicted values (Section C.II.d), it does not seem possible to differentiate between the two types of model. All that can be said, on the basis of the data available at present, is that the three-strand-rope model is not inconsistent with the observed infrared dichroism.

d. *Other Structural Studies*

Three structurally different types of NH group occur in the three-strand-rope models for collagen (Chapter 11, Section B.II) and these might be expected to differ with regard to the rate of exchange of hydrogen for deuterium or tritium in the presence of the respective oxide. The results obtained by Bradbury *et al.* (1958), which were interpreted as favoring a structure with one interchain hydrogen bond per triplet, have already been discussed in Section C.II.b. Later studies (Bensusan and Neilsen, 1964; Englander and von Hippel, 1962; von Hippel, 1967; Kingham and Brisbin, 1968) were, however, generally held to favor a structure with two interchain hydrogen bonds per triplet. A similar conclusion was reached (Harrington, 1964; Rao and Harrington, 1966) from a consideration of data on thermal transition temperatures, and these studies have been held to support the RK type of structure. In every case, however, some aspect of the interpretation of the experimental data has been open to criticism (von Hippel, 1967; Katz, 1970; Cooper, 1971) and in the most recent study of hydrogen exchange (Katz, 1970) it was found that the proportion of slowly exchanging hydrogens approached that calculated for a structure in which only one interchain hydrogen bond was formed per triplet.

Chapman et al. (1970) have used nuclear magnetic resonance spectrometry to study the hydrogen-bonded states of the amide groups in deuterated tendon and obtained spectra which they considered could only be explained in terms of the structure proposed for [Gly-Pro-Pro]$_n$ by Yonath and Traub (1969). Until more precise methods are available for predicting the effects of hydrogen bonding on the spectra it is difficult to make a critical assessment of this claim.

Calculations of intramolecular potential energy have been carried out for various collagen models and for polytripeptides (Ramachandran and Venkatachalam, 1966; Hopfinger and Walton, 1970a,b,c; Tumanyan, 1970) but it is difficult to know what reliance can be placed on the numerical values obtained in view of the large number of assumptions involved in the calculation. Intermolecular and solvent interactions have generally been disregarded and in addition relatively minor adjustments of the molecular parameters of a model or the parameters used in potential functions are capable of producing appreciable variations in the calculated energy. The problem in accepting the various assertions which have been made on the basis of such calculations has been nicely stated by Traub and Piez (1971): "We feel that such calculations may be suggestive of interesting structural possibilities, but that, in the present state of the art, these should be treated with caution unless supported by firm experimental evidence."

Nuclear magnetic resonance spectrometry has also been used to study interactions between collagen and water (Berendsen, 1962, 1968; Berendsen and Migchelsen, 1965, 1966; Migchelsen and Berendsen, 1967; Dehl and Hoeve, 1969; Chapman and McLauchlan, 1969; Khanagov and Gabuda, 1969; Chapman et al., 1971) and evidence obtained for ordering of water molecules. Various theories have been advanced to account for the observed ordering but there is no general agreement at present concerning the detailed interpretation of the data in terms of specific models.

V. SUMMARY

The data so far obtained from X-ray diffraction studies of collagen are consistent with a three-strand-rope model in which the chain conformation is a coiled-coil derived from the helical conformations present in form II of $[Gly]_n$ and $[Pro]_n$. At present, however, insufficient data are available to enable the exact details of the conformation to be determined.

Studies of sequential polypeptides indicate that the RC II type of hydrogen bond topology is present in $[Gly-Pro-Pro]_n$ and probably also in four polyhexapeptides containing a pair of glycine-led triplets (Chapter 11). The RC II type of model is therefore the only one which has been directly established to be a physically possible type of structure.

Two distinct views have been expressed regarding the chain conformation in collagen. According to Yonath and Traub (1969), the whole of the molecule is visualized as having a structure similar to that found by them for $[Gly-Pro-Pro]_n$, whereas Ramachandran (1963a, 1967) supposes that both the conformation and the hydrogen bond topology are variable and are determined by the local distribution of imino acid residues. In the latter case the structure is supposed to oscillate between the RK model and two alternative structures, one with five hydrogen bonds per three triplets and another which has the RC II type of hydrogen bond topology. However, Kühn et al. (1968) have pointed out that conditions favorable to the formation of the alternative with five hydrogen bonds per three triplets are unlikely to be present in collagen.

The original model proposed by Ramachandran and Kartha (1955a), in which two hydrogen bonds were formed per triplet, appears to have been designed in such a way that the maximum number of interchain hydrogen bonds were formed. However, in an aqueous environment, where opportunities exist for the formation of hydrogen bonds to water molecules, this requirement may not be of overriding importance (Klotz and Franzen, 1962; Klotz and Farnham, 1968; Kresheck and Klotz, 1969; Lumry and Biltonen, 1969; Cooper, 1971). Cooper (1971), in a consideration of the thermal stability of collagen, comments:

> We are forced to the conclusion that interchain hydrogen bonds cannot be the dominant interaction in the stabilization of the collagen triple helix. This is not to deny the importance of hydrogen bonding in structural considerations of the protein but it would

seem in the case of collagen, at least, that peptide–peptide hydrogen bonds differ little in stability from the peptide–water interaction [p. 125].

In the final form of the RK model (Ramachandran, 1967) the angle τ(H, N, O) is at the limit of the observed range for both hydrogen bonds (Fig. 14.10), and short CH···O contacts are also present. Thus if the criterion of maximizing the number of *interchain* hydrogen bonds is not applicable to the structure of native, hydrated collagen, there are no obvious stereochemical grounds for preferring the RK model to the RC II type of model, which is intrinsically better able to form structures with near-linear hydrogen bonds and without short contacts. There remains the possibility that in parts of the molecule where the sequence would allow an RK type of structure to form, a transition to this type of hydrogen bond topology occurs when collagen-containing tissue is dehydrated. The available evidence, however, suggests that the structural changes which take place on drying are more in the nature of a dis-ordering phenomenon.

Although no definite decision can be arrived at on the basis of the presently available data regarding the RK and RC II types of model, the evidence, on balance, appears to favor the latter alternative. The refinement of the RC II model, based on a study of [Gly-Pro-Pro]$_n$ (Table 11.3), would seem to be the closest approach that has so far been attained to a model for the chain conformation in collagen but it must be emphasized that this model cannot be regarded as being completely refined, even for the sequential polytripeptide, due to the paucity of the available diffraction data.

E. MOLECULAR STRUCTURE AND PACKING

Following biosynthesis of the collagen molecule, extracellular aggrega-tion takes place with the formation of collagen fibrils which vary in diameter, depending upon the tissue and the stage of development, from about 100 to 2000 Å. In most tissues the fibrils are approximately circular but occasionally sheet structures are formed as in the elastoidin of elasmobranch fishes (McGavin, 1962). The fibrils in turn are arranged in a geometric pattern adapted to the function of the tissue. In tendon they are aligned parallel to one another in bundles and the collagen fibers so formed are a convenient source of experimental material for structural studies. In the dermis the fibers form a complicated woven pattern (Rumplestiltskin, 1957) and many other variations have been noted (Fitton Jackson, 1968; Harkness, 1968; Bailey, 1968). In the present section attention will be restricted to a consideration of the way in which the collagen molecules are arranged in fibrils of the type found in tendon and of features of the molecular structure that are important in fibril formation.

I. LOW-ANGLE X-RAY PATTERN

a. *Meridian*

Low-angle meridional reflections indicating axial regularity of large period in collagen were observed in early studies (Clark *et al.*, 1935; Wyckoff *et al.*, 1935; Clark and Schaad, 1936; Corey and Wyckoff, 1936) and later it was established that the reflections in dry tendon could be indexed on an axial period of 640 Å (Bear, 1942; Kratky and Sekora, 1943). Subsequently, the nature and fine structure of the pattern and its dependence on the physical state of the specimen have been investigated in considerable detail (Bear, 1944a, 1951, 1952; Kaesberg *et al.*, 1948; Wright, 1948; Marks *et al.*, 1949; Bear and Bolduan, 1950a,b, 1951;

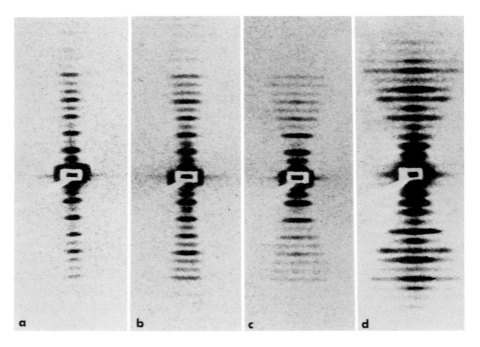

FIG. 14.16.   Progressive change in the low-angle X-ray diffraction pattern of kangaroo-tail tendon in going from wet (a) to dry (d) state (Rougvie and Bear, 1953).

Bolduan and Bear, 1950, 1951; Bear *et al.*, 1951; Bolduan *et al.*, 1951; Kaesberg and Shurman, 1953; Rougvie and Bear, 1953; North *et al.*, 1954; Burge and Randall, 1955; Cowan *et al.*, 1955a; Tomlin and Worthington, 1956; Bear and Morgan, 1957; Ericson and Tomlin, 1959; Tomlin and Ericson, 1960; McGavin, 1962, 1964; Ellis and McGavin, 1970).

The pattern obtained from wet tendon (Fig. 14.16a) consists of a series of layer-line streaks of approximately constant profile with maxima on the meridian. When the tendon is dried the lateral widths of the streaks increase progressively with layer-line index and off-meridional maxima appear (Fig. 14.16d); this effect has been described as a "fanning" of the pattern. The extent of the fanning depends upon the source of the collagen, its water content, and its physical and chemical history (Bear *et al.*, 1951; Bear, 1952; Rougvie and Bear, 1953).

The axial period decreases continuously from about 670 to 635 Å when the tendon is dried at room temperature (Bear, 1944a; Wright, 1948; Rougvie and Bear, 1953; Tomlin and Worthington, 1956) and a further reduction is observed (Table 14.7) if elevated temperatures are used (Rougvie and Bear, 1953; Ericson and Tomlin, 1959). This further reduction is associated, at least in part, with the fact that drying *in vacuo* at elevated temperatures extracts additional moisture from the tendon

TABLE 14.7

Effect of Heating on the Axial Period of Tendon Collagen[a]

Treatment	Period (Å)
Kangaroo tail tendon	
Air-dried and rewetted	673
Dried *in vacuo* at about 20°C	636
Dried *in vacuo*, 108°C	625
Dried *in vacuo*, 150°C	608
Dried *in vacuo*, 200°C	483
Sample dried at 150°C and rewetted	660
Sample dried at 150°C, rewetted, and dried at about 20°C	633
Rat tail tendon	
Air-dried and rewetted	671
Dried *in vacuo* at about 20°C	636
Bovine achilles tendon	
Air-dried and rewetted	670
Dried *in vacuo* at about 20°C	635

[a] Ericson and Tomlin (1959).

and is accompanied by a loss in weight which amounts to as much as
10% at 150°C (Ericson and Tomlin, 1959). The transition from the
hydrated to the dry state is accompanied by a redistribution of intensity
among the layer lines (Fig. 14.16).

Rougvie and Bear (1953) found that the decrease in axial period with
drying at room temperature was linearly related to the decrease in the
spacing of the strong high-angle equatorial reflection (Fig. 14.6) but
when elevated temperatures were used to extract further water from the
tendon the additional decrease in axial period took place without any
significant change in the spacing of the high-angle equatorial reflection.
Although the axial period is appreciably reduced, the high-angle meridional reflection of about 2.9 Å does not appear to be affected by drying
(Section C.I.c).

The axial period of air-dried tendon collagen increases when the
tendon is stretched (Cowan *et al.*, 1955a) and the increase was found to
be linearly related to the increase in the length of the tendon (Fig. 14.17).
Data were also obtained on the spacing of the meridional arc at about
2.9 Å, which was found to increase when the tendon was stretched by
more than about 2% (Fig. 14.17). The rate at which the spacing of the
high-angle arc increased was, however, appreciably less than the rate at
which the axial period increased.

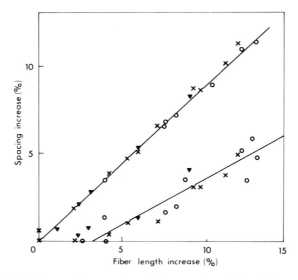

FIG. 14.17. Variation with extension of the spacing of the high-angle meridional
reflection at about 2.9 Å (lower set of points) compared with the variation in the layer-line
spacing at low angles (upper set of points). Data of Cowan *et al.* (1955a).

Many factors other than water content and stretching have been shown to affect the axial period of collagen, including treatment with mineral and organic acids and tanning agents (Bear, 1944a; Bolduan *et al.*, 1951). However, a number of heavy-atom derivatives have been prepared in which the axial period was unchanged from its value in the native material (Wright, 1948; Bear *et al.*, 1951; Bolduan *et al.*, 1951; Kaesberg and Shurman, 1953; Tomlin and Worthington, 1956; Ericson and Tomlin, 1959; Ellis and McGavin, 1970), thus providing opportunities for the application of crystallographic methods which depend on the treated and untreated materials being isomorphous.

Marks *et al.* (1949) obtained low-angle meridional diffraction patterns from a wide variety of invertebrate collagens and found that the axial period for air-dried specimens ranged from 650 to 676 Å and the distribution of intensity among the orders differed somewhat from that observed in mammalian tendon collagen under similar conditions of hydration. The axial period of collagen from the sea cucumber *Thyone* was studied in some detail and found to vary from 670 Å in the wet condition to 640 Å when oven-dried and was also affected by other treatments.

Kaesberg *et al.* (1948) measured the breadth of the first layer line in the diffraction pattern of wet beef tendon and the value obtained can be interpreted, using the theory outlined in Chapter 1, Section A.I.e, as indicating a coherent length of about 20,000 Å, corresponding to about 30 repeats of the observed 670 Å period. The estimate of coherent length so obtained is a minimum value and it is clear that, in one dimension at least, the collagen fibril is ordered to an extent which is comparable to that present in crystals of globular proteins. In most early studies collagen specimens were allowed to dry in air and the water content was subsequently adjusted to the required value. The number of low-angle meridional reflections in patterns obtained from such specimens is usually only about 30–35 and the gradual decay of intensity with increase in layer-line index can be attributed to the combined effects of various types of disorder in the axial arrangement of the collagen molecules (Chapter 1, Section A.IV). Later, North *et al.* (1954) showed that drying and rewetting tendon did not entirely restore the order present in the native structure and Miller and Wray (1971) found that if appropriate precautions are taken to preserve the native structure, the low-angle series of meridional or near-meridional reflections persists at least to spacings of about 10 Å (Fig. 14.4).

b. *Equator*

Low-angle and medium-angle reflections were observed in a number of early studies of the X-ray pattern of collagen-containing tissues but these have generally been attributed to radiation artefacts or to the presence of lipids (Clark and Schaad, 1936; Bear, 1952). An extensive series of equatorial and near-equatorial reflections extending to low angles was reported to be present in patterns obtained from tendon fibers maintained in the wet state (North *et al.*, 1954; Randall, 1954; Cowan *et al.*, 1955a). Since these reflections disappeared when the tendon was dried and rewetted, they were assumed to originate from the native collagen structure. Later Nemetschek (1968) found that they could, in fact, be restored if the fibers were rewetted with appropriate solvent mixtures.

North *et al.* (1954) indexed the observed series of reflections on a monoclinic cell with a base of dimensions $a = 76$ Å, $c = 60$ Å, and $\beta = 125°$ and Bear (1956) suggested that this corresponded to a specific grouping of 16 molecules. A number of alternative explanations of the series of reflections reported by North *et al.* (1954) have been suggested based on their assignment to subsidiary maxima (Chapter 1, Section A.I.e) of a limited cylindrical or spiral lattice (Ramachandran and Sasisekharan, 1956; Sasisekharan and Ramachandran, 1957) or a limited hexagonal lattice (Burge, 1963). However, Miller and Wray (1971) have obtained evidence which suggests that certain of the reflections recorded by North *et al.* (1954) did not originate from collagen and so the various suggestions with regard to their significance are unlikely to be correct and therefore will not be discussed in detail. In a further study Miller and Parry (1973) found that the observed equatorial and near-equatorial reflections could be indexed on a square-based cell with side $a = 76.9$ Å, but it was noted that diffuse scattering was present between the sharp reflections, suggesting some cell-to-cell variation in the scattering material.

c. *Direct Interpretation*

Bear and Bolduan (1950a,b, 1951) attempted to find a model distribution of scattering matter which would account for the fanning of the meridional intensity maxima which occurs when a collagen fiber is dried. The transforms of various types of cylindrical objects with periodic axial variations of electron density were examined and it was concluded that the effects could best be explained by supposing that the collagen fibril consisted of bundles of filaments which were subject to periodic distortion. An idealized model for the collagen fibril was described (Bear *et al.*, 1951; Bear, 1951, 1952) in which it was supposed that segments with perfect order (interbands) and segments with disorder (bands) alternated along the length of the filament (Fig. 14.18). The interbands in neighboring filaments were supposed to be in lateral register and to

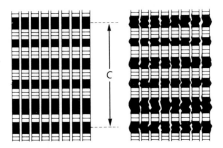

FIG. 14.18. Idealized model for the collagen fibril proposed by Bear and co-workers to account for the changes observed in the low-angle meridional X-ray diffraction pattern with varying water content (Fig. 14.16). Segment with perfect order (interbands) were supposed to alternate with segments of less perfect order (bands) along the length of the filaments. The extent of disorder in the band was supposed to increase in going from the wet state (left) to the dry state (right) (after Bear *et al.*, 1951).

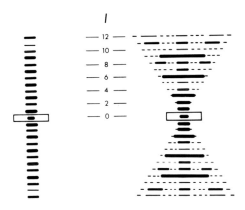

FIG. 14.19. Diffraction patterns predicted for the wet (left) and dry (right) fibril models illustrated in Fig. 14.18 (after Bear *et al.*, 1951).

be responsible for the high-angle pattern. The extent of the disorder in the bands was supposed to be small in the wet fibril and to increase with drying. The appearance of off-meridional maxima on drying (Fig. 14.19) was attributed to an increase in the radial component of the disorder in the bands.

Tomlin and Ericson (1960) offered an alternative explanation of the fanning of the pattern on drying which was based on the observation that the banding pattern of collagen fibers, visible in the electron microscope, frequently shows evidence of longitudinal shearing. The effect of such shearing on the diffraction pattern was examined in detail and it was shown that the fanning could also be explained on this basis.

Quantitative intensity data have been collected for the low-angle meridional patterns from wet and dry collagen and from a number of axially isomorphous derivatives and used to calculate various types of one-dimensional Patterson syntheses (Kaesberg and Shurman, 1953; Tomlin and Worthington, 1956; Burge et al., 1958; Ericson and Tomlin, 1959; McGavin, 1962; Ellis and McGavin, 1970). These syntheses give information about prominent interpeak vectors in the projection of the electron density onto the fiber axis. Slit cameras have frequently been used for this purpose but it does not appear to have been generally appreciated that these are unsuitable for data collection from fanned patterns since the measured intensity is a summation of meridional and off-meridional contributions. Only meridional intensity should be included in the calculation of one-dimensional syntheses and thus syntheses based on data obtained with slit cameras from dry fibers are subject to errors of unknown magnitude.

One-dimensional Patterson functions have been used to devise models for the distribution of scattering matter in collagen fibrils in the wet and dry states and also after treatment with heavy atoms. On the basis of these models the reflections can be phased and in some instances one-dimensional Fourier syntheses have been calculated. Unfortunately, a one-dimensional Patterson function of a structure as complex as a collagen fibril is extremely difficult to interpret in an unequivocal manner and the model distributions deduced from them are often gross oversimplifications of the actual situation. If such models are used to phase the reflections, a Fourier synthesis calculated on this basis inevitably "agrees" with the proposed model since the synthesis is dominated by the phase information (Ramachandran, 1964; Fraser and MacRae, 1969; Ramachandran and Srinivasan, 1970).

Ellis and McGavin (1970) collected intensity data from elastoidin and from several heavy-atom derivatives of this material and attempted to use the method of isomorphous replacement (Holmes and Blow, 1965; Phillips, 1966) to phase the reflections. Models were devised for the distribution of heavy atoms on the basis of difference syntheses and refined on the assumption that the distribution of electron density was centrosymmetric. These models were then used to phase the reflections and Fourier syntheses were calculated for both elastoidin and tendon collagen.

In view of the difficulties involved in the interpretation of Patterson syntheses and the assumption, not supported by electron microscope studies, that the structure is centrosymmetric, it is doubtful whether much significance can be attached to the fine detail in the Fourier synthesis calculated for tendon collagen. However, there appears to be general agreement between this and other Fourier syntheses with regard to the presence of a rectangular step function in the electron density distribution in wet tendon collagen with a bandwidth of either about 0.46 (Kaesberg and Schurman, 1953; Tomlin and Worthington, 1956; Ericson and Tomlin, 1959) or $1 - 0.46 = 0.54$ (Bear and Morgan, 1957) of the period.

The Patterson function for dry tendon is appreciably different from that for wet tendon but there is no general agreement regarding the nature of the corresponding differences in the electron density distribution (Bear and Morgan, 1957; Ericson and Tomlin, 1959). Burge and Randall (1955) attempted to determine the axial density distribution by microphotometry of electron micrographs of unstained collagen. The Fourier transform of the density distribution so obtained was found to be in reasonable agreement with the observed structure amplitudes in the low-angle meridional X-ray pattern. The density function did not exhibit a step function of the type believed to be present in the density function for wet tendon.

II. ELECTRON MICROSCOPE STUDIES

a. *Native Collagen*

At about the same time that Bear (1942) successfully indexed the low-angle meridional X-ray pattern of dry collagen in terms of a 640-Å periodicity, electron microscope studies furnished direct evidence of a transverse banding pattern of similar period in isolated collagen fibrils (Hall *et al.*, 1942; Schmitt *et al.*, 1942; Wolpers, 1943). Shadow-cast specimens of dried fibrils were found to have a characteristically corrugated appearance but replicas of moist fibrils showed relatively little intraperiod fluctuation in diameter (Gross and Schmitt, 1948). This suggested that the water content of the moist fibril varied in a periodic manner.

The use of heavy-atom staining methods has shown that within each period there exists a series of transverse bands which stain preferentially and as many as 13 such bands have been identified (Nemetschek *et al.*, 1955). The distribution of bands within the repeat is asymmetric so that attempts to interpret the low-angle meridional X-ray diffraction pattern on the basis of a centrosymmetric distribution of electron density are of questionable validity. Convincing evidence has been obtained that the band pattern extends over the whole cross section of the fibril, rather than being confined to the surface (Borysko, 1963; Grover, 1965).

Fine filaments have been detected in collagen fibers (Olsen, 1963a,b; Tromans *et al.*, 1963; Grant *et al.*, 1965; Smith and Frame, 1969;

Steven, 1970; Bouteille and Pease, 1971) and these have been assumed to correspond either to individual molecules or to a small grouping of molecules. In the preparations examined by the last-named authors the filaments had a diameter of 30–35 Å and presented a beaded appearance related to stain-combining sites. The sites in neighboring filaments appeared to be in lateral register and the banding pattern of the fibril was thus presumed to reside in the individual filaments.

b. *Regenerated Collagen*

The axial period of collagen fibrils is not obviously related to the molecular length of 2800 Å (Section B.I) but studies of ordered aggregates prepared from soluble collagen preparations have provided information on this relationship. When acid-soluble collagen is precipitated from solution, fibrous material is obtained which gives an X-ray diffraction pattern (Wyckoff and Corey, 1936) and exhibits a banding pattern (Schmitt *et al.*, 1942) which are similar to those of native collagen. Further studies (Highberger *et al.*, 1950, 1951; Gross *et al.*, 1952, 1954; Schmitt *et al.*, 1953, 1955a) showed that in addition to native-type fibrils, other modes of fibrillar aggregation were possible. In addition, a segment long-spacing (SLS) form was observed (Fig. 14.20) in which

FIG. 14.20.   Electron micrographs of SLS aggregates of native collagen molecules with two α1 chains and one α2 chain (top), molecules renatured from three α1 chains (center), and molecules renatured from three α2 chains (bottom) (Tkocz and Kühn, 1969).

growth proceeded by lateral aggregation. The length of the segment in this form was similar to that estimated for the collagen molecule and the banding pattern was polarized, thus suggesting that the SLS form was an aggregate of collagen molecules in lateral register. From a considera-tion of the symmetry or lack of symmetry in the banding patterns and of the repeat distances, the various types of aggregates were rationalized in terms of different packing modes of the collagen molecules and the important conclusion was reached that the banding pattern in native collagen fibrils arises through an axially staggered arrangement of collagen molecules (Schmitt *et al.*, 1955b; Gross, 1956) as depicted in Fig. 14.21. A similar conclusion was reached independently by Tomlin

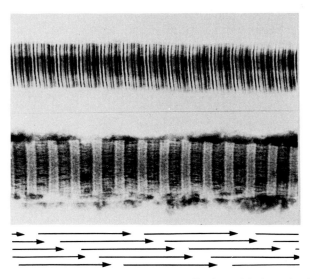

FIG. 14.21. Electron micrographs of collagen fibrils positively stained with phosphotungstic acid and uranyl acetate (upper) and negatively stained with phosphotungstic acid (middle). The lower diagram is a schematic representation of the arrangement of molecules in the fibril (after Kühn, 1967).

(1956) in an attempt to reconcile the results obtained from X-ray diffraction and from electron microscopy. Subsequent studies have shown that further varieties of *in vitro* aggregates can be obtained under appropriate conditions of precipitation (Kühn and Zimmer, 1961; Olsen, 1963b; Bard and Chapman, 1968; Kühn, 1969; Armitage and Chapman, 1971).

Convincing support for the scheme outlined in Fig. 14.21 was obtained by the demonstration that the banding pattern in native-type fibrils stained with phosphotungstic acid or with uranyl acetate could be reproduced from the corresponding SLS pattern by combining four of the latter patterns displaced by zero, one, two, and three times the native fibril period (Hodge and Schmitt, 1960; Kühn *et al.*, 1960; Kühn and Zimmer, 1961). The important conclusion was also reached that each stained band in the native fibril corresponded to the superposition of four of the bands visible in the SLS form. Lateral interactions between segments of the molecules rich in basic and acidic amino acid residues were therefore presumed to play an important part in determining the molecular packing arrangement in the native fibril.

A further implication of the superposition experiments was that a pseudoperiod equal to the native period must be present in the distribution of staining sites along the length of the collagen molecule. Renaturation experiments in which SLS-type aggregates were prepared from molecules with three $\alpha 1$ chains or with three $\alpha 2$ chains (Kühn *et al.*, 1965; Tkocz and Kühn, 1969) showed that the distribution of staining sites closely resembled that of the normal molecule with two $\alpha 1$ chains and one $\alpha 2$ chain (Fig. 14.20). From this it can be concluded that the pseudoperiodicity in staining sites is a property of the individual α chains and direct correlations have been established with the occurrence of charged groups in the amino acid sequence (von der Mark *et al.*, 1970).

The segment length in the SLS type of aggregate was estimated, by negative staining, to be 4.40 ± 0.02 times the axial period in the native-type fibril (Hodge and Petruska, 1963) and since this does not correspond to an integral number of periods, it was suggested that there must be voids in the distribution of protein in the native material. The length of the void (Fig. 14.21) was presumed to be $5 - 4.4 = 0.6$ times the axial period, that is, approximately 400 Å in the native wet fibril. It seems likely that the step function in the axial distribution of electron density in wet tendon, deduced from X-ray studies (Section E.I.c), is a manifestation of such voids (Burge, 1965). The voids are presumably occupied by water, which has a lower electron density than hydrated protein. Electron microscope studies of negatively stained fibrils (Hodge and Petruska, 1963; Olsen, 1963a,b; Tromans *et al.*, 1963) showed that the uptake of stain over a zone a little in excess of one-half of the repeat was appreciably greater than in the remainder and it was assumed (Hodge and Petruska, 1963) that the densely stained zone corresponded to the voids aligned in lateral register (Fig. 14.21). Although this interpretation is attractive in its simplicity, later studies by Grant *et al.* (1965) showed that the length of the densely stained zone was reduced in fibers treated with glutaraldehyde and Spadaro (1970) found that in unstained preparations the "void zone" appeared to be more electron-dense than the remainder of the repeat. Caution must be exercised, however, in relating this observation to the situation in native tendon since various lines of evidence suggest that the voids, which are presumably filled with water in the native material, shrink and possibly disappear altogether in completely dehydrated material.

In addition to providing evidence on the molecular packing arrangement in collagen, studies of the *in vitro* aggregation of native and partially degraded molecules have also yielded information on the distribution of chemical groups in the polypeptide chains. Bands in SLS-type aggregates corresponding to acidic and basic staining loci were found to exhibit a close correspondence (Hodge and Schmitt, 1960), leading to the view that regions rich in acidic and basic side chains alternated with regions devoid of such side chains along the length of the molecule. Basic groups were found to predominate in some regions, while relative parity or an excess of acidic groups existed in others and it was suggested that the "charge profile" is such that under physiological conditions maximum complementation of charge occurs when the stagger observed in the native fibril is present.

Another important finding (Gross and Nagai, 1965; Nagai *et al.*, 1965) based on the study of SLS-type aggregates was that the directions, with respect to the sequence $-NH-CHR-CO\rightarrow$, are the same for the three α chains in the collagen molecule. The N-terminal and C-terminal ends of the SLS aggregate were also identified in these experiments. Other applications of *in vitro* aggregation studies have been reviewed by Hodge (1967), Kühn (1969), and Traub and Piez (1971).

III. MOLECULAR PACKING

The evidence discussed in Sections E.I and E.II indicates that the molecules in a collagen fibril are packed together with a precisely determined axial stagger, but several authors have pointed out that it is not possible to form a regular three-dimensional array of molecules, according to the scheme depicted in Fig. 14.21, which preserves a constant relationship between neighboring molecules (Ross and Benditt, 1961; McGavin, 1964; Smith, 1965). In view of this, it was suggested (Grant *et al.*, 1965, 1967; Cox *et al.*, 1967) that the packing arrangement might be one in which the molecules are only statistically equivalent. A specific model of this type was proposed in which each molecule was assumed to consist of five bonding zones about 265 Å long separated by four "nonbonding" zones about 375 Å long and the molecules were supposed to aggregate with the bonding zones in register. The axial stagger between bonded molecules was therefore limited to a multiple of 640 Å but was otherwise assumed to be random. This model does not appear to be consistent with the presence of filaments 30–35 Å in diameter (Section E.II.a) or with a lateral periodicity of about 38 Å (Section E.I). Other difficulties have been discussed by Veis *et al.* (1967a) and Smith (1968).

An alternative scheme, based on studies of cross-linking, has been suggested (Veis *et al.*, 1967a, 1970) in which it is supposed that tetramers of collagen molecules are packed end to end to form a filament of precisely determined structure (Fig. 14.22a) which may conveniently be

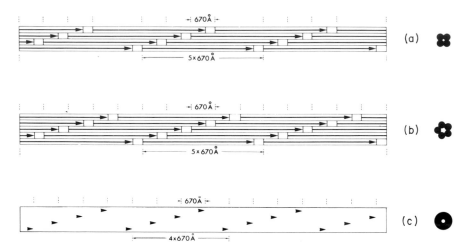

FIG. 14.22. Schematic representation of the arrangement of collagen molecules in models for the microfibril. Radial projections are shown on the left and axial projections on the right. The upper diagram (a) shows the arrangement proposed by Veis *et al.* (1970); this arrangement has a period of 3350 Å when projected onto the axis and does not preserve a constant relationship between molecules. The arrangement proposed by Smith (1968) shown in the middle diagram (b) has a period of 670 Å when projected onto the axis and also preserves a constant relationship between molecules. The lower diagram (c) shows a refinement of Smith's model proposed by Miller and Parry (1973) in which the structure has been given a helical twist to give the microfibril a fourfold screw axis. The exact path followed by the molecule is uncertain.

referred to as a microfibril. The four molecules in the tetramer are related by a rotation of 90° and an axial displacement of 695 Å and successive tetramers are packed end to end in the same azimuthal orientation with gaps of $0.6 \times 695 = 415$ Å between the ends of the molecules. The asymmetric unit of the microfibril is a tetramer and so the individual molecules in the tetramer are not equivalent. The axial period of the microfibril is $5 \times 695 = 3475$ Å (or 3350 Å if the axial displacement is taken as 670 Å), and if the microfibrils were packed in lateral register, this would be the periodicity of the banding pattern. However, no evidence for a periodicity of this magnitude has been obtained either from electron microscope studies or from X-ray studies (Huxley and Brown, 1967). It was suggested (Veis *et al.*, 1970) that the observed periodicity might result from a random staggering of the microfibrils by multiples of the observed period but there is no obvious reason why this should occur. Unless some highly specific packing arrangement of the microfibrils is assumed, this model does not appear to be capable of accounting for the observed banding pattern (Miller and Wray, 1971). It is also at variance with the observation (Bouteille and Pease, 1971) that this pattern appears to reside in the microfibril.

Smith (1968) pointed out that a constant relationship between adjacent molecules of the type indicated by *in vitro* reaggregation studies (Fig. 14.21) existed in a helical arrangement with five molecules per turn (Fig. 14.22b). The microfibril so formed would have a diameter close to the value of 30–35 Å estimated by Bouteille and Pease (1971). The axial period is the same as in the model proposed by Veis *et al.* (1967a) but the asymmetric unit is now a single molecule and the projection of the structure onto the axis of the microfibril has a period which is inherently the same as that deduced for the fibril from X-ray and electron microscope studies. The bands observed in collagen fibrils extend across the entire thickness (Section E.II) and so it is necessary to assume that the microfibrils are packed with the void regions in lateral register. Smith (1968) pointed out that random axial displacements equal to the axial rise per molecule would not destroy the banding pattern of the fibril and suggested that this situation obtained in the native material. Because of the helical symmetry, this type of disorder could equally well be described as being generated from a monoclinic crystalline arrangement by rotations of the microfibrils about their axes.

Miller and Wray (1971) have discussed the implications of the very detailed X-ray diffraction pattern which they obtained from native tendon fibers and conclude that the fibril is a highly ordered structure in which groupings of collagen molecules are arranged on a lattice which extends laterally over distances comparable with the diameter of a fibril. The equatorial and near-equatorial data were interpreted as indicating that the collagen molecules were grouped into small bundles and this was supported by the fact that the row line corresponding to a lateral periodicity of 38 Å was intensified when the tendon was treated with heavy-metal stain.

The models for molecular packing proposed by Sasisekharan and Ramachandran (1957), Burge (1963, 1965), Grant *et al.* (1965), and Veis *et al.* (1967a) were all considered to be incompatible with some feature of the observed data, but the five-stranded microfibril model proposed by Smith (1968) appeared to be inherently capable of accounting for much of the data. In order to explain the splitting of the strong high-angle equatorial reflection into equatorial and near-equatorial components (Section E.I.a), it was suggested that the molecules were not straight but followed a helical path, so that the microfibril became a five-strand rope. Since the individual molecules are themselves three-strand ropes of coiled-coil chains, the final chain conformation becomes that of a "coiled coiled-coil." The possibility that the splitting of the strong high-angle equatorial reflection might be due to a conformation of this type had been suggested earlier by Burge (1965) who, however, proposed a specific model for a microfibril consisting of a two-strand rope. An alternative explanation for the splitting has been offered by McFarlane (1971) based on the assumption that the fibril is a triply twinned crystal of hexagonally close-packed molecules.

Since the equatorial and near-equatorial reflections can be indexed on a tetragonal cell (Section E.I.b), Miller and Parry (1973) argued that the microfibril must therefore have a fourfold screw axis. A model for the microfibril having the required symmetry can be obtained by supposing that the model proposed by Smith (1968), which has five molecular ends per turn (Fig. 14.22b), is twisted about its axis until there are only four molecular ends per turn (Fig. 14.22c). The desirable features of the original model relating to the equivalence of the molecules and the 670-Å repeat in the axial projection are not affected by this operation. The axial period of the microfibril is, however, reduced to $4 \times 670 = 2680$ Å and the molecular axes follow a helical path. The conversion from fivefold to fourfold screw symmetry is not a unique operation and so the paths followed by the molecular axes are not indicated in Fig. 14.22c. Miller and Parry (1973) estimated the pitch of the molecular path to be approximately 700 Å. The square-based cell with $a = 76.9$ Å was assumed to contain four microfibrils related by a four-fold axis but various features of the diffraction pattern were consistent with the presence of azimuthal disordering whereby a proportion of the molecules were rotated by $\pi/2$, π, or $3\pi/2$ from their idealized positions.

The concept that the collagen fibril is an ordered aggregate of micro-fibrils, which in turn are helical assemblies of collagen molecules, appears to provide a basis for the integration of a large body of experimental data. Great stress was placed in early studies on the differences between the X-ray diffraction patterns of wet and dry tendon and of stretched and unstretched tendon and any model for the collagen fibril should provide a rational explanation for these differences. In hydrated collagen fibrils the voids in the microfibrils would be filled with water, and since the microfibrils are packed together with the void regions in register, the water content in the native fibril would vary periodically as deduced from the development of corrugations in dried fibrils (Section E.II.a). In addition to a radial contraction some contraction in the length of the void zone might also be expected as the chains in this section of the microfibril reorganize in order to fill the space previously occupied by water molecules. In the absence of a radial contraction, this would entail a reduction in axial period from 670 Å in the wet state to about $670 \times 4.54/5 = 608$ Å in the completely dehydrated state, assuming that the void region is 0.46 times the period (Section E.I.c). This figure can be compared with the value of 603 Å observed by Rougvie and Bear (1953). The model proposed by Bear et al. (1951) (Fig. 14.18) on the basis of X-ray studies envisaged the filaments as individual molecules and the disordered zones as the staining sites visible in the electron microscope. The alternative possibility might be considered that the filaments with alternating regions of order and disorder are the microfibrils, and the disordered regions are to be identified with the collapsed void zones.

As noted earlier (Section E.I.a), the pseudoperiod associated with the chain conformation increases at a smaller rate than the axial period when air-dried collagen is stretched and this observation was interpreted in terms of an axial alternation of regions of different order (Cowan et al., 1955a). According to Bear and co-workers, these disordered regions are also to be identified with the staining sites in the banding pattern but again an alternative possibility is that the disordered regions correspond to the partially collapsed void zones of the microfibrils.

In a microfibril of the type proposed by Smith (1968) there is a central lumen (Fig. 14.22c) which will presumably be occupied by water in the native state. It is highly improbable that this lumen will persist in the dried material and the microfibril must therefore contract laterally when water is removed from the specimen. The spacing of the hydration-sensitive equatorial reflection studied by Rougvie and Bear (1953) must be largely determined by interference between the scattering from molecules in the same microfibril, and the reduction in spacing from 14.6 to 10.6 Å on drying is an indication of the extent of the lateral contraction. It is unlikely that this contraction could take place without some disturbance of the geometric regularity of the microfibril.

Support for the belief that both the void zone (Fig. 14.22b) and the microfibril lumen (Fig. 14.22c) are eliminated in dry collagen is obtained from density measurements (Pomeroy and Mitton, 1951; Rougvie and Bear, 1953). The value of 1.34 g cm^{-3} obtained by the latter authors for dried kangaroo-tail tendon fibers is similar to that of other dried fibrous proteins. Rougvie and Bear (1953), in their study of the effects of hydration on the X-ray pattern, found that the period decreased from 670 to 637 Å when the water content was reduced from about 120 to 7 g/100 g of collagen, but the removal of the last 7 g was accompanied by a further reduction of 34 Å in the period without significant change in the spacing of the high-angle equatorial reflection. This suggests that the microfibril lumen has been virtually eliminated by drying to a water content of 7 g/100 g of collagen and that further dehydration leads to a pronounced crumpling of the chains in certain zones of the microfibril with an attendant axial contraction.

The void zones are involved in the mineralization of bone collagen (see, for example, Glimcher and Krane, 1968). The elucidation of the precise form of the packing of the molecules in the microfibrils should enable the nature of the internal surface of the voids to be determined and thus provide a basis for understanding their role in mineralization. Great interest also attaches to the elucidation of the spatial distribution of residues along the polypeptide chains in the microfibril since this will determine the potentialities for the establishment of cross-linkages. In addition to intramolecular cross-links, two types of intermolecular cross-links are possible depending on whether the bond is intramicrofibrillar or intermicrofibrillar. Some preliminary ideas concerning the possible interrelationship between chain conformation and microfibril geometry have been discussed by Segrest and Cunningham (1971). Cassel (1966) has stressed the important part played by hydrophobic bonding in the formation of the native-type of molecular aggregate and it will be of interest to determine whether this is a particular feature of the formation of the microfibril or also plays a part in the specific lateral aggregation of the microfibrils. Other outstanding questions relate to the parts played by the nonhelical "telopeptide" ends of the α chains and the hydroxyproline residues in determining the properties of the fibril.

Chapter 15

Myofibrillar Proteins

A. INTRODUCTION

The protein constituents of the contractile systems of muscles have been extensively characterized and considerable progress has been made in the determination of their distribution in the muscle fiber and their role in muscular contraction (Huxley, 1966, 1971; Young, 1969; Pepe, 1971a,b). The study of the conformation of the polypeptide chain in these proteins in their native state is complicated by the fact that they occur together in intimate contact and the X-ray diffraction pattern of striated muscle, for example, is determined by the whole sarcomere rather than the distribution of any particular protein. Some progress has been made in assigning particular features of the diffraction pattern at low and medium angles to specific components of the contractile system but little attention has been given so far to the high-angle pattern. Current ideas concerning the chain conformation of the individual proteins rest heavily upon studies of extracted proteins and on the assumption that the conformation is preserved or recovered after extraction.

In this chapter we shall be concerned with the chain conformation in actin, tropomyosin, myosin, and paramyosin. The compositions and molecular properties of these proteins are summarized in Tables 15.1 and 15.2, respectively. It would be beyond the scope of the present treatment to discuss the interactions between these proteins in resting muscle and the changes in conformation and interaction which occur during muscular contraction. Detailed discussions of these topics are

TABLE 15.1

Amino Acid Compositions of Fibrous Proteins Isolated from the
Contractile Systems of Muscles[a]

Amino acid	Actin[b]	Myosin[c]	Tropomyosin[d]	Paramyosin[e]
Alanine	8.0	8.8	12.1	12.2
Arginine	4.8	5.3	4.8	9.8
Aspartic acid	9.0	9.9	10.1	12.8
Glutamic acid	11.3	19.1	24.7	21.3
Glycine	7.5	4.7	1.3	2.0
Half-cystine[f]	1.4	1.1	1.0	0.6
Histidine	2.0	1.8	0.7	1.3
Isoleucine	7.6	4.7	3.8	3.3
Leucine	7.0	9.4	10.9	12.4
Lysine	5.4	10.9	13.2	6.9
Methionine	4.1	2.5	4.4	1.5
Phenylalanine	3.1	3.3	0.5	1.0
Proline	4.9	2.5	0.2	0.1
Serine	5.9	4.5	4.4	5.3
Threonine	6.9	4.8	2.7	4.5
Tryptophan	1.2	—	—	—
Tyrosine	4.0	2.0	1.9	1.6
Valine	5.4	4.9	3.4	3.6

[a] Given in residues/100 residues, rounded to nearest decimal place.

[b] Rabbit skeletal (Johnson and Perry, 1968); small amount of 3-methylhistidine also detected.

[c] Rabbit skeletal (Huszar and Elzinger, 1971); small amounts of 3-methylhistidine and ϵ-N-methyllysine also detected.

[d] Rabbit skeletal (Hodges and Smillie, 1970).

[e] Oyster adductor muscle (Woods and Pont, 1971).

[f] Estimated, in the case of modified proteins, from the cysteine derivatives.

available in reviews by Huxley and Brown (1967), Huxley (1966, 1971), Young (1969), and Pepe (1971a,b).

B. ACTIN

Actin is a water-soluble protein first isolated by Straub (1942, 1943) which may be extracted from muscle with neutral or slightly alkaline solutions. Such preparations are contaminated with variable amounts of tropomyosin and troponin, which may be removed by appropriate purification procedures. The physicochemical properties of actin in aqueous solution are those of a globular protein and the available evidence indicates that the molecule contains only one chain of weight about

TABLE 15.2

Molecular Properties of Myofibrillar Proteins in Solution

Protein	Helix content[a] (%)	Molecular weight[b]	Chain structure
Actin	30[c]	47,000[d]	Single chain; globular structure
Myosin	60[e]	460,000[f]	Two heavy chains and several light chains; rodlike molecule with two globular heads at same end; overall length about 1500 Å
Tropomyosin	100[g]	68,000[h]	Two chains; rodlike with length about 450 Å
Paramyosin	100[i]	210,000[j]	Two chains; rodlike with length about 1400 Å

[a] Approximate value based on measurements of the optical rotatory dispersion parameter b_0, assuming % helix $= -100\, b_0/630$.

[b] Approximate value based on data given in references cited.

[c] Nagy and Jencks (1962), Nagy (1966).

[d] Rees and Young (1967), Johnson et al. (1967).

[e] Cohen and Szent-Györgyi (1957), Lowey and Cohen (1962), McCubbin et al. (1966), Lowey et al. (1969), Rainford and Rice (1970).

[f] Lowey et al. (1969), Godfrey and Harrington (1970).

[g] Cohen and Szent-Györgyi (1957), McCubbin and Kay (1969), Woods (1969a), Rainford and Rice (1970).

[h] Holtzer et al. (1965), E. F. Woods (1967).

[i] Cohen and Szent-Györgyi (1957), Olander (1971).

[j] Lowey et al. (1963), Woods (1969b).

47,000 daltons (Table 15.2). As isolated, actin contains bound ATP in the proportion of one mole per mole (Rees and Young, 1967; Tsuboi, 1968).

Upon acidification of solutions of the globular form of ATP-containing actin or upon the addition of 0.1 M KCl the actin polymerizes and the two states are distinguished by the prefixes G (globular) and F (fibrous), respectively. During the process of polymerization the ATP is dephosphorylated to ADP but the presence of ATP is not essential for the G-actin to F-actin transformation since ATP-free G-actin can be polymerized by the addition of magnesium ion.

The optical rotatory dispersion parameter b_0 has been determined for solutions of G-actin–ATP (Nagy and Jencks, 1962; Nagy, 1966) and values of -180 to -184 deg cm^2 dmole^{-1} obtained. If it is assumed that a value of -630 deg cm^2 dmole^{-1} corresponds to 100% α-helix (Chapter 8, Section B), this would imply a helix content of 29%. A somewhat greater value of $-b_0$ was obtained from partially polymerized actin which, if interpreted as a conformational change, would indicate an increase of 10% in the helix content. Murphy (1971) measured the circular dichroism of solutions of G-actin and F-actin in the range 200–250 nm, and, using spectra obtained from the α-helix, β-pleated-sheet, and random-coil forms of $[Lys]_n$ as standards (Greenfield and Fasman, 1969), estimated the contents of these conformations to be, respectively, 26%, 26%, and 48% for G-actin and 29%, 22%, and 49% for F-actin. Using the ellipticity of $[Glu]_n$ at 210 nm as a standard (Wu and Yang, 1970), the α-helix contents were estimated to be 27% and 28%, respectively, for G-actin and F-actin. These results suggest that there is little change in secondary structure in the G → F transformation. The amino acid composition of actin (Table 15.1) shows that it contains appreciably greater amounts of proline and glycine than the highly helical proteins tropomyosin and paramyosin and this presumably accounts (Chapter 12, Section C) for the lower helix content of actin.

The molecular shape and mode of aggregation of the subunits in F-actin appear to be very similar to those which obtain in the native state (Hanson and Lowy, 1963; Moore et al., 1970) and so the study of this material offers one possibility of determining the chain conformation of the actin molecule in a near-native state. Physicochemical studies of F-actin have been reviewed by Oosawa and Kasai (1971) and in Section B.I an account is given of physical studies of this material. In Section B.II evidence relating to the arrangement and conformation of the actin molecules in the thin filaments of muscle is discussed.

I. Physical Studies of F-Actin

Very little work has been reported on either the high-angle X-ray pattern or infrared spectrum of F-actin and our knowledge of the chain conformation in this material is only rudimentary. Infrared spectra have been obtained from films of actin (Morales et al., 1951; Kominz, 1965) but these have not been suitable for conformation analysis. High-resolution studies of dichroism in oriented films would undoubtedly be rewarding. Information on the geometric form of the F-actin polymer has been obtained mainly from electron microscope studies.

a. Electron Microscope Studies

Jakus and Hall (1947) and Rozsa et al. (1949) demonstrated that F-actin contained fine filaments about 100 Å in diameter and later studies (Hanson and Lowy, 1963, 1964a), in which the negative staining technique was used, showed that the filaments consist of two helical strands that present a beaded appearance (Fig. 15.1). The beaded appearance

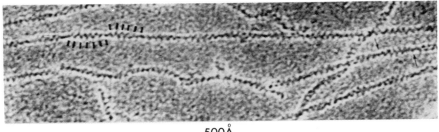

500Å

FIG. 15.1. Electron micrograph of F-actin filament, negatively stained with uranyl acetate, showing globular subunits. The arrows indicate positions in which the strands cross over one another. Individual globular units are visible as indicated by the short lines. Underfocus micrograph, reverse-contrast print (Hanson and Lowy, 1963).

was interpreted as indicating that F-actin was composed of approximately spherical subunits about 55 Å in diameter. A spherical molecule of molecular weight 47,000 (Table 15.2) and density 1.3 g cm^{-3} would have a diameter around 48 Å and from a similar calculation, together with a consideration of the actin content of skeletal muscle, it was concluded that each subunit corresponded to one actin molecule.

The pitch of the individual strands was estimated to be about 700 Å and a model was proposed for the geometric arrangements of the subunits (Fig. 15.2). The number of subunits per turn of each helical strand was estimated to be approximately 13 and the subunits in the two strands were supposed to be staggered relative to one another by one-half of the subunit height. The overall diameter of the F-actin filament was estimated to be about 80 Å.

Huxley (1963) also studied F-actin filaments and complexes formed with myosin and with heavy meromyosin. The manner in which these molecules attached to the filaments showed that the filaments were structurally polarized and suggested that all the actin molecules were similarly oriented relative to the axis of the helix. Thus the line group of the F-actin helix appears to be s (Chapter 1, Section A.II.d). Depue and Rice (1965) established that the screw sense of the strands was right-handed, and if all the actin molecules are assumed to be equivalent, the basic helix (Chapter 1, Section A.II.c) is left-handed with about 13 molecules in six turns of pitch -58 Å (Fig. 15.2). The unit height is $h \simeq 350/13 = 26.9$ Å and the unit twist is $t \simeq 6 \times 360/13 = -166.2°$. In the event that there were exactly 13 residues in six turns, the period of the filament would be around 350 Å.

Under appropriate conditions paracrystals of F-actin may be obtained and a number of polymorphic forms have been described (Hanson,

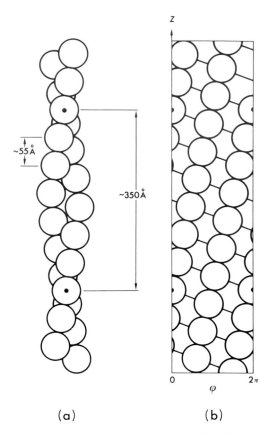

(a)   (b)

FIG. 15.2.   (a) Model for the arrangement of the globular subunits in an actin filament (Hanson and Lowy, 1963). (b) Radial projection of the model showing the basic helix.

1967, 1968; Kawamura and Maruyama, 1970; O'Brien *et al.*, 1971). In the so-called type I paracrystals (Fig. 15.3) a regular lattice is formed in which the projection of the repeating unit has a side of length 320 Å. This presumably is related to the repeat (or near-repeat) in the structure of isolated F-actin filaments (Fig. 15.2).

Moore *et al.* (1970) have applied the technique of three-dimensional image reconstruction (Chapter 3, Section D.IV) to F-actin paracrystals of the type described by Hanson (1967, 1968). The paracrystals were found to contain filaments, having a structure similar to that shown in Fig. 15.2, which were packed in lateral register so that equivalent points in neighboring filaments were related by a simple translation perpendicular to the filament-axis direction. The method used for reconstruction

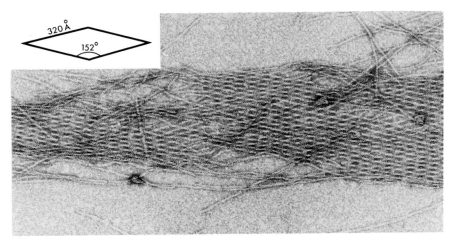

Fig. 15.3. Electron micrograph of paracrystals of F-actin formed at pH 5.0. The inset shows the lattice dimensions deduced from the micrograph (Kawamura and Maruyama, 1970).

was only applicable to helical objects and an additional operation of deconvolution was required in order to obtain the transform of a single filament. The result is illustrated in Fig. 15.4 and the stain-excluding volume measures approximately 55 Å axially, 35 Å radially, and 50 Å tangentially. The calculated volume is of the order expected for a single actin molecule.

b. *X-Ray Studies*

Astbury and Spark (1947) examined thin films of F-actin preparations by X-ray diffraction and obtained a fiber-type pattern (Chapter 1, Section A.I.d) when the X-ray beam was oriented parallel to the surface of the film. It was concluded that the aggregates were fibrous and that uniplanar orientation was produced by forces generated during drying. In terms of the orientation parameter discussed in Chapter 5, Section B.II.b this type of preferred uniaxial orientation corresponds to $f < 0$. A series of meridional arcs was observed with spacings ranging from 3.6 to 27 Å together with an imperfectly resolved lower-angle reflection of about 54 Å, a number of equatorial reflections, and two diffuse rings of spacing about 4.5 and 9 Å. The thesis was developed that the pattern originated from a linear polymer of corpuscular units (Astbury, 1947a, 1949a; Astbury and Spark, 1947). Somewhat improved patterns were obtained in a later study by Cohen and Hanson (1956) and the spacings recorded by these authors and by Astbury (1949a) are

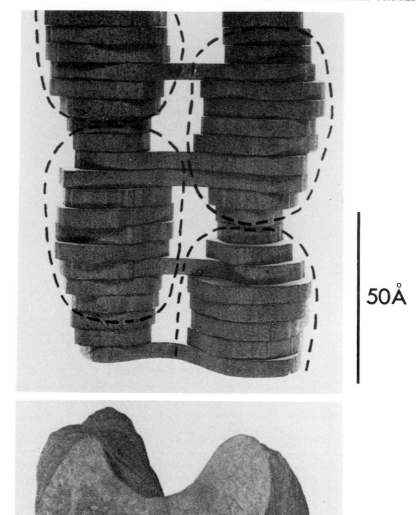

FIG. 15.4.   Model of a portion of an F-actin filament, built from data obtained by three-dimensional reconstruction from an electron micrograph. The probable outlines of the actin subunits are indicated by broken lines. The lower photograph shows an end-on view of the model (Moore *et al.*, 1970).

given in Table 15.3. Vazina *et al.* (1965) have also given a list of spacings for fibers of F-actin but these differ somewhat from the values given in Table 15.3.

TABLE 15.3

Spacings Recorded in the X-Ray Diffraction Pattern of F-Actin

Layer line[a]		Bessel functions order[b]	Observed spacings	
Index *l*	Spacing Z^{-1} (Å)	*n*	Film[c] (Å)	Dry fiber[d] (Å)
6, 7	58.3, 50.0	−1, 1	54[e] vs	53[f]
13	26.9	0	26.8 m	26.7
19	18.4	−1	18.1 m	18.3
26	13.5	0	13.5 s	13.4
32	10.9	−1	10.9 s	10.9
39	9.0	0	9.1 m	9.0
45	7.8	−1	7.9 s	7.7
52	6.7	0	6.8 vw	6.8
65	5.4	0	5.3 m	—
78	4.5	0	4.5 m	—
84	4.2	−1	4.2 mw	—
91	3.8	0	3.9 m	—
97, 98	3.6	−1, 1	3.6 vw	—
110, 111	3.2	−1, 1	3.1 vw	—
123, 124	2.8	−1, 1	2.8 vw	—

Equatorial reflections[d]: 55 Å, 25 Å, 12 Å, 9.8 Å[g]

[a] Tentative assignment based on a value of $c = 350$ Å; see also Selby and Bear (1956).
[b] Derived from Eq. (15.1).
[c] Data of Astbury (1947a, 1949a), relative intensities indicated by w (weak), m (medium), s (strong), v (very).
[d] Data of Cohen and Hanson (1956).
[e] Imperfectly resolved.
[f] Average value of a pair of closely spaced reflections.
[g] Rather diffuse and poorly oriented.

Cohen and Hanson (1956) found that the meridional pattern sharpened with increase in humidity and noted changes in the equatorial pattern that indicated an increase in the lateral lattice dimension. Vazina *et al.* (1965) and Spencer (1969) recorded the low-angle scattering from concentrated sols of F-actin and the latter author showed that the spacing of a strong equatorial maximum varied, with increasing concentration, from 260 to 120 Å in the manner expected for a regular hexagonal array of cylinders. If the formula used by Spencer is extrapolated to a concentration of 1.3 g cm^{-3}, corresponding to the usual value for dried protein, an intercylinder distance of about 50 Å is predicted. This confirms the suggestion (Cohen and Hanson, 1956) that the prominent equatorial reflection of spacing 55 Å in dried fibers of F-actin (Table 15.3) corresponds to an interfilament plane.

Although a number of high-angle reflections have been recorded from F-actin (Table 15.3), none can be directly related to regular conformations observed in synthetic polypeptides (Chapters 9–11) or in other fibrous proteins. Studies of optical rotatory dispersion and circular dichroism suggest that F-actin has an α-helix content of about 30% (Section B). If this is correct, it follows that the helical segments cannot all be oriented parallel to the axis of the F-actin filament, otherwise diffraction effects characteristic of the α-helix would be discernible in the high-angle pattern.

The series of meridional and near-meridional reflections observed in the diffraction pattern of F-actin (Table 15.3) can be accounted for on the basis of a helical structure for the F-actin filament of the type depicted in Fig. 15.2. For a helix with $u = 13$ units in $v = -6$ turns the formula given in Eq. (1.23) yields

$$l = 13m - 6n \qquad (15.1)$$

for the selection rule governing the appearance of Bessel functions of order n on a layer line of index l, where m is an integer. Meridional reflections only occur if $n = 0$ and so the spacings of the meridional reflections will be $c/13m$, where $c \simeq 350$ Å. Clearly the reflection at 26.7 Å corresponds to $m = 1$ and reflections corresponding to values of m from 2 to 7 are also observed (Table 15.3). Near-meridional reflections would most likely be attributable to Bessel functions with $n = \pm 1$ and from Eq. (15.1) these occur on layer lines with $l = 13m \pm 6$. The pair of closely spaced reflections of mean spacing 53 Å clearly corresponds to the first-order Bessel functions on $l = 6$ and 7 and all the remaining observed near-meridional reflections can be similarly assigned to layer lines on which first-order Bessel functions are predicted.

The fact that sharp reflections appear in positions determined by the symmetry in the F-actin helix out to spacings of 2.8 Å implies that the regularity with which the actin molecules are arranged in the filament is comparable with that which occurs in crystals of globular proteins. The F-actin filament is in fact a one-dimensional crystal. There is no *a priori* reason for supposing that the F-actin helix repeats exactly after six turns of the basic helix but the diffraction data obtained so far are not of sufficiently good quality to allow the true situation to be determined.

This point is discussed in more detail in Section B.II in connection with the actin helix in muscle.

II. ACTIN IN MYOFILAMENTS

Hanson and Huxley (1953, 1955) established that the actin in striated muscle is contained in the thin myofilaments and subsequently Hanson and Lowy (1963) showed that these filaments contain a two-strand-rope structure which is essentially similar to that found in the F-actin filaments discussed in Section I. Reedy (1968) showed that in insect flight muscle the screw sense of the actin strands in the thin filaments is right-handed, which is the same as in F-actin (Section B.I.a). Bear (1945) recorded a series of X-ray reflections, now usually referred to as the "actin" reflections, which appear in every type of muscle so far examined. Their origin was established through the study of films of F-actin (Astbury, 1947a, 1949a; Astbury and Spark, 1947). Due to the effects of imperfect orientation, it is difficult to measure the lateral coordinate R of the actin reflections, particularly at high angles. In some studies values of $d = (R^2 + Z^2)^{-1/2}$ are recorded, while in others, values of Z^{-1} are recorded. Unfortunately, it is not always made clear as to which quantity is being estimated.

a. *Arrangement of Molecules*

A detailed study of the actin pattern was reported by Selby and Bear (1956) and it was pointed out that the observed reflections (Table 15.4) could be indexed on the basis of either a two-dimensional net or a helical structure. In the light of subsequent electron microscope studies, the latter interpretation is the appropriate one. Two possible types of helical model were considered. In the first the repeating units were supposed to be distributed on a helix with 15 units in seven turns having an axial repeat $c = 406$ Å. An alternative indexing of the reflections gave a second possibility with 13 units in six turns and an axial repeat $c = 351$ Å. These two helices are, in fact, very similar, as is shown by their description in terms of unit height h, unit twist t, and pitch P (Table 15.5).

Although the determination of accurate helical parameters for the actin helix is a difficult task, knowledge of the precise values is essential for an understanding of the molecular basis of muscular contraction. Hanson and Lowy (1963) suggested, on the basis of electron microscope studies, that the number of units per turn, if integral, was 13 and estimated values of $h = 56.5$ Å and $c = 349$ Å. A later study (Hanson, 1967) showed, however, that the values obtained for the helical parameters varied with the preparative methods used and so particular interest

TABLE 15.4

X-Ray Reflections Associated with Actin-Containing Filaments

(a)   Spacings of Medium- and High-Angle Reflections (Å)

m^a	Bessel function order n	Calculated layer-line spacing[b] Z^{-1}	Muscle Adductor of Venus mercenaria[c]	Muscle Frog sartorius[d]
0	−1	58.95	59	59.1
1	1	50.84	51	50.97
				(39.0)
0	−2	29.47	28.2	29.8
1	0	27.30	26.9	27.30
0	−3	19.65	19.0	(20.4)
1	−1	18.66	18.8	18.75
2	0	13.65	13.6	13.66
3	2	13.16	12.9	
2	−1	11.08	10.9	11.06
3	0	9.10	9.0	9.08
3	1	7.88	7.9	7.85
4	0	6.83	6.8	6.87

(b)   Spacings (Å) of Layer Lines with Small Z Values[e]

m^a	Bessel function order n	Muscle Anterior byssus retractor of Mytilus edulis	Muscle Guinea pig Taenia coli	Muscle Frog sartorius
1	2	381	372	400[f]
2	4	196	184	180
3	6	132	124	125
−1	−3	70.0	70.9	70.0
0	−1	59.2	59.4	59.1
1	1	51.4	51.1	51.1

[a] See Eq. (15.2).
[b] Calculated on the basis of $h = 27.3$, $P = 58.95$ Å.
[c] Selby and Bear (1956); pattern obtained from tinted portion of dried muscle.
[d] Huxley and Brown (1967); pattern obtained from living, resting muscle.
[e] Vibert et al. (1972); spacings represent summary of observations on muscles in resting, contracting, and rigor states.
[f] Spacing cannot be determined accurately; see text.

TABLE 15.5

Estimates of the Parameters of the Actin Helix in the Thin Filaments of Muscles

Type of muscle	Pitch of strand $\mid P_s \mid$	Filament helix	
		Unit height h	Unit twist[a] $\mid t \mid$
Electron microscopy of isolated filaments			
Rabbit psoas, indirect flight muscles of *Dytiseus* and *Calliphora*, striated adductor of *Pecten*, adductor of *Crassostrea*, anterior byssus retractor of *Mytilus*, mantle of *Loligo*, body wall of *Lumbricus*, taenia coli of guinea pig[b]	698	28.25	165.4
X-ray diffraction from whole muscle			
Adductor of *Venus mercenaria* (dried)[c]	812	27.1	168.0
	or 702	27.0	166.2
Frog sartorius (living, resting)[d]	710–740	27.3	166.2–166.7
Toad sartorius (living, resting)[e]	720–760	—	—
Anterior byssus retractor of *Mytilus edulis*[e]			
Living, resting	738	—	—
Dried	712	—	—
Smooth adductor of *Pecten chlamys* (dried)[d]	734	—	—
Pharynx retractor of *Helix pomatia* (dried)	805[f]	27.4	167.8
	746[e]	—	—
Flight muscle of *Lethocerus* (in rigor)[g]	775	27.5	167.2

[a] Calculated from Eq. (15.3) or from $t = 2\pi h/P$.

[b] Hanson and Lowy (1963). A total of 156 measurements of $\frac{1}{2}P_s$ gave values in the range 330–363 Å with a mean of 349 Å and standard deviation of the mean of 1.1 Å; 132 measurements of $2h$ gave values in the range 52–58 Å with a mean of 56.5 Å and standard deviation of the mean of 0.25 Å.

[c] Selby and Bear (1956); pattern from tinted portion of muscle. The first set of values corresponds to a helix with 15 units in seven turns and the second set to a helix with 13 units in six turns.

[d] Huxley and Brown (1967).

[e] Lowy and Vibert (1967).

[f] Worthington (1959).

[g] Miller and Tregear (1972), see also Reedy (1968).

attaches to the possibility of determining these parameters from the X-ray diffraction pattern (Fig. 15.5b). In the discussion of diffraction by helical structures given in Chapter 1, Section A.III it was shown that the Z coordinates of the observed layer lines are related to the helical parameters by the expression

$$Z(m, n) = (m/h) + (n/P) \tag{15.2}$$

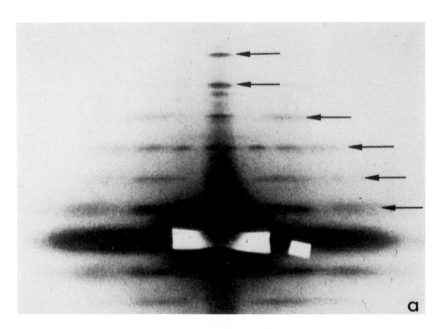

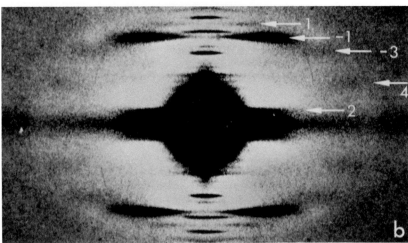

FIG. 15.5. (a) X-ray diffraction patterns from live frog sartorius muscle; fiber axis vertical (Huxley and Brown, 1967). The pattern of low-angle layer lines (arrowed) corresponding to a repeat of 429 Å is believed to arise from the helical array of cross-bridges on the thick (myosin) filaments. (b) Diffraction pattern from living *taenia coli* muscle of the guinea pig (Vibert *et al.*, 1972). The arrows indicate layer lines associated with the thin actin-containing filaments and the numbers indicate the order of the Bessel function contributing to the layer line.

where h is the unit height, P is the pitch of the basic helix (Chapter 1, Section A.II.c), and m and n are integers. The values of m and n appropriate to the actin pattern are indicated in Table 15.4 and are based on assignments suggested by Selby and Bear (1956). The values of h and $|P|$ can be determined with reasonable precision simply by taking the reciprocals of the values of $Z(1, 0)$ and $Z(0, -1)$, respectively. It can be shown that the pitch of the individual strands P_s is related to h and P by the expression

$$2/P_s = (1/h) + (2/P) \qquad (15.3)$$

Values of P_s obtained by the application of Eq. (15.3) are unlikely to be particularly reliable since $h \sim -\frac{1}{2}P$ and so the quantity $(1/h) + (2/P)$ cannot be determined with high precision. An alternative method, used by Huxley and Brown (1967) and Lowy and Vibert (1967), depends upon the fact that, from Eq. (15.2), $Z(1, 1) = (1/h) + (1/P)$ and $Z(0, -1) = -1/P$. Thus we have

$$Z(1, 1) - Z(0, -1) = (1/h) + (2/P) \qquad (15.4)$$

and by applying Eq. (15.3), we obtain

$$Z(1, 1) - Z(0, -1) = 2/P_s \qquad (15.5)$$

Estimates of P_s obtained by the application of Eq. (15.5) are included in Table 15.5 and the values obtained for living, resting frog sartorius muscle by Huxley and Brown (1967) appear to preclude a helix with 15 units in seven turns in this material since this would imply a value of $P_s = (27.3 \pm 0.12) \times 30 = 823\text{--}815$ Å compared with the estimated value of 710–740 Å. A helix with 13 units in six turns would, on the other hand, give a value of $P_s = (27.3 \pm 0.12) \times 26 = 713\text{--}707$ Å and thus cannot be excluded on the basis of the available data. Miller and Tregear (1972) suggest that the data obtained from an insect flight muscle in rigor are consistent with a helical structure with 28 units in 13 turns.

According to Eq. (15.2), a near-equatorial layer line is predicted to be present in the actin pattern corresponding to the set of values $m = 1$, $n = 2$. From Eq. (15.3) it can be seen that the Z coordinate of this layer line will be $2/P_s$ and this offers the possibility of determining P_s directly. Layer lines with a Z coordinate of this order have been observed in the diffraction patterns of a wide variety of muscles (Selby and Bear, 1956;

Reedy *et al.*, 1965; Huxley *et al.*, 1966; Huxley and Brown, 1967; Lowy and Vibert, 1967; Millman *et al.*, 1967; Vibert *et al.*, 1972). In the case of vertebrate striated muscle the Z coordinate cannot be determined accurately due to the presence of a strong layer line nearby associated with the thick filaments. However, the Z coordinates for molluscan and mammalian smooth muscle (Vibert *et al.*, 1972) are generally consistent with the values for P_s given in Table 15.5.

According to Huxley and Brown (1967), the helical symmetry of the arrangement of the actin molecules in the thin filaments of frog sartorius muscle remains relatively invariant under different physiological conditions. Some indication of the possibility of a small change in the value of P_s was, however, noted between the relaxed state and iodoacetate-induced rigor. Vibert *et al.* (1972) consider that the differences noted by them (Table 15.4b) between the spacings of the actin layer lines in various types of muscle represent genuine differences in the helical parameter of the actin helices present in these materials.

b. *Nature of the Repeating Unit*

If the near-meridional reflections in the diffraction pattern of the actin-containing filament (Table 15.4) are assumed to arise from a helical distribution of points at a constant radius r, the lateral coordinate R of the intensity maximum can be used to obtain an estimate of r by equating the quantity $2\pi Rr$ to the argument corresponding to the first maximum of the appropriate Bessel function. Selby and Bear (1956) applied this method to the reflections of spacing 51 and 59 Å in a dried actin-rich muscle and obtained an estimate of about 25 Å for the effective radius on which the scattering matter of the structural units could be considered to be concentrated. The diameter of the filament was therefore assumed to be in the range 50–100 Å. A similar calculation was made by Vibert *et al.* (1972) using the R coordinates of maxima observed on layer lines with Z in the range 0–0.02 Å^{-1} in the pattern from resting molluscan smooth muscle. A value of 24 Å was obtained for the effective radius at which the scattering matter could be considered to be concentrated. From electron microscope studies the diameter of the thin filaments was estimated by Hanson and Lowy (1963) to be 80 Å and by Huxley (1963) to be 60–70 Å.

Moore *et al.* (1970) applied the technique of three-dimensional image reconstruction to isolated thin filaments from muscle and the results obtained were essentially similar to those obtained from F-actin filaments apart from the appearance of a small additional stain-excluding volume situated between the main stain-excluding volumes in the two strands.

C. TROPOMYOSIN

Tropomyosin is a water-soluble protein first isolated by Bailey (1948), which appears to be present as a minor component, on a weight basis, of all types of muscle. As mentioned in Section B, tropomyosin commonly occurs as a contaminant of crude preparations of actin and it has been established that in the native state tropomyosin occurs in association with actin in the thin filaments (Pepe, 1966). The molecular weight, as determined by hydrodynamic studies, is about 68,000 (Table 15.2) but in the presence of denaturants and reducing agents the molecule dissociates into two chains of very similar weight and composition (E. F. Woods, 1967). Measurements of the optical rotatory dispersion parameter b_0 indicate that in aqueous solution the molecule has an α-helix content approaching 100% (Table 15.2) and this may be correlated with the high content of helix-favoring residues and the low content of proline, cysteine, glycine, and other non-helix-favoring residues (Table 15.1).

I. MOLECULAR STRUCTURE OF EXTRACTED TROPOMYOSIN

a. *Chain Conformation*

Astbury *et al.* (1948a) showed that films prepared from tropomyosin gave the so-called α pattern characterized by a meridional arc of spacing 5.15 Å and a diffuse equatorial reflection centred at 9.8 Å. A meridional arc of spacing 1.49 Å has also been recorded (Perutz, 1951c). These features, which are discussed in detail in Section E, are generally regarded as indicating that the chain conformation is that of an α-helix distorted into a coiled-coil (Crick, 1953b). Quantitative studies of the distribution of scattered intensity on the equator and on the near-equatorial layer lines were carried out by Fraser *et al.* (1965e) and the results obtained are included in Fig. 15.6. The intensity profiles are similar to those observed in the "paramyosin pattern" obtained from dried oyster catch muscle which had been treated with ethanol in order to eliminate the actin reflections (Bear and Selby, 1956). Since paramyosin appears to contain two-strand coiled-coil ropes (Section E), it was concluded that so does tropomyosin.

Oriented films of tropomyosin suitable for infrared studies were prepared by Ambrose *et al.* (1949b) and parallel dichroism of the amide A and B bands was observed. This character of the dichroism is consistent with the presence of α-helices having a preferred orientation parallel to the direction of stretching.

Crick (1953a) suggested that the tropomyosin molecule consisted of

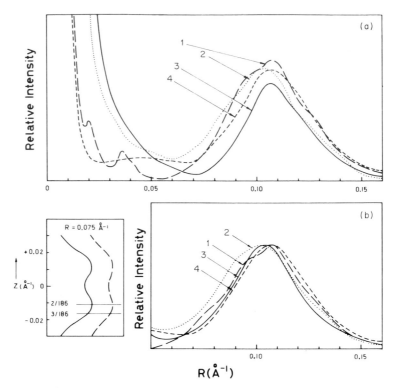

FIG. 15.6. Comparison of the intensity distributions on (a) the equatorial and (b) near-equatorial layer lines in the X-ray diffraction patterns of various α-fibrous proteins in the dried state: (1) Porcupine quill, (2) rabbit psoas muscle, (3) oyster muscle, (4) tropomyosin. The inset (bottom left) shows the variation in intensity (plotted horizontally) in the Z direction at $R = 0.075$ Å^{-1} for oyster muscle (full line) and porcupine quill (broken line). The layer line spacings appropriate to two-strand and three-strand coiled coils with pitch 186 Å are indicated (Fraser *et al.*, 1965e).

a ropelike aggregate of α-helices having a coiled-coil conformation. He also pointed out that it would be energetically favorable if the nonpolar sidechains were concentrated in the interior of the rope. This would lead to a periodic distribution of nonpolar residues in the sequence, and in the case of a two-strand rope in which the main-chain atoms are related by a dyad (Section E.II.e) the sequence would be expected to contain repeating heptads of the type [X—N—X—X—N—X—X], where N is a nonpolar residue and X is any residue (Fig. 15.7a). A sufficient length of the tropomyosin chain has been sequenced (Hodges and Smillie, 1970, 1972a,b; Hodges *et al.*, 1973) to test this expectation and the result is shown in Fig. 15.7b. It can be seen that positions 2

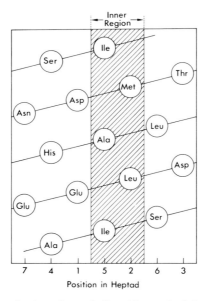

FIG. 15.7a.   Radial projection of an α-helix with exactly 3.5 residues per turn. When two such helices are placed in contact so that they are related by a parallel dyad, two residues (in this case the second and fifth) in each consecutive heptad are directed toward the inner region of the assembly. A similar situation obtains when two α-helices with other than 3.5 residues per turn are twisted around each other to form a two-strand coiled-coil rope. Crick (1953b) suggested that the inner positions would be occupied by nonpolar residues and this expectation is realized in the sequence of the tropomyosin chain, a portion of which is illustrated.

and 5, which correspond to the interior of the rope, are very frequently occupied by Ala, Leu, Ile, or Val residues and thus the occurrence of the coiled-coil conformation in tropomyosin can be directly correlated with the disposition of nonpolar residues in the sequence.

b. *Molecular Length*

Rodlike particles believed to correspond to individual tropomyosin molecules have been observed by electron microscopy (Rowe, 1964; Ooi and Fujime-Higashi, 1971) and the mean length of the particles was estimated to be about 400 Å. Tropomyosin forms a variety of ordered aggregates, including tactoids with regular polar banding patterns of period about 400 Å, as in the example shown in Fig. 15.8 (Cohen and Longley, 1966; Caspar *et al.*, 1969; Millward and Woods, 1970) and also tactoids with banding patterns of period about 800 Å, having dihedral symmetry (Tsao *et al.*, 1965; Cohen and Longley, 1966). These observations suggest a minimum length for the tropomyosin

Position in Heptad

1	2	3	4	5	6	7
-	Leu-	Lys-	Glu-	Ala-	Lys-	His-
				Leu		
Ile-	Ala-	Glu-	Asp-	Ala-	Asp-	Arg-
Lys-	Tyr-	Glx-	Glx-	Val-	Ala-	Arg-
Lys-	Leu-	Val-	Ile-	Leu-	Glx-	Ser-
				Ile-		Gly
Asx-	Leu-	Glx-	Arg-	Ala-	Glx-	Glx-
Glx				Ser		
Arg-	Ala-	Glx-	Leu-	Ser-	Glx-	Gly-
			Val	Ala		Ser
Lys-	Cys-	Ala-	Glx-	Leu-	Glx-	Glx-
		Gly	Asx			
		Thr				
Glx-	Leu-	Lys-	Ile-	Val-	Thr-	Asn-
			Lys			
Asx-	Leu-	Lys-	Ser-	Leu-	Glx-	Ala-
Glx-	Ala-	Glx-	Lys-	Tyr-	Ser-	Glx-
Val						
Lys-	Glx-	Asx-	Lys-	Tyr-	Glx-	Glx-
Glx-	Ile-	Lys-	Val-	Leu-	Ser-	Asx-
			Leu			
Lys-	Leu-	Lys-	Glx-	Ala-	Glx-	Thr-
Arg-	Ala-	Glx-	Phe-	Ala-	Glx-	Arg-
Ser-	Val-	Ala-	Lys-	Leu-	Glx-	Lys-
		Thr				
Ser-	Ile-	Asx-	Asx-	Leu-	Glx-	Asx-
Glx-	Leu-	Tyr-	Ala-	Glx-	Lys-	Leu-
Val						
Lys-	Tyr-	Lys-	Ala-	Ile-	Ser-	Glu-
Glu-	Leu-	Asp-	His-	Ala-	Leu-	Asn-
Asp-	Met-	Thr-	Ser-	Ile-	COOH	

	1	2	3	4	5	6	7
Ala + Leu + Ile + Val	$1\frac{1}{2}$	15	2	$5\frac{1}{3}$	16	2	2
Asp + Asn + Glu + Gln	$7\frac{1}{2}$	1	9	6	1	11	9
Arg + His + Lys	8	0	6	$5\frac{1}{3}$	0	2	5

FIG. 15.7b. Provisional sequence of 137 residues from the C-terminal portion of the tropomyosin chain (Hodges *et al.*, 1973). The sequence has been arranged as a series of heptads and phased so that positions 2 and 5 are selectively occupied by nonpolar residues. As illustrated in Fig. 15.7a, these positions correspond to the inner region of a two-strand coiled-coil rope. The tabulation in the lower part of the figure illustrates the high correlation between position in the heptad and residue type in tropomyosin. Asx = Asp or Asn, Glx = Glu or Gln.

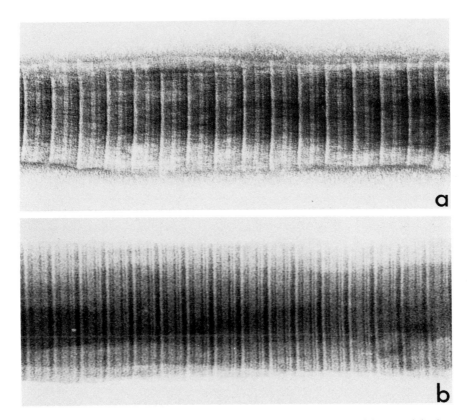

F IG. 15.8. Electron micrographs of tropomyosin tactoids formed by precipitation with lead acetate, stained with uranyl acetate. (a) Negative stain, (b) positive stain. The polar banding pattern has an axial period of 395 Å (Caspar *et al.*, 1969).

molecule in the vicinity of 400 Å. A similar conclusion was arrived at through the study of tropomyosin crystals (Section C.I.c).

The weight of the individual chains in the tropomyosin molecule has been determined to be 33,500 ± 2000 daltons. (E. F. Woods, 1967) and the mean residue weight, calculated from the analysis given in Table 15.1, is 115.6 daltons. Thus the number of residues in each chain is 272.5–307 and assuming a mean axial rise per residue in the coiled coil of 1.485 Å (Fraser *et al.*, 1964e; Elliott *et al.*, 1968a), a length of 405–456 Å would be predicted for a completely helical molecule. This is compatible with the minimum lengths of 405 ± 8 Å estimated by electron microscopy and 402 ± 5 Å estimated from X-ray diffraction studies of crystalline tropomyosin (Caspar *et al.*, 1969). When allowance is made for the supercoiling present in the crystal the latter estimate is increased to 410 Å (Cohen *et al.*, 1971b).

c. *Studies of Crystals*

Macroscopic crystals of tropomyosin (Fig. 15.9) can be prepared (Bailey, 1948) and these have been studied by both electron microscopy (Hodge, 1959; Huxley, 1963; Caspar *et al.*, 1969; Fujime-Higashi and Ooi, 1969; Ooi and Fujime-Higashi, 1971) and X-ray diffraction (Caspar *et al.*, 1969; Cohen *et al.*, 1971b). The crystals contain about 95% by volume of water and collapse if removed from the mother liquor. The high degree of hydration leads to appreciable disorder in the lattice and reflections with spacings less than about 20 Å are faint and diffuse. The observed X-ray reflections were indexed on an orthorhombic unit cell with dimensions $a = 126$ Å, $b = 243$ Å, $c = 295$ Å (Caspar *et al.*, 1969). The unit-cell dimensions were found to vary somewhat from crystal to crystal but the length of the diagonal remained relatively constant, with a mean value of 401.8 Å (Cohen *et al.*, 1971b).

The space group was originally thought to be $P2_12_12$ (see Henry and

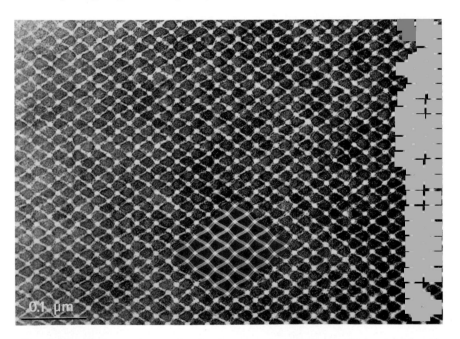

FIG. 15.9. Electron micrograph of a tropomyosin crystal, negatively stained with uranyl acetate. In one portion of the micrograph the electron density map calculated from the X-ray pattern of the wet crystal has been superimposed (Caspar *et al.*, 1969).

Lonsdale, 1952) and following crystallographic convention, the unique twofold axis was chosen as the c axis (Caspar *et al.*, 1969). This is the convention used earlier to list the unit-cell dimensions. Later studies (Cohen *et al.*, 1971b) showed, however, that the space group was in fact $P222_1$. Conventionally, the unique twofold screw axis would be taken as c, but to avoid changing the original assignment of axes, the space group may be rewritten $P22_12$. The a-axis projection of structure has a plane group symmetry *pmg* (see Henry and Lonsdale, 1952) with glide planes parallel to b and mirror planes parallel to c (Fig. 15.10). Interpretation of the diffraction data related to this projection was carried out on the basis of the netlike appearance visible in electron micrographs where the viewing direction is parallel to the a axis (Fig. 15.9). It was concluded that the crystal is built up from filaments formed by the head-to-tail aggregation of tropomyosin molecules and that the mean direction of the filaments is parallel to the body diagonals of the unit cell. Thus the length of the diagonal (402 Å) gives an estimate for the minimum length of the molecule. The filaments are not straight but have a conformation which is approximately sinusoidal in a-axis projection (Fig. 15.9) with a mean amplitude of about 12 Å.

An important feature to emerge from studies of ordered aggregates of tropomyosin is that the molecule is polar. The available evidence on

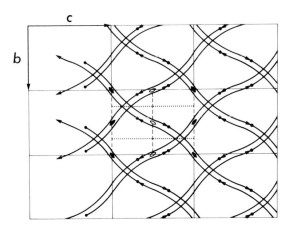

FIG. 15.10. The a-axis projection of the tropomyosin crystal lattice showing the symmetry relationships between the tropomyosin molecules. The cell is orthorhombic with dimensions $a = 126$ Å, $b = 243$ Å, and $c = 295$ Å. The molecules aggregate end-to-end to form filaments as shown but the positions of the molecular ends, marked by arrowheads, are unknown. The two filaments of opposite polarity which appear to run parallel in this projection actually run along opposite diagonals in the a-axis direction (Caspar *et al.*, 1969).

molecular length, chain weight, and chain conformation lead to a picture of the molecule as a two-strand coiled-coil rope and the polarity of the molecule could be explained simply on the basis that the two chains have different amino acid sequences. However, it seems likely that the two chains are very similar chemically (E. F. Woods, 1967) and on balance the evidence seems to favor a two-strand parallel-chain structure, where the term parallel refers to the sense of the main-chain sequence $-NH-CHR-CO\rightarrow$.

In the structure proposed for tropomyosin crystals by Cohen *et al.* (1971b) the filament runs parallel to the body diagonals of the unit cell and the net appearance in electron micrographs (Fig. 15.9) is attributed to a collapse of the structure parallel to the *a* axis. Ooi and Fujime-Higashi (1971), on the other hand, suggest that the crystal is formed by the piling up, in register, of two-dimensional nets. In this case the filaments would run parallel to the *bc* plane. The evidence quoted in support of this model is largely circumstantial and there does not appear to be any compelling reason for preferring it to the model proposed by Cohen *et al.* (1971b).

II. TROPOMYOSIN IN MUSCLE

As mentioned earlier, tropomyosin has been shown to be located in the thin filaments (Pepe, 1966) and a periodicity of about 400 Å in the axial projection of these filaments that cannot be accounted for by the actin helix (Section B) has been detected by both electron microscopy (Draper and Hodge, 1949; Carlsen *et al.*, 1961; Page and Huxley, 1963) and X-ray diffraction (Worthington, 1959; Huxley and Brown, 1967). Hanson and Lowy (1963) suggested that the tropomyosin molecules might be located in the grooves in the helical array of actin molecules (Fig. 15.2) and that the periodicity of about 400 Å could be accounted for by the arrangement of the tropomyosin molecules. Subsequent studies suggest that the periodicity is accentuated by the presence of troponin in association with the tropomyosin (Ohtsuki and Wakabayashi, 1972).

Some support for this view has been obtained from studies of optical diffraction patterns from electron micrographs of paracrystals formed by actin and by actin contaminated with tropomyosin (O'Brien *et al.*, 1971). Evidence has been obtained for a periodicity of about 28 Å in the staining properties along the length of the tropomyosin molecule (Caspar *et al.*, 1969; Parry and Squire, 1973) and the latter authors have drawn attention to the fact that the globular units in the F-actin filament occur at similarly spaced axial intervals (Fig. 15.2).

Huxley and Brown (1967) estimated the "non-actin" periodicity in living, resting frog sartorius muscle to be 385 ± 2 Å and assuming head-to-tail assembly of the tropomyosin molecules along a helical groove of radius r_s, the molecular length would be $385/\cos[\operatorname{atan}(2\pi r_s/P_s)]$, where P_s is the pitch of the groove. Assuming $P_s = 730$ Å (Section B) and $r_s = 30$ Å, for example, the predicted molecular length is 398 Å.

D. MYOSIN

Myosin is the major protein component of most muscles and is the principal, if not the only, component of the thick filaments of striated muscle. Reviews of various aspects of the chemical and physicochemical properties of myosin have been given by Gergely (1966), Perry (1967), Lowey *et al.* (1969), Young (1969), and Lowey (1971). Limited proteolysis with trypsin cleaves the molecule into two fragments termed light meromyosin (LMM) and heavy meromyosin (HMM) and further proteolysis leads to the appearance of two subfragments of HMM termed S1 and S2. By digesting myosin in a precipitated state with papain, it is also possible to isolate a fragment which corresponds to LMM plus the S2 subfragment of HMM. The molecular weights and helix contents of these fragments and subfragments are given in Table 15.6. These data, taken in conjunction with electron microscope

TABLE 15.6

Molecular Weights and Helix Contents of Myosin and
Its Enzymatic Subfragments S1 and S2[a]

Parameter	Myosin	HMM			LMM	Papain rod
		Whole	S1	S2		
Molecular weight $\times 10^{-3}$	510 ± 10	340 ± 10	115 ± 5	62 ± 2	140 ± 5	220 ± 10
Dispersion constant b_0	400 ± 10	320 ± 10	230 ± 10	610 ± 10	630 ± 10	660 ± 10
Helix content[b]	63	51	37	97	100	105

[a] Lowey *et al.* (1969).
[b] Calculated assuming $b_0 = -630$ is equivalent to 100% helix.

studies of individual myosin molecules and of enzymatic fragments (Rice, 1961; Zobel and Carlson, 1963; Huxley, 1963; Slayter and Lowey,

1967), lead to the model for the extracted myosin molecule depicted in
Fig. 15.11 (Lowey *et al.*, 1969). The molecule has a rodlike "tail" with

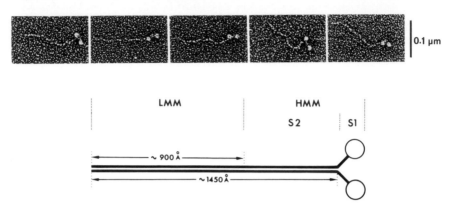

FIG. 15.11.   Top: electron micrographs of individual myosin molecules (Lowey *et al.*,
1969). Bottom: schematic representation of the chain structure of the myosin molecule.
Limited proteolysis with trypsin cleaves the molecule into fragments termed light
meromyosin (LMM), which is highly α-helical, and heavy meromyosin (HMM) which
is only partly α-helical. Further proteolysis cleaves the HMM fragment into subfragments
S1 and S2. In addition to the two so-called heavy chains depicted in the diagram, the
myosin molecule contains so-called light chains, which are bound in the S1 region.
(After Lowey *et al.*, 1969 and Cohen *et al.*, 1970.)

a very high content of α-helix and a "head" composed of two globular
regions, about 100 Å in diameter, with a relatively low helix content. It
has been shown that the enzymatic activity of myosin resides in the
globular portions of the molecule. The fragment obtained by digestion
of precipitated myosin with papain appears to constitute the rodlike tail
portion of the molecule and is conveniently referred to as the "papain
rod."

Considerable difficulty was experienced in elucidating the chain
structure of the myosin molecule (Lowey, 1971) but it is now reasonably
well established that the molecule consists of two "heavy" chains which
run side by side throughout the rodlike portion of the molecule and then
terminate in the two globular units which form the head region. In
addition to these heavy chains, there appear, in the case of mammalian
myosin, to be four light chains with molecular weights about 20,000
which are noncovalently linked to the head region (Lowey, 1971).

From a consideration of the form of the myosin molecule it would
seem likely that the two chains in the rodlike portion of the molecule
run in a parallel sense with respect to the sequence $-NH-CHR-CO\rightarrow$
of main-chain atoms. Various lines of evidence from chemical studies

support this conclusion, including the identification of the expected number of N-terminal residues in the HMM region (Offer, 1965; Offer and Starr, 1968) and the absence of C-terminal homoserine in a fragment resembling LMM obtained by cleavage of the molecule with cyanogen bromide (Young *et al.*, 1968).

I. STUDIES OF EXTRACTED MYOSIN

a. *Chain Conformation*

Films cast from solutions of myosin were studied by Astbury and Dickinson (1940) and shown to possess uniplanar orientation (Chapter 5, Section B.II). A further type of preferred orientation was induced in the plane of the sheet by stretching, and up to 50% stretch, the high-angle X-ray pattern was of the α type, subsequently attributed to the coiled-coil modification of the α-helix (Section E.II.e). In addition, a meridional reflection of spacing 1.50 Å, as predicted for the α-helix, was observed by Perutz (1951b). Infrared spectra were obtained from similar specimens using polarized radiation (Ambrose *et al.*, 1949b) and the dichroism exhibited by the amide A and B bands was of the character expected for a preferred orientation of α-helices parallel to the direction of stretching.

The fragments obtained by limited enzymatic proteolysis have also been examined by X-ray diffraction. LMM can readily be oriented and gives a high-angle α pattern (Section E.II.e) together with a series of meridional and near-meridional reflections which can be indexed on an axial repeat of 430 Å (Cohen and Szent-Györgyi, 1960; Szent-Györgyi *et al.*, 1960). The HMM fragment, on the other hand, is rather easily denatured and the high-angle pattern obtained from this material is a function of specimen treatment (Cohen and Szent-Györgyi, 1960). The helical subfragment S2 of HMM gives a high-angle α pattern but so far no well-developed low-angle pattern has been reported (Lowey *et al.*, 1967).

The observations on LMM and subfragment 2 of HMM support the results obtained from studies of optical rotatory dispersion (Table 15.6) which suggest that the α-helix content of the myosin molecule resides mainly in the rodlike tail region. A comparison of the amino acid compositions of the papain rod and of the HMM subfragment S1 (Table 15.7) shows that no proline is present in the rod and that the contents of glycine and aromatic amino acids are lower than in the head region. The principal difference between the two regions is the greatly increased content of glutamic acid in the rod. The overall composition of the rod is similar to that of the highly helical proteins tropomyosin and para-

TABLE 15.7

Amino Acid Compositions of Myosin and Enzymatic Fragments of Myosin[a]

Amino acid	Myosin	HMM Subfragment 1	Papain rod
Alanine	9.0	7.8	9.7
Arginine	5.0	3.8	6.4
Aspartic acid[b]	9.9	9.4	10.1
Glutamic acid[b]	18.2	13.0	25.0
Glycine	4.6	6.8	2.3
Half-cystine[c]	1.0	1.2	0.5
Histidine	1.9	2.0	1.7
Isoleucine	4.9	5.9	4.1
Leucine	9.4	8.3	11.3
Lysine	10.7	9.2	12.2
Methionine	2.7	3.1	2.6
Phenylalanine	3.4	5.8	0.8
Proline	2.5	4.1	0
Serine	4.5	4.5	4.5
Threonine	5.1	5.4	4.3
Tyrosine	2.3	3.8	0.7
Valine	5.0	6.1	3.8

[a] Lowey et al. (1969). Given in residues/100 residues, rounded to nearest 0.1 residue.
[b] Includes amide.
[c] Estimated as cysteic acid.

myosin (Table 15.2) and the high proportion of α-helix can be correlated with high content of glutamic acid and low contents of glycine and proline (Chapter 12, Section C).

b. *Ordered Aggregates*

Myosin and fragments of myosin form a wide variety of ordered aggregates and their study has provided information on the mode of assembly of the myosin molecules in the thick filaments of striated muscle.

Huxley (1963) showed that if purified myosin is precipitated at low ionic strength, spindle-shaped aggregates are formed which are similar in many respects to the thick filaments of striated muscle. From the appearance of the synthetic filaments and from studies of individual molecules it was concluded that the aggregates consist of a bipolar array of myosin molecules in which the molecules in either half of the filament are oriented with opposite polarity, leaving a bare region devoid of heads in the center of the filament (Fig. 15.12).

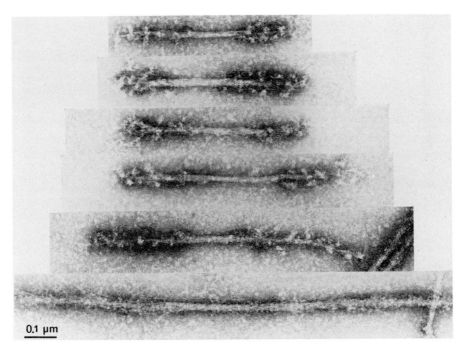

FIG. 15.12. Electron micrographs of filamentous aggregates of extracted myosin; all show the same characteristic of a smooth central zone and projections all the way along the rest of the length of the filament (Huxley, 1963).

A different type of bipolar aggregate has been obtained by precipitating myosin with high concentrations of divalent cations (Harrison *et al.*, 1971; King and Young, 1972a). The aggregates have a segmented form in which the molecules are packed in an antiparallel array with an overlap of 900 Å. Similar types of bipolar segment aggregate have been obtained from enzymatic fragments of myosin and overlaps of 1300, 900, and 430 Å (Fig. 15.13) have been observed (Cohen *et al.*, 1970; Harrison *et al.*, 1971; Kendrick-Jones *et al.*, 1971). The length of the rodlike portion of the myosin molecule was estimated from these studies to be 1450 Å in myosin from vertebrate striated muscle and 1560 Å in myosin from vertebrate smooth muscle.

LMM forms tactoids with banding patterns of period 430 Å (Philpott and Szent-Györgyi, 1954; Szent-Györgyi *et al.*, 1960; Huxley, 1963; Cohen *et al.*, 1970; Nakamura *et al.*, 1971) and also of period 145 Å (Cohen *et al.*, 1970). The patterns have dihedral symmetry (Fig. 15.14), making it difficult to determine the precise nature of the molecular packing, but it is likely that the observed periodicities arise through a

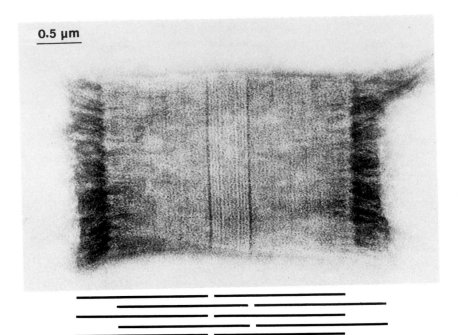

FIG. 15.13. Electron micrograph of a bipolar segment formed by the rod portion of myosin, isolated from vertebrate smooth muscle, when precipitated with calcium ions. A model of the molecular packing is shown below; the overlap region was estimated to be about 430 Å in length (Kendrick-Jones et al., 1971).

staggered arrangement of molecules. The two periodicities correlate with the period and unit height, respectively, of the helical arrangement of projections from the thick filament in living, resting vertebrate skeletal muscle (Section D.II.b). Other aggregates have the form of square nets with side equal to about 400 Å (Philpott and Szent-Györgyi, 1954; Lowey et al., 1967), hexagonal nets with side about 600 Å (Huxley, 1963; Lowey et al., 1967), and segments about 900 Å in width (Cohen et al., 1970).

The fragment of the myosin molecule termed LMM-C, prepared by treatment of myosin with cyanogen bromide (Young et al., 1968), forms tactoids with periods of 143 and 429 Å (King and Young, 1970), which are similar to the periods of the enzymatic LMM tactoids. In addition, square nets of side 685 and 390 Å were observed, the latter type apparently originating from a three-dimensionally crystalline form. Bipolar segment-type aggregates have been obtained from LMM-C (King and Young, 1972a) and their form suggests that the fragments are about

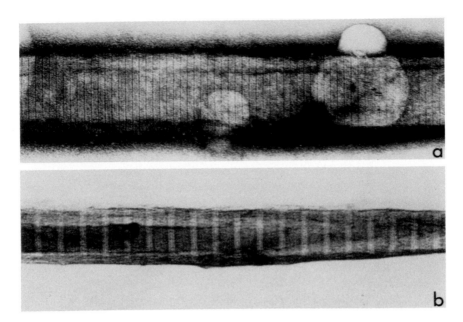

FIG. 15.14. Electron micrographs of paracrystals formed by (a) heavy meromyosin subfragment 2, axial periodicity 145 Å, and (b) light meromyosin, periodicity 430 Å (Lowey *et al.*, 1967).

1025 Å long and are packed in an antiparallel manner with an overlap of 890 Å. As mentioned earlier, both myosin and LMM form bipolar aggregates with overlaps of about 900 Å. A further type of aggregate, so far only observed with LMM-C, appears to be a tubular assembly in which the molecular fragments follow a helical path (King and Young, 1972b).

Tactoids have also been prepared from subfragment 2 of HMM (Lowey *et al.*, 1967) and from the papain rod (Lowey, 1971). In both cases the banding period was 145 Å. The ability to form tactoids with a 145-Å periodicity is a property shared by all the helical fragments and subfragments of the myosin molecule.

II. NATIVE AGGREGATES

a. *Electron Microscope Studies*

In striated muscle, myosin exists as thick filaments of well-defined length and diameter which are packed in hexagonal array (Fig. 15.15). The appearance of the filaments depends upon the origin and physiological state of the muscle, on the position along the length of the A band,

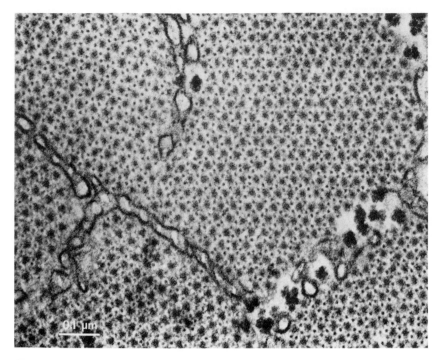

FIG. 15.15. Electron micrograph of cross section of live frog sartorius muscle fixed in glutaraldehyde and treated with osmium tetroxide. The thin, actin-containing filaments are situated at the trigonal points of the hexagonal lattice formed by the thick, myosin-containing filaments (Huxley, 1968).

and on the method of fixation and staining. In vertebrate striated muscle the thick filaments are about 1.5–1.6 μm long (Huxley, 1963) and appear to consist of a cylindrical core around 100 Å in diameter (Huxley, 1968) with a series of regularly disposed projections (Huxley, 1957; Huxley and Brown, 1967). At the center of the A band the filaments are linked together to form the M band and for a short distance each side of the M band the filaments are free of projections. In this so-called pseudo-H-zone the filament profile has been reported to be triangular (Sjöstrand and Andersson, 1956; Franzini-Armstrong and Porter, 1964; Pepe, 1967, 1971b; Afzelius, 1969).

In other parts of the A band the thick filaments have been reported to present the appearance in cross section of either hollow or solid circles or triangles, with overall diameters variously estimated to be in the range 100–300 Å (Huxley, 1966; Halvarson and Afzelius, 1969; Hayes *et al.*, 1971; Pepe, 1971a). Substructure within the filament profile has been reported (Auber and Couteaux, 1963; Baccetti, 1965; Gilëv, 1966a,b, 1970; Huxley, 1968; Pepe and Drucker, 1972) but the degree of specimen preservation and effective resolution so far achieved have been insufficient for the precise nature of the substructure to be elucidated.

Hanson et al. (1971) isolated thick-filament assemblies from vertebrate striated muscle (Fig. 15.16) and these show the central M band about 495 Å wide flanked on either side by heavily stained zones about 395 Å wide. On each side of the central region there are ten well-defined repeats of a 420 ± 15 Å period, after which the filaments appear to taper. The periodic portions of the thick-filament assembly are presumably responsible for the 429-Å axial periodicity observed in the X-ray diffraction pattern of the living, resting muscle (Elliott, 1964a; Huxley and Brown, 1967).

The geometric form of the myosin-containing filaments in vertebrate smooth muscle is not well established and useful summaries of existing knowledge have been given by Burnstock (1970) and Lowy et al. (1972). According to Lowy and Small (1970), the filaments are ribbonlike, about 80 Å thick, 200–1100 Å wide, and up to 30,000 Å long. Small et al. (1971) have shown there is a regular array of projections from the flat surfaces of the filament; the projections are arranged in transverse rows and the rows are spaced at axial intervals of about 140 Å. Optical diffraction patterns obtained from edge-on views of the filaments show that the arrangement of projections is of opposite polarity on the two sides of the filament. In studies by other workers approximately cylindrical filaments with diameters about 150–200 Å have generally been observed and the appearance of one type of filament or the other seems to depend on the method of specimen preparation (Rice et al., 1971; Somlyo et al., 1971a,b; Cooke and Fay, 1972). Which form of filament exists in vivo remains to be established. In paramyosin-containing smooth muscle the myosin is believed to be located at the surface of the thick filaments (Hanson and Lowy, 1964b; Squire, 1971; Szent-Györgyi et al., 1971) but the precise arrangement is not known.

b. *Low-Angle X-Ray Studies*

The low-angle X-ray patterns of striated muscles have been studied extensively but it would be beyond the scope of the present treatment to discuss these in detail. Valuable reviews have been given by Huxley and Brown (1967) and Hanson (1969). Various lines of evidence (Worthington, 1961; Elliott, 1964a; Huxley and Brown, 1967) suggest that the pattern of meridional and near-meridional reflections observed at low angles in the diffraction pattern of living, resting striated muscle

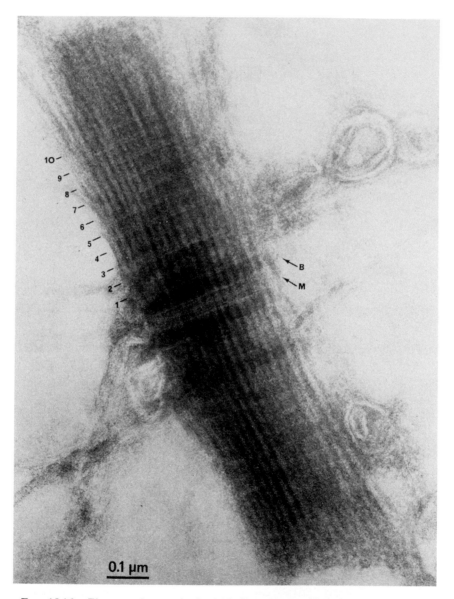

FIG. 15.16. Electron micrograph of a thick-filament assembly isolated from vertebrate striated muscle. The central M band about 495 Å wide is flanked by densely stained zones (B) about 395 Å wide. A periodicity of 420 Å is visible in the remainder of the filament (Hanson *et al.*, 1971).

(Fig. 15.5) originate from the projections from the core of the thick filament (Section D.II.a). In vertebrate striated muscle these reflections lie on a series of layer lines corresponding to a periodicity of 429 Å and from a consideration of the distribution of intensity and earlier electron microscope studies Huxley and Brown (1967) suggested that the projections were arranged in pairs, related by a twofold axis parallel to the axis of the filament, on a helix with three pairs per turn of pitch 429 Å and unit height 143 Å (Fig. 15.17a). Squire (1971) has suggested an alternative model in which each unit consists of four projections related by a parallel fourfold rotation axis.

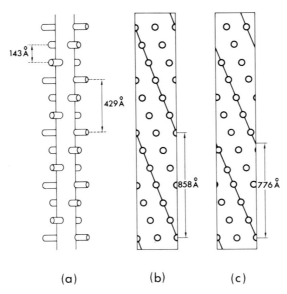

(a)   (b)   (c)

Fig. 15.17. (a) Model for the distribution of cross-bridges on the surface of the thick filaments of vertebrate striated muscle (Huxley and Brown, 1967). (b) Radial projection of the model showing the basic helix. (c) Radial projection of the distribution of cross-bridges in insect flight muscle (Reedy, 1968).

The diffraction patterns predicted for the two models will be similar except for the lateral distribution of intensity. In the second model the Bessel functions that contribute to a particular layer line have orders that are twice those which contribute to the diffraction pattern of the first (Chapter 1, Section A.III.c). If the nature of the projection was known, a choice could therefore be made on the basis of a comparison of the observed and calculated distributions of intensity.

Although the general distribution of intensity is consistent with a helical symmetry in the arrangement of projections, meridional reflections are observed on layer lines other than those for which zeroth-order Bessel functions are predicted by Eq. (1.23). This is likely to be associated with departures from strict helical symmetry (Chapter 1, Section A.IV.c).

The pattern obtained from insect flight muscle differs from that obtained from other striated muscle in that a series of layer lines corresponding to a periodicity of 388 Å appears at low angles accompanied by meridional orders of a 146-Å periodicity (Reedy et al., 1965). These periodicities are not in the ratio 3:1 as in vertebrate striated muscle and it has been suggested (Reedy, 1968) that the arrangement of projections in insect flight muscle has a slightly different helical symmetry which can be described in terms of a helix with 16 units in three turns (Fig. 15.17c). Miller and Tregear (1972) estimated values of 145.1 ± 0.3 Å for the unit height and 389 ± 2 Å for the pitch. The screw sense of this helix was determined by electron microscopy to be right-handed (Reedy, 1968). An alternative description of this model in terms of the basic helix (Chapter 1, Section A.II.c) is a unit height of 146 Å and a unit twist of $-67.5°$ leading to a left-handed helix of pitch $2 \times 388 = -776$ Å. As before, the unit consists of projections related by a parallel twofold axis.

In the wet state, vertebrate striated muscle gives a characteristic set of low-angle equatorial reflections which is associated with the hexagonal array of myofilaments, and studies of these reflections have been reviewed by Hanson (1969). Measurements of the relative intensities have been used to determine the axial projection of electron density in the muscle (Hanson and Huxley, 1955; Huxley, 1968). The resolution obtained in these syntheses is very low but a substantial amount of material, corresponding to about 30% of the original mass of myosin, was found to move from the vicinity of the thick filaments to the vicinity of the thin filaments in the transition from the resting state to rigor (Fig. 15.18). These observations were correlated with the appearance of the thick filaments in cross sections examined in the electron microscope (Huxley, 1968), where the diameter was found to be substantially less in the rigor state. The diameter of the thick filaments in muscle in rigor was estimated to be about 100 Å and this presumably represents the size of the rodlike core of the thick filament.

c. *Chain Conformation*

It has generally been assumed that the chain conformation in extracted myosin and in enzymatic fragments of myosin are the same as in the native state but direct evidence in support of this assumption is difficult to obtain. Both native and dried vertebrate striated muscle give a high-

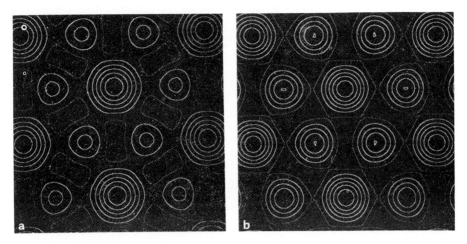

FIG. 15.18. Axial projections of electron density in the sarcomere of live frog sartorius muscle, calculated from the low-angle equatorial X-ray diffraction pattern (Huxley, 1968). (a) In resting muscle the concentration of electron density at the trigonal positions (corresponding to the positions of the thin, actin-containing filaments) is small. (b) In rigor muscle the electron density is increased around the trigonal positions and this may be attributed to a change in the distribution of the projections (cross-bridges) from the thick filaments.

angle α pattern when examined by X-ray diffraction, with a meridional arc at 5.15 Å and a diffuse equatorial reflection at 9.8 Å (Astbury, 1947b). References to earlier studies have been given by Astbury and Dickinson (1940) and Bear (1944b). In addition, a meridional reflection of spacing 1.5 Å has been recorded (Huxley and Perutz, 1951). On the basis of studies of the high-angle patterns of isolated actin (Section B.I.b) and myosin (Section D.I.a) it seems likely that the α pattern observed in striated muscle is to be attributed to those portions of the myosin rod that are aligned parallel to the length of the muscle fiber. Infrared spectra have also been obtained from muscle fibers using polarized radiation (Morales and Cecchini, 1951; Malcolm, 1955) but the observed dichroism, while consistent with the X-ray data, was very low, suggesting that the α-helical material aligned parallel to the fiber axis represented only a small portion of the total protein material present in the muscle.

The high-angle diffraction pattern of dried frog sartorius muscle was investigated in detail by MacArthur (1943) and Fraser and MacRae (1961a) and reflections out to spacings of about 1 Å were recorded (Fig. 15.19). The general distribution of intensity was found to be similar to that expected for a coiled-coil rope structure of the type proposed by Crick (1953a,b). The diffuse equatorial reflection at 9.8 Å

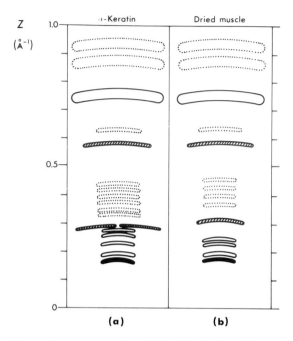

FIG. 15.19. Schematic representation of the distribution of intensity in the high-angle X-ray diffraction pattern of (a) dried α-keratin and (b) dried toad sartorius muscle (Fraser and MacRae, 1961a).

is partially resolved into equatorial and near-equatorial components (Cohen and Holmes, 1963) and the intensity distribution has been determined for dry muscle (Fraser *et al.*, 1965e) and is included in Fig. 15.6. All the high-angle X-ray data are consistent with a coiled-coil rope structure for the oriented α-helical portions of myosin in muscle.

d. *Models for the Thick Filaments*

No detailed information is available at present on the arrangement of the myosin molecules in the thick filaments of striated muscle but several models have been put forward. The core of the filament, which appears to about 100 Å in diameter (Huxley, 1968), is generally considered to consist of some or all of the rodlike portions of the molecules arranged in a staggered helical array with some or all of the HMM portion forming the projections (Huxley and Hanson, 1960; Lowey and Cohen, 1962; Huxley and Brown, 1967; Lowey 1971; Pepe, 1971a,b). There have been suggestions that myosin might be confined to a peripheral zone of the thick filament, leaving a central zone which was either empty or occupied by another type of protein (Lowey and Cohen, 1962; Squire, 1971).

Evidence has been obtained from the study of cross sections of muscle fibers which suggests that the thick filaments contain finer filaments about 20–30 Å in diameter arranged in a specific geometric pattern. Baccetti (1965) supposed that nine fine filaments were arranged peripherally about a central pair of filaments, while Gilëv (1966b) put forward a scheme appropriate to crab muscle involving 18 fine filaments. In view of the difficulties of preserving and observing structures of these dimensions in the electron microscope (Thon, 1966b; Haydon, 1968, 1969), independent confirmation of these proposals by other methods is desirable.

Pepe (1967, 1971a,b) has attempted to devise a model for the thick filaments which would be consistent with the triangular profile observed in the pseudo-H-zone (Section D.II.a) and various other features of the observed data. Myosin molecules (or groups of myosin molecules) were supposed to form linear aggregates with a 430-Å overlap (Fig. 15.20a) and the thick filament was considered to consist of nine such aggregates surrounding a central group of three aggregates. At the commencement of the overlap region the HMM portion was thought to pass out to the surface of the filament. In section, therefore, the 12 central subunits would consist of LMM rods or groups of LMM rods (Fig. 15.20b). It was estimated that the diameter of the LMM core would be 90–120 Å for an aggregate of individual myosin molecules, 120–160 Å for two molecules, and 150–200 Å for three molecules. Some support for this type of model has been obtained from electron microscope studies of cross sections (Pepe and Drucker, 1972) which show some evidence of the expected geometric pattern of subunits. The diameter of the core appears to about 160–180 Å in these preparations.

Harrison et al. (1971) devised a packing scheme for the myosin molecules (Fig. 15.20e) based on the helical symmetry deduced by Huxley and Brown (1967) for the projections in living, resting vertebrate striated muscle (Fig. 15.17). The rodlike portions of the molecules were assumed to be 1450 Å in length and inclined to the filament axis. The antiparallel packing at the center of the filament leads to a bipolar structure and the molecular overlaps are of the same dimensions as observed in certain types of segment aggregates. The length of the head-free zone is predicted to be about 1300 Å, which is of the correct order of magnitude (Huxley, 1967; Pepe, 1967; Knappeis and Carlsen, 1968). This is also close to the dimensions of the central region in the isolated filament assemblies (Section D.I.b). The model proposed by Harrison et al. (1971) is essentially a two-dimensional one since the radial coordinates of the molecules are not specified. King and Young (1972b) have suggested the possibility that the rodlike portions of the myosin molecules

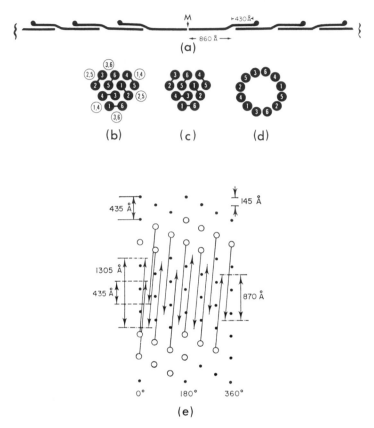

FIG. 15.20. Models for the disposition of the myosin molecules in the thick filaments of striated muscle. In the model proposed by Pepe (1966, 1967, 1971a, b) the myosin molecules are supposed to form linear aggregates (a) with a 430-Å overlap and the thick filament is visualized as a group of 12 such aggregates with the LMM portions of the molecules forming a central core (black circles). (b), (c), (d) Cross sections through the thick filament at different levels. Harrison *et al.* (1971) suggested that the rod portions of the molecules, assumed to be 1450 Å long, were packed as shown in (e) in the central zone of the thick filament.

may be twisted together to form cables, but no specific model was proposed.

E. PARAMYOSIN

Paramyosin is the principal constituent of specialized filaments found in certain molluscan and annelid muscles that are able to maintain tension for prolonged periods without a large expenditure of energy. The occurrence of these paramyosin-containing filaments was detected by low-angle X-ray studies (Bear, 1944b, 1945) and by electron microscopy (Jakus *et al.*, 1944; Hall *et al.*, 1945). Subsequently Hodge (1952) isolated and characterized the filament protein and highly purified paramyosin was prepared by Bailey (1956, 1957). No enzymatic properties have been detected and the precise function of paramyosin in the specialized filaments is unknown but appears to be mainly structural.

Paramyosin has a molecular weight of about 200,000 (Lowey *et al.*, 1963; McCubbin and Kay, 1968; Woods, 1969b) but in denaturing solvents of appropriate composition the weight is reduced to around half that of the native molecule (Olander *et al.*, 1967; McCubbin and Kay, 1968; Weber and Osborn, 1969; Woods, 1969b), suggesting that the molecule has a two-chain structure. Measurements of the optical rotatory dispersion parameter b_0 (Cohen and Szent-Györgyi, 1957, 1960; Simmons *et al.*, 1961; Riddiford, 1966; Olander, 1971) indicate that the protein has a very high α-helix content and this can be correlated with the amino acid composition (Table 15.1). The composition is similar to that of tropomyosin and also the rod portion of myosin (Table 15.7), with a high content of the α-helix-favoring residues glutamic acid, leucine, and alanine and a low content of the non-α-helix-favoring residues proline, glycine, and cysteine (Finkelstein and Ptitsyn, 1971).

Hodge (1952, 1959) established by means of hydrodynamic and light scattering studies that the molecule was highly asymmetric and, assuming a rodlike form, the length was estimated to be between 1200 and 1500 Å and the axial ratio to be about 70:1. Subsequent investigations on more highly purified preparations (Lowey *et al.*, 1963) gave values of 1330 Å and 64:1, respectively. When combined with the data on molecular weight and helix content mentioned earlier these measurements were interpreted as indicating a two-strand α-helical rope structure for the paramyosin molecule in solution.

Very little work has been reported on the chain conformation of extracted paramyosin in the solid state. Beighton (1956) showed that oriented films of paramyosin gave a high-angle X-ray diffraction pattern of the α type and Loeb (1969) reported that the frequency of the amide I band in infrared spectra obtained from compressed monolayers of paramyosin was consistent with an α-helical conformation.

I. *In Vitro* Aggregates

A variety of fibrous aggregates have been obtained from paramyosin solutions under different conditions of precipitation. By variation of pH or ionic strength, regularly banded aggregates have been obtained with periods of 1400 Å (Hodge, 1952), 360 and 1800 Å (Locker and Schmitt,

1957), 145 and 725 Å (Hanson *et al.*, 1957), and 70, 145, 725, and 1800 Å (Hodge, 1959). All these patterns appear to have dihedral symmetry, which makes it difficult to interpret the mode of molecular packing, but all the observed periods appear to be multiples of about 72 Å (Table 15.8). In a number of instances the longer repeats were found to contain

TABLE 15.8

Comparison of Periods Observed in *In Vitro* Aggregates of Paramyosin

Observed period c (Å)	n	c/n (Å)	Ref.[a]
70	1	70	1
145	2	73	1, 2
360	5	72	3, 4
725	10	73	1, 4
1000	14	71	4
1400	20	70	5
1800	25	72	1, 3

[a] 1, Hodge (1959); 2, Hanson *et al.* (1957); 3, Locker and Schmitt (1957); 4, Cohen *et al.* (1971a); 5, Hodge (1952).

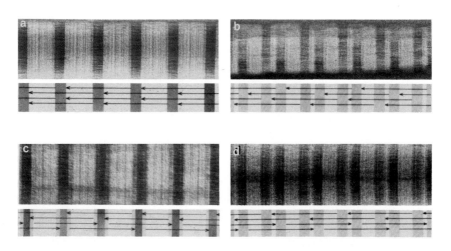

FIG. 15.21. Electron micrographs illustrating various forms of ordered aggregates obtained by precipitating paramyosin with divalent cations. (a) The PI form; (b) the PI(2/5) form; (c, d) forms with banding patterns having dihedral symmetry. Two dimensional representations of the molecular packing are given below each micrograph (Cohen *et al.*, 1971a).

regularly spaced subbands separated by a distance which was also a multiple of 72 Å.

Cohen *et al.* (1971a) obtained a series of aggregates by precipitation with divalent cations and two of these, when negatively stained, showed polar banding patterns with periods of 725 Å (Fig. 15.21). In addition, three types of pattern with dihedral symmetry were observed, also with periods of 725 Å. It was assumed that the observed patterns were determined by specific lateral interactions rather than by end-to-end aggregation and the darkly stained zones were identified with gaps in the assembly. The simpler of the two polar banding patterns, termed PI, had one dark zone and was interpreted in terms of a packing arrangement in which molecules 1275 Å long associated laterally with an axial stagger of 725 Å (Fig. 15.21a). The second polar banding pattern, termed PI(2/5) was interpreted as a superposition of two PI arrays displaced axially by two-fifths of the period (Fig. 15.21b). A frequently observed feature was the occurrence of a dihedral banding pattern in the central region of a paracrystal with a transition to oppositely directed PI forms

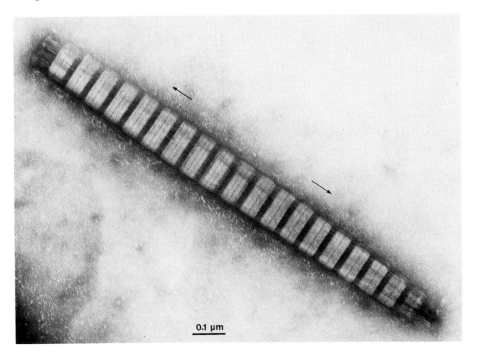

FIG. 15.22. Electron micrograph of bipolar paracrystal of paramyosin (Cohen *et al.*, 1971a). There is one period of the dihedral pattern shown in Fig. 15.21c at the center and a transition to oppositely directed polar PI forms (Fig. 15.21a) on each side.

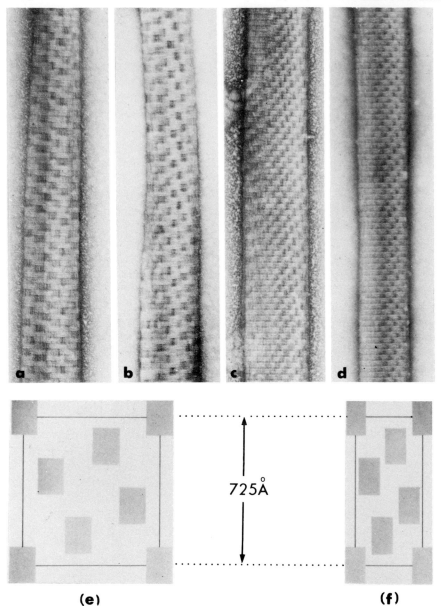

FIG. 15.23. Electron micrographs of synthetic and native paramyosin filaments. (a) Synthetic filament from *Atrina rigida* translucent muscle paramyosin; (b) synthetic filament from *Placopecten magellanicus* smooth muscle paramyosin; (c) native filament from *Mercenaria mercenaria* white muscle; (d) native filament from *Crassostrea viginica* translucent muscle; (e) unit cell of lattice in synthetic filament; (f) unit cell of lattice in native filament (Cohen *et al.*, 1971a).

on either side (Fig. 15.22). The three types of dihedral banding pattern observed could be accounted for in terms of the superposition of oppositely directed PI or PI(2/5) forms (Fig. 15.21c,d). The molecular length deduced from the banding pattern varied with species in the range 1250–1308 Å but the period remained sensibly constant.

A further type of aggregate obtained by Cohen *et al.* (1971a) exhibited an array of darkly stained "gap" regions arranged on a lattice similar, in many respects, to that observed in the native filament (Section II.a). In projection the gap regions appear to be arranged on a two-dimensional lattice, or net, which is conveniently specified in terms of a nonprimitive rectangular cell containing five gap regions (Fig. 15.23). The cell edge parallel to the fiber axis has a dimension of 725 Å and it was suggested that the pattern of gap regions could be generated by the assembly of subfilaments with a PI arrangement of molecules in which adjacent subfilaments were axially displaced by 290 Å. This displacement represents two-fifths of the period and so the net pattern and PI(2/5) forms are closely related.

II. PARAMYOSIN IN MUSCLE

a. *Electron Microscope Studies*

The thin filaments in the paramyosin-containing muscles of molluscs and annelids are very similar in form to the actin-containing filaments of striated muscle (Hanson and Lowy, 1963) but the thick filaments, which consist largely of paramyosin, are appreciably longer (up to 60 μm) and of greater diameter (200–1600 Å) than the thick filaments of striated muscle. Paramyosin-containing filaments have been isolated from a variety of molluscan and annelid muscles by mechanical disintegration and, when appropriately treated and stained or shadowed, exhibit, in projection, a two-dimensionally periodic structure as illustrated in Fig. 15.23 and 15.24 (Hall *et al.*, 1945; Hanson and Lowy, 1964b; Elliott and Lowy, 1970; Szent-Györgyi *et al.*, 1971; Cohen *et al.*, 1971a). The last-named authors have suggested that treatments that reveal this periodic structure remove a layer of myosin molecules from the surface of the filament.

In the preparations of Hall *et al.* (1945) the nonprimitive cell illustrated in Fig. 15.23 was found to have dimensions $a = 193$ Å and $c = 720$ Å, with $\beta = 90°$, and from the observed constancy of the lattice across the filaments it was assumed that they were flat ribbons rather than cylindrical structures with helical symmetry. Support for this view was obtained from the observation of lattices, in different filaments, which were mirror images of each other. This is to be expected for a planar net viewed from

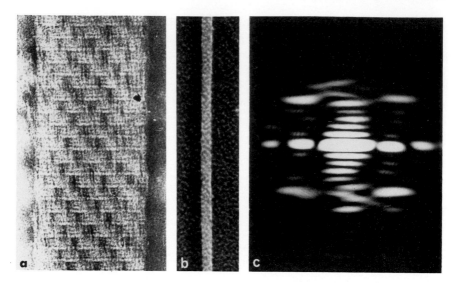

Fig. 15.24.   (a) Electron micrograph of negatively stained paramyosin filament from the adductor muscle of *Crassostrea angulata* showing surface lattice. (b) Electron micrograph of filament shadowed on both sides. (c) Optical diffraction pattern, which is essentially symmetric about a line parallel to the filament axis, from (b) (Elliott, 1971, with permission of the Royal Society, London).

opposite sides. Enantiomorphic images have been observed in other studies (Elliott, 1964b; Elliott and Lowy, 1970; Cohen *et al.*, 1971a) and the last-named authors drew attention to the possibility that the filament was in fact helical but that selective staining of the top surface occurred in some filaments and of the bottom surface in others. This would also lead to the appearance of enantiomorphic images (Finch and Holmes, 1967). Compelling evidence in favor of a helical arrangement of repeating units was obtained by Elliott (1971), who devised a means of shadowing both the front and rear surfaces of a filament. The optical diffraction pattern of the image (Fig. 15.24) was that expected for a helical structure. Elliott and Lowy (1970) observed an apparent change in the screw sense of a filament in a shadowed preparation but this was also accompanied by an appreciable change in the *a* dimension of the lattice.

The repeating unit, as revealed by negative staining, is asymmetric (Fig. 15.23) and Cohen *et al.* (1971a) observed that in a number of cases an abrupt transition occurred at a certain point along the length of the filament in the appearance of the repeating unit. The repeating units appeared to be oppositely directed with respect to the length of the filament on each side of the transition point, giving the filament a bipolar character. The screw sense of the helix remained the same along the entire length of the filament.

Elliott and Lowy (1970) measured the a dimension of the cell in negatively stained preparations and obtained a wide range of values. For filaments with diameters above 800 Å the a values were in the range 200–400 Å, but for smaller filament diameters the upper limit increased to about 660 Å. It was suggested that this increase might be due to distortion at the outer surface of a layer of finite thickness bent into a cylindrical shape.

The fine structure of paramyosin-containing filaments has also been studied by thin-sectioning of muscle (Hodge *et al.*, 1954; Elliott, 1964b). The latter author found a variety of banding patterns in longitudinal sections and, occasionally, a striated appearance in transverse section. However, some doubt has been cast (Elliott and Lowy, 1970) on the relevance of the latter observation to the internal structure of the filament.

b. *Low-Angle X-Ray Studies*

The low-angle X-ray diffraction patterns obtained from paramyosin-containing muscles (Fig. 15.25a) contain a series of meridional and near-meridional reflections which can be indexed in terms of a two-dimensional lattice of the type shown in Fig. 15.25b. For the opaque portion of the adductor of the clam *Venus mercenaria*, Bear and Selby (1956) determined the cell parameters for the dried muscle to be $a = 250 \pm 10$ Å, $c = 720 \pm 5$ Å, and $\beta = 90.5 \pm 0.2°$ and for rewetted muscle to be $a = 325 \pm 20$ Å, $c = 725 \pm 5$ Å, and β somewhat greater than in the dried muscle. G. F. Elliott (1963, 1964b) obtained values of $a = 420 \pm 20$ Å, $c = 720 \pm 5$ Å, and $\beta = 90°$ for specimens of the opaque adductor of the Portugese oyster *Crassostrea angulata* which had been fixed with formaldehyde, stained with phosphotungstic acid, and embedded in Araldite. Estimates of the cell parameters in different materials and in different states are summarized in Table 15.9.

The c dimension of the cell is relatively constant, decreasing by only about 1 % between wet and dry muscle, and in the wet state does not appear to depend upon the physiological state of the muscle (Elliott and Lowy, 1961; Millman and Elliott, 1965). The values quoted for a are based on measurements of the lateral coordinates of the off-meridional reflections and the assumption that the lattice is planar. If instead the lattice is assumed to be curved, the dimensions of the curved lattice can be related to the estimates of a in a simple manner (Bear and Selby, 1956; Elliott and Lowy, 1970). In contrast to the relative constancy of c, the value of a varies with species and is also sensitive to the water content and to the chemical and physical history of the specimen

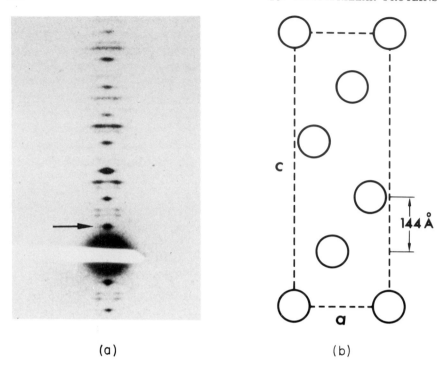

(a)   (b)

FIG. 15.25.   (a) X-ray diffraction pattern obtained from the opaque adductor muscle of *Crassostrea angulata* after fixation with formaldehyde and staining with phosphotungstic acid. The 144-Å meridional reflection is arrowed (Elliott, 1964). (b) Unit cell of two-dimensional surface lattice deduced from the distribution of the "paramyosin" reflections. By permission of the Royal Society, London.

(Table 15.9). Elliott and Lowy (1970) found that in the opaque adductor of *C. angulata* the *a* dimension varied in a continuous manner between a value of about 320 Å for acetone-dried muscle to about 400 Å for muscle in a 1:3 acetone–water mixture. The *a* dimensions, in the dried state, for three species listed in Table 15.9 are almost exactly in the ratio 3:4:5 (Elliott and Lowy, 1970) and appear to be multiples of an elementary unit with an equivalent planar width of about 82 Å in the dry state. In the wet state the *a* dimension increases by about 30% (Bear and Selby, 1956) so that the equivalent planar width of the elementary unit would be about 106 Å in the native state.

Analysis of the lateral widths of the low-angle meridional and near-meridional reflections led to the conclusion (Bear and Selby, 1956) that the native filaments were not ribbonlike, but no evidence could be found for a periodic organization in a direction normal to the observed lattice. Elliott and Lowy (1961) observed an equatorial reflection of spacing

TABLE 15.9

Dimensions of Equivalent Planar Net Cell in the Paramyosin-Containing Filaments of
Molluscan Muscles, Deduced from Low-Angle X-Ray Studies

| Species and muscle | Cell dimensions[a] | | | Ref.[b] |
	a (Å)	c (Å)	β (deg)	
Venus mercenaria				
Opaque adductor, dried	250	720	90.5	1
	250	—	—	2
rewetted	325	725	>90.5	1
Crassostrea angulata				
Opaque adductor, dried	340	—	—	3
	330	—	—	2
rewetted in acetone–water (1:3)	405	—	—	2
fixed and embedded	420	720	90	3
Mytilus edulis				
Posterior adductor, dried	420	—	—	2
	—	719–722	—	4
Anterior byssus retractor, dried	—	721	—	4
living	—	727	—	4

[a] See Fig. 15.25.
[b] 1, Bear and Selby (1956); 2, Elliott and Lowy (1970); 3, Elliott (1964b); 4, Elliott
(1968).

120 Å which was interpreted in terms of a filament composed of stacked
sheets, but this reflection was subsequently shown to be attributable to
arrays of the thin, actin-containing filaments (Lowy and Vibert, 1967).

c. *Models of Filament Structure*

A number of arguments have been advanced (Elliott, 1964b) in favor
of a filament structure consisting of a stack of ribbons so organized as to
give a three-dimensionally regular array, but the absence of reflections
associated with the dimension perpendicular to the plane of the ribbon,
and the results obtained by shadowing both sides of a filament (Section II.a), appear to preclude such a possibility. Elliott and Lowy (1970)
suggested an alternative model (Fig. 15.26) in which it was supposed
that a planar net with a two-dimensional lattice of the type envisaged
by Bear and Selby (1956) was rolled up to form a spiral cylindrical
lattice (Whittaker, 1955). Much of the observed data could be accounted
for by such a structure but so far no equatorial reflection attributable to

Axis of filament

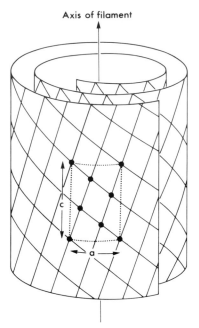

FIG. 15.26.   Model for the paramyosin filament suggested by Elliott and Lowy (1970). A two-dimensional lattice with a cell like that depicted in Fig. 15.25b is rolled up to form a spiral cylindrical lattice.

the intersheet spacing has been observed, nor has the free edge of the outermost portion of the rolled-up sheet been visualized by electron microscopy.

A detailed model for the arrangement of the individual molecules in filaments from *C. angulata* was put forward based on a structural unit consisting of a rectangular lamina formed from eight molecules. These were envisaged as aggregating side by side with a stagger of 145 Å. However, studies of *in vitro* aggregates, discussed in Section I, suggest that a stagger of 2 × 145 Å is the important factor in the formation of the native aggregate.

d. *High-Angle X-Ray Studies*

Paramyosin-rich molluscan muscles yield typical α patterns at high angles both in the dried state (Bear, 1944b) and in the living state (Astbury, 1947b) and this is generally attributed to the paramyosin component of the muscle. Cohen and Holmes (1963) examined the high-angle pattern of the anterior byssus retractor of *M. edulis* in some

detail and estimated the spacing of the prominent meridional reflection
in the 5-Å region to be 5.15 ± 0.06 Å in the wet state. Off-meridional
maxima on a layer line of spacing 5.25 Å were also observed (Fig. 15.27).
Elliott (1968) showed that the system of layer lines corresponding to
the axial periodicity of about 720 Å (Section E.II.b) extends meridionally
at least as far as the 5-Å region and in both living and dried muscle the
prominent reflection occurred on the layer line with $l = 142$. With the
cell shown in Fig. 15.25 meridional reflections only occur when l is a
multiple of five and so the strong arc is not a true meridional reflection.
The c dimension of the cell was determined to be 727 Å in the living
muscle and 721 Å in the dried muscle, giving spacings for the strong
layer line of 5.123 Å and 5.076 Å, respectively. An important implication

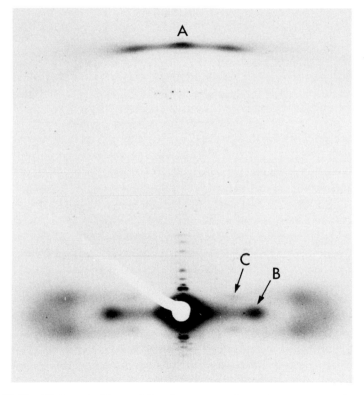

Fig. 15.27. High-angle X-ray diffraction pattern from the opaque portion of the
adductor muscle of *Crassostrea angulata* in 50% aqueous acetone. The specimen has
been tilted in order to show the detail around the 5.1-Å meridional reflection (A).
Equatorial reflection B has a spacing of 19.3 Å and reflection C is situated on a near-
equatorial layer line of spacing about 70 Å (Elliott and Lowy, 1970).

of these observations is that the lattice extends throughout the whole volume of the paramyosin filament and is not confined to a surface layer. A weak meridional arc occurs in the 1.5-Å region (Cohen and Holmes, 1963) and the spacing has been estimated to be 1.485 ± 0.005 Å (Elliott *et al.*, 1968a).

Bear (1944b) noted that the 10-Å equatorial reflection had a characteristic shape with "horns" directed toward the center of the pattern and subsequent studies have shown that this is due to the presence of intensity on a pair of near-equatorial layer lines (Fig. 15.27). The spacing of this layer line in patterns obtained from the anterior byssus retractor of *Mytilus edulis* in the native state was determined to be 89 ± 5 Å (Cohen and Holmes, 1963) and an essentially similar value was observed in the dried opaque adductor of *Crassostrea commercialis* (Fraser *et al.*, 1965e). Somewhat different values were given by Elliott *et al.* (1968a) for the opaque adductor of *Crassostrea angulata* soaked in sea water after fixation with glutaraldehyde (68–69 Å) or rewetted with acetone–water mixtures after drying (70 Å). Quantitative estimates of the diffracted

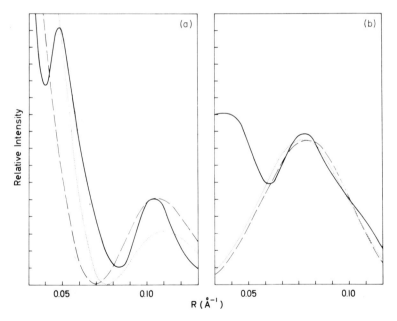

FIG. 15.28. Comparison of the observed equatorial intensity (a) and near-equatorial intensity (b) in live molluscan catch muscle (———) (after Cohen and Holmes, 1963) with the intensity transforms calculated for a two-strand coiled-coil rope structure with $r_0 = 5$ Å (···) and for a two-strand rope with $r_0 = 5.5$ Å surrounded by water (– – –) (Fraser *et al.*, 1965e).

intensity on the equator and the near-equatorial layer line have been reported for native muscle (Cohen and Holmes, 1963) and for dried muscle (Fraser *et al.*, 1965e) and the results are included in Figs. 15.28 and 15.29.

In native muscle there is an interference maximum on the equator with a spacing of about 20 Å (Cohen and Holmes, 1963) and when dried muscle is soaked in acetone–water mixtures interference maxima also appear on the near-equatorial layer line (Fig. 15.27). These can be interpreted in terms of a body-centered tetragonal packing arrangement (Elliott *et al.*, 1968a). The spacing of the equatorial interference maximum, corresponding to the lateral distance between molecules, varies from 17 to 20 Å depending on the composition of the mixture (Elliott and Lowy, 1970).

The half-width of the near-equatorial layer line in wet muscle was estimated by Cohen and Holmes (1963) to be 0.0067 Å^{-1}, implying a coherent length of about 150 Å. This is very much shorter than the length of the molecule (Section E.I) and the most likely explanation of the discrepancy is that the pitch of the coiled-coil, which is highly

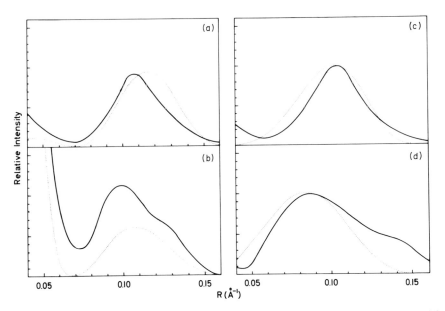

FIG. 15.29. Comparison of the observed equatorial intensity (a, b) and near-equatorial intensity (c, d) in the X-ray diffraction patterns of wet (b, d) and dry (a, c), ethanol-extracted adductor muscle from *Crassostrea commercialis* (——) with the intensity transforms calculated for two-strand ropes with $r_0 = 5.5$ Å and appropriate corrections for background (···) (Fraser *et al.*, 1965e).

dependent on the parameters of the undistorted α-helix (Section E.II.e), varies along the length of the molecule.

e. *The Coiled-Coil Model*

As discussed in Chapter 9, the α-helix model proposed by Pauling and Corey (1951a) accounts satisfactorily for the X-ray diffraction pattern of synthetic polypeptides in the α form but in its simplest form does not predict a meridional reflection of spacing about 5.15 Å as observed in the α pattern (Astbury and Woods, 1930, 1933) obtained from the so-called keratin-myosin-epidermin-fibrinogen (k-m-e-f) group of fibrous proteins (Astbury, 1947b, 1949b). Crick (1952, 1953a,b) and Pauling and Corey (1953b) independently suggested that the observed reflection could be accounted for if the axes of the α-helices in these materials were distorted so as to follow a long-pitch helix. The resulting polypeptide chain conformation was termed a coiled-coil.

The distortion was envisaged by Pauling and Corey (1953b) as resulting from a repeating sequence in which the individual residues formed hydrogen bonds of slightly different lengths. The origin of the distortion

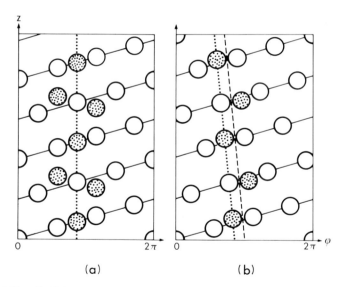

FIG. 15.30. Radial projections of α-helices with (a) 3.5 residues per turn and (b) 3.6 residues per turn. The dotted circles show the pattern of contacts formed by a second α-helix when brought into contact with first. The rotation and translation of the helices has been chosen for optimum "knob–hole" packing. In (b) "knob–hole" packing can be attained over a short length if the axis of the second helix is inclined to the axis of the first helix. The dotted and dashed lines represent two possibilities considered by Crick (1953b) for the line of contact between two coiled-coils in a two-strand rope.

was thus considered to be intrahelical. Crick (1952, 1953b), on the other hand, attributed the distortion to interhelical interactions.

The radial projection of an α-helix with a unit twist $t = 4\pi/7$, having exactly 3.5 residues per turn, is shown in Fig. 15.30 and if the sidechains are considered as "knobs," it can be seen that the pattern of knobs and holes on the surface of the helix would allow two such helices to be brought together, with their axes parallel, in such a manner that the knobs on one helix mesh with the holes in the other. Crick pointed out that the same type of knob–hole packing could be achieved with α-helices having other values of t if the axes of the helices were mutually inclined at an appropriate angle and coiled around each other (Fig. 15.31). The greater the departure of t from the value of $4\pi/7$, the greater is the tilt

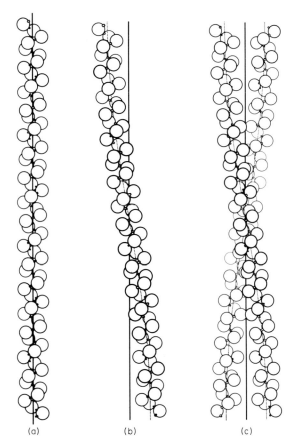

(a)   (b)   (c)

FIG. 15.31.   Diagrammatic representations of the distribution of residues in (a) an α-helix, (b) a coiled-coil, and (c) a two-strand rope.

required to maintain knob–hole packing and there is a simple relationship connecting $\Delta t = t - (4\pi/7)$, the unit height h, the radius of the coiled coil r_0, and its pitch P given by

$$P = (2\pi/\Delta t)[h^2 - (r_0\Delta t)^2]^{1/2} \tag{15.6}$$

In all known cases t is less than $4\pi/7$ and so Δt and P are negative. Equation (15.6) was formulated for right-handed α-helices and a negative value for P corresponds to a left-handed major helix. The dependence of P on h and t is illustrated in Fig. 15.32 and it can be seen that the value of P is very sensitive to the value of Δt.

It seems likely that the knobs on two chains forming a two-

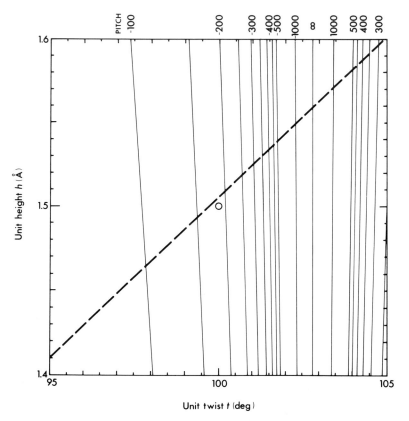

FIG. 15.32. Relationship connecting the unit height h and unit twist t of an α-helix and the pitch of the coiled-coil predicted from Eq. (15.6) with $r_0 = 5.2$ Å. The broken line corresponds to the fitted curve $t = 52.43h + 21.08$ (see Fig. 9.7) and the circle corresponds to an α-helix with $h = 1.5$ Å and 3.6 residues per turn ($t = 100°$).

strand coiled-coil rope would be at least quasiequivalent and Crick (1953b) suggested that the most probable packing arrangement would be that illustrated in Figs. 15.30 and 15.31 where the two sets of knobs are related by a dyad parallel to the axis of the major helix. When the main chain is taken into consideration it is necessary to distinguish between the parallel-chain, two-strand rope, in which the sequences $-NH-CHR-CO\rightarrow$ run in the same direction in both chains, and the antiparallel-chain, two-strand rope, in which they run in opposite directions. In the first arrangement it is possible to construct a model in which the main-chain atoms are related by a dyad parallel to the axis of the major helix, and in the second a model can be constructed in which the main-chain atoms are related by a dyad perpendicular to the axis of the major helix. In order to define the models more closely, it is necessary, in both cases, to specify the azimuthal relationship between the two chains. This will depend on where the effective "centers" of the knobs formed by side chains are located. As a first approximation, it may be assumed that if the axis of the major helix is coincident with the z axis of a set of Cartesian coordinates $Oxyz$ and the axis of an α-helix intercepts the x axis at r_0, then a C^β atom occupies the position $(r_0 + r_\beta, 0, 0)$, where r_β is the radial coordinate of C^β in the undistorted α-helix. If this atom is designated $C_0{}^\beta$, then the atoms in the "inside" of the rope will be $C_2{}^\beta$ and $C_5{}^\beta$, and $C_7{}^\beta$ will occupy an equivalent position to $C_0{}^\beta$ on the "outside" of the rope. In the parallel-chain, two-strand rope the second chain is generated by the operation of a dyad along Oz, and in the antiparallel-chain rope by a dyad along Ox.

Crick (1953a,b) has described mathematical transformations which produce a coiled-coil distortion of the α-helix and coordinates for a coiled-coil formed from the α-helix of Table 9.1 are given in Table 15.10. The distortion of individual bond lengths and interbond angles is small but it is important to realize that in an actual coiled-coil the distortion would not be uniformly distributed and it is likely that most of the distortion is accommodated by changes in the torsion angles ϕ and ψ and the interbond angle $\tau(C^\alpha)$ rather than by changes in bond lengths (Chapter 6). Crick (1952) estimated that the energy of deformation would be very small and Parry and Suzuki (1969a) confirmed this by carrying out potential energy calculations for the interactions in straight and deformed α-helices. In view of the fact that the precise manner in which the distortion is accommodated is unknown, the significance of the numerical values obtained is uncertain.

The next simplest type of rope is that formed by taking three α-helices which are quasiequivalently related with regard to the "knobs" formed by the side chains. A parallel-chain, three-strand rope in which the three

TABLE 15.10

Atomic Coordinates of the Repeating Unit in a Coiled-Coil with
$r_0 = 5.5$ Å and $P = 186$ Å[a]

Atom	x (Å)	y (Å)	z (Å)	r (Å)	ϕ (deg)
N_4	6.950	1.165	-4.451	7.047	9.52
H_4	6.892	1.559	-5.378	7.066	12.75
C_4^α	7.173	2.154	-3.378	7.489	16.71
H_4^α	8.176	2.015	-2.943	8.421	13.84
C_4'	6.118	1.964	-2.276	6.426	17.80
O_4	6.438	1.828	-1.079	6.692	15.85
C_4^β	7.054	3.565	-3.969	7.904	26.81
$H_4^{\beta 1}$	6.075	3.699	-4.382	7.113	31.34
$H_4^{\beta 2}$	7.216	4.284	-3.186	8.392	30.70
$H_4^{\beta 3}$	7.783	3.698	-4.753	8.617	25.41
N_5	4.877	1.961	-2.723	5.256	21.90
H_5	4.644	2.072	-3.686	5.085	24.04
C_5^α	3.708	1.794	-1.854	4.119	25.82
H_5^α	3.597	2.682	-1.218	4.487	36.71
C_5'	3.893	0.554	-0.989	3.932	8.10
O_5	3.760	0.619	0.231	3.811	9.35
C_5^β	2.451	1.625	-2.697	2.941	33.54
$H_5^{\beta 1}$	2.555	0.761	-3.313	2.666	16.59
$H_5^{\beta 2}$	1.600	1.504	-2.073	2.196	43.23
$H_5^{\beta 3}$	2.315	2.491	-3.296	3.401	47.10
N_6	4.187	-0.541	-1.659	4.222	352.64
H_6	4.277	-0.560	-2.647	4.314	352.54
C_6^α	4.405	-1.844	-1.021	4.775	337.28
H_6^α	3.453	-2.211	-0.617	4.100	327.37
C_6'	5.420	-1.692	0.110	5.678	342.67
O_6	5.155	-2.091	1.253	5.563	337.92
C_6^β	4.932	-2.842	-2.049	5.692	330.05
$H_6^{\beta 1}$	5.854	-2.484	-2.454	6.359	337.01
$H_6^{\beta 2}$	5.091	-3.791	-1.583	6.347	323.33
$H_6^{\beta 3}$	4.213	-2.950	-2.830	5.143	325.00
N_0	6.550	-1.113	-0.251	6.644	350.36
H_0	6.724	-0.801	-1.191	6.772	353.21
C_0^α	7.658	-0.869	0.694	7.707	353.53
H_0^α	8.104	-1.830	1.000	8.308	347.27
C_0'	7.120	-0.149	1.944	7.122	358.80
O_0	7.328	-0.588	3.097	7.352	355.41
C_0^β	8.718	-0.000	0.000	8.718	0.00
$H_0^{\beta 1}$	8.282	0.934	-0.296	8.334	6.43
$H_0^{\beta 2}$	9.524	0.177	0.695	9.526	1.06
$H_0^{\beta 3}$	9.085	-0.503	-0.885	9.099	356.83
N_1	6.433	0.944	1.667	6.502	8.35
H_1	6.275	1.272	0.731	6.403	11.46

TABLE 15.10 (*continued*)

Atom	x (Å)	y (Å)	z (Å)	r (Å)	ϕ (deg)
C_1^{α}	5.823	1.790	2.707	6.092	17.09
H_1^{α}	6.613	2.290	3.288	6.998	19.10
C_1'	4.974	0.922	3.638	5.059	10.50
O_1	5.125	0.962	4.864	5.215	10.63
C_1^{β}	4.932	2.842	2.049	5.692	29.95
$H_1^{\beta 1}$	4.160	2.355	1.494	4.780	29.51
$H_1^{\beta 2}$	4.488	3.458	2.801	5.666	37.61
$H_1^{\beta 3}$	5.519	3.450	1.393	6.509	32.01
N_2	4.096	0.163	3.015	4.099	2.28
H_2	3.990	0.154	2.030	3.993	2.21
C_2^{α}	3.178	−0.746	3.706	3.264	346.79
H_2^{α}	2.439	−0.157	4.263	2.444	356.32
C_2'	3.967	−1.631	4.660	4.289	337.65
O_2	3.639	−1.714	5.842	4.022	334.78
C_2^{β}	2.451	−1.625	2.697	2.941	326.46
$H_2^{\beta 1}$	3.165	−2.197	2.149	3.853	325.23
$H_2^{\beta 2}$	1.784	−2.286	3.195	2.900	307.97
$H_2^{\beta 3}$	1.891	−1.006	2.043	2.142	331.99
N_3	4.981	−2.266	4.107	5.472	335.54
H_3	5.210	−2.175	3.139	5.646	337.34
C_3^{α}	5.869	−3.169	4.857	6.670	331.63
H_3^{α}	5.310	−4.071	5.150	6.691	322.52
C_3'	6.374	−2.457	6.120	6.831	338.92
O_3	6.245	−2.973	7.250	6.917	334.54
C_3^{β}	7.054	−3.565	3.969	7.904	333.19
$H_3^{\beta 1}$	7.594	−2.686	3.677	8.055	340.52
$H_3^{\beta 2}$	7.700	−4.223	4.523	8.782	331.26
$H_3^{\beta 3}$	6.691	−4.061	3.085	7.827	328.74

[a] Parry and Suzuki (1969a). Coordinates refer to alanyl residues; equivalent units of seven residues at $(r, \phi - 20n, z + 10.333n)$, where n is an integer.

main chains are equivalent can be generated by the same procedure as that described earlier for the two-strand rope except that the three chains are generated by a triad rather than a dyad along Oz. Detailed descriptions of three-strand rope models have been given in connection with studies both of paramyosin and α-keratin (Crick, 1953b; Pauling and Corey, 1953b; Lang, 1956a; Fraser and MacRae, 1961b; Cohen and Holmes, 1963; Fraser *et al.*, 1964a, 1965g).

The helical sections of proteins belonging to the k-m-e-f group have a high proportion of negatively charged side chains (Tables 15.1, 15.7, and 16.2) and this is consistent with the adoption of an extended rather than a globular form in an aqueous environment (Fisher, 1964). The stability of a coiled-coil rope in such an environment is likely to derive from interactions between nonpolar residues in the interior of the rope (Crick, 1953b; Cohen, 1966a) and this leads, in the case of ropes in which the main chains are related by a dyad or a triad, to the prediction of a repeating pattern of residue types which may be represented symbolically by $[X-N-X-X-N-X-X]_n$, where N is a nonpolar residue and X is any residue. The terminal residue in the repeating sequence is on the outermost surface of the rope.

Parry and Suzuki (1969b) calculated the interaction energies for various coiled-coil rope models of poly(L-alanine) and found that they were generally more stable than the corresponding straight-chain assemblies. So far poly(L-alanine) has not been prepared in a coiled-coil conformation in the solid state and this stems from the fact that a three-dimensionally crystalline arrangement will generally be more stable, since coiled-coil ropes do not pack neatly together and so the packing density is very low. Coiled-coil ropes are unlikely to be favored in any polypeptide structure except when they occur in conjunction with appreciable amounts of liquid or other relatively unorganized material.

f. *Intensity Transform of a Coiled-Coil*

Crick (1953a,b) described a method for calculating the Fourier transform of a coiled-coil and this was generalized by Lang (1956a). Subsequently Pardon (1967) showed that certain of the assumptions made in these treatments were not strictly correct but it was considered that the errors involved in using Crick's method would be small since the distortion of the α-helix was relatively slight. An alternative method of calculating the Fourier transform, which is not subject to these uncertainties, depends on the fact that the coiled-coil contains a repeating unit of seven residues which is arranged with helical symmetry (Cohen and Holmes, 1963; Fraser *et al.*, 1964a, 1965e). Exact calculations can therefore be carried out using the formula given in Eq. (1.29) for a simple discontinuous helix.

The calculation of the cylindrically averaged intensity transform predicted for a coiled-coil rope in a fibrous protein specimen presents a number of difficulties which have not so far been completely resolved. The problems are similar to those encountered with tests of collagen models (Chapter 14, Section D.IV.b). The appearance of strong reflections at high angles is due to the presence of a helical array of a pseudo-repeat containing seven residues and the difficulties stem from the fact that the side chain is variable. The C^β atom probably occupies a fixed position relative to the main-chain atoms but the number and positions of the remaining atoms are variable and constitute a background of electron density which must be allowed for. In hydrated specimens the contribution from the water molecules must also be included. Methods of approximating the contribution from backgrounds have been discussed in Chapter 1, Section D.III and the application of such methods to paramyosin models has been described in detail by Fraser *et al.* (1965e). To a first approximation the α-helix appears, in projection, as a main chain

$$[\mathrm{NH-\overset{|}{C}-CO-}]_n$$

which excludes water (if present) and surrounding side chains from a cylindrical volume of radius $r_\mathrm{b} \simeq 3.3$ Å (Fig. 15.33). It can be shown

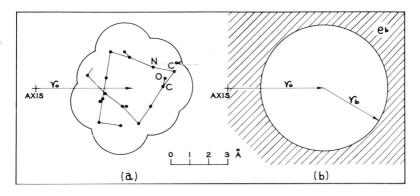

Fig. 15.33. Approximation used to represent a coiled-coil α-helix in an aqueous environment. (a) Axial projection of a short length of the distorted α-helix main chain; (b) continuum with empty volume element (unshaded) which approximates the volume occupied by the main chain. The electron density is represented as the sum of (a) and (b) (Fraser *et al.*, 1965e).

that with this representation of the electron density the cylindrically averaged intensity distribution in the near-equatorial region for a coiled-coil in water is given by

$$\langle I(R, n/P)\rangle_\psi = A^2 + B^2 \tag{15.7}$$

where

$$A = \sum_j f_j J_n(2\pi R r_j) \cos[n(-\phi_j + 2\pi z_j/P)] - \gamma N_\mathrm{e} J_n(2\pi R r_0) \, J_1(2\pi R r_\mathrm{b})/\pi R r_\mathrm{b}$$

$$\tag{15.8}$$

and

$$B = \sum_j f_j J_n(2\pi R r_j) \sin[n(-\phi_j + 2\pi z_j/P)] \qquad (15.9)$$

In these formulas the symbols are as the same as those used in Chapter 1, Section A.III.d. Also P is the pitch of the major helix, γ is the ratio of the electron density of the background to the mean electron density in the volume occupied by the main chain, N_e is the number of electrons in the asymmetric unit, and the summation extends over one asymmetric unit consisting of seven

$$[\text{NH}-\overset{|}{\underset{|}{\text{C}}}-\text{CO}-]$$

groupings. The value of γ was estimated to be 0.6.

In regions remote from the equator the contribution from the background will be less important and to a first approximation the transform can be calculated on the basis of an asymmetric unit consisting of seven [NH—CH(C)—CO] groupings with no allowance for background.

The nature of the intensity transform from a coiled-coil was discussed in a qualitative way by Crick (1953b) and studies of optical diffraction patterns confirmed his conclusions (Fraser and MacRae, 1961a,b; Fraser et al., 1962b). Intensity transforms have also been calculated for a number of coiled-coil models, generally without background correction (Lang, 1956b; Cohen and Holmes, 1963; Fraser et al., 1964a,e, 1965e, 1971b; Holmes and Blow, 1965). Strong reflections are predicted in the 1.5- and 5-Å regions of the meridian and in the 10-Å region of the equator. To this extent the coiled-coil model provides a satisfactory explanation of the α pattern given by the k-m-e-f group of fibrous proteins.

In the case of an infinite coiled-coil with perfect helical symmetry the diffraction pattern would be confined to a set of layer lines with Z coordinates given by

$$Z(m, n) = (m/h) + (n/P) \qquad (15.10)$$

where P is the pitch of the coiled coil, h is the height of the seven-residue asymmetric unit ($\sim$10.3 Å), and m and n are integers (Chapter 1, Section A.III.b). The regions of strong diffraction (Fig. 15.34) correspond to $m = 0$ (equator and near-equatorial layer lines), $m = 2$ (5-Å region of the meridian), and $m = 7$ (1.5-Å region of the meridian). Since $P \gg h$, each region is split into a closely spaced set of layer lines corresponding to $n = 0, \pm 1, \pm 2$, etc. When coiled-coils are combined to form ropes the distribution of intensity is modified by interference and expressions for the modified intensity transforms have been given by Fraser *et al.* (1964a). When a parallel $\mathcal{N}$-fold rotation axis is present the intensity on layer lines for which n is not a multiple of $\mathcal{N}$ is reduced to zero (Chapter 1, Section A.III.c). In ropes with mixed chain directions this is also true to a good approximation for $m = 0$ (equatorial and near-equatorial regions) and so the spacing of the first near-equatorial layer line, in the absence of complicating factors, can be taken to be equal to the pitch of the coiled-coil divided by the number of strands in the rope. The effect of mixed chain directions on the rest of the transform is more complicated and systematic absences do not occur (Fig. 15.34).

Potentially, the transforms of the various types of rope are sufficiently different for the form of the rope in a particular material to be determined from the diffraction in the 5-Å region of the meridian, but in actual structures various factors operate to complicate this procedure. These relate to departures from perfect helical symmetry and to the modulation of the intensity transform due to regularity in the disposition of the coiled-coil ropes associated with higher levels of structural organization (Fraser *et al.*, 1964b).

g. *Tests of the Coiled-Coil Model for Paramyosin*

Cohen and Holmes (1963) collected quantitative intensity data for the equator and for the near-equatorial layer line in the pattern obtained from the anterior byssus retractor of *Mytilus edulis* in the wet state and these data were corrected for actin scattering and for overlap. The corrected observed intensity (Fig. 15.28) was compared with the intensity distribution calculated for two-strand and three-strand ropes with various assumptions about the side-chain contributions. Of the models tested, the best fit was obtained for a two-strand rope with $r_0 = 5$ Å and contributions from the main-chain atoms and a C^β atom. The contributions of the remaining side-chain atoms and water were neglected. Fraser *et al.* (1965e) showed that a marginally better fit could be obtained (Fig. 15.28) if these contributions were taken into account and the value of r_0 increased to 5.5 Å. Exact comparison of the observed and calculated intensities was complicated in both cases by the presence of an interference function with a maximum at a spacing of 20 Å, and probably also at 10 Å.

Data from both wet and dried opaque adductor muscle of the oyster *Crassostrea commercialis* were collected by Fraser *et al.* (1965e) and compared with the intensity transform for a two-strand rope (Fig. 15.29). Allowance for background was made and it was found that reasonable agreement was obtained for a value of $r_0 = 5.5$ Å. In particular the

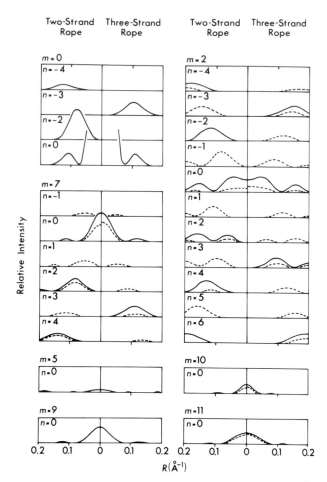

Fig. 15.34.   Intensity transforms calculated for coiled-coil α-helix ropes. The group of layer lines with $m = 0$ corresponds to the equator and near-equatorial region. The layer line with $m = 2$, $n = 0$ is the 5.15-Å meridional reflection, and the group of layer lines with $m = 7$ corresponds to the 1.5 Å meridional arc. Full lines, chain directions parallel; broken lines, mixed chain directions (Fraser *et al.*, 1964a).

observed shift in the R coordinate of the maximum on the near-equatorial layer line on drying (Fig. 15.29) was correctly reproduced by the different background corrections applied for wet and dried specimens.

The spacing of the near-equatorial layer line in the wet muscle was estimated to be 89 ± 5 Å (Cohen and Holmes, 1963), suggesting a value of 178 ± 10 Å for the pitch of the coiled-coil. In the dried muscle the pitch appeared to be very similar (Fraser *et al.*, 1965e). Elliott *et al.*

(1968a) measured the layer-line spacing in the pattern obtained from the opaque adductor of *C. angulata* and obtained values of 68–69 Å for the wet state and 70 Å for acetone-dried material. These would lead to values of 134 and 140 Å for the pitch of the coiled-coil in this material. The origin of the difference between these and the values obtained with other muscles is not clear but could be due to the effect of an overlying interference function which was periodic in *Z*. The molecular packing giving rise to this interference function may be related to the stagger, by multiples of about 70 Å, which is observed with *in vitro* aggregates (Table 15.8).

The results obtained in the 1.5- and 5-Å regions of the meridian have already been described in Section E.II.d. It can be shown (Cohen and Holmes, 1963) that in an *N*-strand rope the number of residues per turn in the undistorted α-helix ($= 360/t$, where *t* is measured in degrees) is related to the spacings $d(0, N)$ and $d(2, 0)$ of the layer lines with $m = 0$, $n = N$ and $m = 2$, $n = 0$, respectively, by the expression

$$360/t = 3.5Nd(0, N)/[Nd(0, N) - d(2, 0)] \qquad (15.11)$$

Using the values $d(0, N) = 89$ Å, $d(2, 0) = 5.15$ Å, and $N = 2$ determined by Cohen and Holmes (1963), this gives a value of $t = 99.9°$, which can be compared with the value of $t = 99.6°$ in poly(L-alanine) (Table 9.3). This calculation is based on the assumption that the spacings of the layer lines are not appreciably affected by interference functions due to larger-scale organization. The evidence obtained by Elliott *et al.* (1968a) indicates that such functions exert a powerful influence under certain conditions and it is uncertain how literally the results obtained by applying Eq. (15.11) can be interpreted.

In the 5-Å region of the meridian the intensity in the transform of a coiled-coil rope is distributed on a series of closely spaced layer lines and the precise distribution differs for the parallel-chain and antiparallel-chain two-strand ropes (Fig. 15.34). The observed pattern, however, did not correspond to either of these possibilities (Cohen and Holmes, 1963) and various explanations were considered including a type of screw disorder in which the line of contact (Fig. 15.30) was variable.

The spacings of the layer lines with $m = 2$, $n = 0$ and $m = 7$, $n = 0$ should, from Eq. (15.10), be in the ratio 7:2 but Elliott *et al.* (1968a) have pointed out that their measured values of 5.084–5.157 Å and 1.485 ± 0.005 Å, respectively, for these two spacings are not exactly in this ratio. It was suggested that the side chains might have a different symmetry to the main chain. Since atoms at a large radius make an important contribution to the intensity in the 5-Å region and a minor contribution to the intensity in the 1.5-Å region (Crick, 1953a, Fraser and MacRae, 1961b), this explanation is a plausible one. Crick (1953a) pointed out that if the relationship between the two coiled-coils is not symmetric, this can lead to a departure from the simple 7:2 ratio and this is a further possibility.

In summary, the two-strand coiled-coil rope model for paramyosin appears to be capable of accounting for the main features of the high-angle X-ray diffraction pattern but departures from the idealized model prevent the determination of precise helix parameters or the determination of the parallel or antiparallel nature of the two chains. Evidence obtained from electron microscope studies of *in vitro* aggregates of paramyosin (Section E.I) indicate that the molecule is polar, and if the two chains have similar sequences this would imply that the parallel-chain two-strand rope model is appropriate. In the event that the sequences are not identical, the polarity does not imply a particular choice of model (Cohen *et al.*, 1971a).

Chapter 16

Keratins

A. INTRODUCTION

Keratin is a proteinaceous material found in the outer layers of the epidermis of vertebrates and keratin-containing tissues are typically unreactive toward the natural environment and also mechanically strong and durable. It has long been recognized that keratin is not a single material but a complex of sulfur-containing proteins that is stabilized by disulfide linkages. No precise definition of keratin exists at present (Rudall, 1968b, Mercer and Matoltsy, 1969) and in this chapter the term will be used simply to denote an insoluble complex of sulfur-containing epidermal proteins. A distinction has been drawn between the so-called soft keratin found in stratum corneum, corns, callouses, and the eponychium of nails, and the hard keratin found in hair, nails, claws, beaks, horns, and quills (Giroud *et al.*, 1934). Although the classification is based on tactile sensation, the two varieties of keratin contain different types of proteins and are produced by different modes of biosynthesis (Giroud and Leblond, 1951; Fraser *et al.*, 1972). Typically, soft keratin contains less sulfur than does hard keratin (Table 16.1) and is produced in desquamating tissue.

Keratins and keratin-containing tissues have also been classified on the basis of their high-angle X-ray diffraction patterns (Rudall, 1947; Fraser *et al.*, 1972). Mammalian hard keratins generally give an α pattern (Fig. 16.1a), while avian and reptilian hard keratins give a quite different pattern which is conveniently referred to as the "feather pattern" (Figure 16.1c). Soft keratin generally gives a poorly developed

469

TABLE 16.1

Classification and Origin of Proteins Isolated from Mammalian Keratins[a]

	Hard keratins (hair, nail, horn, hoof, claw)			Soft keratins (stratum corneum)	
	Low-sulfur proteins	High-sulfur proteins	Glycine-rich proteins	Lower-sulfur proteins	Higher-sulfur proteins
Proportion (%)	40–85	5–45	1–30	~93[b]	~7[b]
Chain weight	~50,000	10–30,000	5–10,000	~60,000	20–40,000[c]
% Cys residues	4	10–30	5–11	1	2
% Gly residues	3	3–9	20–40	14	17
Helix content	~50%	0	0	~45%	—
Origin	Microfibrils	Matrix	Matrix	Filaments	Unknown

[a] The authors are indebted to Dr. J. M. Gillespie for assistance in the preparation of this table.

[b] Proportion of extract (O'Donnell, 1971).

[c] Value determined for the precursor fraction (O'Donnell, 1971).

α pattern and certain hard keratins, such as the cuticle of hair, give a pattern consisting of two broad halos with maxima at spacings of about 4.5 and 9.5 Å (Fig. 16.1d), indicating an absence of regular secondary structure. A fourth pattern, not so far observed in native keratins, is obtained when keratins yielding the α pattern are mechanically deformed (Fig. 16.1b); this has been termed the β pattern. The terms α-keratin, β-keratin, feather keratin, and amorphous keratin are used to denote classification on the basis of the X-ray diffraction pattern.

Keratins are produced in specialized cells termed keratinocytes and the cysteine residues in the newly synthesized proteins exist in the thiol form. During the process of keratinization the thiol groups are oxidized in pairs leading to the formation of the diamino acid cystine. Although no direct evidence has been obtained for the formation of intermolecular disulfide linkages, it is generally believed that such linkages are largely responsible for the natural insolubility of keratinized tissues and their resistance to enzymatic proteolysis. The task of preparing and purifying soluble derivatives of keratin proteins has proved to be one of great complexity and reviews of this topic have been given by Crewther et al. (1965) and Fraser et al. (1972). The most widely studied type of derivative has been that prepared by reduction of the disulfide linkages followed by alkylation with iodoacetic acid to give an S-carboxymethyl derivative. In some instances this derivative has been prepared directly from the newly synthesized proteins which, as mentioned earlier, occur in the thiol form (Rogers, 1959a; Downes *et al.*, 1966; Frater, 1966; I. E. B. Fraser, 1969).

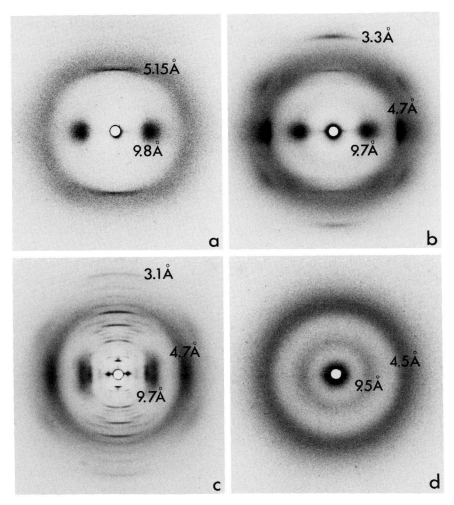

FIG. 16.1.   Classification of keratins on the basis of their high-angle X-ray diffraction patterns: (a) the α pattern obtained from porcupine quill; (b) the β pattern obtained from stretched porcupine quill; (c) the "feather pattern" obtained from seagull feather rachis; (d) the "amorphous pattern" obtained from keratins that are devoid of regular secondary structure. From Fraser, R. D. B., MacRae, T. P., and Rogers, G. E., "Keratins: Their Composition, Structure and Biosynthesis," 1972. Courtesy of Charles C. Thomas, Publisher, Springfield, Illinois, U.S.A.

B. MAMMALIAN HARD KERATIN

I. CHEMICAL AND PHYSICOCHEMICAL STUDIES

The S-carboxymethyl protein derivatives isolated from mammalian hard keratins can be classified into three groups on the basis of amino acid composition, conformation, and structural origin. The native keratin consists of filaments, termed microfibrils, embedded in a ground substance or matrix (Section B.IV) and many lines of evidence suggest that the microfibrils are ordered aggregates of a group of proteins that have a lower sulfur content than the whole keratin (Birbeck and Mercer, 1957; Rogers, 1959b,c; Bendit, 1968b; Swift, 1969; Fraser *et al.*, 1972). These so-called low-sulfur proteins were first isolated by Goddard and Michaelis (1935). The matrix contains two distinct groups of proteins, one characterized by having a higher sulfur content than whole keratin (high-sulfur proteins) and the other by a high content of glycyl residues (glycine-rich proteins). All the glycine-rich proteins so far studied contain cystine and also an unusually high proportion of aromatic residues, particularly tyrosine (Table 16.3). Comprehensive accounts of the preparation and properties of the three groups of proteins have been given by Crewther *et al.* (1965) and Fraser *et al.* (1972).

a. *Low-Sulfur Proteins*

The low-sulfur proteins isolated from different keratins appear to be very similar and to comprise two subgroups, termed components 7 and 8, which are present in a ratio of about 2:1 (Thompson and O'Donnell, 1965). The molecular weights are, respectively, about 53,000 and 45,000 (Jeffrey, 1972) and the helix contents in solution at pH 9, estimated from the optical rotatory dispersion parameter b_0 , are about 56 and 62% (Crewther and Dowling, 1971). Oriented films and fibers of low-sulfur protein derivatives have been prepared and shown to give an α pattern when examined by X-ray diffraction (Happey, 1950) but so far ordered aggregates of the type obtained with most other fibrous proteins have not been observed.

Crewther *et al.* (1968) studied the properties of solutions containing mixtures of components 7 and 8 and found that both the helix contents and the resistance to enzymatic hydrolysis of the mixtures were greater than those of the pure components. Maximum enhancement of these properties occurred at a molar concentration of approximately 2:1 and on this basis it was suggested that a specific aggregate was formed containing two molecules of component 7 and one of component 8. The molecular weight of the aggregate would be predicted to be 151,000 but its existence has not so far been demonstrated directly by hydrodynamic or other measurements.

A helix-rich fragment has been isolated from unfractionated preparations of the S-carboxymethyl derivatives of the low-sulfur proteins after partial hydrolysis with enzymes (Crewther and Harrap, 1967). Hydrodynamic and electron microscope studies suggested that the fragment was rodlike and about 170 Å in length (Crewther et al., 1968, 1973). The helix content of the fragment depends upon the method of preparation but has been estimated to be in the range 77–90% (Crewther and Harrap, 1967; Crewther and Dowling, 1971).

Oriented films of the helix-rich fragment have been prepared from formic acid solution and found to give an α pattern and a strong meridional arc of spacing 1.49 Å when examined by X-ray diffraction (Suzuki et al., 1973). A sharp low-angle meridional pattern was also observed (Fig. 16.2) indicating the formation of ordered aggregates of extended length. The reflections can be indexed on a periodicity of 160 Å, which

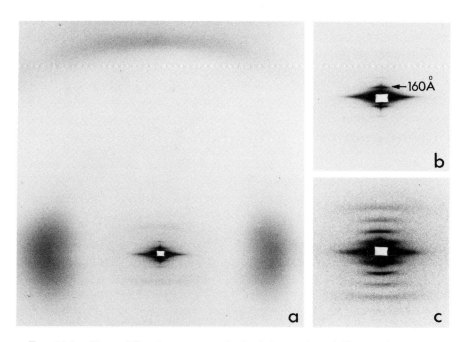

FIG. 16.2.   X-ray diffraction patterns obtained from oriented films of the helix-rich fragment obtained by partial proteolysis of the low-sulfur protein fraction: (a, b) untreated film; (c) film treated with osmium tetroxide. The system of layer lines at low angles corresponds to an axial periodicity of 160 Å in the arrangement of the fragments (Suzuki et al., 1973).

is close to the value estimated by Crewther *et al.* (1968, 1973) for the length of the particle. The films also exhibit infrared dichroism (Fig. 16.3) and the frequencies and dichroic character of the amide

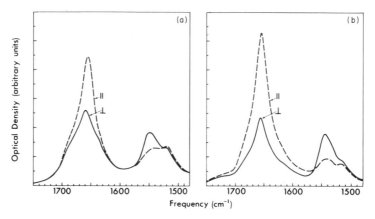

FIG. 16.3.   Amide I and II bands in the infrared spectrum of (a) a thin section of porcupine quill; (b) an oriented film of the helix-rich fragment from the low-sulfur proteins of wool keratin. Spectra were measured with the electric vector vibrating perpendicular (full line) and parallel (broken line) to the fiber axis (Suzuki *et al.*, 1973).

bands indicate that a high proportion of the material consists of α-helices preferentially oriented parallel to the direction of stroking.

The helix-rich fragment obtained from chymotryptic digests of the low-sulfur proteins has been shown to be dissociable into subfragments, termed ChB and ChC, with molecular weights of about 25,000 and 12,500, respectively. The amino acid compositions of the subfragments are very similar (Table 16.2) and comparison with the compositions of components 7 and 8 shows that the helix-favoring residues Glu and Leu are more abundant in the helix-rich material while the non-α-helix-favoring residues Pro, Gly, Ser, and Cys are less abundant. On the basis of chemical and physicochemical studies of subfragments ChB and ChC, Crewther and Dowling (1971) have suggested that the low-sulfur proteins contain three sections of α-helix around 128 Å in length separated by nonhelical sections. It has also been suggested that in keratin two molecules of component 7 and one molecule of component 8 aggregate laterally with the helical sections aligned to give segments of three-strand rope (Crewther *et al.*, 1968; Crewther and Dowling, 1971). The helix-rich fragments are identified with the segments of three-strand rope.

O'Donnell (1969) investigated the possibility of sequencing a fragment of molecular weight about 6600, containing about 60 residues, obtained

TABLE 16.2

Amino Acid Analyses of Purified Fractions of the Low-Sulfur Proteins from
Wool and of Subfragments Present in Chymotryptic Digests[a]

Amino acid	Low-sulfur proteins[b]		Chymotryptic fragments[c]	
	Component 7	Component 8	ChB	ChC
Alanine	7.8	6.0	8.1	8.3
Arginine	7.5	7.8	7.6	6.3
Aspartic acid[d]	9.0	10.8	10.4	10.6
Glutamic acid[e]	15.6	18.2	21.9	21.4
Glycine	7.4	4.2	2.9	3.2
Half-cystine[f]	6.2	6.0	3.0	3.5
Histidine	0.5	0.6	1.1	0.8
Isoleucine	3.9	3.8	4.3	4.0
Leucine	9.9	11.8	14.0	12.7
Lysine	4.4	3.1	5.8	6.2
Methionine	0.6	0.3	0	Trace
Phenylalanine	2.4	2.1	2.2	1.8
Proline	2.9	3.7	1.1	1.4
Serine	8.3	7.6	5.5	6.3
Threonine	4.4	5.3	3.2	4.0
Tyrosine	2.9	2.6	3.4	2.7
Valine	6.4	6.3	5.8	6.6

[a] Given in residues/100 residues, rounded to nearest 0.1 residue.
[b] Thompson and O'Donnell (1965).
[c] Crewther and Dowling (1971).
[d] Includes asparagine.
[e] Includes glutamine.
[f] Content of half-cystine plus cysteine in native material.

by cyanogen bromide cleavage from the N-terminus of component 8. Several examples of substitutions were found in the sequence near the acetylated N-terminus, indicating that the material was not homogeneous. The data obtained were combined with earlier results (Corfield *et al.*, 1968; Hosken *et al.*, 1968) to give the partial provisional sequence shown in Fig. 16.4. The distribution of residues is nonuniform and the first 50 positions contain disproportionately large amounts of Cys, Pro, Gly, and Ser and presumably constitutes part of the nonhelical material that is removed during the preparation of the helix-rich fragment. The positions 51–76 contain no Cys or Pro and are enriched in Glu and Leu, suggesting that they form part of an α-helical segment.

```
      1    Pro        Leu   5                        10
  Acetyl-Ser- Phe- Asp- Phe- Cys- Leu- Pro- Asp- Leu- Ser-
             Tyr
      Phe- Arg ————————————————————— (Cys₇, Ser₆,
                             Lys
      Asx₃, Pro₃, Ala₂, Thr₂, Val₂, Arg, Glx, Leu, Ile, Gly₃.₄,
                                  Glu      50
      Phe₀.₆) ———————————— Phe- Cys- Gly- Gly- Phe-
                       Ser                      60
      — (Asx₂, Glx, Gly, Thr) —Lys- Glu- Thr- Met- Gln-
      61                   65                   70
      Phe- Leu- Asp- Asp- Arg- Leu- Ala- Ser- Tyr- Leu-
      71               75                ~397
      Glu- Lys- Val- Arg- Glu- Leu- (———————————)
```

FIG. 16.4. Partial provisional sequence of component 8 of the low-sulfur proteins from wool, based on the studies of Corfield *et al.* (1968), Hosken *et al.* (1968), and O'Donnell (1969). Glx, glutamine or glutamic acid; Asx, asparagine or aspartic acid; Cys, *S*-carboxymethyl cysteine. Substitutions occur at several positions, indicating that component 8 is not a pure protein.

b. *High-Sulfur Proteins*

Although there is some evidence of microheterogeneity and species to species variation in composition in the low-sulfur proteins, they are a relatively constant feature of keratin compared with the high-sulfur proteins. These proteins constitute an extremely heterogeneous group with regard to molecular weight and composition (Gillespie, 1963; Joubert *et al.*, 1968; Darskus *et al.*, 1969; Swart *et al.*, 1969; Haylett *et al.*, 1971) and variations in type and proportion have been reported from species to species (Darskus and Gillespie, 1971) and with changes in nutritional status (Gillespie *et al.*, 1964; Gillespie, 1965; Lindley *et al.*, 1971).

The molecular weights of the high-sulfur proteins range from about 10,000 to 28,000 and some correlation between molecular weight and per cent cystine content has been noted (Lindley *et al.*, 1970; Gillespie and Broad, 1972). The S-carboxymethyl derivatives of the high-sulfur proteins appear to be devoid of α-helix in aqueous solution (Gillespie and Harrap, 1963) and this can be correlated with the high contents of Pro and Cys (Table 16.3). The wide range of composition is illustrated by the two examples given in Table 16.3 and no obvious relationship exists with the composition of the low-sulfur proteins (Table 16.2). When allowance is made for amide content, the high-sulfur proteins are found, in their native state, to be basic in character, whereas the low-sulfur proteins are found to be acidic (Crewther and Dowling, 1960).

TABLE 16.3

Amino Acid Analyses of Purified Fractions of the High-Sulfur and
Glycine-Rich Proteins from Keratins[a]

Amino acid	High-sulfur protein fractions		Glycine-rich protein fractions	
	Wool III B2[b]	Wool UHS[c]	Porcupine quill type I[d]	Echidna quill type II[d]
Alanine	3.2	2.0	1.0	0.8
Arginine	4.0	7.4	3.9	3.8
Aspartic acid[e]	8.3	0.8	3.8	2.9
Glutamic acid[f]	4.5	8.2	1.3	0.7
Glycine	4.9	4.9	36.2	40.1
Half-cystine[g]	16.2	28.3	1.4	9.1
Histidine	2.0	1.1	0.3	0.2
Isoleucine	4.0	1.8	0.3	0.1
Leucine	7.4	1.6	6.7	5.3
Lysine	1.0	0.8	0.3	0.3
Methionine	0	0	0	0
Phenylalanine	3.0	0.5	8.1	5.0
Proline	13.5	13.7	3.8	3.9
Serine	9.4	12.7	6.8	5.3
Threonine	10.4	10.4	1.6	1.2
Tyrosine	1.9	1.7	22.5	18.9
Valine	6.2	4.2	1.4	0.9

[a] Given in residues/100 residues, rounded to nearest 0.1 residue.
[b] Swart et al. (1969).
[c] Lindley et al. (1971).
[d] Gillespie (1972).
[e] Includes asparagine.
[f] Includes glutamine.
[g] Content of half-cystine plus cysteine in native material.

The sequences of several high-sulfur proteins have been determined (Haylett and Swart, 1969; Elleman, 1971, 1972a,b; Haylett et al., 1971; Elleman and Dopheide, 1972) and two examples are given in Figs. 16.5 and 16.6. There are few similarities between the two proteins apart from the N-terminal sequence Acetyl-Ala-Cys-Cys- and the frequency with which like residues occur in pairs. Lindley and Haylett (1967) have drawn attention to the fact that Cys-Cys sequences are a common feature of hard mammalian keratin but are not found in feather keratin. The sequence given in Fig. 16.5 shows little evidence of any systematic distribution of residues but in the sequence given in Fig. 16.6 there is a

```
         1               5                  10
2 Acetyl-Ala-  Cys- Cys- Ala-  Pro- Arg- Cys-Cys- Ser- Val-
3 Acetyl-Ala-  Cys- Cys- Ala-  Arg- Leu- Cys-Cys- Ser- Val-
4 Acetyl-Ala-  Cys- Cys- Ala-  Arg- Leu- Cys-Cys- Ser- Val-

         11              15                 20
         Arg- Thr- Gly- Pro-  Ala- Thr- Thr- Ile-  Cys- Ser-
         Pro- Thr- Ser- Pro-  Ala- Thr- Thr- Ile-  Cys- Ser-
         Pro- Thr- Ser- Pro-  Ala- Thr- Thr- Ile-  Cys- Ser-

         21              25                 30
         Ser- Asp- Lys- Phe-  Cys- Arg- Cys- Gly- Val-  Cys-
         Ser- Asp- Lys- Phe-  Cys- Arg- Cys- Gly- Val-  Cys-
         Ser- Asp- Lys- Phe-  Cys- Arg- Cys- Gly- Val-  Cys-

         31              35                 40
         Leu-Pro- Ser-  Thr- Cys- Pro-  His- Asn- Ile-  Ser-
         Leu-Pro- Ser-  Thr- Cys- Pro-  His- Thr- Val-  Trp-
         Leu-Pro- Ser-  Thr- Cys- Pro-  His- Thr- Val-  Trp-

         41              45                 50
         Leu- Leu- Gln- Pro-  Thr- Cys- Cys ——— Asp- Asn-
         Leu- Leu- Gln- Pro-  Thr- Cys- Cys- Cys- Asp- Asn-
         Phe- Leu- Gln- Pro-  Thr- Cys- Cys- Cys- Asp- Asn-

         51              55                 60
         Ser- Pro- Val- Pro-  Cys- Val- Tyr-Pro- Asp- Thr-
         Arg- Pro- Pro- Pro-  Tyr- His- Val- Pro- Gln- Pro-
         Arg- Pro- Pro- Pro-  Cys- His- Ile-  Pro- Gln- Pro-

         61              65                 70
         Tyr- Val- Pro-  Thr- Cys- Phe-  Leu- Leu- Asn- Ser-
         Ser- Val- Pro-  Thr- Cys- Phe-  Leu- Leu- Asn- Ser-
         Ser- Val- Pro-  Thr- Cys- Phe-  Leu- Leu- Asn- Ser-

         71              75                 80
         Ser- His- Pro-  Thr- Pro-  Gly- Leu- Ser-  Gly- Ile-
         Ser- Gln- Pro-  Thr- Pro-  Gly- Leu- Glu- Ser-  Ile-
         Ser- Gln- Pro-  Thr- Pro-  Gly- Leu- Glu- Ser-  Ile-

         81              85                 90
         Asn- Leu- Thr- Thr- Phe-  Ile-  Gln- Pro-  Gly- Cys-
         Asn- Leu- Thr- Thr- Tyr-  Thr- Gln- Ser-  Ser-  Cys-
         Asn- Leu- Thr- Thr- Tyr-  Thr- Gln- Pro-  Ser-  Cys-

         91              95
         Glu- Asn- Val-  Cys- Glu-Pro-  Arg- Cys-
         Glu ——— Pro-  Cys- Ile-  Pro-  Ser- Cys- Cys-
         Glu ——— Pro-  Cys- Ile-  Pro-  Ser- Cys- Cys-
```

FIG. 16.5.   Amino acid sequences of the purified high-sulfur wool proteins SCMKB-IIIB 2, 3, and 4 (Haylett and Swart, 1969; Haylett *et al.*, 1971); Cys, *S*-carboxymethyl cysteine.

```
            1                  5                      10
Acetyl-Ala-  Cys- Cys- Ser-  Thr- Ser- Phe- Cys- Gly- Phe-
           11                 15                      20
Pro-  Ile- Cys- Ser-  Thr- Ala- Gly- Thr- Cys- Gly-
           21                 25                      30
Ser-  Ser- Cys- Cys-  Arg- Ser- Thr- Cys- Ser- Gln-
           31                 35                      40
Thr- Ser- Cys- Cys-  Gln- Pro- Thr- Ser- Ile- Gln-
           41                 45                      50
Thr- Ser- Cys- Cys-  Gln- Pro- Thr- Cys- Leu- Gln-
           51                 55                      60
Thr- Ser- Gly- Cys-  Glu- Thr- Gly- Cys- Gly- Ile-
           61            Ile  65                      70
Gly- Gly- Ser-  Thr- Gly- Tyr- Gly- Gln- Val- Gly-
           71                 75                      80
Ser-  Ser- Gly- Ala-  Val- Ser- Ser- Arg- Thr- Arg-
           81                 85                      90
Trp- Cys- Arg- Pro-  Asp- Cys- Arg- Val- Glu- Gly-
           91
Thr- Ser- Leu-
                95                   100
           Pro-  Pro- Cys- Cys-  Val- Val- Ser- Cys- Thr- Ser-
               105                  110
           Pro-  Ser- Cys- Cys-  Gln- Leu- Tyr- Tyr- Ala- Gln-
               115                  120
           Ala-  Ser- Cys- Cys-  Arg- Pro- Ser- Tyr- Cys- Gly-
               125                  130
           Gln-  Ser- Cys- Cys-  Arg- Pro- Ala- Cys-
                                   135
                        Cys- Cys- Gln- Pro- Thr- Cys- Thr- Glu-
               140                  145
           Pro-  Val- Cys- Glu- Pro- Thr- Cys- Ser- Gln-
               150
           Pro-  Ile- Cys-
```

FIG. 16.6.   Amino acid sequence of the purified high-sulfur wool protein SCMK-B2C aligned to show repitition of a decapeptide (italics) (Elleman, 1972a).

repetition, with some replacements and deletions, of a decapeptide sequence Thr-Ser-Cys-Cys-Gln-Pro-Thr-Cys-Ser-Gln in much of the molecule. There is an internal section from 59–73 that contains seven Gly residues and no Cys residues and it has been suggested (Elleman, 1972a) that this region may form a flexible link between the two cystine-rich sections, which were visualized as forming two compact, globular units.

The cuticle of mammalian hair contains a keratin which gives an amorphous, rather than an α-type, diffraction pattern (Woods, 1938; Fraser *et al.*, 1971b) and it is believed that the constituent proteins resemble the high-sulfur proteins present in the matrix of α-keratin (Mercer, 1961; Birbeck, 1964; Asquith and Parkinson, 1966; Bradbury *et al.*, 1970).

c. *Glycine-Rich Proteins*

The proportion of glycine-rich proteins in mammalian hard keratins is highly variable, ranging from about 1% in horn and nail to about 30% in echidna quill (Gillespie, 1972), and as with the high-sulfur proteins, the compositions and molecular weights vary appreciably even among the proteins obtained from a single keratin. The extent of the variation in composition is illustrated in Table 16.3 by the analyses obtained by Gillespie (1972) for a so-called type I fraction from porcupine quill and a type II fraction from echidna quill. In both cases Gly + Tyr accounts for about 60% of the residues. Other data on the composition of glycine-rich proteins have been given by Zahn and Biela (1968a,b) and Gillespie and Darskus (1971). As might be expected from the high content of glycyl residues, these proteins are devoid of α-helix in aqueous solution (Harrap, 1970).

The sequence of a glycine-rich protein with 62 residues has been

```
1              5                    10                    15
Ser- Tyr- Cys- Phe- Ser- Ser- Thr- Val- Phe- Pro- Gly- Cys- Tyr- Trp- Gly-
16             20                   25                    30
Ser- Tyr- Gly- Tyr- Pro- Leu- Gly- Gly- Tyr- Ser- Val- Gly- Cys- Gly- Tyr-
31             35                   40                    45
Gly- Ser- Thr- Tyr- Ser- Pro- Val- Gly- Tyr- Gly- Phe- Gly- Tyr- Gly- Tyr-
46             50                   55                    60
Asp- Gly- Gly- Ser- Ala- Phe- Gly- Cys- Arg- Arg- Phe- Trp- Pro- Phe- Ala-
61
Leu- Tyr-
```

FIG. 16.7. Amino acid sequence of a purified glycine-rich protein from wool (Dopheide, 1973).

determined by Dopheide (1973) and is given in Fig. 16.7. Little regularity is evident in the distribution of residues apart from the frequency with which Gly alternates with other types of residue. The sequences -Gly-Cys-Gly-Tyr-Gly-Ser- and Gly-Tyr-Gly-Phe-Gly-Thr-Gly-Tyr- together account for almost one-quarter of the molecule. It may be recalled that the alternation of Gly with other types of residue is a feature of the sequence in the crystalline regions of the arthropod silks belonging to group 1 (Chapter 13, Section B.V.a). Infrared spectra obtained from films of the glycine-rich protein, cast from dichloroacetic acid, show an amide I band with components at frequencies associated with a random conformation and with the antiparallel-chain, pleated-sheet conformation, the former component being the more prominent (Suzuki, 1972).

In keratins that contain an appreciable proportion of glycine-rich protein it seems likely that the major part is accommodated in the matrix (Fraser *et al.*, 1973) but the possibility cannot be excluded that a small part originates from the cell membrane complex, which is modified during keratinization by the deposition of an intercellular layer (De Deurwaerder *et al.*, 1964; Andrews *et al.*, 1966; Bradbury *et al.*, 1966).

II. Physical Studies of Chain Conformation

a. *Infrared Dichroism*

In early studies of infrared dichroism in α-keratins it was found that the amide A and I bands exhibited parallel dichroism with respect to the fiber axis, and it was concluded that the main-chain $C=O$ and $N-H$ bonds were preferentially oriented parallel to the molecular axis (Ambrose and Hanby, 1949; Ambrose *et al.*, 1949b; Ambrose and Elliott, 1951b; Beer *et al.*, 1959). Quantitative analysis of the spectra was complicated by band overlap and by uncertainties regarding the transition moment directions associated with the amide bands. Parker (1955, 1956) studied the changes which occurred in the spectrum when the specimen was exposed to deuterium oxide vapor and found that as deuteration of the NH and OH groups proceeded, the intensity of the absorption at about 3300 cm^{-1} decreased in the expected manner. However, the dichroic ratio of the residual band increased and it was concluded that the low dichroic ratio in the native material was due, in part, to the presence of nondichroic absorption associated with side chains and with the amide groups of the amorphous material. In a later study Fraser and Suzuki (1964) found that in some specimens the dichroic ratio of the amide A band attained a value as high as 5.5 after partial deuteration (Fig. 16.8). Bendit (1962, 1966b) studied the progress of the deuteration process in some detail and found that exchange commenced in the oriented regions before it was complete in the unoriented regions. Thus the dichroic ratio of 5.5 mentioned earlier represents a minimum value for the amide A band of the oriented material.

The spectrum of porcupine quill tip in the region of the amide I and II bands is shown in Fig. 16.3 and the frequencies of the main dichroic

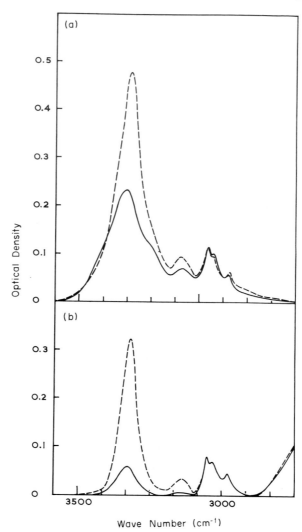

Fig. 16.8.   (a) Infrared spectra of a thin section from porcupine quill measured with the electric vector vibrating perpendicular (full line) and parallel (broken line) to the fiber axis. (b) Corresponding spectra obtained after exposing the section to deuterium oxide vapor for 16 hr (Fraser and Suzuki, 1964).

components are close to those predicted by Miyazawa and Blout (1961) for the $\nu_\parallel(0)$ and $\nu_\perp(100°)$ modes of the α-helix (Chapter 9, Section C.I). Bendit (1966a) carried out a detailed study of the effects of moisture content on the infrared spectrum of horse hair and noted small changes in the frequencies of the amide A, I, and II bands. When the relative humidity was changed from 0 to 93% some increase in the dichroism of the amide A absorption was observed and this was attributed to an increase in the orientation of the α-helices upon hydration. A small amount of dichroism was observed in the broad water band centered at 3400 cm^{-1} and attributed to a preferred orientation of the water molecules in the specimen. However, the effect was slight and may have been due to frequency-dependent form dichroism.

As discussed in Chapter 8, there is no satisfactory method available at present for determining the α-helix content of anisotropic solids and, in view of the conclusions reached by Bendit (1966b) regarding the deuteration process in α-keratin, methods for estimating α-helix content based on deuterium exchange (Jordan and Speakman, 1965) or tritium exchange (Leach *et al.*, 1964) must also be regarded as inherently unsatisfactory. In view of these difficulties Fraser *et al.* (1971b) attempted to obtain an estimate of the content of oriented α-helices from quantitative measurements of dichroism in the amide I region of the infrared spectrum. The problem is complicated by the fact that the amide I components of the α-helix and of unordered material absorb at approximately the same frequency (Miyazawa and Blout, 1961) and cannot readily be separated by the methods outlined in Chapter 5, Section C. However, if certain conditions regarding band overlap are satisfied, it is possible to estimate the fraction of oriented α-helices f_α from the integrated intensities A_π and A_σ of the amide I band, measured for radiation vibrating, respectively, parallel and perpendicular to the fiber axis, from the relationship

$$f_\alpha = (A_\pi - A_\sigma)/[f(A_\pi + 2A_\sigma)(1 - \tfrac{3}{2}\sin^2\beta)(1 - \tfrac{3}{2}\sin^2\gamma)] \qquad (16.1)$$

where f is the orientation parameter defined in Chapter 5, Section B.II.b, β is the inclination of the transition moment direction of the amide I mode to the axis of the α-helix for an unperturbed amide group, and γ is the inclination of the α-helix axis to the coiled-coil axis. Before Eq. (16.1) can be applied in practice the measured values of the integrated areas must be corrected for contributions from side-chain bands.

When applied to the spectrum illustrated in Fig. 16.3a this method yielded a value of $ff_\alpha = 0.191$ assuming $\beta = 33.9°$ and $\gamma = 12.6°$. The value of β was calculated from the atomic coordinates for the α-helix given in Table 9.1 and the transition moment data given by Suzuki (1967). The value of γ is appropriate to a coiled-coil with $P = 155$ Å and $r_0 = 5.5$ Å.

Corrections for absorption due to acidic and amide side chains were made on the basis of the known amino acid composition and the spectra of model compounds. The value of the orientation parameter f was estimated from the azimuthal distribution of intensity in the equatorial low-angle X-ray diffraction pattern according to the procedure set out in Chapter 5, Section B.II.d. The azimuthal width at half-height was estimated to be 9.4°, giving a value of $f = 0.825$ (Fig. 5.6b) and so the estimated value of the fraction of oriented α-helical material in quill was $f_\alpha = 0.232$.

A crude estimate of the proportion of oriented α-helical material present in the microfibrils in α-keratin can be obtained by combining the value of f_α obtained from the infrared spectrum with estimates of the proportions of different keratin proteins in porcupine quill (Fraser et al., 1973). Of a total extract of 84%, it was found that 31% consisted of high-sulfur plus glycine-rich proteins, so that f for the microfibril would be $0.232 \times 100/(84 - 31) = 0.44$, assuming that no low-sulfur protein remains in the residue. This is somewhat lower than the helix contents of the S-carboxymethyl derivatives in solution, as determined from measurements of optical rotatory dispersion (Section B.I.a).

The dichroism of the combination bands in the infrared spectrum of α-keratin has also been studied in some detail (Ambrose and Elliott, 1951a; Elliott, 1952a; Elliott et al., 1954; Fraser, 1955, 1956b; Fraser and MacRae, 1958b,c, 1959c; Bendit et al., 1959). The interpretation of dichroism in spectra obtained from fiber bundles is complicated by the presence of form dichroism but this can be corrected empirically (Fraser, 1956c) and the residual dichroisms of the amide combination bands are then found to resemble those observed in the α form of $[Ala]_n$ (Elliott et al., 1954). Of particular interest is the presence of an additional dichroic component in α-keratin at a frequency of about 5060 cm^{-1} (Fig. 16.9), which has been shown to be associated with primary amide groups (Elliott, 1952a,b; Fraser, 1955). The character of the observed dichroism suggests that the $-CO \cdot NH_2$ groups of glutamine and asparagine are preferentially oriented so that the line joining the two hydrogen atoms tends to be perpendicular to the fiber axis (Fraser, 1955).

The prominent band at 4860 cm^{-1} in Fig. 16.9 has been shown to be due to a combination of the amide A and amide II modes (Glatt and Ellis, 1948), and the intensity decreases in the expected manner when the N—H groups in the specimen are converted to N—D groups by exposure to deuterium oxide. In a quantitative study of the reaction (Fraser and MacRae, 1958b, 1959c) it was shown that about 30% of the peptide linkages were not readily accessible for exchange, but a precise estimate could not be made due to errors of unknown magnitude associated with the presence of weak underlying absorption of different origin.

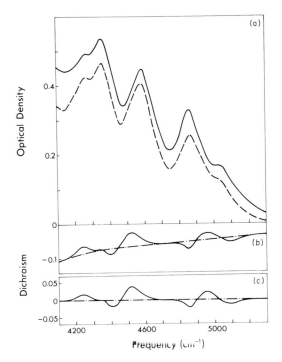

FIG. 16.9. (a) Combination bands in infrared spectra of a bundle of wool fibers obtained with the electric vector vibrating perpendicular (full line) and parallel (broken line) to the fiber axis; (b) the observed dichroism (equal to the difference between parallel and perpendicular spectra) and the estimated form dichroism (– · –); (c) the intrinsic dichroism (Fraser and MacRae, 1958b).

b. *High-Angle X-Ray Pattern*

Herzog and Jancke (1921) showed that α-keratin gave a high-angle X-ray diffraction pattern containing discrete reflections and a detailed study of the pattern was carried out by Astbury and co-workers (Astbury and Woods, 1930, 1933; Astbury and Street, 1931). Eleven reflections were recorded and indexed on an orthogonal cell with a side of length 10.3 Å parallel to the fiber axis and sides of length 9.8 and 27 Å perpendicular to the fiber axis. The prefix α was introduced to distinguish this pattern from the quite different pattern (β pattern) given by stretched specimens. The distinguishing features of the two types of pattern are illustrated in Fig. 16.1.

Hair consists of a cortex, composed of spindle-shaped cells, which is surrounded by a cuticle composed of flattened cells. Astbury and Street (1931) found that the α pattern remained after removal of the cuticle

and Woods (1938) showed that oriented films of cortical cells released
by partial proteolysis also gave an α pattern. More recently this has
been confirmed by observations on individual cortical cells using electron
diffraction (Fraser *et al.*, 1971b). Similar studies of cuticle cells show no
evidence of an α pattern.

Further studies (MacArthur, 1943; Perutz, 1951a; Astbury, 1953;

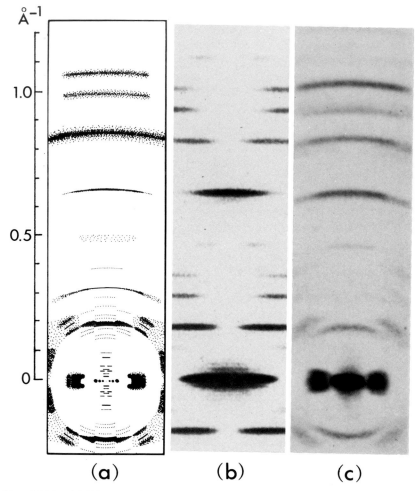

(a)   (b)   (c)

Fɪɢ. 16.10.   (a) Diagram illustrating the overall distribution of intensity in the X-ray
diffraction pattern of α-keratin. (b) Optical diffraction pattern obtained from a mask
representing the distribution of atoms in an α-helix with alanyl residues. (c) Optical
diffraction pattern obtained from a mask representing a 66-Å length of a three-strand,
parallel-chain α-helix rope with $r_0 = 5.2$ Å, oscillated $\pm 10°$ (Fraser and MacRae,
1961b; Fraser, 1969).

Astbury and Haggith, 1953; Fraser and MacRae, 1961a,b; Fraser *et al.*, 1964e) extended the number and range of observed X-ray data and the principal features are summarized in Fig. 16.10a. A number of quantitative measurements of the intensity transform have been reported: Lang (1956b) mapped the complete transform out to spacings of 1.33 Å (Fig. 16.11) and the equatorial region has been investigated in detail

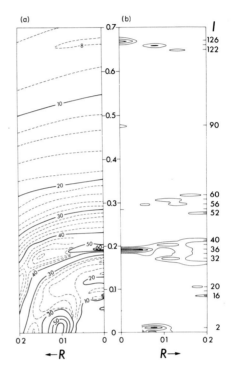

FIG. 16.11. Comparison of (a) the intensity distribution in the high-angle X-ray diffraction pattern of porcupine quill tip (Lang 1956b) with (b) the cylindrically averaged intensity transform calculated for a 100-Å length of two-strand, parallel-chain α-helix rope with $r_0 = 5.5$ Å (Fraser and MacRae, 1969).

both at low resolution (Bendit, 1957, 1960; Skertchly and Woods, 1960; Heidemann and Halboth, 1970) and at high resolution (Fraser *et al.*, 1964e; 1965e; 1971b). The results obtained in the high-resolution studies are illustrated in Fig. 16.12.

III. MODELS FOR CHAIN CONFORMATION

a. *Early Suggestions*

Attempts to devise a model conformation that would account for the high-angle X-ray diffraction pattern obtained from α-keratin have provided a stimulus to conformational studies of proteins for more than forty years. Astbury and Woods (1933) suggested, on the basis of their studies

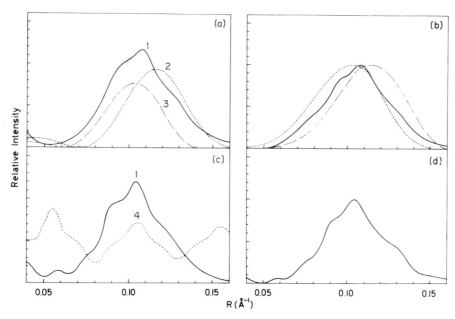

FIG. 16.12. X-ray diffraction pattern of porcupine quill tip (curve 1); (a) equator, dry; (b) near-equatorial layer line, dry; (c) equator, wet; (d) near-equatorial layer line, wet. Curve 2 is the intensity calculated for a two-strand rope with $r_0 = 5.5$ Å; curve 3 is the intensity calculated for a three-strand rope with $r_0 = 6.25$ Å. In both cases a correction has been made for background material. Curve 4 [in(c)] is an inter-coiled-coil rope interference function (Fraser *et al.*, 1965e).

of the $\alpha \rightarrow \beta$ transformation, that the polypeptide chain in α-keratin was folded so that three residues were accommodated in an axial distance of 5.1 Å. The axial length per residue in the β form was estimated to be 3.4 Å so that the molecular extensibility was $100(3 \times 3.4 - 5.1)/5.1 = 100\%$, and this agreed exactly with the macroscopic extensibility. However, some doubt has been cast on the assumption that the transformation is intramolecular (Bendit, 1957, 1960) and the possibility must be considered that sections of chain without regular secondary structure in the native material contribute to the β pattern in the stretched state. The so-called Astbury structure for α-keratin was further elaborated (Astbury and Bell, 1941) but as new data accumulated on the stereochemistry of the polypeptide chain and on the general principles of polymer structure, it became apparent that this model was untenable. The search for an alternative model, culminating in the proposals of a coiled-coil α-helix, has been well documented (Kendrew, 1954; Dickerson, 1964) and the evolution of thought leading to this model can be traced through the contributions of Astbury and Street (1931), Taylor (1941), Huggins (1943), Bragg *et al.* (1950), Pauling *et al.* (1951), and Crick (1952).

b. *The Coiled-Coil Model*

A detailed description of the coiled-coil α-helix model for the chain conformation in the k-m-e-f group of proteins has been given in Chapter 15, Section E.II.e and in the present section the extent to which this model accounts for the observed diffraction and spectroscopic data obtained from α-keratin will be examined.

According to Eq. (15.10), an infinite coiled-coil should give rise to a series of meridional reflections that are orders of a periodicity of about 10.3 Å, corresponding to the length of the seven-residue repeating unit. As mentioned earlier, Astbury and Street (1931) indexed the high-angle diffraction pattern on an axial repeat of this magnitude and identified maxima in the intensity transform corresponding to $m = 1, 2, 3$, and 4. The last two maxima were also observed by Astbury and Haggith (1953). The strong meridional arc of spacing about 1.49 Å (MacArthur, 1943; Perutz, 1951b; Fraser *et al.*, 1964e) occurs close to the position predicted for $m = 7$ and meridional scatter in the vicinity of the positions predicted for $m = 5, 9$, and 11 has also been reported (Fraser and MacRae, 1961a,b). When the pattern is examined at high resolution the situation is found to be more complex than these low-resolution studies would indicate, due to the operation of interference functions associated with higher levels of organization.

The spacings of the meridional reflections corresponding to $m = 2$ and $m = 7$ should be in the ratio 3.5:1 but the measured values of 5.165 and 1.485 Å (Fraser *et al.*, 1964e) give a somewhat different ratio. A similar discrepancy exists in the pattern obtained from paramyosin (Chapter 15, Section E.II.g). It was suggested (Fraser *et al.*, 1964e) that the reflection at 1.485 Å may be a superposition of meridional ($n = 0$) and near-meridional components ($n \neq 0$) of the $m = 7$ branch of the pattern (Fig. 15.34), in which case the observed maximum would occur at a larger spacing than predicted. Other factors which could destroy a precise ratio include the possibility that the rope strands are not equivalent (Crick, 1953b) and the likelihood that meridional reflections are restricted to orders of 470 Å (Fraser and MacRae, 1971, 1973).

According to the treatment of diffraction by coiled-coil α-helix ropes given by Crick (1953a), equatorial and near-equatorial layer-line streaks separated by $\Delta Z = N/P_c$ should be observed, where Z is the reciprocal space coordinate parallel to the meridian, P_c is the pitch of the coiled-coil, and N is the number of strands in the rope. These layer lines are part of the branch of the helix pattern corresponding to $m = 0$ in

Eq. (15.10). In α-keratin the observed high-angle equatorial region is only very poorly separated into layer lines (Figs. 16.1 and 16.11) and this has been attributed to a combination of the effects of disorientation and limited coherent length of rope (Fraser and MacRae, 1961b; Fraser *et al.*, 1964e). The coherent length, estimated from the half-width, is around 50–70 Å but this cannot be equated directly with the actual length since the presence of any distortion from the idealized form would reduce the coherent length. The poor separation of the layer lines makes the estimation of the spacing of the near-equatorial layer line more difficult than in the case of paramyosin (Fig. 15.27) or honeybee silk (Fig. 13.1b). The apparent spacing in α-keratin varies with distance from the meridian (Fraser *et al.*, 1964e) and has been estimated to be in the range 70–85 Å (Fraser *et al.*, 1965e). The implied range of pitch values is 140–170 Å for a two-strand rope and 210–255 Å for a three-strand rope. In common with other features of the transform attributable to coiled-coil ropes, it is likely that the precise values of these spacings are determined by the macroperiod of the microfibril (Section B.V.c).

Since the pitch of a coiled-coil is very sensitive to the parameters of the α-helix from which it is formed (Chapter 15, Section E.II.e), the spacing of the near-equatorial layer line does not provide a reliable indication of the number of strands in a coiled-coil rope. General studies of the transform (Crick, 1953b; Lang, 1956b; Fraser and MacRae, 1961a,b; Fraser *et al.*, 1964e) indicated that many features predicted for a coiled-coil model were present in the diffraction pattern, and the three-strand model was generally favored although no very compelling evidence was found for the rejection of a two-strand model. It would appear to be possible, on the basis of the calculated transforms depicted in Fig. 15.34, to determine both the number of strands in the rope and the chain arrangement from the observed pattern in the 5.15-Å region of the meridian. No quantitative data are available but a preliminary mapping of the features in this region was reported by Fraser *et al.* (1964e) and is reproduced in Fig. 16.13. None of the calculated transforms agrees with the observed transform and it is clear that the region is overlaid by an interference function associated with higher levels of organization and that there is a rapid falloff in intensity away from the meridian. As with paramyosin (Chapter 15, Section E.II.g), this is likely to be associated with departures from the idealized coiled-coil geometry.

It is possible, in principle, to determine the number of strands in the rope from the distribution of intensity on the equatorial and near-equatorial layer lines (Chapter 15, Section E.II.g). Quantitative intensity data for this region have been collected for porcupine quills from various species (Lang, 1956b; Fraser *et al.*, 1964d,e, 1965e, 1971b) and when

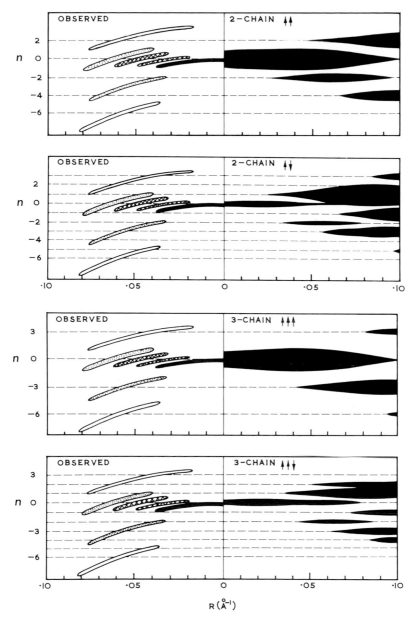

FIG. 16.13. Comparison of observed intensity distribution in the 5-Å meridional region of the X-ray pattern of α-keratin with cylindrically averaged intensity transforms of various coiled-coil α-helix rope models (Fraser *et al.*, 1964e).

the results obtained with dried specimens (Fig. 16.12) were compared with the intensity transforms calculated for two-strand and three-strand ropes (Fig. 15.34) it was noted that somewhat better agreement was obtained for a three-strand-rope model with a major helix radius of about 5.2 Å (Fraser *et al.*, 1964d,e). In a further study (Fraser *et al.*, 1965e) it was found, however, that when allowance is made for the presence of background material in a solid specimen the differences between the transforms of two-strand and three-strand ropes are less marked, particularly in dry specimens (Fig. 16.14). Taken together with the unknown modification introduced by higher levels of organization (Section B.V), the differences between the calculated transforms are insufficient to readily determine which model is appropriate to α-keratin (Fig. 16.12).

According to the calculated transforms shown in Fig. 16.14, the differences between two-strand and three-strand ropes should be emphasized if the rope is surrounded by water rather than dried protein and this effect was confirmed by measurements on wet and dried oyster muscle (Fig. 15.29). In the case of α-keratin the observed changes are much smaller (Fraser *et al.*, 1965e) and this may be due to the restricted entry of water into the microfibril.

In summary, the measurements that have been carried out so far on the near-equatorial layer line do not enable a definite decision to be made about the number of strands in the coiled-coil ropes but the general similarity between the distributions of intensity in porcupine quill and in tropomyosin, myosin, and paramyosin (Fig. 15.6), all of which contain two-strand ropes, would seem to favor a two-strand-rope model for α-keratin. If this is in fact the case, some explanation would need to be found for the apparent conflict with the evidence for three-strand ropes obtained by Crewther and co-workers (Section B.I.a). This evidence is based mainly on estimates of the molecular weights and helix contents of fragments and subfragments of the low-sulfur proteins and involves the assumption that the helix contents of molecular fragments equals, in aggregate, the helix content of the original molecule. Although a three-strand-rope model would account satisfactorily for many aspects of the chemical and physicochemical data, it is doubtful whether this is sufficiently precise at present to rule out the possibility of a two-strand-rope model for the native molecule. In order to reconcile the data with a two-strand-rope model, one possibility would be to suppose that each low-sulfur protein molecule contained two 160-Å lengths of α-helix which formed either intramolecular or intermolecular two-strand ropes. This would account for about 50% of the molecule, in agreement with the observed helix content, and the subfragments B and

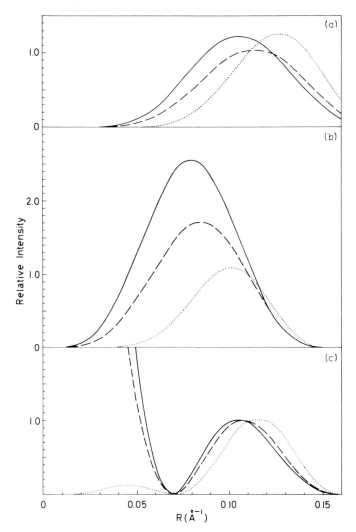

FIG. 16.14.   Intensities calculated for an isolated coiled-coil rope (———) compared with
those calculated for ropes surrounded by water (– – –) and by dried protein (···). (a) Near-
equatorial layer line, three-strand rope; (b) near-equatorial layer line, two-strand rope;
(c) equator, all ropes (Fraser et al., 1965e).

C would correspond, respectively, to two or one of the 160-Å lengths
of α-helix.

   The infrared spectrum of an α-helix distorted into a coiled-coil would
not be expected to differ greatly from that of a straight α-helix provided

that the distortion was not too great. However, the dichroic ratios predicted for the amide bands would be modified by the tilt of the α-helix relative to the axis of the coiled-coil. The predicted dichroic ratio for the amide A band of an α-helix depends upon the transition moment direction in the main-chain amide group and the orientation of the main-chain amide group relative to the axis of the α-helix but is in the vicinity of 12.1. The inclination of the axis of the α-helix to the coiled-coil axis has been estimated to be about 12.6° (Section B.II.a) and this would reduce the predicted dichroic ratio to 9.1. This prediction cannot be compared directly with the observed spectrum, due to the presence of strong overlapping absorption associated with nonhelical material, but much of this absorption can be removed by treatment of the specimen with deuterium oxide (Fig. 16.8). The residual dichroic ratio of 5.5 would be consistent with the predicted value for a coiled-coil if it were assumed that about 18% of nonhelical material had not exchanged or that the coiled-coils were themselves inclined to the microfibril axis by an angle of 20.4°. In view of the finding by Bendit (1966b) that part of the nonordered material exchanges slowly, the first possibility is likely to account for at least part of the discrepancy between the observed and calculated dichroic ratios. Thus there does not seem to be any serious conflict between the available data on infrared dichroism and the coiled-coil model.

c. *Other α-Helical Models*

In the years immediately following the proposal of the coiled-coil model for α-keratin a number of authors drew attention to the topological difficulties which would be encountered in attempting to transform a bundle of coiled-coil ropes into a pleated-sheet β structure. Huggins (1957) proposed a model based on a three-chain unit in which segments of α-helix were inclined to the axis of the unit and interspersed with connecting regions in which the twist acquired through tilting was removed. Skertchly and Woods (1960) suggested that the difficulty could be overcome by supposing that the coiled-coils existed as independent entities rather than being grouped into ropes and proposed a specific packing scheme. Subsequent studies have shown that the lengths of the regions possessing regular secondary structure are much shorter than had been supposed originally and are unlikely to exceed the pitch length of the major helix. Thus steric hindrance to an $\alpha \rightarrow \beta$ transformation would be less important than originally envisaged.

Swanbeck (1961, 1963) proposed a model in which a three-strand coiled-coil rope was surrounded by a series of concentric layers each containing more chains, and having a greater tilt, than the previous layer. The three inner layers consisted of α-helices of pitch 5.25, 5.65, and 6.40 Å followed by two further layers having a 3_{10} helical structure (Donohue, 1953) and a β structure.

A possible modification of the coiled-coil model was suggested by Fraser and MacRae (1961a) in an attempt to improve agreement between the optical diffraction pattern of the model and the observed high-angle X-ray diffraction pattern. It was thought possible that the distortion required to produce a rope structure, instead of being continuous, might be concentrated at particular residues. This would lead to a "segmented rope" consisting of short, straight sections of α-helix with axes tangential to the path of the major helix. It was shown that the segment length was unlikely to exceed 20–30 Å since this would lead to a significant departure from the "knob–hole" packing scheme suggested by Crick (1953b). A variant of the segmented-rope model, in which the segment length was envisaged as being around 70 Å, was proposed by Lundgren and Ward (1962), and Parry (1970) proposed an alternative modification in which it was assumed that a repeating sequence of hydrophobic residues was present. The intensity transform of the latter model was calculated but did not give satisfactory agreement with the observed pattern either in the near-equatorial or 5.1-Å meridional regions.

Originally, Pauling and Corey (1951e) suggested that α-keratin consisted of a close-packed array of α-helices but this was later abandoned in favor of a coiled-coil model (Pauling and Corey, 1953b). Parry (1969) suggested the possibility that the α-helices were in fact straight and that the 5.165-Å reflection was due to a repeating sequence of residues.

Although it has not proved possible so far to provide an unequivocal proof of the existence of coiled-coil α-helix ropes in α-keratin, none of the alternative models that have been put forward appears to be capable of providing a simpler or a more satisfactory explanation of the observed diffraction data.

IV. ELECTRON MICROSCOPE STUDIES

Birbeck and Mercer (1957) showed that during the biosynthesis of hair the cortical cells become densely packed with fine filaments of uniform diameter and indefinite length, preferentially aligned parallel to the direction of growth, and Rogers (1959a) demonstrated that the filaments are also present in fully keratinized hair. In both cases osmium tetroxide was used as a stain (in the latter case after reduction of the disulfide linkages) and the filaments in cross section appeared as lightly stained areas about 70 Å in diameter in an osmiophilic background. On

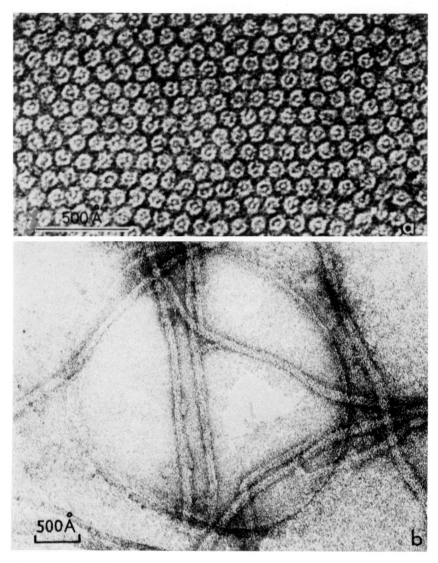

FIG. 16.15. (a) Electron micrograph of a cross section of the paracortex of Merino wool stained, following reduction, with osmium tetroxide and lead hydroxide showing ring–core substructure in the microfibrils (Filshie and Rogers, 1961). (b) Individual microfibrils isolated from hair root after treatment with chymotrypsin (Whitmore, 1972), stained with uranyl acetate.

the basis of the affinity of osmium for thiol groups it was concluded that the microfibrils consisted of low-sulfur proteins and the matrix consisted of high-sulfur proteins. Although the premises on which these conclusions were based are questionable, they have never been seriously contested.

Further studies established that the microfibrils in different types of α-keratin were essentially similar, but the mode of packing and the proportion of matrix were found to be highly variable (Rogers, 1959a; Rogers and Filshie, 1963; Fraser et al., 1972). Evidence of a substructure within the microfibril was noted by Rogers (1959a) and later Filshie and Rogers (1961) obtained sections in which a lightly stained ring about 20 Å wide and 60 Å in diameter could be discerned in each microfibril and in many instances a lightly stained central core about 20 Å in diameter was also visible (Fig. 16.15). The outer ring appeared to be broken up into unstained areas about 20 Å in diameter and it was suggested that the microfibril was composed of protofibrils about 20 Å in diameter, possibly arranged in a 9 + 2 pattern of the type found in certain cilia and flagella. This interpretation of the image was criticized by Sikorski and co-workers (Johnson and Sikorski, 1962, 1964, 1965; Dobb et al., 1962), who drew attention to the fact that a periodic axial deposition of heavy atoms would produce diffraction effects in the image, and concluded that the image did not therefore represent an axial projection of the microfibril. An alternative explanation of the 20-Å substructure observed by Filshie and Rogers (1961), which appears more likely in the light of later studies (Thon, 1966b; Haydon, 1968, 1969), is that it is associated with the granular phase image which is superimposed on the amplitude image of a stained biological thin section. Millward (1970) examined the appearance of microfibril images taken with the specimen very close to focus under conditions where the phase image granularity would be very small. The contrast was reduced, as expected, and most of the evidence for a 20-Å substructure disappeared (Fig. 16.16).

Since the granular phase image is not correlated with the microfibril lattice, an average of many microfibril images should lead to a reduction of granularity. Using the method outlined in Chapter 4, Section C.III, an averaged image was obtained (Fraser et al., 1969b) and this is compared in Fig. 16.16 with the near-focus image.

A number of investigators have reported that protofibrils about 20 Å in diameter were present in chemically modified keratins that had been fragmented by ultrasonic irradiation (Dobb, 1964b; Rogers and Clarke, 1965; Johnson and Speakman, 1965a,b). However, there is strong circumstantial evidence (Millward, 1969; Fraser et al., 1969c) that the

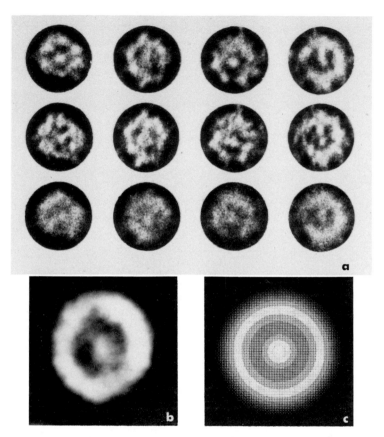

FIG. 16.16. (a) Electron micrographs obtained from cross sections of poodle hair, stained with osmium tetroxide following reduction, showing the effects of focal position on the appearance of the fine structure in four individual images. Upper set, 0.66 μm underfocus; middle set, 0.44 μm underfocus; lower set, close to focus (Millward, 1970). (b) Averaged microfibril image obtained from the electron micrograph of Fig. 16.15a using the method outlined in Chapter 4, Section C.III (Fraser et al., 1969b). (c) Low-resolution synthesis of the rotationally averaged electron density calculated from the electron diffraction pattern of the specimen used for the micrograph in (a) (Fraser et al., 1971b).

observed protofibrils originated from cellulosic contaminants, which were shown to produce copious amounts of similar protofibrils.

In summary, the evidence obtained from electron microscope studies of cross sections suggests that the microfibril has a ring–core substructure but gives no reliable indication at present of the way in which the ring is subdivided. Studies of the longitudinal appearance either in sections (Rogers, 1959a) or in isolated microfibrils (Whitmore, 1972) have failed to reveal any evidence relating to the axial organization of the low-sulfur proteins. A report of a 200-Å periodicity in fragmented α-keratin (Dobb, 1964a) is likely to be associated with cellulosic contaminants (Millward, 1969).

V. LOW-ANGLE X-RAY PATTERN

Electron microscope studies of α-keratin and of keratin proteins have so far failed to provide any detailed information on the form or arrangement of the molecules in either the microfibril or the matrix. As a result, the interpretation of the low-angle X-ray pattern is less advanced than is the case with collagen and with the myofibrillar proteins. In the present section an account will be given of the available data and of the progress that has been made in its interpretation. The interference functions which are prominent in the low-angle pattern also extend into the high-angle region and these functions will also be discussed in this section.

a. General Features

The low-angle pattern given by dry porcupine quill, shown in Fig. 16.17, consists of a well-defined equatorial reflection of spacing

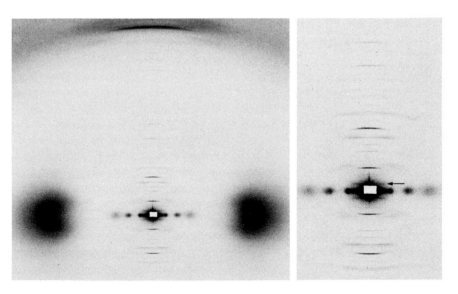

FIG. 16.17.   X-ray diffraction pattern obtained from a chromium derivative of porcupine quill (left). The low-angle pattern (right) shows a layer line of spacing 220 Å (arrowed) (Fraser and MacRae, 1973).

75 Å together with broader maxima of spacing around 45 and 27 Å and a series of meridional and near-meridional reflections which are extremely sharp in a direction parallel to the meridian but comparatively broad in the lateral direction. Lists of spacings for reflections in various α-keratins were given by Corey and Wyckoff (1936), MacArthur (1943), Bear (1944b), Bear and Rugo (1951), and Lang (1956b), and the meridional and near-meridional reflections were indexed on an axial repeat of structure of 198 Å. Later studies (Onions *et al.*, 1960; Fraser and MacRae, 1961b; Dobb *et al.*, 1965) suggested that the true axial period might be greater than 198 Å and in a detailed investigation of the spacings of the meridional reflections in native α-keratin and in several heavy-atom derivatives of α-keratin (Fraser and MacRae, 1971) it was found that the smallest periodicity which could account for the observed meridional reflections was 470 Å (Table 16.4). Crewther (1972) suggested some alternative possibilities.

TABLE 16.4

Z Values of Meridional Reflections in the X-Ray Pattern of α-Keratin $(10^{-3}\ \text{Å}^{-1})^a$

African porcupine quill										
Dry	—	14.8	—	25.4	—	—	40.4	55.3	80.9	95.7
Hydrated	10.8	14.9	—	25.5	—	—	40.4	55.1	80.5	95.2
Silver derivative	10.7	15.0	21.3	—	29.9	—	40.4	—	—	—
Osmium derivative	10.8	14.7	—	25.4	29.8	34.1	40.4	—	80.7	—
Chromium derivative	—	14.9	—	25.6	29.7	—	40.4	55.1	80.8	95.7
South American										
porcupine quill	10.7	14.9	—	25.6	29.9	—	40.4	55.3	80.7	95.4
Lincoln wool fibers	—	15.0	—	25.6	29.8	—	40.4	—	—	—
Mean Z value	10.8	14.9	21.3	25.5	29.8	34.1	40.4	55.2	80.7	95.5
Index with 198-Å period	2	3	4	5	6	7	8	11	16	19
Calculated Z value	10.1	15.2	20.2	25.3	30.3	35.4	40.4	55.6	80.8	96.0
Error	0.7	−0.3	1.1	0.2	−0.5	−1.3	0	−0.4	−0.1	−0.5
Index with 470-Å period	5	7	10	12	14	16	19	26	38	45
Calculated Z value	10.6	14.9	21.3	25.5	29.8	34.0	40.4	55.3	80.8	95.7
Error	0.2	0	0	0	0	0.1	0	−0.1	−0.1	−0.2

[a] Fraser and MacRae (1971). All patterns scaled to make the Z value of the 19th order of 470 Å equal to 0.0404 Å⁻¹.

The layer-line translations of the near-meridional reflections are difficult to assess in fiber patterns but the spacings of two layer lines have been measured with sufficient precision (Fraser and MacRae, 1973) to show that they cannot be indexed on a 470-Å periodicity, indicating that the true axial repeat of structure c must be some multiple of 470 Å. The spacing of the innermost near-meridional reflection was estimated to be 220 $\pm$ 4 Å, which suggests a minimum value for c in the range $1/[(1/216) - (1/235)]$ to $1/[(1/224) - (1/235)]$, that is, 2700–4800 Å. Thus the minimum multiple of 470 Å would be in the range 6–10.

It may be recalled that there is no *a priori* reason why a helical structure should have an integral number of units in some integral number of turns. If the number is nonintegral, the value of c becomes infinite and a more useful method of description is in terms of the unit height h and the pitch P.

b. *Interpretation of the Equatorial Pattern*

The low-angle equatorial maxima in α-keratin were provisionally indexed by Bear and Rugo (1951) as orders of a macroperiod about 80–90 Å transverse to the fiber axis, and Hosemann (1951) analyzed the pattern obtained by these authors in terms of a crystal lattice subject to random deviations of the cell parameters. The root-mean-square deviation of the lateral macroperiod was estimated to 3.3 Å. Following their discovery of a microfibril–matrix texture in developing hair, Birbeck and Mercer (1957) made the suggestion that the innermost reflection in fully keratinized α-keratin originated from the microfibrillar substructure and the problem of determining the precise nature of the interference function due to the microfibril lattice has been investigated in some detail (Fraser and MacRae, 1958a; Fraser *et al.*, 1960, 1964c).

To the extent to which the Fourier transform of the microfibril F_m is cylindrically symmetric for $Z = 0$, the cylindrically averaged equatorial intensity transform of a large group of parallel microfibrils will be given by the expression

$$\langle I(R, 0)\rangle_\psi = F_m(R, 0)\, F_m{}^*(R, 0) \left\{1 + 2\pi\rho_0 \int_0^\infty [g(r) - 1]\, r J_0(2\pi Rr)\, dr\right\} \quad (16.2)$$

where (R, ψ, Z) are cylindrical polar coordinates in reciprocal space, ρ_0 is the mean density of microfibrils per unit area, $2\pi\rho_0 g(r)r\, dr$ is the probability of finding a microfibril center at a distance between r and $r + dr$ from a reference microfibril, and $J_0(x)$ is a Bessel function of the first kind of order zero and argument x (Oster and Riley, 1952). The function $g(r)$, for a specimen of porcupine quill, was determined from an electron micrograph of a cross section and a means devised of correcting for finite sample size (Fraser *et al.*, 1964c). The function has a sharp peak at a value around $r = 100$ Å and tends to unity for large r (Fig. 16.18). The cylindrically averaged equatorial intensity transform of the lattice, $I_l(R) = \langle I(R, 0)\rangle_\psi/[F_m(R, 0)\, F_m{}^*(R, 0)]$, was calculated from

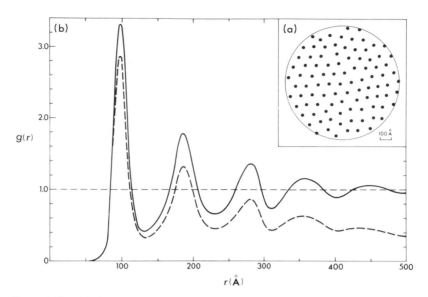

FIG. 16.18. (a) Distribution of microfibril centers in an electron micrograph of a cross section of porcupine quill. (b) The function $g(r)$ defined in the text; (———) corrected for finite sample size, (– – –) uncorrected (Fraser *et al.*, 1964c).

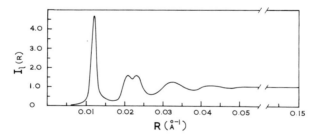

FIG. 16.19. Equatorial interference function of the microfibril lattice calculated from the function $g(r)$ (Fig. 16.18b) by means of Eq. (16.2).

Eq. (16.2) by numerical integration and is shown in Fig. 16.19. There is a pronounced maximum at a value of R close to the value predicted for a hexagonal lattice with a cell edge around 100 Å in length but the resolution of succeeding maxima is poor and beyond $R = 0.02$ Å^{-1} the function is within 0.78–1.26 of 1.0, the value for an isolated microfibril. The results depicted in Figs. 16.18 and 16.19 refer to specimens containing heavy metal stains and the corresponding curves for native keratin would be somewhat displaced.

Fraser and MacRae (1958a) suggested that the Fourier transform F_m

of a microfibril could be approximated at low angles, in the equatorial plane, by the Fourier transform of a cylinder of uniform density

$$F_m(R, 0) = k(\rho_m - \rho_b)[J_1(2\pi Rr)/2\pi Rr] \qquad (16.3)$$

where ρ_m and ρ_b are the mean electron densities of the microfibril and matrix, respectively, and k is a constant of proportionality. When F_m is normalized to give unit amplitude for $R = 0$ the expression for the intensity transform becomes

$$I_m(R, 0) = 4[J_1(2\pi Rr)/2\pi Rr]^2 \qquad (16.4)$$

(Oster and Riley, 1952). This function is illustrated in Fig. 16.20, and it was pointed out by Fraser and MacRae (1958a) that if a value of r of

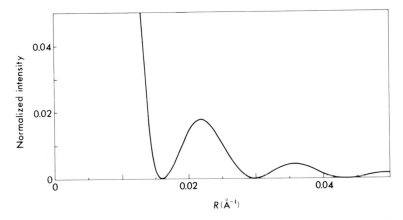

Fig. 16.20. Normalized equatorial intensity transform of a cylinder with a diameter of 75 Å.

about 37 Å is assumed for the cylinder radius, then the first two subsidiary maxima of the intensity transform correspond closely in position to the maxima at 45 and 27 Å in the pattern of α-keratin. A recent study of porcupine quill dried at 50°C *in vacuo* gave an estimate of 36 Å for the effective radius of the microfibril (Fraser *et al.*, 1971b). If the cylinder approximation is used in conjunction with the interference function shown in Fig. 16.19, the predicted intensity transform shows a general correspondence to the observed transform (Fraser *et al.*, 1964e) although the relative intensities of the maxima are not in quantitative agreement (Woods and Tyson, 1966; Bailey *et al.*, 1965). Support for the interpretation of the low-angle equatorial pattern in terms of an electron density difference between the microfibrils and the matrix was obtained from the observation that the pattern is greatly intensified when α-keratin is reacted with osmium tetroxide (Fraser and MacRae, 1958a), which is known from electron microscope studies to be deposited preferentially in the matrix. Thus ρ_b in Eq. (16.3) would be greatly increased in osmium-stained α-keratin.

Although the representation of the microfibril as a cylinder accounts in a general way for the low-angle pattern, it is too crude an approximation to account for the finer details of the low-angle pattern or for the fine structure in the 10-Å region (Fig. 16.21). These depend upon the

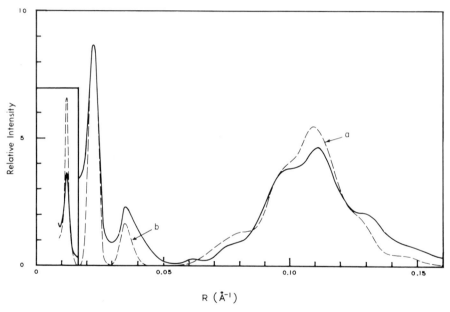

R (Å⁻¹)

FIG. 16.21. A comparison of the observed equatorial intensity distribution in dry α-keratin (——) with: (curve a) the product of the microfibril lattice intensity transform (Fig. 16.19) and the intensity transform of a cylinder (Fig. 16.20); (curve b) the intensity transform of coiled-coil α-helix ropes arranged in a ring structure (Fraser et al., 1964e).

internal structure of the microfibril, and the suggestion (Filshie and Rogers, 1961; Fraser et al., 1962b) that the microfibril might be made up of a 9 + 2 arrangement of coiled-coil α-helix ropes prompted a series of attempts to find a model of this type that would account for both the low-angle and high-angle patterns (Wilson, 1963; Fraser et al., 1964d,e, 1965e,f; Bailey et al., 1965). The last-named authors focused attention on the positions and relative intensities of the 45- and 27-Å maxima and showed that a measure of agreement could be obtained with a ring–core type of arrangement of protofibrils. The 9 + 2 structure, as such, did not give satisfactory agreement with the observed intensity transform in the 10-Å region (Fraser *et al.*, 1964d). Calculations were also carried out (Fraser *et al.*, 1964e) for an alternative model in which the scattering matter in the ring was much less sharply divided than in the 9 + 2 model and this was shown to give better agreement with the observed data (Fig. 16.21). Much additional information has been obtained on intensity changes that take place when heavy atoms are deposited in the structure (H. J. Woods, 1960, 1967; Bailey *et al.*, 1965; Bailey and Woods, 1968; Wilson, 1972a,b) but only a rudimentary interpretation in structural terms has so far been undertaken.

The attempts to devise a model for the axial projection of the microfibril by trial-and-error methods from the equatorial pattern have obvious limitations and the possibility of obtaining an axial projection by direct Fourier synthesis has been investigated (Fraser *et al.*, 1968b). The method used was based on the fact that the axial projection of a fibril with an *m*-fold screw axis or rotation axis can be represented by a Fourier series

$$\rho(r, \phi) - \sum_{k=-\infty}^{\infty} a_k(r) \exp(ikm\phi) \tag{16.5}$$

where ρ is the projected electron density distribution and r and ϕ are cylindrical polar coordinates. Klug *et al.* (1958) have shown that the Fourier transform, in the equatorial plane, of the distribution given in Eq. (16.5) is

$$F(R, \psi) = \sum_{k=-\infty}^{\infty} A_k(R) \exp[ikm(\psi + \tfrac{1}{2}\pi)] \tag{16.6}$$

where

$$A_k(R) = 2\pi \int_0^{\infty} a_k(r) \, J_{km}(2\pi Rr)r \, dr \tag{16.7}$$

Since ρ is also the Fourier transform of F (Chapter 1, Section A.I.f), it follows that

$$a_k(r) = 2\pi \int_0^{\infty} A_k(R) \, J_{km}(2\pi Rr)R \, dR \tag{16.8}$$

and so Eq. (16.5) can be used to calculate the axial projection of electron density if m and the components A_k can be determined. The observed intensity transform corresponds to

$$\langle F(R) F^*(R) \rangle_\psi = \sum_{k=-\infty}^{\infty} A_k(R) A_k^*(R) \tag{16.9}$$

and so the problem reduces to the resolution of the observed equatorial intensity distribution into its components $|A_k|^2$, $k = 0, \pm 1, \pm 2, ...,$ and the determination of the phases of the A_k components at each value of R. No direct means for determining the A_k components is available but it is possible to obtain a low-resolution synthesis by multiplying the observed data by a term $\exp(-\tfrac{1}{2}BR^2)$ in which B is chosen so that all components other than A_0 are negligible. This is possible since the microfibril has a finite radius and a range of R can be defined in which $J_0(2\pi Rr)$ is the only Bessel function term in Eq. (16.6) that has a significant value (Fig. 1.10). When this method was applied to the data obtained from dry porcupine quill, using a value of $B = 920 \text{ Å}^2$ (i.e., $\exp(-\tfrac{1}{2}BR^2) = 0.01$ for $R = 0.1 \text{ Å}^{-1}$), the result shown in Fig. 16.22 was obtained (Fraser $et\ al.$, 1971b). The synthesis supports the concept of a ring–core structure and similar distributions of electron density were obtained from the low-angle patterns of several other keratins. The relative intensities of the 45- and 27-Å maxima vary from keratin

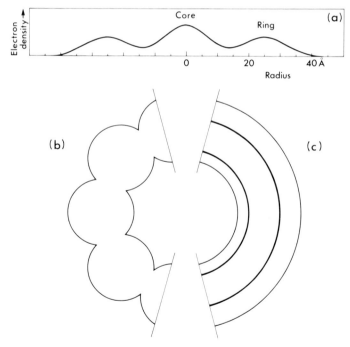

FIG. 16.22.   (a) Cylindrically averaged electron density distribution in dry porcupine quill obtained by Fourier synthesis from the equatorial X-ray diffraction pattern (Fraser $et\ al.$, 1971b). (b) Section and (c) projection of the model shown in Fig. 16.24b for the ring portion of the microfibril.

to keratin (Woods and Tyson, 1966; Bailey and Woods, 1968) and this appears to be related to the relative electron densities in the ring and in the annulus between the ring and core (Fig. 16.22). The relative intensities of the 45- and 27-Å maxima are a function of water content and this suggests that the radial distribution of water in a hydrated microfibril is nonuniform. A similar type of synthesis has been carried out on the basis of electron diffraction data obtained from the specimens depicted in the electron micrographs shown in Fig. 16.16 (Fraser *et al.*, 1971b). The electron density distribution, also shown in Fig. 16.16, is dominated in this case by the contribution of the osmium stain. The distribution closely resembles that deduced from the electron microscope studies.

c. *Interpretation of Meridional and Near-Meridional Patterns*

The layer-line spacings of the meridional and near-meridional reflections in α-keratin (Section B.V.a) indicate that the axial repeat of structure is very long and may even be infinite. The pattern of reflections does not appear to correspond to that of a simple discontinuous helix (Chapter 1, Section A.III.b) and it has been suggested (Fraser and MacRae, 1971, 1973) that the microfibril might have a helical structure that is subject to periodic distortion. The pattern observed at low angles (Table 16.5) is capable of explanation in terms of a helix with a unit height $h = 67$ Å and pitch $P = 220 \pm 4$ Å that is subject to an axial distortion of period 235 Å. The effects of a periodic distortion have been discussed in Chapter 1, Section A.IV.c and it will be seen from Fig. 16.23 that all the observed meridional reflections are clustered around the meridional reflections predicted for the undistorted helix in the manner expected for an axially periodic distortion. A crude estimate of the magnitude of the distortion amplitude δz can be made from the Z value at which the satellite layer lines corresponding to a particular value of s in Eq. (1.48) first appear, and the value of δz would seem to be about 10 Å and the corresponding helix is depicted in Fig. 16.24a. It may be noted that "forbidden" meridional reflections are also observed in the patterns obtained from striated muscle (Huxley and Brown, 1967) and from feather keratin (Fig. 16.35).

Some ambiguity exists in the prediction of helical symmetry from the Z distribution of the meridional and near-meridional reflections due to the possibility that an $\mathcal{N}$-fold parallel rotation axis of symmetry may be present (Chapter 1, Section A.III.c) and if this is so, the near-meridional reflections depend on factors of the type $J_{\mathcal{N}}(2\pi Rr)$ rather than $J_1(2\pi Rr)$. The positions of these reflections in the pattern obtained from α-keratin correspond to a value of r of about 30–40 Å for $\mathcal{N} = 1$ and

TABLE 16.5

Layer-Line Translations of the Meridional and
Near-Meridional Reflections in the X-Ray Diffraction Patterns Obtained from
Porcupine Quill and from Certain Heavy-Atom Derivatives

Selection rule parameters[a]			Z value of layer line[b] (10^{-3} Å^{-1})	Observed Z values (10^{-3} Å^{-1})[c]				
				Untreated		Silver derivative	Chromium derivative	Mercury derivative
m	n	s		Dry	Wet			
0	0	1	4.3	—	—	—	—	4.2
0	1	0	4.6	—	—	—	4.6	—
0	0	3	12.8	—	12.8	—	—	—
0	3	0	13.6	—	13.5	—	—	—
1	0	−2	6.4	—	—	—	—	6.3
1	0	−1	10.6	—	10.8	10.7	10.7	10.8
1	0	0	14.9	14.8	14.9	15.0	14.9	—
1	1	0	19.4	—	—	19.4	—	19.5
2	−3	0	16.2	—	—	—	16.2	—
2	0	−3	17.0	—	—	—	17.1	—
2	0	−2	21.3	—	—	21.3	—	—
2	0	−1	25.5	25.4	25.5	—	25.6	25.6
2	0	0	29.8	—	—	29.9	29.7	—
2	1	0	34.3	—	34.3	—	34.3	34.5
3	0	−3	31.9	—	—	—	31.9	—
3	0	−2	36.2	—	—	—	36.0	36.4
3	0	−1	40.4	40.4	40.4	40.4	40.4	40.4
3	0	0	44.7	—	—	—	44.9	—
3	1	0	49.2	—	—	—	49.5	—
4	0	−2	51.1	—	—	—	50.9	—
4	0	−1	55.3	55.3	55.1	—	55.1	—

[a] Based on the diffraction pattern predicted for a helix with unit height $h = 67.1$ Å and pitch $P = 220$ Å, and subject to an axial distortion of period $P_d = 235$ Å. The Z values of the layer lines are given by $Z = (m/h) + (n/P) + (s/P_d)$. Reflections with $n = 0$ are meridional, those with $n \neq 0$ are off-meridional.

[b] Calculated from $Z = [(7m + 2s)/470] + (n/220)$.

[c] Relative values based on $Z(m = 3, n = 0, s = -1) = 19/470$ Å^{-1}.

to much greater values of r for $\mathcal{N} > 1$. Since the effective radius of the microfibril in dry porcupine quill is around 36 Å (Fraser et al., 1971b, 1973), it is therefore unlikely that the microfibril has an $\mathcal{N}$-fold rotation axis. It is noteworthy that the low-angle meridional and near-meridional patterns appear to originate mainly from the surface of the microfibril. A similar situation obtains with the pattern obtained from the thick filaments of striated muscle (Chapter 15, Section D.II.b).

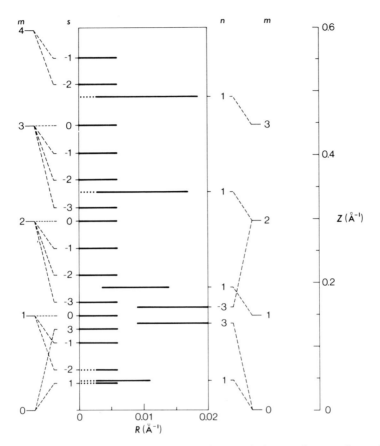

FIG. 16.23. Composite diagram showing the translations and approximate lateral extents of layer lines observed in the low-angle X-ray diffraction pattern of α-keratin and of certain heavy-atom derivatives of α-keratin. The pattern is indexed in terms of the diffraction pattern of a helix with unit height h and pitch P which is subject to an axial distortion of period P_d. The layer-line translations are given by $Z = (m/h) + (n/P) + (s/P_d)$, where $h = 67.1$ Å, $P = 220$ Å, and $P_d = 235$ Å.

The strong reflection in the 5-Å region of the meridian, believed to be associated with the coiled-coil α-helical conformation, is broken up into components that lie on layer lines which would be expected to be an extension of the low-angle pattern. As far as can be judged, the main component is meridional and corresponds in the terminology of Table 16.5 to $m = 13$, $n = s = 0$, and a weak satellite reflection corresponding to $m = 13$, $n = 0$, $s = 1$ also appears to be present (Fraser and MacRae, 1973). Since the coiled-coil segments are believed to be located in the ring and core of the microfibril, it seems likely that the

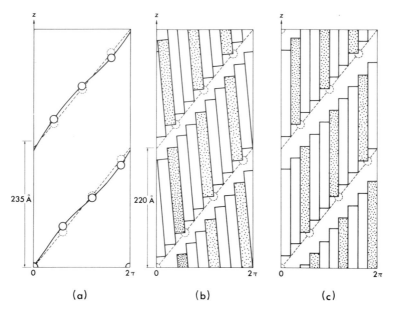

FIG. 16.24. (a) Radial projection of the positions of the structural units (represented by circles) in the ring portion of the microfibril as deduced from the meridional and near-meridional X-ray data. The distribution is derived from a basic helix with a unit height of 67 Å and a pitch about 220 Å (broken line) by the operation of a periodic axial distortion. The distortion period is around 235 Å and the amplitude about 10 Å (full line). (b) Possible distribution of coiled-coil α-helix rope segments in the form of a close-packed spiral band of axial dimension approximately equal to 160 Å. The molecular boundaries are not specified and the tilt of segments is not known. The tilt shown leads to a packing at the rate of about nine ropes per 360° in section; in the absence of tilt (c), the number would be around ten.

helical symmetry deduced from the low-angle pattern extends inward at least to include the ring.

The microfibril volume per unit of the undistorted helix depends rather critically on the value assumed for the effective radius but if this is taken as 36 Å, the volume per unit is $2.71 \cdot 10^5$ Å^3. The weight-average molecular weight of the low-sulfur proteins is about 50,000 (Section B.I.a) and so the average volume per molecule, assuming a density of 1.3 g cm^{-3}, is $1.66 \times 50,000/1.3 = 6.4 \cdot 10^4$ Å^3. Thus there are approximately 4.2 molecules per unit, the actual number depending rather critically upon the assumed radius. If, as suggested by physicochemical studies (Section B.I.a), there is a chemical unit consisting of two molecules of component 7 and one of component 8, the preceding calculation would suggest that the helical unit in the ring consists of one such chemical grouping with the balance of the material being located in the core.

Rudall (1956) has shown that a reasonable close packing of two-strand ropes can be obtained if appropriate combinations of axial displacement and azimuth exist between neighboring ropes (Fig. 16.25a).

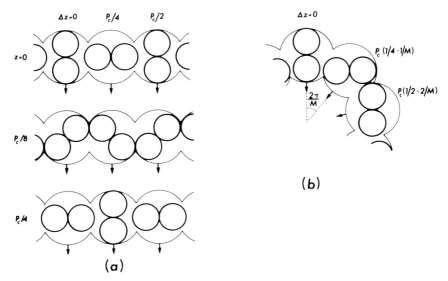

FIG. 16.25.   (a) Close-packing scheme for two-strand ropes showing the appearance in cross section at three different levels of z. (b) The close packing can be maintained when the sheet is bent around a cylinder provided that appropriate adjustment of the phasing is made.

In the case of a ring with M equivalent ropes, close packing will be achieved if there is an axial stagger between neighboring ropes of

$$\Delta z = (P_c/4) - (P_c/M) + (kP_c/2) \tag{16.10}$$

for two-strand ropes, or

$$\Delta z = -(P_c/M) + (kP_c/3) \tag{16.11}$$

for three-strand ropes, where $k = 0, \pm 1, \pm 2,...$ and P_c is the pitch of the coiled-coil (Fig. 16.25b). For infinite lengths of two-strand rope, it is necessary that M be an even integer but in the case of a finite helical structure with pitch P no overlap occurs between the rope segments provided that the length is less than $(M - 1)P/M$, and so odd and nonintegral values of M are feasible.

If the length of the coiled-coil segments is taken to be 160 Å (Section B.I.a), a close-packed ring of such segments could be formed in the manner shown in Fig. 16.24. Although the three rope sections associated with each chemical unit are depicted as having the same axial stagger, this is not required by the helical symmetry but the mean axial stagger per rope $\langle \Delta z \rangle$ must still satisfy Eq. (16.10) or (16.11) as appropriate. The X-ray data available at present are insufficient to be able to test specific suggestions as to the molecular boundaries and the inclination of the rope segments to the microfibril axis. Since the low-angle pattern appears to relate primarily to the outer layers of the microfibril, the X-ray data do not provide a basis for speculation as to the nature of the core. However, it seems likely that the annulus between the ring and the core consists of nonhelical material.

A number of authors have noted that particular meridional or near-meridional reflections are intensified following chemical treatments of α-keratin (Kratky et al., 1955; Fraser and MacRae, 1957a, 1961b, 1971, 1973; Fraser et al., 1960; Sikorski and Woods, 1960; Simpson and Woods, 1960; Dobb et al., 1965; Heidemann and Halboth, 1967; Spei et al., 1968a,b, 1970; Spei, 1970a,b,c, 1971, 1972a; Spei and Zahn, 1971; Bonart and Spei, 1972; Wilson, 1972a,b). A number of detailed models for the distribution of particular amino acids and for the arrangement of the polypeptide chains in the microfibril and the matrix have been made on the basis of these observations. In no case, however, has the model been demonstrated, by the calculation of intensity transforms, to account either qualitatively or quantitatively for the observed effects.

VI. THE MICROFIBRIL–MATRIX TEXTURE

The microfibrils in various types of α-keratin appear to be very similar as judged by the appearance in cross section in electron micrographs (Rogers, 1959b; Fraser et al., 1972) and Fourier syntheses of the axial projection of electron density (Fraser et al., 1971b), and the constancy of the layer-line spacings in the meridional and near-meridional diffraction patterns. In contrast, both the proportion and composition of the matrix are highly variable, as evidenced by differences in the spacing of the innermost low-angle equatorial reflection and the contents of high-sulfur and glycine-rich proteins (Table 16.6).

Little is known about the disposition or conformation of the matrix proteins. Onions et al. (1960) noted that the intensities of the meridional and near-meridional reflections differed among α-keratins from different species and this may be due, in part, to differences in the microfibril–matrix boundary. Sikorski and Woods (1960) obtained micrographs from reduced human hair that had been stained with silver nitrate and disintegrated by grinding in lithium bromide. It was claimed that this

TABLE 16.6

Estimates of the Proportions of Matrix Proteins Obtained from
X-Ray Diffraction and Extraction Studies[a]

Material	X-ray data[b]			Mass fraction of extracted proteins			
	Spacing of innermost reflection (Å)	Estimated microfibril diameter (Å)	Estimated volume fraction of matrix[c] (%)	High-sulfur (%)	Glycine-rich (%)	High-sulfur plus glycine-rich[d] (%)	All proteins (%)
Lincoln wool	72.5	73	28	22	0–2	23	87
Porcupine quill[e]	74.5	72	32	16	15	31	84
Pig bristle[f]	75.5	75	34	32	0–2	33	83
Echidna quill[g]	90.0	—	53	21	30	51	>95
Human hair[h]	92.5	78	56	38	0–2	39	84

[a] Fraser *et al.* (1973).

[b] Specimens dried *in vacuo* for 16 hr at 50°C.

[c] Assuming a microfibril diameter of 74.5 Å and hexagonal packing.

[d] Where the content of glycine-rich proteins is 0–2% a value of 1% has been taken.

[e] Cortex of quills from the African porcupine *Hystrix cristata*.

[f] Domestic pig, variety large white.

[g] Cortex of quills from the Australian spiny anteater *Tachyglossus aculeatus*.

[h] Specimens of hair from a child newly recovered from the protein deficiency disease kwashiorkor.

treatment revealed a "pseudoglobular" structure in the matrix with 50-Å-diameter units arranged in a regular manner with a longitudinal repeat of 95 Å. No confirmation of this conclusion has been obtained so far. An X-ray diffraction pattern obtained from unkeratinized hair that had been treated with silver nitrate showed a ring of spacing about 21 Å superposed on the normal diffraction pattern and it was suggested on this basis that the matrix may contain globular particles about 20 Å in diameter (Fraser, 1961; Fraser *et al.*, 1962b); a similar effect was noted in films of extracted high-sulfur proteins. A much broader ring of similar spacing has been observed in fully keratinized α-keratin after reduction and treatment with silver nitrate (Sikorski and Woods, 1960).

Other evidence which suggests that the matrix proteins have a globular structure has been obtained from studies of the mechanical properties of wool (Crewther and Dowling, 1960) and from studies of the hydration of keratin. When porcupine quill is hydrated the spacing of the innermost equatorial low-angle reflection increases from 75 to 85 Å, which represents an increase in the dimensions of the microfibril lattice of about 13%, but the microfibril radius only increases by about 6% as determined from the positions of the maxima and minima in the equatorial low-angle pattern (Fraser et al., 1971b). Studies of the effects of hydration on the high-angle pattern suggest a similar value for the swelling of the microfibril (Fraser et al., 1965e). The respective volume swellings for the microfibril and matrix were calculated to be 11 and 53% (Fraser et al., 1971b) and the high volume swelling of the matrix implies that it is not a randomly cross-linked network, but suggests rather that the disulfide bonds are largely intramolecular, leading to a globular character. The disparity between the uptake of water by the matrix and microfibrils results in a much greater decrease in rigidity modulus than in stretching modulus when dry keratin is hydrated (Feughelman, 1959). Bendit and Feughelman (1968) have suggested that as biosynthesis occurs in an aqueous environment, the microfibrils and matrix will be in mechanical equilibrium in the fully hydrated fiber and when the fiber is dried, the matrix contracts, resulting in a small longitudinal compression of the microfibrils. According to MacArthur (1943), the fine detail in the X-ray pattern is better developed in wet than in dry porcupine quill and this tends to support the view expressed by Bendit and Feughelman (1968).

The effect of stretching on the low-angle X-ray pattern has also been studied (Kratky, 1951; Spei, 1970b, 1971, 1972b; Spei and Zahn, 1971), and a series of changes noted in the spacings of both the equatorial and meridional patterns. A contraction of about 7% in axial period has been noted when reduced wool is treated with silver nitrate (Wilson, 1972a) and a similar effect has been observed in unkeratinized horsehair when treated with the same reagent (Fraser et al., 1962c).

C. SOFT MAMMALIAN KERATINS

The fibrous proteins of soft mammalian keratins have not been studied as extensively as the proteins from hard keratin and are less well characterized. Rudall (1952) extracted the epidermis of cow's nose with 6 M urea for 24 hr at room temperature and isolated two fractions, one with a lower sulfur content than the fully keratinized stratum corneum (1.16%) and the other with a higher sulfur content. The fraction with the lower sulfur content was termed epidermin and oriented films were prepared which gave an α-type X-ray diffraction pattern and exhibited infrared dichrism of the amide A, I, and II bands of similar character to sections of hair or porcupine quill. The second fraction represented a relatively minor component of the extract and was estimated to be present in about one-tenth of the concentration of the lower-sulfur fraction. The higher-sulfur fraction gave a disoriented β pattern. Other fractionations of unkeratinized epidermal proteins into components of differing sulfur content have been described (Matoltsy, 1965; Matoltsy and Parakkal, 1967; O'Donnell, 1971).

Keratin proteins have also been prepared from fully keratinized stratum corneum by reduction and subsequent alkylation to give the S-carboxymethyl derivatives (Baden and Bonar, 1968; O'Donnell, 1971). The principal component of the extract resembles the epidermin isolated by Rudall (1952) with regard to its physical properties, and the amino acid composition of the purified lower-sulfur fraction isolated by O'Donnell is given in Table 16.7. Apart from the lower content of Cys and higher content of Gly, the composition resembles that of the low-sulfur proteins of hard mammalian keratin (Table 16.2). The distribution of residues in the sequence appears to be highly nonuniform since Baden (1970) was able to isolate a fraction, after treatment with trypsin, that contained appreciably less Gly and more Glu than the original material (Table 16.7).

The molecular weight of the lower-sulfur fraction has been estimated to be about 60,000 (Mercer and Olofsson, 1951; O'Donnell, 1971) and the helix content, estimated from measurements of optical rotatory

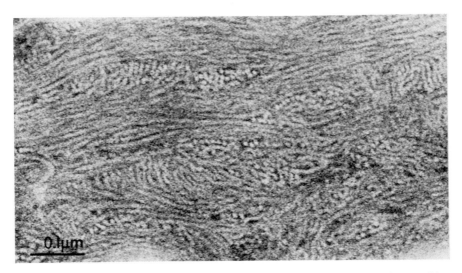

FIG. 16.26. Electron micrograph of a section of stratum corneum from human skin, stained with phosphotungstic acid and uranyl acetate, showing filaments about 70 Å in diameter in interleaved bundles (Brody, 1960).

dispersion, is about 45% (Matoltsy, 1965; Baden and Bonar, 1968). So far the higher-sulfur fraction has not been satisfactorily characterized.

The fine structure of the keratin in the stratum corneum layer of mammalian skin is superficially very similar to that of the keratin of hair cortex (Brody, 1959a,b, 1960, 1964; Patrizi and Munger, 1966) with filaments about 70 Å in diameter that can be differentiated on the basis of their staining properties (Fig. 16.26). Brody (1959a) has suggested that the interfilamentous matrix contains more sulfur than the filaments and Matoltsy and Parakkal (1967) have presented a number of arguments in favor of the view that the matrix is composed of higher-

TABLE 16.7

Amino Acid Analyses of Purified Fractions of Extracts from Stratum Corneum and Feather and of Enzymatic Fragments[a]

Amino acid	Stratum corneum proteins			Emu feather proteins[d]	
	Fraction S2[b]	Helix-rich fragment[c]	Fraction S3[b]	Purified component	β-Rich fragment
Alanine	6.9	8.1	6.1	4.5	4.2
Arginine	6.9	6.1	5.2	3.8	1.5
Aspartic acid[e]	9.1	12.1	8.2	6.2	4.5
Glumatic acid[f]	13.6	20.4	11.0	5.9	7.1
Glycine	13.8	6.8	16.5	10.0	13.5
Half-cystine[g]	0.9	1.1	2.3	7.2	1.5
Histidine	0.9	0.9	1.6	0	0
Isoleucine	4.4	4.4	4.4	4.5	5.9
Leucine	9.9	12.6	8.1	9.1	8.5
Lysine	5.0	6.2	5.8	0	0
Methionine	0.9	1.4	0.5	0	0
Phenylalanine	3.1	1.5	4.1	2.7	3.0
Proline	1.4	0.6	3.9	12.8	10.5
Serine	9.5	7.1	9.2	17.1	20.2
Threonine	4.5	4.3	4.7	4.8	5.5
Tyrosine	3.4	2.7	2.5	1.9	0
Valine	5.8	3.7	6.1	9.3	14.1

[a] Given in residues/100 residues, rounded to nearest 0.1 residue.
[b] O'Donnell (1971).
[c] Baden (1970).
[d] O'Donnell (1973).
[e] Includes asparagine.
[f] Includes glutamine.
[g] Content of half-cystine plus cysteine in native material.

sulfur proteins. So far, however, the presence of a separate matrix protein in soft keratins has not been directly demonstrated.

A number of X-ray diffraction studies of epidermis have been reported (Giroud and Champetier, 1936; Derksen and Heringa, 1936; Derksen *et al.*, 1937; Rudall, 1946, 1947; Swanbeck, 1959; Fraser *et al.*, 1972) and the stratum corneum layer shown to give an α pattern. However, the natural orientation of the filaments is very poor and the patterns are quite unsuitable for detailed analysis. Some improvement in orientation can be effected by stretching, but so far no clearly defined low-angle diffraction pattern has been recorded. It seems likely that the filaments of soft keratin resemble the microfibrils of hard keratin.

D. BETA-KERATIN

In this book attention has been restricted to the conformation of fibrous proteins in their native form, and artificially induced conformational transitions have not generally been considered. An exception will be made, however, in the case of β-keratin since the structure of this material provides a key to the understanding of the structure of feather keratin (Section E). In this section relevant physical studies of the conformation in β-keratin will be reviewed and discussed in terms of the pleated-sheet model. It would be beyond the scope of the present account to include a discussion of the molecular mechanism of the $\alpha \rightarrow \beta$ transformation. This has been the subject of much speculation (see, for example, Chapman, 1969b).

I. INFRARED SPECTRUM

Parker (1956) recorded the infrared spectrum of a steam-stretched horsehair and the observed dichroism of the amide A, I, and II bands was similar to that found in the spectra of synthetic polypeptides in the β conformation (Chapter 10, Section B). Miyazawa and Blout (1961) suggested that the observed frequencies were consistent with a parallel-chain, pleated-sheet structure (Chapter 10, Section A.I) but in a later study (Elliott, 1962; Bradbury and Elliott, 1963b) it was demonstrated that a weak shoulder with parallel dichroism was present at 1692 cm^{-1} in the position expected for the $\nu_{\parallel}(0, \pi)$ amide I mode of the antiparallel-chain pleated sheet (Chapter 5, Section A.II). Bendit (1966a) also recorded the spectrum of stretched horsehair (Fig. 16.27) and by plotting a dichroism spectrum showed that a sharp parallel component of amide I was present at the frequency expected for the $\nu_{\parallel}(0, \pi)$ component.

There is appreciable overlap between the various components of the amide I band in preparations of β-keratin (Fig. 16.27) but an analysis of the dichroism into individual components was carried out, for spectra obtained from steam-stretched porcupine quill, by the methods outlined in Chapter 5, Section C.II (Fraser and Suzuki, 1970c). Components

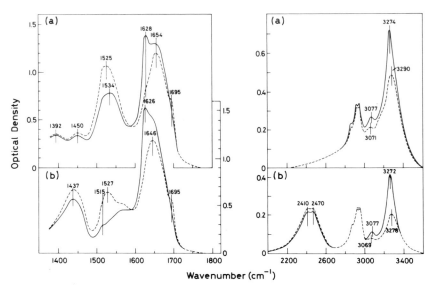

FIG. 16.27. Infrared spectra obtained from a thin section of stretched horsehair (a) untreated and (b) after exposure to deuterium oxide vapor at 35°C for 70 min, obtained with the electric vector vibrating perpendicular (full line) and parallel (broken line) to the fiber axis (Bendit, 1966b).

associated with the $\nu_{\parallel}(0, \pi)$, $\nu_{\perp}(\pi, 0)$, and $\nu_{\perp}(\pi, \pi)$ modes of the antiparallel-chain pleated sheet were identified (Fig. 16.28) and used (Fraser *et al.*, 1969a) to calculate the azimuthal orientation of the chains relative to the *ab* plane of the pleated sheet (Fig. 10.28).

The $\nu_{\perp}(\pi, 0)$ and $\nu_{\perp}(\pi, \pi)$ components of the amide I vibration of an antiparallel-chain pleated sheet have transition moments parallel, respectively, to the *a* and *c* axes and the magnitudes of the transition moments associated with these modes will be proportional to the corresponding components of the transition moment **M** associated with the unperturbed amide I mode of an individual amide group. If M_u and M_w are the components for the azimuthal orientation of the chain defined by $\Delta\phi = 0$ (Fig. 10.2), it can be shown that

$$\Delta D_{\pi 0}/\Delta D_{\pi\pi} = [(M_u \cos \alpha - M_w \sin \alpha)/(M_u \sin \alpha + M_w \cos \alpha)]^2 \qquad (16.12)$$

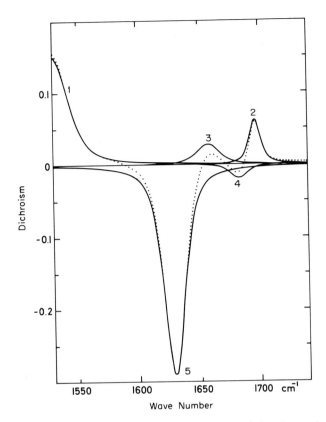

FIG. 16.28. Infrared dichroism in the amide 1 region of the observed spectrum of β-keratin (···) and components determined by least squares analysis: (1) amide II, (2) $\nu_{\parallel}(0, \pi)$, (3) ν_{α}, (4) $\nu_{\perp}(\pi, \pi)$, (5) $\nu_{\perp}(\pi, 0)$ (Fraser and Suzuki, 1970c).

so that

$$\alpha = -\tan^{-1}(M_u/M_w) \pm \tan^{-1}(\Delta D_{\pi 0}/\Delta D_{\pi \pi}) \qquad (16.13)$$

where $\Delta D_{\pi 0}$ and $\Delta D_{\pi \pi}$ are the integrated dichroisms for the $\nu_{\perp}(\pi, 0)$ and $\nu_{\perp}(\pi, \pi)$, bands, respectively (Fig. 16.28). The observed value of $\Delta D_{\pi 0}/\Delta D_{\pi \pi}$ was found to be 18.1 and the value of M_u/M_w was calculated to be -23.8 from the atomic coordinates of a polypeptide chain with an axial repeat of 6.68 Å. When these values are substituted in Eq. (16.13) the solutions $\Delta\phi = 11°$ and $\Delta\phi = -16°$ are obtained. On the basis of energy maps (Section III) the second solution was found to be the appropriate one.

Complementary studies of the combination bands in β-keratin have also been carried out (Elliott, 1952a; Fraser, 1955, 1956c; Fraser and

MacRae, 1958c) and the spectrum obtained from steam-stretched hair (Fig. 16.29) bears many similarities to that obtained from [Ala]$_n$ in the β form (Elliott *et al.*, 1954).

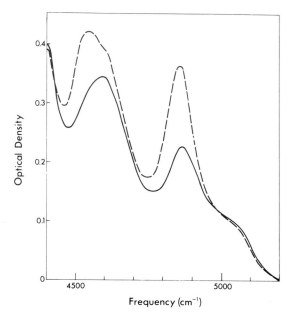

FIG. 16.29. Combination bands in the infrared spectrum of stretched wool measured with the electric vector vibrating perpendicular (full curve) and parallel (broken curve) to the fiber axis (Fraser and MacRae, 1958c).

II. X-RAY PATTERN

Astbury and Street (1931) recorded ten reflections in the high-angle X-ray diffraction pattern obtained from β-keratin and these were indexed on an orthogonal cell, the dimensions of which were later revised to $a = 9.3$ Å, $b = 6.66$ Å (chain axis), and $c = 9.7$ Å (Astbury and Bell, 1939). The indexing of certain reflections was confirmed by the preparation of doubly oriented specimens (Astbury and Sisson, 1935) and it was noted that the $hk0$ reflections were markedly sharper than the two reflections for which $l \neq 0$. The conclusion was reached that "the width, or effective width, of the crystallites in the direction of the backbone spacing (a axis) is considerably greater than in the direction of the side-chain spacing (c axis)" [p. 543].

More extensive data were collected by Fraser and MacRae (1962b) and it was found that the reflections could be indexed on an orthogonal cell with $a = 9.46$ Å, $b = 6.68$ Å, and $c = 9.7$ Å, similar to that suggested by Astbury and Street (1931). It was also found that the reflections could be indexed on a monoclinic cell, of the type suggested by Kratky and Porod (1955), with $a = 9.57$ Å, $b = 6.68$ Å (chain axis), $c = 9.99$ Å, and $\beta = 103.9°$ (Fig. 16.30). An insufficient number of

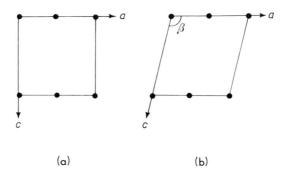

(a)   (b)

FIG. 16.30. Projection down the fiber (b) axis of unit cells proposed for β-keratin by (a) Astbury and Street (1931), and (b) Kratky and Porod (1955). The pleated sheets lie in the ab plane.

reflections of the general hkl type were observed to distinguish between these two possibilities. In a later study (Fraser et al., 1969a) quantitative intensity data were collected for $hk0$ reflections with $0 \leqslant h \leqslant 6$ and $0 \leqslant k \leqslant 3$ (Table 16.8) and these were used, as described in the following section, to test the pleated-sheet model for β-keratin.

III. THE PLEATED-SHEET MODEL

In the unit cell proposed for β-keratin by Astbury and co-workers (Astbury and Street, 1931; Astbury and Bell, 1939) the repeat of structure is two chains wide in the a direction and on this basis Astbury and Woods (1933) proposed an antiparallel-chain model of the type depicted in Fig. 10.1a. Pauling and Corey (1953a), on the other hand, considered that the axial repeat of structure, which is shorter than that of $[\text{Ala}]_n$ or $B.$ $mori$ fibroin, indicated that a parallel-chain structure (Fig. 10.1b) was present. Neither the infrared data (Section I) nor the X-ray diffraction data (Section II) can be accounted for by a parallel-chain structure but it was found (Fraser and MacRae, 1962b) that the intensity transform calculated for the antiparallel-chain pleated sheet was not in agreement with the observed intensity data. In particular the reflections with odd h, which required the a dimension of the cell to be 9.3 Å rather than 4.65 Å, were found to be generally weaker than predicted.

Following the demonstration by Arnott et al. (1967) that the crystal

TABLE 16.8

Comparison of Observed Intensities in the X-Ray Pattern of β-Keratin with Those
Calculated for an Antiparallel-Chain, Pleated-Sheet Model with Intersheet Disorder[a]

	Integrated intensity		Observed structure amplitude[d] F_o	Calculated phase (rad)
hk	Observed[b] I_o	Calculated[c] I_c		
20	100	103	35.3	π
40	9	12	16.5	0
60	4	3	14.4	π
11	0	0	—	—
21	30	26	20.8	0
31	0	0	—	—
41	8	7	16.8	0
51	0	0	—	—
61	0	0	—	—
12	0	1	—	—
22	10	7	13.1	π
32	0	1	—	—
42	0	0	—	—
52	0	0	—	—
62	0	0	—	—
13	3	3	9.5	0
23	13	15	17.3	π
33	2	1	12.4	π
43	0	0	—	—
53	0	0	—	—
63	1	1	11.5	π

[a] Fraser *et al.* (1969a).
[b] Zero in this column indicates below-minimum observable intensity.
[c] Zero in this column indicates $I_c < 0.5$.
[d] Corrected for Lorentz, polarization, and temperature factors.

structure in the β form of $[Ala]_n$ is subject to disorder with regard to sheet packing (Chapter 10, Section B.II), the structure of β-keratin was reinvestigated (Fraser *et al.*, 1969a) to see whether a similar type of disorder might account for the intensity anomalies mentioned earlier. As pointed out by Astbury and Sisson (1935), the diffuse nature of the reflections with $l \neq 0$ indicates that the crystallites are poorly developed in a direction perpendicular to the plane of the sheet and it was estimated, from the observed breadth of the 001 reflection, that the coherent length parallel to the c axis was about 22 Å, which corresponds to 2–3 sheets (Fraser *et al.*, 1969a).

If the sheets are subject to a random displacement of $\pm a/2$ in the direction of the a axis, as in the β form of $[\text{Ala}]_n$, the intensities of the reflections with odd h will be reduced by a factor $1/M$, where M is the number of pleated sheets in the crystallite. In the case of $[\text{Ala}]_n$, M appears to be sufficiently great to reduce the intensities of reflections with odd h to a negligible value, but M is not sufficiently large in β-keratin for this to be so.

In order to provide a starting point for the refinement of the pleated-sheet model on the basis of the rather limited X-ray data, a trial structure was derived by computing the packing energies of a pleated-sheet structure for various dispositions of the chain within the unit cell, using the methods outlined in Chapter 6, Section C. Terms corresponding to van der Waals, electrostatic, and hydrogen bond interactions were included and an "average" side chain $-CH_2-CH_3$ was used. The effects of varying the parameters χ_{12}, a, $\Delta\phi$, and Δz (Fig. 10.2) were explored and the minimum-energy conformation was found to correspond to $\chi_{12} = 300°$, $a = 9.4$ Å, $\Delta\phi = -18°$, and $\Delta z = 0.1$ Å. A sharp rise in energy was found to occur outside the range $\Delta\phi = -22°$ to $-5°$ (Fig. 16.31) due to the close approach of side-chain atoms and so the appropriate choice from the two possible values for this parameter derived from infrared studies (Section I) would appear to be $\Delta\phi = -16°$.

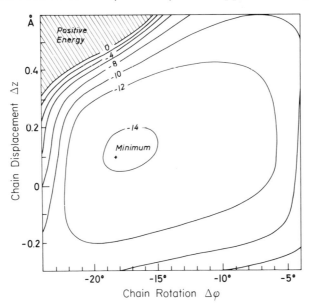

FIG. 16.31.   Packing energy of the pleated-sheet model for β-keratin as a function of the chain rotation $\Delta\phi$ and chain displacement Δz (Fraser et al., 1969a).

An iterative nonlinear least squares refinement of the parameters $\Delta\phi$, Δz, and M was carried out for the trial structure derived from energy calculations, assuming that intersheet disorder was present. Refinements were carried out both for the main chain and for the main chain with various combinations of side-chain atoms. It was found that the best agreement between the calculated and observed intensities was obtained when side-chain atoms were omitted and when $\Delta\phi = -16°$, $\Delta z = 0.28$ Å, and $M = 2.6$. A comparison of the observed and calculated intensities for the $hk0$ reflections is given in Table 16.8 and atomic coordinates for the refined model are given in Table 16.9.

TABLE 16.9

Atomic Coordinates of the Main-Chain Atoms in the Repeating Unit of the Antiparallel-Chain, Pleated-Sheet Structure in β-Keratin[a]

| Atom | "Up" chain | | | Hydrogen bonds | "Down" chain | | |
	u (Å)	v (Å)	w (Å)		u (Å)	v (Å)	w (Å)
N	2.619	−0.954	0.264		6.781	0.954	0.264
H	3.613	−1.049	0.323		5.787	1.049	0.323
C$^\alpha$	2.100	0.280	0.872		7.300	−0.280	0.872
H$^\alpha$	1.029	0.379	0.641		8.371	−0.379	0.641
C′	2.857	1.484	0.307		6.543	−1.484	0.307
O	4.091	1.564	0.400		5.309	−1.564	0.400
N	2.081	2.386	−0.264		7.319	4.294	−0.264
H	1.087	2.291	−0.323		8.313	4.389	−0.323
C$^\alpha$	2.600	3.620	−0.872		6.800	3.060	−0.872
H$^\alpha$	3.671	3.719	−0.641		5.729	2.961	−0.641
C′	1.843	4.824	−0.307		7.557	1.856	−0.307
O	0.609	4.904	−0.400		8.791	1.776	−0.400

[a] Fraser *et al.* (1969a). Equivalent positions (u, v, w), $(4.7 - u, 3.34 + v, -w)$, $(9.4 - u, -v, w)$, $(4.7 + u, 3.34 - v, -w)$.

A Fourier synthesis of the c-axis projection of the structure was carried out on the basis of the observed structure amplitudes and the calculated phases and additional resolution were obtained by using the technique of artificial sharpening (Fig. 16.32). Peaks in the vicinity of the main-chain atoms were resolved by this means. The main difference between the structure for β-keratin derived in these studies and the original pleated-sheet models is in the increased value of Δz, and it may be noted that a Δz shift of the same sense was also introduced during the refinement of the structure of [Ala]$_n$ (Chapter 10, Section B.II).

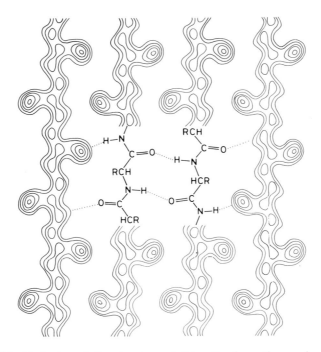

FIG. 16.32. Projection of the structure of β-keratin down the c axis obtained by Fourier synthesis from the $hk0$ reflections in the X-ray diffraction pattern (Fraser, 1969).

From the breadth of the 200 reflection the number of chains in a sheet was estimated to be 10.1, so that on average each β-crystallite would be expected to contain about 26 chains. Each microfibril in α-keratin is believed to contain about this number of α-helices in cross section (Section B.V.c) and this suggests that the chains in an individual β-crystallite probably all originate from the same microfibril. If this is the case, it implies that each microfibril contains equal numbers of oppositely directed chains.

E. FEATHER KERATIN

I. SOLUBLE PROTEINS

The keratin of feathers can be solubilized by a variety of methods (Goddard and Michaelis, 1934; Jones and Mecham, 1943; Ward *et al.*, 1946; Woodin, 1954; Harrap and Woods, 1964a), all of which involve scission of the disulfide linkages. The extract, which may constitute as much as 90% of the mass of the feather, has been shown to consist of a mixture of closely related proteins of uniform molecular weight around 10,400 (Woodin, 1954; Harrap and Woods, 1964b; Woods, 1971). No evidence has been obtained so far for the presence of high-sulfur and low-sulfur fractions as found in hard mammalian keratin (Section B.I).

Comparative studies of the proteins extracted from different feather parts (Harrap and Woods, 1964a) showed that the electrophoretic pattern of the S-carboxymethyl derivatives from the barbs was different from that of the rachis and calamus. Variations from species to species were also found (Harrap and Woods, 1967). A pure protein has been prepared from emu feather rachis and the composition (Table 16.7) and sequence (Fig. 16.33) determined (O'Donnell, 1973). The overall composition of the protein is quite different from that of the low-sulfur proteins of hard mammalian keratin but shows certain affinities with the high-sulfur proteins (Fraser *et al.*, 1972). Gly, Ser, and Pro account for 40% of the residues.

```
      1                   5                      10                  15
Ac-Ser- Cys- Tyr- Asn- Pro- Cys- Leu- Pro- Arg- Ser- Ser- Cys- Gly- Pro- Thr-
     16                  20                     25                  30
     Pro- Leu- Ala- Asn- Ser- Cys- Asn- Glu- Pro- Cys- Leu- Phe- Arg- Gln- Cys-
     31                  35                     40                  45
     Gln- Asp- Ser- Thr- Val- Val- Ile- Glu- Pro- Ser- Pro- Val- Val- Val- Thr-
     46                  50                     55                  60
     Leu- Pro- Gly- Pro- Ile- Leu- Ser- Ser- Phe- Pro- Gln- Asn- Thr- Val- Val-
     61                  65                     70                  75
     Gly- Gly- Ser- Ser- Thr- Ser- Ala- Ala- Val- Gly- Ser- Ile- Leu- Ser- Ser-
     76                  80                     85                  90
     Gln- Gly- Val- Pro- Ile- Ser- Ser- Gly- Gly- Phe- Asn- Leu- Ser- Gly- Leu-
     91                  95                     100
     Ser- Gly- Arg- Tyr- Ser- Gly- Ala- Arg- Cys- Leu- Pro- Cys
```

FIG. 16.33.   Amino acid sequence of a purified feather keratin protein extracted from emu feather rachis; Cys, *S*-carboxymethyl cysteine (O'Donnell, 1973).

The distribution of residues in the sequence is highly nonuniform. The Cys residues only occur in the terminal sections comprising residues 1–30 and 99–102, and the central section comprising residues 39–92 contains no charged side chains. When the S-carboxymethyl derivative of the protein was treated with trypsin a precipitate comprising residues 29–93 was obtained (O'Donnell, 1973) and this was found to have a higher content of β conformation than the original material (Suzuki, 1973). The composition of the β-rich fragment is given in Table 16.7 and is notable for the high contents of Ser, Val, Gly, and Pro. The presence of appreciable amounts of Val, Leu, and Ile can be correlated with the tendency for these residues to occur in β sections of globular proteins (Finkelstein and Ptitsyn, 1971).

In aqueous solution the S-carboxymethyl derivatives of the feather-keratin proteins appear to be devoid of α-helix as judged by measurements of the optical rotatory dispersion parameter b_0 (Harrap and Woods, 1964b) but α-helix contents up to 38% were found in solutions of the protein in mixtures of organic solvents. Films cast from aqueous solutions of feather-keratin proteins appear to contain a proportion of polypeptide chain in the β conformation (Rougvie, 1954; Fraser and MacRae, 1959a, 1963; Filshie et al., 1964; Burke, 1969) but no evidence has been obtained for the occurrence of an α-helical conformation.

II. INFRARED DICHROISM

Early studies of infrared dichroism in feather keratin (Ambrose et al., 1949b; Ambrose and Elliott, 1951b; Hecht and Wood, 1956; Beer et al., 1959) showed that the dichroisms of the amide A, I, and II bands were of the same character as observed in the β form of synthetic polypeptides and in β-keratin. A similar result was found for the combination bands (Hecht and Wood, 1956; Fraser and MacRae, 1958c). The magnitude of the observed dichroism in the amide A band was small but Parker (1955) found that exposure of a specimen to deuterium oxide resulted in a decrease in intensity in this band and at the same time the dichroic ratio changed from 1/1.8 to 1/4.8. This indicates that a proportion of the polypeptide chains in feather keratin are well oriented.

Following the analysis of the effects of interchain coupling on the spectra of polypeptides by Miyazawa and Blout (1961), attempts were made to interpret the spectrum of feather keratin in terms of specific conformations but the results were inconclusive (Krimm, 1962; Bradbury and Elliott, 1963b). In a later study (Fraser and Suzuki, 1965), however, it was found that the amide I band contained components that could be ascribed to the $\nu_{\parallel}(0, \pi)$ and $\nu_{\perp}(\pi, 0)$ modes of an antiparallel-chain pleated sheet together with a broad component with a frequency close to that predicted for material devoid of regular secondary structure.

A detailed analysis of the amide I band was carried out (Fraser et al., 1971a), using the methods outlined in Chapter 5, Section C, and the component bands are illustrated in Fig. 16.34. The fraction of residues in the antiparallel-chain, pleated-sheet conformation was estimated from the expression

$$f_\beta = ([A_{ps}]_\pi + 2[A_{ps}]_\sigma)/([A]_\pi + 2[A]_\sigma) \qquad (16.14)$$

where $[A_{ps}]_\pi$ is the sum of the areas of the amide I pleated-sheet bands,

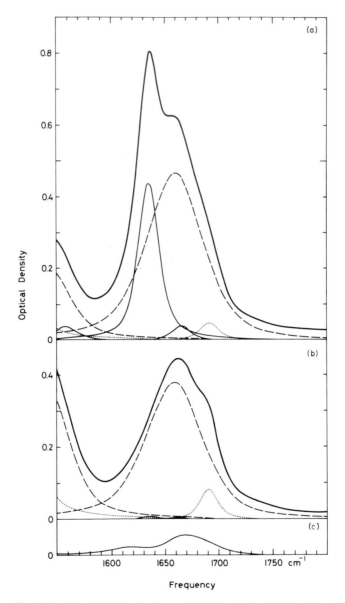

FIG. 16.34. Infrared spectra obtained from seagull feather rachis with the electric
vector vibrating perpendicular (a) and parallel (b) to the fiber axis. The spectra have been
analyzed into a band associated with amide side chains (c) and into amide I components.
The $(\pi, 0)$ and (π, π) components of the antiparallel-chain pleated sheet are shown as full
lines, the $(0, \pi)$ component is shown dotted, and the broad band associated with material
devoid of regular secondary structure is shown as a broken line (Fraser et al., 1971a).

$[A]_\pi$ is the sum of the areas of all the amide I bands in the parallel spectrum, and the subscript σ refers to similar quantities in the perpendicular spectrum. From the band areas found in the analysis (Table 16.10) the value of f_β was determined to be 0.25, and after correction for 10% nonkeratinous material a final estimate of 0.28 was obtained.

TABLE 16.10

Components of the Amide I Band in the Infrared Spectrum of Feather Keratin[a]

Component	Frequency of maximum (cm^{-1})	Optical density at maximum	Half-width (cm^{-1})	Area
Parallel spectrum				
$\nu(\pi, 0)$	1634.8	0.004	23.3	0.14
$\nu(\pi, \pi)$	1665.0	0.005	20.0	0.14
$\nu(0, \pi)$	1691.4	0.083	20.2	2.34
ν_a[b]	1658.3	0.380	Asymmetric	31.68
Perpendicular spectrum				
$\nu(\pi, 0)$	1634.8	0.440	23.2	14.30
$\nu(\pi, \pi)$	1665.0	0.040	20.0	1.11
$\nu(0, \pi)$	1691.4	0.049	20.2	1.37
ν_a	1658.3	0.465	Asymmetric	38.76

[a] Fraser *et al.* (1971a).
[b] Material devoid of regular secondary structure.

III. X-Ray Pattern

The high-angle X-ray diffraction pattern of feather keratin (Fig. 16.1) was first recorded by Marwick (1931), who commented on the similarity to the pattern obtained from β-keratin. From a consideration of the spacings of the reflections, Astbury and Marwick (1932) suggested that feather keratin contained polypeptide chains having a β conformation with an axial repeat of 6.2 Å rather than the 6.8-Å repeat that had been found in β-keratin. This view was supported by the observation that the repeat could be increased to 6.6 Å by stretching the specimen. Lists of the spacings of reflections in the high-angle patterns of various feathers have been given by Corey and Wyckoff (1936), Astbury and Bell (1939), Bear (1944b), Rudall (1947), Rugo (1949, 1950), Kraut (1954), and Schor and Krimm (1961a). The last-named authors showed that the "meridional" reflection of spacing 3.1 Å reported by Astbury and Marwick (1932) was a superposition of a pair of near-meridional reflections of spacing 3.15 Å and a true meridional reflection of spacing 2.96 Å.

Astbury and Marwick (1932) recorded a strong meridional reflection of spacing about 24 Å and Bear (1944b) showed that this was part of a well-developed low-angle pattern (Fig. 16.35a). Layer-line spacings out to 5 Å were catalogued by the latter author and it was shown that these could be indexed on an axial repeat of structure of $c = 94.6$ Å. In addition, it was noted that a lateral periodicity around 34 Å was present.

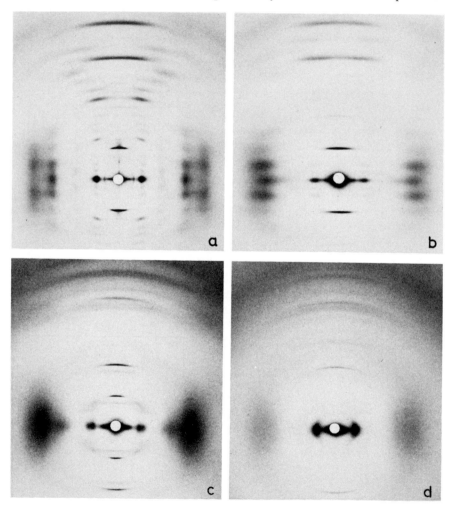

FIG. 16.35. X-ray diffraction patterns obtained from (a) dried feather rachis of the seagull *Larus novae-hollandiae*; (b) feather rachis after pressing in steam; (c) a claw of the lizard *Varanus varius*; (d) a claw of the tuatara (*Sphenodon*). From Fraser, R. D. B., MacRae, T. P., and Rogers, G. E., "Keratins: Their Composition, Structure and Biosynthesis," 1972, Courtesy of Charles C. Thomas, Publisher, Springfield, Illinois, U.S.A.

An extensive study of the diffraction patterns of both wet and dry feather rachis was carried out by Rugo (1949, 1950). The value of c was found not to vary significantly with moisture content but an increase of about 4% was noted in the prominent lateral periodicity in going from the dry to the wet state. An important finding was that in specimens heated in the presence of water or aqueous butanol the low-angle pattern was greatly modified, with the intensity becoming concentrated at positions that suggested the presence of a two-dimensional lattice of the type depicted in Fig. 16.36a. The distribution of intensity on layer lines with $l \geqslant 15$ was little affected by these treatments.

In a later study Fraser and MacRae (1963) showed that when feather

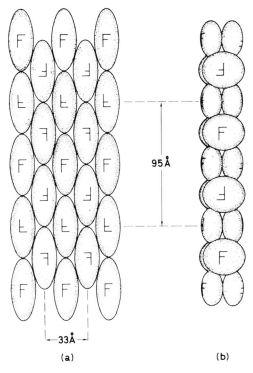

FIG. 16.36. (a) Two-dimensional array of particles suggested as the basis of the molecular structure of feather keratin by Bear and Rugo (1951). In order to satisfy certain aspects of the X-ray data, it was assumed that the particle orientation followed the pattern indicated by the letter F. (b) Alternative model based on a helical array of particles (Fraser and MacRae, 1963). The representation of the scattering material is highly schematic. From Fraser, R. D. B., MacRae, T. P., and Rogers, G. E., "Keratins: Their Composition, Structure and Biosynthesis," 1972, Courtesy of Charles C. Thomas, Publisher, Springfield, Illinois, U.S.A.

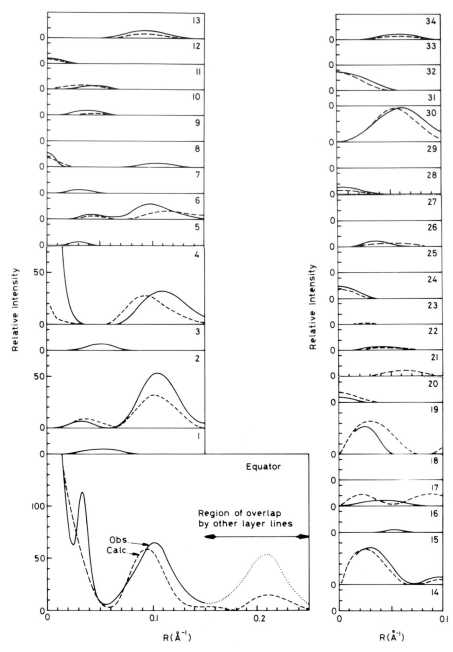

FIG. 16.37. Comparison of observed intensity (———) in the simplified X-ray diffraction pattern obtained from feather keratin after pressing in steam, with the cylindrically averaged intensity transform (– – –) of the model for the central framework of the microfibril illustrated in Fig. 16.41d. The double-headed arrow indicates the region of overlap by other layer lines (Fraser *et al.*, 1971a).

was subjected to lateral pressure while being held under tension in steam the lateral interference function was eliminated except for $l = 0$ (Fig. 16.35b), and the distribution of intensity resembled that expected for a helix with four residues per turn of pitch close to 95 Å (Fig. 16.36b). From a consideration of the lateral distribution of intensity on the layer lines it was concluded that no rotation axis was present parallel to the fiber axis, but the absence of layer lines with odd l in the 10-Å near-equatorial region suggested the possibility that a perpendicular twofold axis was present. In this case the line group would be **s2** (Chapter 1, Section A.II.d). In a further study (Fraser et al., 1971a) quantitative intensity data were collected for the simplified pattern (Fig. 16.37).

Schor and Krimm (1961a) carried out a detailed study of the pattern of native feather and observed certain reflections which, in their view, required a value of $c = 189$ Å rather than 94.6 Å. It was also suggested that an equatorial reflection of spacing about 49 Å was present in the pattern of feather keratin but this was later shown to be associated with lipid impurities (Fraser et al., 1963). Cylindrical Patterson functions were computed from the observed intensities and it was found that these bore no resemblance to the Patterson function calculated for the α-helix (Yakel and Schatz, 1955). It was also noted that the Patterson function calculated for feather keratin had certain peaks in common with that of collagen but there were sufficient differences to rule out a common conformation.

IV. ELECTRON MICROSCOPE STUDIES

When cross sections of hard avian keratins such as feathers, beaks, and claws are suitably stained with heavy atoms a fine structure is visible in the electron microscope (Fig. 16.38) that suggests a microfibril–matrix texture (Filshie and Rogers, 1962; Rogers and Filshie, 1963). The diameter of the lightly stained areas and the distance between them varies somewhat from preparation to preparation, depending on the nature of the staining method, and it is clear that the native structure is appreciably distorted by the staining procedure. The diameter of the lightly stained areas is about 30 Å and this suggests that the prominent lateral periodicity of about 33 Å in native feather keratin (Section E.III) might be associated with the apparent microfibril–matrix texture. In an earlier study of the lateral periodicity (Fraser and MacRae, 1959a) it had been found that treatment of feather rachis with osmium tetroxide greatly enhanced the innermost equatorial reflection in the low-angle X-ray pattern and this had been attributed to the deposition of osmium between the fibrous elements of the structure. Taken together

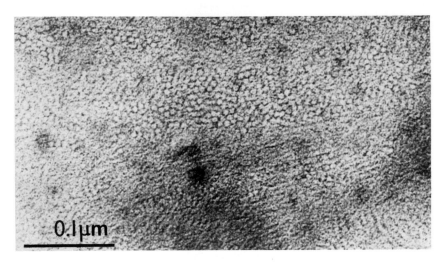

FIG. 16.38.   Electron micrograph of a cross section of fowl claw stained with potassium permanganate (Filshie and Rogers, 1962).

with the electron microscope studies, this confirms that the lateral interference function is associated with the microfibril lattice.

Although the appearance of feather keratin in cross section (Fig. 16.38) gives the impression of a microfibril–matrix texture, there is no evidence for the occurrence of two different types of protein (Section E.I) and the alternative possibility was suggested (Fraser *et al.*, 1971a) that the lightly stained areas represent the portions of the feather keratin molecules that possess regular secondary structure, while the "matrix" consists of the remaining portions of the molecules. Thus feather keratin was visualized as consisting of microfibrils with a central framework, having a β conformation, surrounded by the sections of the molecules that are devoid of regular secondary structure.

Tulloch (1971) prepared longitudinal sections of feather rachis stained with uranyl acetate, which enhances the prominent 24-Å meridional reflection in the X-ray and electron diffraction patterns, and succeeded in visualizing the corresponding lattice planes (Fig. 16.39a). Image averaging (Chapter 4, Section C. III) was used to enhance the periodicity and the result is shown in Fig. 16.39d. The lateral extent of the planes is considerable, indicating that the microfibrils are in precise register over distances of the order of several hundred angstroms. This can be correlated with the appearance of row lines in the diffraction pattern (Fig. 16.35a).

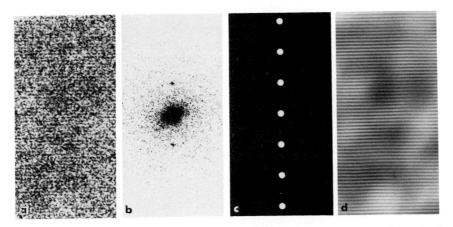

FIG. 16.39.   (a) Electron micrograph of a longitudinal section of feather rachis stained with uranyl acetate resynthesized from optical diffraction pattern (b) without filtering. (c) Mask used to filter the diffraction pattern. (d) Averaged image. (Tulloch, 1971.)

V. MODEL STRUCTURES

Early attempts to devise a model for feather keratin (Fraser and MacRae, 1959a; Astbury and Beighton, 1961) were formulated in terms of the two-dimensional lattice (Fig. 16.36a) proposed by Bear and Rugo (1951). A fresh approach to the problem followed the recognition that the structure might be helical (Schor and Krimm, 1961b; Ramachandran and Dweltz, 1962; Fraser and MacRae, 1962b).

The model proposed by Schor and Krimm (1961b) was based on the assumption that every eighth residue along the polypeptide chain was Pro and that the chain conformation was similar to that of the polar pleated sheet (Pauling and Corey, 1951c,d). Ten chains were supposed to be wrapped around a cylinder to give a tubular structure. In the model proposed by Ramachandran and Dweltz (1962) it was supposed that the polypeptide chains had a collagenlike conformation. Neither of these models appears to be capable of accounting satisfactorily for the observed X-ray diffraction pattern (Fraser and MacRae, 1963) and more recent studies have centered on the elaboration of the suggestion (Fraser and MacRae, 1962b, 1963) that the structure might be based on a helical array of β-crystallites.

The use of the X-ray diffraction pattern as a criterion for testing models of the structure of feather keratin is complicated by the presence of an interference function associated with the microfibril lattice (Fig. 16.35a) that leads to the appearance of row lines. As discussed in Section E.III, these can be almost completely removed by suitable specimen treatment (Fig. 16.35b) and the simplified pattern obtained refers essentially to a single microfibril. The persistence of the high-angle pattern in a virtually unchanged form after treatment of specimens with a variety of denaturing agents (Rugo, 1949, 1950; Bear and Rugo, 1951; Fraser and MacRae, 1963) suggests that this stems primarily from the central β framework of the microfibril, while the low-angle pattern, which is much changed by such treatments, stems primarily from the outer layers of the microfibril. A greater stability of the β framework would be expected, due to the regular array of hydrogen bonds and to the essentially nonpolar character of this portion of the molecule (Section E.I).

A helical model for the central framework of the microfibril was developed (Fraser *et al.*, 1971a) on the basis of the simplified X-ray pattern and from considerations of the observed density and the fraction

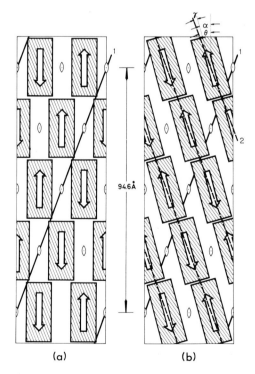

FIG. 16.40. Diagrammatic radial projection of the disposition of the pleated sheets (shown shaded) in (a) the trial structure and (b) the refined structure for the pleated-sheet framework of the feather keratin microfibril: (1) primitive helix, (2) generating helix for ruled surface (Fraser *et al.*, 1971a).

f_β of the molecule having the β conformation. A 94.6-Å length of micro-fibril will have a volume of about $33 \times 33 \times 94.6$ Å $= 1.03 \cdot 10^5$ Å^3 and if the density is taken as 1.27 g cm^{-3} (Fraser and MacRae, 1957b), this leads to an estimate of 78,800 daltons for the mass. If the line group is **s2** and a fourfold screw axis is present (Fig. 16.40), the number of asymmetric units per pitch length will be eight and so the mass of the asymmetric unit will be about 9900 daltons. This is close to the value of 10,400 measured for the molecular weight of the feather keratin mole-cule (Section E.I) and so it seems likely that the asymmetric unit of structure is a single molecule.

A striking feature of the intensity data collected from the simplified pattern (Fig. 16.37b) is the way in which strong diffraction occurs on even layer lines in the near-equatorial and 3-Å meridional regions and on odd layer lines in the 6-Å meridional region. It was found that this observation could be explained in a very simple manner by supposing that the microfibril contained a helical array of β-crystallites each con-taining two sheets related by a horizontal dyad (Fig. 16.40a). Earlier studies of possible models using optical diffraction patterns (Fraser and MacRae, 1959a) had suggested that the axial length of the crystallite was about one-quarter of the period, and using a value of 3.08 Å for the mean axial projection of a residue (Astbury and Bell, 1939), it was estimated that the mean number of residues per chain was about 7.7. The number of pleated-sheet residues per molecule was calculated to be 29.3 from the value of f_β (Section E.II) and so an estimate of $29.3/7.7 = 3.8$ was obtained for the number of chains in a sheet.

In order to reduce the number of parameters, a low-resolution model was used in which each residue was represented by a spherically sym-metric distribution of electron density given by

$$\rho = \rho_0 \exp(-d^2/d_e^2) \tag{16.15}$$

where d is distance from the symmetry center of the distribution and d_e is the radius at which the electron density falls to $1/e$ of its central value. The normalized transform of the distribution, referred to the symmetry center, is

$$F = \exp[-\pi^2 d_e^2 (R^2 + Z^2)] \tag{16.16}$$

and an additional factor f_c equal to the atomic scattering factor for carbon was incorporated, so that the normalized function for representing the scattering from a residue at low resolution became

$$F(R, Z) = [f_c(R, Z)/f_c(0, 0)] \exp[-\pi^2 d_e^2 (R^2 + Z^2)] \tag{16.17}$$

The cylindrically averaged transform for the trial structure was calculated, with this approximation, using the expression for helical
structures given in Eq. (1.33) and an additional term was included for
$l = 0$ to account for the background scattering from the non-β material.
The agreement between the observed and calculated transforms was
very poor but it was found that a greatly improved fit could be obtained
by twisting the pleated sheets so that they followed a helical ruled surface of opposite screw sense to the basic helix (Fig. 16.41b). The parameters of the ruled surface and of the pleated sheet were optimized and
the residue coordinates for the final model are given in Table 16.11.
The intensity transform calculated for this model is compared with the
observed intensity data in Fig. 16.37. The agreement at high angles is
generally good but some differences are apparent at low angles and
account would need to be taken of the contribution of the non-β portions before satisfactory agreement in this region could be expected.

No information was obtained on the course of the polypeptide chain
in individual molecules but one possibility is that each sheet corresponds

TABLE 16.11

Coordinates of the Residues in the Asymmetric Unit of the Model for the
Pleated-Sheet Framework in Feather Keratin[a]

r (Å)	ϕ (deg)	z (Å)	r (Å)	ϕ (deg)	z (Å)
	Chain 1 (up)			Chain 2 (down)	
7.92	169.5	−9.49	5.70	97.3	10.84
6.99	171.2	−5.94	7.01	94.0	7.44
8.58	153.1	−3.67	5.01	111.2	4.94
7.86	153.4	−0.05	6.57	107.9	1.64
9.30	136.0	2.15	4.45	122.9	−0.97
8.76	134.8	5.81	6.26	120.9	−4.14
10.07	118.3	7.99	4.07	132.1	−6.89
9.70	115.6	11.67	6.10	133.3	−9.93
	Chain 3 (up)			Chain 4 (down)	
7.27	93.4	−10.34	6.57	0.3	8.88
5.35	75.5	−7.89	8.24	18.6	6.58
6.78	78.9	−4.54	7.42	17.6	2.99
4.71	62.7	−1.98	8.94	35.3	0.76
6.40	65.5	1.25	8.31	35.8	−2.88
4.23	52.2	3.93	9.68	52.8	−5.07
6.16	52.8	7.03	9.23	54.8	−8.74
3.97	44.0	9.85	10.46	70.8	−10.92

[a] Fraser et al. (1971a). Equivalent positions (r, ϕ, z), $(r, -\phi, -z)$.

to the β portion of a single molecule. If this is so, there is a striking similarity to the type of structure envisaged for the *Chrysopa* egg stalk (Fig. 13.15). At the other extreme a molecule might span four sheets so that the four β sections of the molecule occupied an axial length of 94.6 Å. The disposition of the pleated sheets (Fig. 16.41c) would readily accommodate such an arrangement.

Films cast from solutions of feather-keratin proteins have been reported to give two types of X-ray diffraction pattern. In the first, many features of the original high-angle pattern are reproduced (Rougvie, 1954; Fraser and MacRae, 1959a, 1963) and it may be concluded that

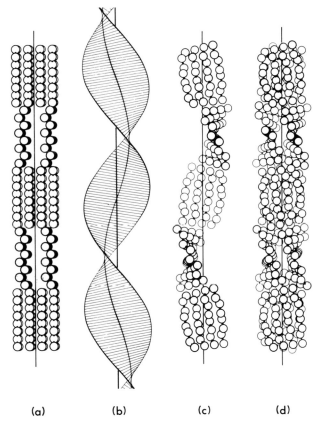

(a)   (b)   (c)   (d)

FIG. 16.41. (a) Trial structure for the pleated-sheet framework of the microfibril in feather keratin; pairs of pleated sheets related by a horizontal dyad are related by a right-handed fourfold screw axis; (b) a left-hand helical ruled surface; (c) a strand of pleated sheets distorted so as to conform to the ruled surface in (b); (d) complete model for the framework consisting of two strands of pleated sheets (Fraser *et al.*, 1971a).

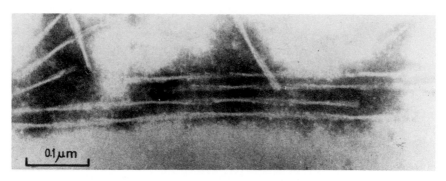

FIG. 16.42. Filaments regenerated from a purified fraction of keratin proteins extracted from fowl feather (Filshie *et al.*, 1964).

the original pleated-sheet framework is regenerated. In contrast, fibers and films obtained from certain purified fractions contain ordered aggregates (Fig. 16.42) which exhibit a cross-β pattern (Filshie *et al.*, 1964; Burke, 1969) and it has been suggested that in this form the pleated-sheet portions of the molecules aggregate side-to-side rather than end-to-end (Fraser *et al.*, 1971a).

Rudall (1947) showed that heavy atoms could be used as markers for studying the structure of feather keratin and a number of other examples of the selective deposition of heavy atoms in the structure have been reported (Fraser and MacRae, 1959a; Dweltz, 1962; Fraser *et al.*, 1971a). In some instances the change in the pattern appears to be confined to intensifications of the group of low-angle reflections that are eliminated when the feather keratin is partially denatured. It seems likely, therefore, that the heavy atoms are being preferentially deposited in the non-β portions of the molecules, which constitute the outer layers of the microfibril.

In conclusion, it should be recalled that Astbury predicted, from a general consideration of the X-ray diffraction pattern, that feather keratin was built up by the polymerization of corpuscular units having a β conformation (Astbury and Lomax, 1934; Astbury, 1938, 1949a).

F. OTHER KERATINS

Very few studies have been made of conformation in keratins other than those discussed in Sections B, C, and E. Marwick (1931) reported that reptilian hard keratin gave a diffraction pattern similar to that obtained from feather, and Rudall (1947) carried out a survey of the X-ray diffraction patterns yielded by a wide variety of keratins. Apart from hard avian and reptilian keratin, all other structures of epidermal origin were found to give an α pattern, and electron microscope studies, which have been summarized by Fraser *et al.* (1972), show that these α structures all contain filaments about 70 Å in diameter. In contrast, materials such as lizard claw (Whitmore, 1971) and the hard outer surface of snake scales (Roth and Jones, 1970), which give a featherlike pattern, contain filaments about 30–40 Å in diameter.

Rudall (1947) reported that the axial period, measured from the X-ray diffraction pattern, of a claw from the lizard *Varanus niloticus* was 98.5 Å, compared with 94.6 Å in feather rachis, while the period in the hard layer of a snake scale was found to be 92.8 Å. The low-angle diffraction patterns obtained from claws of the lizard *Varanus varius* and the tuatara (*Sphenodon*) are shown in Fig. 16.35c and d, respectively, and a number of differences between these patterns and that of feather keratin (Fig. 16.35a) are evident. In particular the prominent meridional reflection of spacing around 24 Å in feather keratin is weaker in the lizard claw and absent in the tuatara claw. In contrast, the high-angle patterns are more similar. It seems likely, therefore, that the microfibrils in these materials have a pleated-sheet core similar to that present in feather keratin but the non-β portions of the molecules are different. No chemical data are available on the proteins from these materials but the overall composition of lizard claw (Table 16.12) is considerably different from that of feather.

TABLE 16.12

Amino Acid Analyses of Fowl Feather[a,c] and Lizard Claw[b,c]

Residue	Fowl feather	Lizard claw	Residue	Fowl feather	Lizard claw
Alanine	8.7	5.2	Lysine	0.6	0.9
Arginine	3.8	2.9	Methionine	0.1	1.0
Aspartic acid[d]	5.6	5.5	Phenylalanine	3.1	2.8
Glutamic acid[e]	6.9	3.5	Proline	9.8	10.2
Glycine	13.7	27.6	Serine	14.1	6.3
Half-cystine	7.8	13.1	Threonine	4.1	3.5
Histidine	0.2	2.1	Tryptophan	0.7	0
Isoleucine	3.2	2.5	Tyrosine	1.4	5.6
Leucine	8.3	2.6	Valine	7.8	4.8

[a] Calculated from the data of Harrap and Woods (1964a).
[b] Hard keratin dissected from the claw of *Varanus varius*.
[c] Given in residues/100 residues, rounded to nearest 0.1 residue.
[d] Includes asparagine.
[e] Includes glutamine.

Chapter 17

Conformation in Other Fibrous Proteins

The conformations of the fibrous proteins considered in the preceding chapters have been investigated in some detail but there are many other fiber-forming or filament-forming proteins in which the conformation has been less well characterized. These include the extracellular structure proteins of molluscs (Hunt, 1970; Flower *et al.*, 1969), the proteins of insect egg shells (Kawasaki *et al.*, 1971; Furneaux and MacKay, 1972), the tubulins that form the structural framework of microtubules (Grimstone and Klug, 1966; Grimstone, 1968; Schmitt, 1968; Davison and Huneeus, 1970; Huneeus and Davison, 1970; Cohen *et al.*, 1971c; Stephens, 1971; Yamaguchi *et al.*, 1972), and the proteins that form filamentary structures in rodlike viruses such as TMV (Lauffer, 1971) and bacteriophages (Marvin, 1966; Moody, 1967). It would be beyond the scope of the present treatment to discuss structural studies of all these materials in detail but brief summaries are given in the following sections of the information that is presently available on fibrinogen and flagellin, both of which appear to contain a proportion of regular secondary structure, and on elastin and resilin, both of which appear to be devoid of regular secondary structure.

A. FIBRIN AND FIBRINOGEN

Fibrinogen is a soluble blood plasma protein which during the process of clotting is modified to give a derived product termed fibrin. Conversion of fibrinogen to fibrin takes place under the influence of the enzyme thrombin and results in the release of two peptides, termed fibrinopeptides A and B, which contain, respectively, 18 and 20 residues. The fibrin aggregates to form a fibrous network that is subsequently stabilized by the formation of intermolecular cross-linkages. The molecular weight of fibrinogen has been estimated to be around 330,000 (Shulman, 1953; Fantl and Ward, 1965) and the molecule has been shown to consist of three pairs of polypeptide chains (α, β, γ) leading to the chain formula $\alpha_2\beta_2\gamma_2$ (Blombäck and Yamashina, 1958; Clegg and Bailey, 1962). The chains have been shown to be covalently linked through disulfide bonds and to differ both in sequence and molecular weight (McKee et al., 1966). The molecular weights of the α, β, and γ chains were estimated to be 63,500, 56,000, and 47,000, respectively. The extensive literature on the fibrinogen molecule and on its conversion to fibrin has been reviewed by Scheraga and Laskowski (1957), MacFarlane (1960), and Mihalyi (1968, 1970).

Individual fibrinogen molecules have been visualized in the electron

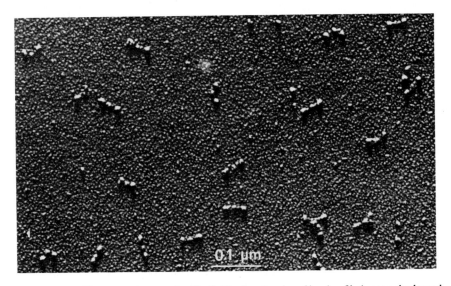

FIG. 17.1. Electron micrograph of individual molecules of bovine fibrinogen shadowed with platinum. The molecules appear to consist of three globular portions connected by thin, rodlike sections (Slayter, 1969).

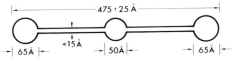

FIG. 17.2. Model for the fibrinogen molecule proposed by Hall and Slayter (1959).

microscope (Hall, 1956; Hall and Slayter, 1959; Slayter, 1969) and
appear to consist of a linear array of three globular units (Fig. 17.1).
The overall length of the molecule in the dried state was found to be a
function of the pH of the solution from which the specimen was prepared (Slayter and Hall, 1962). A model for the fibrinogen molecule,
based on the electron microscope and hydrodynamic data, was proposed
by Hall and Slayter (1959) and this is illustrated in Fig. 17.2. More
recently Köppel (1967), on the basis of an electron microscope study,
has suggested a cagelike pentagonal dodecahedral model for the
fibrinogen molecule, but Krakow et al. (1972) could find no support
for this model in their electron microscope studies.

The helix content of the molecule in solution has been estimated by
optical rotatory dispersion to be about 33% (Cohen and Szent-Györgyi,
1957; Mihalyi, 1965) and high-angle X-ray diffraction patterns obtained
from dried films of fibrinogen exhibit an α pattern (Bailey et al., 1943).
It has been suggested (Cohen, 1961) that the rodlike region connecting
the globular sections of the molecule (Fig. 17.2) has a coiled-coil conformation. The overall amino acid composition is given in Table 17.1
but it seems likely that the compositions of the globular and rodlike
regions will differ appreciably from one another.

Ordered aggregates of fibrin can be prepared by the action of thrombin
on fibrinogen solutions and these exhibit a regular banding pattern with
a periodicity of about 230 Å (Ruska and Wolpers, 1940; Hawn and
Porter, 1947; Hall, 1949; Köppel, 1958; Bang, 1963; Cohen et al., 1966;
Kay and Cuddigan, 1967; Karges and Kühn, 1970). Two types of banding have been observed depending on the method of staining and the
last-named authors have suggested that the dark bands in Fig. 17.3a
represent accumulations of polar groups whereas the dark bands in
Fig. 17.3b represent regions of lower density in the original fibril.
Similar banding patterns have been observed in tactoids of fibrinogen
formed under certain conditions of precipitation (Cohen et al., 1963,
1966). Both types of aggregate yield a series of sharp low-angle X-ray
reflections which can be indexed on an axial repeat of structure of
226 ± 3 Å (Stryer et al., 1963).

A second type of ordered aggregate (Fig. 17.4a) has been prepared
from fibrinogen treated with a protease isolated from Pseudomonas
aeruginosa (Tooney and Cohen, 1972). The modified molecule, termed
fibrinogen Ps, is about 10–15% lower in molecular weight than the
original fibrinogen and forms rod-shaped crystallites when precipitated
at low ionic strength. A related aggregate observed in precipitates of
fibrinogen contaminated with Pseudomonas is shown in Fig. 17.4b. In
both cases the projection of the structure onto an axis parallel to the

TABLE 17.1

Amino Acid Compositions of Certain Fibrous Proteins[a]

Amino acid	Fibrinogen[b]	Salmonella[c] flagellin	Elastin[d]	Resilin[e]
Alanine	4.9	15.0	21.3	11.2
Arginine	5.4	2.7	0.7	3.6
Aspartic acid	12.4	16.0	0.7	10.2
Desmosine	0	0	0.2	0
Glutamic acid	11.4	9.9	1.7	4.3
Glycine	10.0	8.5	31.6	41.1
Half-cystine	2.5	0	—	—
Histidine	2.0	0.3	0.1	0.8
Hydroxyproline	0	0	0.7	0
Isodesmosine	0	0	0.1	0
Isoleucine	4.0	5.0	2.7	1.3
Leucine	6.1	7.6	6.5	2.2
Lysine	7.4	3.6	0.4	0.4
Lysinorleucine	0	0	0.1	0
Methionine	2.2	0.6	0	—
ϵ-N-Methylysine	0	2.9	0	0
Phenylalanine	3.4	1.5	3.4	2.5
Proline	4.8	1.2	12.5	7.5
Serine	7.0	6.0	1.0	7.7
Threonine	6.1	10.4	1.0	2.9
Tryptophan	1.8	0	—	—
Tyrosine	3.5	2.4	0.6	2.1
Valine	5.0	6.5	13.4	2.3
Ammonia	—	15.5	—	—

[a] Residues per 100 residues, rounded to the nearest decimal place.
[b] McKee *et al.* (1966).
[c] McDonough (1965).
[d] Franzblau *et al.* (1965). Bovine ligamentum nuchae.
[e] Andersen (1971). Wing hinge of *Schistocerca gregaria.*

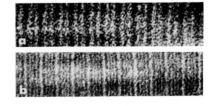

FIG. 17.3. Electron micrographs of fibrin fibrils (a) stained with phosphotungstic acid, pH 4; (b) stained with uranyl acetate, pH 4.3 (Karges and Kühn, 1970).

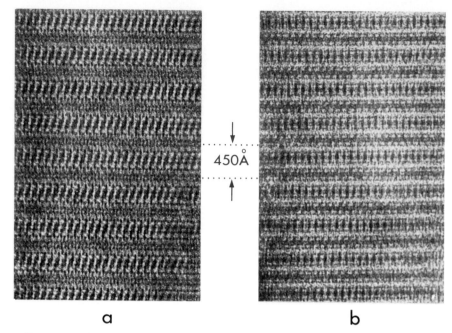

a   b

FIG. 17.4. Electron micrographs of fibrinogen microcrystals: (a) obliquely striated form; (b) orthogonal sheet form. In both forms the projected axial repeat is 450 Å (Tooney and Cohen, 1972).

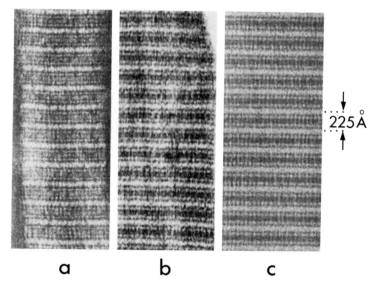

a   b   c

FIG. 17.5. Electron micrographs of fibrinlike aggregate from (a) fibrinogen precipitated at high ionic strength and (b) fibrinogen Ps clotted with thrombin. (c) Photographic superposition of four orthogonal sheet forms (Fig. 17.4b) each shifted 225 Å axially and 46 Å laterally (Tooney and Cohen, 1972).

length of the microcrystal has a period of 450 Å and it was suggested that this corresponds to the length of the molecule.

Fibrinogen Ps can be clotted with thrombin and the precipitate exhibits a banding pattern with a period of 225 Å (Fig. 17.5b) which closely resembles the pattern obtained from fibrin filaments under similar conditions of staining. The interrelationship between the patterns observed in the microcrystals and in the fibrils was demonstrated by superposing four images of the orthogonal sheet form (Fig. 17.4b) with an axial translation of 225 Å (Fig. 17.5c). Originally it was suggested that the 225-Å period observed in fibrin fibrils corresponded to the molecular length (Hall and Slayter, 1959; Hall, 1963), but the alternative suggestion that it arises from a staggered arrangement (Stryer et al., 1963) is supported by the superposition experiment.

The general similarity between the ordered aggregates obtained from fibrinogen and from fibrin suggests that no major change takes place in the form of the molecule during the conversion of fibrinogen to fibrin (Cohen, 1966b). A similar conclusion was reached by Bailey et al. (1943) on the basis of similarities between the high-angle X-ray patterns of the two materials, and by Krakow et al. (1972) on the basis of an electron microscope study of fibrinogen and fibrin monomer.

B. FLAGELLINS

Bacterial flagella can be detached from the cell by mechanical agitation and isolated in a purified form by fractional centrifugation or by a combination of centrifugation and precipitation with ammonium sulfate. When held at a pH below 3–4 the flagellar filaments disintegrate to give a solution of protein molecules of uniform size and composition. Proteins obtained in this way have been termed flagellins (Astbury et al., 1955) and under suitable conditions they can be precipitated in the form of filaments which closely resemble the filamentous portions of original flagella. Studies of flagellins have been reviewed by Koffler (1957) and Seifter and Gallop (1966).

The flagellins isolated from different species differ somewhat in composition and differences have also been noted between different strains from the same species (McDonough, 1965). A representative analysis of a *Salmonella* flagellin is given in Table 17.1. The molecular weights of flagellins from a variety of species have been determined and values around 40,000 have generally been obtained. In the case of flagellin from *Proteus vulgaris*, however, Erlander et al. (1960) found particles of molecular weight 19,000 below pH 3.8 that dimerized above this pH value.

Yaguchi *et al.* (1964) reported that certain flagellins undergo a transition from a random to a partially α-helical structure as the pH of the solution is increased from 2 to 4 and the helix content was estimated, from measurements of optical rotatory dispersion, to be between 25 and 40% at pH 4 depending on the type of organism. There is strong diffraction in the high-angle X-ray pattern of intact flagella in the 1.5- and 5-Å regions of the meridian and the 10-Å region of the equator (Fig. 17.6) and Astbury and co-workers (Astbury and Weibull, 1949;

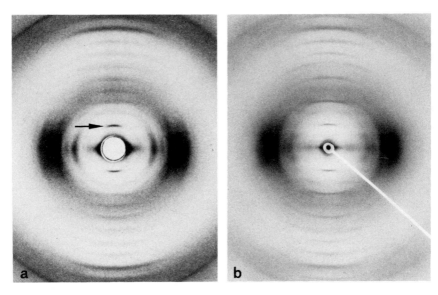

FIG. 17.6. X-ray diffraction patterns obtained from oriented fibers of (a) flagella from *Salmonella typhimurium* (Champness, 1971) and (b) reconstituted filaments from the flagellin of *S. typhimurium* (Wakabayashi and Mitsui, 1970). The arrow indicates the 26-Å meridional reflection.

Weibull, 1950; Astbury *et al.*, 1955) suggested that a conformation was present that was essentially similar to that found in the k-m-e-f group of fibrous proteins (Chapter 15, Section E.II.e). It was also suggested that a meridional reflection of spacing 4.65 Å indicated the presence of chains having the cross-β conformation, but studies of infrared spectra of flagella in the amide I region have failed to confirm the presence of a band near 1630 cm⁻¹, as anticipated for a β conformation (Bradbury, 1961; Champness, 1971).

Detailed interpretation of the high-angle X-ray diffraction pattern of bacterial flagella in terms of regular secondary structure is complicated by the fact that the molecular transform is sampled by an interference function associated with the packing arrangement of the flagellin molecules in the flagellar filament. Details of this packing arrangement have been studied by electron microscopy (Kerridge *et al.*, 1962; Lowy and Hanson, 1964, 1965; Lowy and Spencer, 1968; Champness, 1971) and X-ray diffraction (Astbury *et al.*, 1955; Beighton *et al.*, 1958; Burge, 1961; Swanbeck and Forslind, 1964; Champness and Lowy, 1968; Lowy and Spencer, 1968; Wakabayashi and Mitsui, 1970; Burge and Draper, 1971; Champness, 1971).

Negatively stained preparations of bacterial flagella present two surface appearances when examined in the electron miscroscope (Lowy and Hanson, 1965). In the first, the surface has a beaded appearance and is suggestive of globular units about 50 Å in diameter packed in an approximately hexagonal pattern (Fig. 17.7a), and in the second, the flagellum

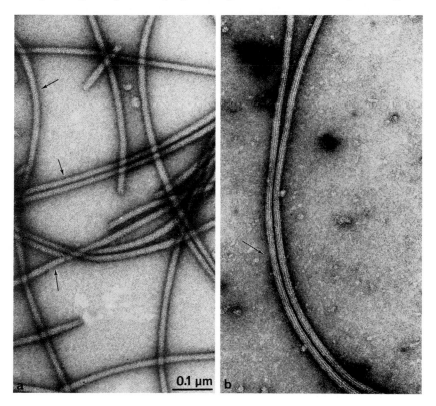

0.1 μm

Fig. 17.7. Electron micrographs of (a) flagella from *Salmonella typhimurium* and (b) flagella from *Pseudomonas fluorescens*. Arrows indicate points where the subunit arrangement is visible (Lowy and Spencer, 1968).

appears to have an outer layer composed of eight or more longitudinal strands (Fig. 17.7b). X-ray diffraction patterns obtained from both types of flagella contain a series of meridional and near-meridional reflections which can be indexed on an axial period of 52 Å (Champness and Lowy, 1968; Lowy and Spencer, 1968; Champness, 1971) and it seems likely that the helical symmetries of the two varieties of filament are essentially similar. Lowy and Hanson (1965) proposed a model for the flagellum of *Salmonella* on the basis of the beaded appearance visible in the electron microscope (Fig. 17.7a). The model consists of eight strands of globular particles that are aligned parallel to the axis of the flagellum and the particles in adjacent strands are staggered by one-half the axial length of the particle. The meridional and near-meridional regions of the X-ray diffraction pattern are consistent with a structure of this type (Champness and Lowy, 1968; Lowy and Spencer, 1968; Champness, 1971) but there are indications that some refinement may be required.

It seems likely that the globular units visible in electron micrographs of "beaded" flagella are individual flagellin molecules and great interest attaches to the observation that flagella-like filaments can be prepared by the precipitation of purified flagellin under appropriate conditions (Ada *et al.*, 1963; Abram and Koffler, 1964; Lowy and McDonough, 1964; Lowy *et al.*, 1966; Oosawa *et al.*, 1966; Wakabayashi and Mitsui, 1970). The last-named authors demonstrated that the X-ray diffraction pattern given by the reconstituted filaments (Fig. 17.6b) was essentially similar to that of the native filaments.

The low-angle equatorial X-ray pattern of flagella is dominated by the interfilament interference but at higher angles the effect is insignificant and there is a striking similarity between the patterns obtained from flagella of different species of bacteria. Possible interpretations of the equatorial pattern have been considered by Swanbeck and Forslind (1964), Burge and Draper (1971), and Champness (1971) but so far no definitive model for the axial projection of electron density has been derived. Estimates of 130 Å (Champness, 1971) and 140 Å (Burge and Draper, 1971) were obtained for the diameter of flagella from *Salmonella typhimurium*, compared with estimates obtained from electron microscope studies of 140 Å for flagella stained with phosphotungstic acid and 180 Å for flagella stained with uranyl acetate (Champness, 1971).

C. ELASTIN AND RESILIN

The proteins elastin and resilin occur in highly elastic structures found in vertebrates and in insects, respectively. The two proteins are not known to be evolutionarily related but their functions are similar.

Elastin is generally a minor component of the proteins present in vertebrate connective tissue but is the major component of the nuchal ligament found in the necks of grazing animals. Elastin is also an important constituent of the walls of large blood vessels, particularly the aorta. In its native state elastin appears to exist in a three-dimensionally cross-linked form (Partridge et al., 1963) and to be devoid of regular secondary structure (Astbury, 1940; Bear, 1952; Ramachandran and Santhanam, 1957). The mechanical properties are rubberlike (Burton, 1954). The amino acid composition is similar in some respects to that of collagen, with Gly constituting about one-third of the residues and Pro about 12%. Elastin is rich in hydrophobic residues and an unusual feature is the high content of Val residues. Structural studies of elastins have been reviewed by Seifter and Gallop (1966), Gotte et al. (1968), Partridge (1968), Franzblau (1971), and Ross and Bornstein (1971).

The occurrence of resilin in insects has been reviewed by Andersen (1971). It forms the major constituent of a number of structures which possess rubberlike elastic properties (Weis-Fogh, 1961a,b; Jensen and Weis-Fogh, 1962) and the polypeptide chains appear to be three-dimensionally cross-linked and to be devoid of regular secondary structure (G. F. Elliott et al., 1965). The amino acid composition (Table 17.1) resembles that of elastin insofar as the Gly content is high, but otherwise there is little in common between the two analyses. Structural studies of resilin have been reviewed by Seifter and Gallop (1966) and Andersen (1971).

Chapter 18

Some General Observations on Conformation

In the foregoing chapters a survey has been given of present-day knowledge of conformation in proteins that form fibrous structures *in vivo*. In some cases, as, for example, with collagen, the fibrous character stems from the development of a highly anisometric shape at the molecular level, while in others, as, for example, with actin, the molecule is globular and the fibrous character has its origin in the mode of aggregation of the molecules. Fibrinogen (Chapter 17, Section A) represents an intermediate case. The unifying feature in the development of a fibrous texture in these materials is the tendency for like residues, like chains, or like molecules to occupy equivalent positions in the structure. The significance and consequences of this tendency have been discussed in detail by a number of authors (Huggins, 1943, 1945, 1958; Crane, 1950; Pauling, 1953; Caspar and Klug, 1962; Caspar, 1966a,b; Klug, 1968, 1969; Caspar and Cohen, 1969; Kushner, 1969) and an important outcome of these considerations was the recognition that in the case of an enantiomorphic repeating unit, such as an L-residue or a protein molecule composed of L-residues, equivalence in a string of interconnected units is preserved only if they are related by helical symmetry. Linear arrays, involving only translational symmetry, and cyclic arrays, involving only rotational symmetry, may be regarded as special cases in which the unit twist and unit height, respectively, are zero. In general, therefore, equivalence leads to the formation of aggregates which are elongated parallel to the axis of helical symmetry.

There is a wide variation among fibrous proteins with regard to the content of regular secondary structures and materials such as collagen, tropomyosin, and paramyosin with highly elongated molecules tend to have a high content of regular secondary structure (generally greater than 90%), while materials with globular molecules, such as actin, feather keratin, and flagellin, generally have much lower contents (often less than 50%). The low-sulfur proteins of mammalian keratins and many of the insect silks form an intermediate group in which the molecules are elongated but a significant proportion of the polypeptide chain does not have a regular secondary structure. Where detailed studies have been carried out it has been found that the two types of region are associated with markedly different types of amino acid composition and sequence. In another group of proteins, comprising elastin, resilin, and the matrix proteins of mammalian keratins, the content of regular secondary structure is very small. In each case the materials are cross-linked and have rubberlike elastic properties.

Any attempt to correlate sequence with structure and function in fibrous proteins must, in view of the very limited data available, necessarily be highly speculative. In many cases the molecule can be subdivided into regions in which there is some type of periodicity in the sequence of residues and regions without such regularity. The latter type may only represent a small fraction of the molecule, as in collagen, or may be as great as two-thirds, as in feather keratin. Possible functions of the nonregular portions include the provision of means for temporary solubilization, transport, post-synthetic modification, and the regulation of aggregation. In materials such as silks, keratins, and elastin the presence of appreciable proportions of chain without regular secondary structure appears to be associated with the regulation of the physical properties of the final aggregate. The simplest examples of repeating sequences are found among the silks (Chapter 13) where, for example, runs of $[Gly]_n$ occur in the silk from *Phymatocera aterrima* and $[Ala]_n$ in the Tussah silks. The corresponding sections of the chains adopt conformations which appear to be identical to those observed in the synthetic homopolypeptides. Synthetic polymers with sequences related to those found in *B. mori* fibroin (Chapter 13, Section B.II) and in collagen (Chapter 14, Section B) have been synthesized and shown to have conformations similar to the parent protein but so far there has been no direct demonstration that particular sequences lead to the formation of the coiled-coil α-helical conformation.

In addition to repeating sequences, the four main conformation-determining factors in fibrous proteins appear to be (i) interactions involving charged side chains, (ii) interactions between apolar side chains, (iii) hydrogen bond formation, and (iv) the content and distribution of glycyl and prolyl residues. Studies of the structure of globular proteins have shown that in an aqueous environment a polypeptide chain adopts a conformation in which the charged side chains are exposed to the solvent and so the relative concentrations of charged and uncharged side chains must exert an important influence on molecular shape. Molecules with a high proportion of charged residues would be expected to assume a shape with a high surface area/volume ratio and attempts have been made to place such considerations on a quantitative basis (Waugh, 1954; Fisher, 1964; Bigelow, 1967; Gates and Fisher, 1971). The higher-than-average concentration of charged residues in the coiled-coil α-helical sections of proteins of the k-m-e-f group would appear to be functionally related to the high ratio of surface area to volume in a coiled-coil rope. The difficulty of attempting to relate overall amino acid composition to conformation is well illustrated by this example, however, since the coiled-coil rope structure will only be stable if the charged residues occupy positions in the sequence such that they are directed away from the axis. By the same token the contacting surfaces of the coiled-coils will be stabilized by van der Waals forces between the apolar side chains and by hydrophobic bonding associated with the exclusion of a structured water layer. The directive influences of hydrophobic bonding and the hydration of charged groups also appear to be important in the formation of ordered aggregates.

The importance of hydrogen bond formation as a conformation-determining factor was recognized at a very early stage and considerable emphasis was placed on the development of models in which all potential intramolecular hydrogen bonds were realized. While this emphasis served a useful purpose in initial studies, it tended to obscure other possibilities and in an aqueous environment, at least, the conformation of minimum free energy may well involve the formation of hydrogen bonds to partly immobilized water molecules.

The influences of glycyl and prolyl residues on conformation stem primarily from their effects on the permissible combinations of the torsional angles ϕ and ψ about these residues. The "permitted" area of the (ϕ, ψ) map is significantly greater for a glycyl residue and significantly less for a prolyl residue than for other types of residue. Again, sequence affects the end result of the incorporation of high proportions of glycyl or prolyl residues into a polypeptide chain; in the case of regular distributions such as occur in collagen and the crystalline regions of *Bombyx mori* fibroin a regular secondary structure results, but other proteins, such as elastin and the glycine-rich proteins of mammalian keratins, appear to be almost devoid of regular secondary structure.

In considering possible conformations of the polypeptide chain, Huggins (1943) and Pauling *et al.* (1951) introduced the simplifying assumption that all residues should be structurally equivalent with regard to the main-chain atoms and as explained earlier, this excludes all conformations in which the residues are not related by helical symmetry. This assumption, combined with other postulates, led to the evolution of models for a number of regular secondary structures, but two additional factors must be taken into consideration in adapting these models to fibrous proteins. The first factor has already been touched upon and relates to the fact that the side chains are not identical; the second concerns the concept of quasiequivalence (Caspar and Klug, 1962; Caspar, 1966a,b). This concept takes cognizance of the possibility that by systematically distorting a regular structure, it may be possible to arrive at a structure of lower free energy. There is abundant evidence from crystallographic studies of synthetic polypeptides for the occurrence of systematic distortion of otherwise helical structures and these stem mainly from packing considerations in three-dimensional crystals. Examples of quasiequivalence in fibrous proteins are found in the coiled-coil conformations where the strict equivalence between the main-chain portions of each residue in the α-helix is replaced by equivalence between groups of seven such portions in the coiled-coil.

Reaggregation studies of collagen, flagellin, and the muscle proteins illustrate the capacity of fibrous proteins for self-assembly into filaments and other organized aggregates and here again the tendency for molecules to occupy equivalent or quasiequivalent positions leads to the development of helical symmetry. The mode of aggregation is generally found to be sensitive to changes in pH, ionic strength, and the presence of other solutes, indicating that the assembly process depends upon a fine balance among various types of weak secondary bonding. Packing considerations are also clearly involved, but in the case of assembly in an aqueous environment these will be less important than in a solvent-free crystalline phase. As a consequence, it is possible to have protein-free volumes, as in the collagen microfibril, which would be most unlikely to occur in a solvent-free system.

A feature of the self-assembly process *in vivo* which is not well understood is the mechanism whereby aggregates of precisely defined dimensions are formed. Two possibilities can be envisaged; in the first, the process may be intrinsically self-limiting so that the information resides in the individual protein molecules, while in the second, the size of the aggregate is potentially unlimited but regulation of size is achieved through some external agency. In the first process it is assumed that up to a certain size the free energy per unit decreases but after that it increases so that the size is determined solely by energy considerations (Caspar, 1963; Kellenberger, 1969). Some specific suggestions regarding ways in which the stability of an ordered aggregate might attain a maximum at some critical size have been made by Cohen (1966b) and Huggins (1966). An example of the second type of process is found in tobacco mosaic virus, where aggregates of indeterminate length may be formed from the protein component *in vitro*, whereas polymerization *in vivo* in the presence of RNA leads to a precisely defined length of the composite aggregate. In this example the length-regulating component forms part of the final aggregate but this need not necessarily be the case. Other possible mechanisms of size regulation include the "poisoning" of the bonding sites (Crane, 1950) and the exhaustion of monomer in a spatially limited volume. In the latter case the mean size of aggregate would be a function of the rate of nucleation. A further feature of the self-assembly process, which is intimately related in some instances to physiological function, is the spontaneous formation of polar or bipolar aggregates. In the case of myosin (Chapter 15, Section D.I.b) it would appear from *in vitro* studies that the ability to form bipolar aggregates is an inherent property of the molecule.

References

Abbott, N. B., and Elliott, A. (1956). *Proc. Roy. Soc. Ser. A* **234**, 247.

Abram, D., and Koffler, H. (1964). *J. Mol. Biol.* **9**, 168.

Ada, G. L., Nossal, G. J. V., Pye, J., and Abbot, A. (1963). *Nature (London)* **199**, 1257.

Adams, M. J. *et al.* (1970). *Nature (London)* **227**, 1098.

Afzelius, B. A. (1969). *In* "Symmetry and Function of Biological Systems at the Macro-molecular Level" (A. Engström and B. Strandberg, eds.), p. 415, Wiley, New York.

Agrawal, H. O., Kent, J. W., and MacKay, D. M. (1965). *Science* **148**, 638.

Alderson, R. H., and Halliday, J. S. (1965). *In* "Techniques for Electron Microscopy" (D. H. Kay, ed.), p. 478. Blackwell, Oxford.

Alexander, L. E. (1969). "X-Ray Diffraction Methods in Polymer Science." Wiley, New York.

Allerhand, A., and Schleyer, P. (1963). *J. Amer. Chem. Soc.* **85**, 1715.

Ambrose, E. J., and Elliott, A. (1951a). *Proc. Roy. Soc. Ser. A* **205**, 47.

Ambrose, E. J., and Elliott, A. (1951b). *Proc. Roy. Soc. Ser. A* **206**, 206.

Ambrose, E. J., and Elliott, A. (1951c). *Proc. Roy. Soc. Ser. A* **208**, 75.

Ambrose, E. J., and Hanby, W. E. (1949). *Nature (London)* **163**, 483.

Ambrose, E. J., Elliott, A., and Temple, R. B. (1949a). *Proc. Roy. Soc. Ser. A* **199**, 183.

Ambrose, E. J., Elliott, A., and Temple, R. B. (1949b). *Nature (London)* **163**, 859.

Ambrose, E. J., Bamford, C. H., Elliott, A., and Hanby, W. E. (1951). *Nature (London)* **167**, 264.

Andersen, S. O. (1971). *In* "Comprehensive Biochemistry" (M. Florkin and E. H. Stotz, eds.), Vol. 26C, p. 633. Elsevier, Amsterdam.

Anderson, A. H., Gibb, T. C., and Littlewood, A. B. (1970). *Anal. Chem.* **42**, 434.

Andreeva, N. S., and Millionova, M. I. (1964). *Sov. Phys-Crystallogr.* **8**, 464.

Andreeva, N. S., Millionova, M. I., and Chirgadze, Y. N. (1963). *In* "Aspects of Protein Structure" (G. N. Ramachandran, ed.), p. 137. Academic Press, New York.

Andreeva, N. S., Esipova, N. G., Millionova, M. I., Rogulenkova, V. N., Shibnev, V. A. (1967). *In* "Conformation of Biopolymers" (G. N. Ramachandran, ed.), Vol. 2, p. 469. Academic Press, New York.

Andreeva, N. S., Esipova, N. G., Millionova, M. I., Rogulenkova, V. N., Tumanyan, V. G., and Shibnev, V. A. (1970). *Biophysics (USSR)* **15**, 204.

Andrews. M. W., Inglis, A. S., and Williams, V. A. (1966). *Textile Res. J.* **36**, 407.

Andrews, P. (1970). *In* "Methods of Biochemical Analysis" (D. Glick, ed.), Vol. 18. Wiley (Interscience), New York.

Andries, J. C., and Walton, A. G. (1969). *Biopolymers* **8**, 523.

Andries, J. C., and Walton, A. G. (1970). *J. Mol. Biol.* **54**, 579.

Andries, J. C., Anderson, J. M., and Walton, A. G. (1971). *Biopolymers* **10**, 1049.

Applequist, J., and Doty, P. (1962). *In* "Polyamino Acids, Polypeptides and Proteins" (M. A. Stahmann, ed.), p. 161. Univ. of Wisconsin Press, Madison, Wisconsin.

Armitage, P. M., and Chapman, J. A. (1971). *Nature (London) New Biol.* **229**, 151.

Arnone, A. *et al.* (1971). *J. Biol. Chem.* **246**, 2302.

Arnott, S. (1965). *Polymer* **6**, 478.

Arnott, S. (1968). *In* "Symposium on Fibrous Proteins, Australia 1967" (W. G. Crewther, ed.), p. 26. Butterworths, London.

Arnott, S., and Dover, S. D. (1967). *J. Mol. Biol.* **30**, 209.

Arnott, S., and Dover, S. D. (1968). *Acta Cryst.* **B24**, 599.

Arnott, S., and Wonacott, A. J. (1966a). *Polymer* **7**, 157.

Arnott, S., and Wonacott, A. J. (1966b). *J. Mol. Biol.* **21**, 371.

Arnott, S., Dover, S. D., and Elliott, A. (1967). *J. Mol. Biol.* **30**, 201.

Ascoli, F., and De Cupis, M. (1964). *Biochim. Biophys. Acta* **79**, 407.

Asquith, R. S., and Parkinson, D. C. (1966). *Textile Res. J.* **36**, 1064.

Astbury, W. T. (1933). *Trans. Faraday Soc.* **29**, 193.

Astbury, W. T. (1938). *Trans. Faraday Soc.* **34**, 378.

Astbury, W. T. (1940). *J. Int. Soc. Leather Trades' Chem.* **24**, 69.

Astbury, W. T. (1947a). *Nature (London)* **160**, 388.

Astbury, W. T. (1947b). *Proc. Roy. Soc. Ser. B* **134**, 303.

Astbury, W. T. (1949a). *Exp. Cell Res., Suppl.* **1**, 234.

Astbury, W. T. (1949b). *Brit. J. Radiol.* **22**, 355.

Astbury, W. T. (1949c). *Nature (London)* **163**, 722.

Astbury, W. T. (1953). *Proc. Roy. Soc. Ser. B* **141**, 1.

Astbury, W. T., and Beighton, E. (1961). *Nature (London)* **191**, 171.

Astbury, W. T., and Bell, F. O. (1939). *Tabulæ Biol.* **17**, 90.

Astbury, W. T., and Bell, F. O. (1940). *Nature (London)* **145**, 421.

Astbury, W. T., and Bell, F. O. (1941). *Nature (London)* **147**, 696.

Astbury, W. T., and Dickinson, S. (1940). *Proc. Roy. Soc. Ser. B* **129**, 307.

Astbury, W. T., and Haggith, J. W. (1953). *Biochim. Biophys. Acta* **10**, 483.

Astbury, W. T., and Lomax, R. (1934). *Nature (London)* **133**, 795.

Astbury, W. T., and Marwick, T. C. (1932). *Nature (London)* **130**, 309.

Astbury, W. T., and Sisson, W. A. (1935). *Proc. Roy. Soc. Ser. A* **150**, 533.

Astbury, W. T., and Spark, L. C. (1947). *Biochim. Biophys. Acta* **1**, 388.

Astbury, W. T., and Street, A. (1931). *Phil. Trans. Roy. Soc. London Ser. A* **230**, 75.

Astbury, W. T., and Weibull, C. (1949). *Nature (London)* **163**, 280.

Astbury, W. T., and Woods, H. J. (1930). *Nature (London)* **126**, 913.

Astbury, W. T., and Woods, H. J. (1933). *Phil. Trans. Roy. Soc. London Ser. A* **232**, 333.

Astbury, W. T., Dickinson, S., and Bailey, K. (1935). *Biochem. J.* **29**, 2351.

Astbury, W. T., Reed, R., and Spark, L. C. (1948a). *Biochem. J.* **43**, 282.

Astbury, W. T., Dalgleish, C. H., Darmon, S. E., and Sutherland, G. B. B. M. (1948b). *Nature (London)* **162**, 596.

Astbury, W. T., Beighton, E., and Weibull, C. (1955). *Symp. Soc. Exp. Biol.* **9**, 282.

Astbury, W. T., Beighton, E., and Parker, K. D. (1959). *Biochim. Biophys. Acta* **35**, 17.

Atassi, M. Z., and Singhal, R. P. (1970). *J. Biol. Chem.* **245**, 5122.

Atkins, E. D. T. (1967). *J. Mol. Biol.* **24**, 139.

Atkins, E. D. T., Flower, N. E., and Kenchington, W. (1966). *J. Roy. Microsc. Soc.* **86**, 123.

Auber, J., and Couteaux, R. (1963). *J. Microsc.* **2**, 309.

Baccetti, B. (1965). *J. Ultrastruct. Res.* **13**, 245.

Baden, H. P. (1970). *J. Invest. Derm.* **55**, 184.

Baden, H. P., and Bonar, L. (1968). *J. Invest. Derm.* **51**, 478.

Badger, R. M., and Pullin, A. D. E. (1954). *J. Chem. Phys.* **22**, 1142.

Bailey, A. J. (1968). *In* "Comprehensive Biochemistry" (M. Florkin and E. H. Stotz, eds.), Vol. 26B, p. 297. Elsevier, Amsterdam.

Bailey, C. J., and Woods, H. J. (1968). *Nature (London)* **218**, 765.

Bailey, C. J., Tyson, C. N., and Woods, H. J. (1965). *Proc. Int. Wool Text. Res. Conf., 3rd* **1**, 21. Inst. Textile de France, Paris.

Bailey, K. (1948). *Biochem. J.* **43**, 271.

Bailey, K. (1956). *Publ. Staz. Zool. Napoli* **29**, 96.

Bailey, K. (1957). *Biochim. Biophys. Acta* **24**, 612.

Bailey, K., Astbury, W. T., and Rudall, K. M. (1943). *Nature (London)* **151**, 716.

Balasubramanian, R., and Ramakrishnan, C. (1972). *Int. J. Peptide Protein Res.* **4**, 91.

Balasubramanian, R., Chidambaram, R., and Ramachandran, G. N. (1970). *Biochim. Biophys. Acta* **221**, 196.

Balasubramanian, R., Lakshminarayanan, A. V., Sabesan, M. N., Tegoni, G., Venkatesan, K., and Ramachandran, G. N. (1971). *Int. J. Protein Res.* **3**, 25.

Bamford, C. H., Hanby, W. E., and Happey, F. (1951a). *Proc. Roy. Soc. Ser. A* **205**, 30.

Bamford, C. H., Hanby, W. E., and Happey, F. (1951b). *Proc. Roy. Soc. Ser. A* **206**, 407.

Bamford, C. H., Brown, L., Elliott, A., Hanby, W. E., and Trotter, I. F. (1952). *Nature (London)* **169**, 357.

Bamford, C. H., Brown, L., Elliott, A., Hanby, W. E., and Trotter, I. F. (1953a). *Nature (London)* **171**, 1149.

Bamford, C. H., Brown, L., Elliott, A., Hanby, W. E., and Trotter, I. F. (1953b). *Proc. Roy. Soc. Ser. B* **141**, 49.

Bamford, C. H., Brown, L., Elliott, A., Hanby, W. E., and Trotter, I. F. (1954). *Nature (London)* **173**, 27.

Bamford, C. H., Brown, L., Cant, E. M., Elliott, A., Hanby, W. E., and Malcolm, B. R. (1955). *Nature (London)* **176**, 396.

Bamford, C. H., Elliott, A., and Hanby, W. E. (1956). "Synthetic Polypeptides." Academic Press, New York.

Bang, U. (1963). *Thromb. Diath. Hæmorrh. Suppl.* **13**, 73.

Bard, J. B. L., and Chapman, J. A. (1968). *Nature (London)* **219**, 1279.

Basu, B. C., Ramanathan, N., and Nayudamma, Y. (1962). *In* "Collagen" (N. Ramanathan, ed.), p. 523. Wiley (Interscience), New York.

Bath, J., and Ellis, J. W. (1941). *J. Phys. Chem.* **45**, 204.

Bear, R. S. (1942). *J. Amer. Chem. Soc.* **64**, 727.

Bear, R. S. (1944a). *J. Amer. Chem. Soc.* **66**, 1297.

Bear, R. S. (1944b). *J. Amer. Chem. Soc.* **66**, 2043.

Bear, R. S. (1945). *J. Amer. Chem. Soc.* **67**, 1625.

Bear, R. S. (1951). *J. Amer. Leather Chem. Ass.* **46**, 438.

Bear, R. S. (1952). *Advan. Protein Chem.* **7**, 69.

Bear, R. S. (1955a). *Symp. Soc. Exp. Biol.* **9**, 97.

Bear, R. S. (1955b). *In* "Fibrous Proteins and their Biological Significance" (R. Brown and J. F. Danielli, eds.), p. 97. Cambridge Univ. Press, London and New York.

Bear, R. S. (1956). *J. Biophys. Biochem. Cytol.* **2**, 363.

Bear, R. S., and Bolduan, O. E. A. (1950a). *Acta Crystallogr.* **3**, 230.

Bear, R. S., and Bolduan, O. E. A. (1950b). *Acta Crystallogr.* **3**, 236.

Bear, R. S., and Bolduan, O. E. A. (1951). *J. Appl. Phys.* **22**, 191.

Bear, R. S., and Morgan, R. S. (1957). *In* "Connective Tissue" (R. E. Tunbridge ed.), p. 321. Blackwell, Oxford.

Bear, R. S., and Rugo, H. J. (1951). *Ann. N.Y. Acad. Sci.* **53**, 627.

Bear, R. S., and Selby, C. C. (1956). *J. Biophys. Biochem. Cytol.* **2**, 55.

Bear, R. S., Bolduan, O. E. A., and Salo, T. P. (1951). *J. Amer. Leather Chem. Ass.* **46**, 107.

Beattie, I. R. (1967). *Chem. Brit.* **3**, 347.

Beer, M. (1956). *Proc. Roy. Soc. Ser. A* **236**, 136.

Beer, M., Sutherland, G. B. B. M., Tanner, K. N., and Wood, D. L. (1959). *Proc. Roy. Soc. Ser. A* **249**, 147.

Beighton, E. (1956). Quoted by Bailey (1956).

Beighton, E., Porter, A. M., and Stocker, B. A. D. (1958). *Biochim. Biophys. Acta* **29**, 8.

Bellamy, L. J. (1958). "The Infra-Red Spectra of Complex Molecules." Methuen, London.

Bellamy, L. J. (1968). "Advances in Infrared Group Frequencies." Methuen, London.

Bendit, E. G. (1957). *Nature (London)* **179**, 535.

Bendit, E. G. (1960). *Textile Res. J.* **30**, 547.

Bendit, E. G. (1962). *Nature (London)* **193**, 236.

Bendit, E. G. (1966a). *Biopolymers* **4**, 539.

Bendit, E. G. (1966b). *Biopolymers* **4**, 561.

Bendit, E. G. (1967). *Biopolymers* **5**, 525.

Bendit, E. G. (1968a). *In* "Symposium on Fibrous Proteins, Australia 1967" (W. G. Crewther, ed.), p. 386. Butterworths, London.

Bendit, E. G. (1968b). *Textile Res. J.* **38**, 15.

Bendit, E. G., and Feughelman, M. (1968). *In* "Encyclopedia of Polymer Science and Technology," Vol. 8, p. 1. Wiley, New York.

Bendit, E. G., Feughelman, M., Fraser, R. D. B., and MacRae, T. P. (1959). *Textile Res. J.* **29**, 284.

Bensing, J. L., and Pysh, E. S. (1971). *Macromolecules* **4**, 659.

Bensusan, H. B., and Nielsen, S. O. (1964). *Biochemistry* **3**, 1367.

Berendsen, H. J. C. (1962). *J. Chem. Phys.* **36**, 3297.

Berendsen, H. J. C. (1968). *In* "Biology of the Mouth" (P. Person, ed.), p. 145. Amer. Ass. for the Advan. of Sci., Washington, D.C.

Berendsen, H. J. C., and Migchelsen, C. (1965). *Ann. N.Y. Acad. Sci.* **125**, 365.

Berendsen, H. J. C., and Migchelsen, C. (1966). *Fed. Proc. Fed. Amer. Soc. Exp. Biol.* **25**, 998.

Berg, R. A., Olsen, B. R., and Prockop, D. J. (1970). *J. Biol. Chem.* **245**, 5759.

Berger, J. E. (1969). *J. Cell Biol.* **43**, 442.

Beychok, S. (1967). *In* "Poly-α-Amino Acids" (G. D. Fasman, ed.), p. 293. Dekker, New York.

Bigelow, C. C. (1967). *J. Theor. Biol.* **16**, 187.

Bikales, N. M. (1966–70). "Encyclopedia of Polymer Science and Technology," Vols. 1–11. Wiley (Interscience), New York.

Birbeck, M. S. C. (1964). *In* "Progress in the Biological Sciences in Relation to Dermatology" (A. Rook and R. H. Champion, eds.), Vol. 2, p. 193. Cambridge Univ. Press, London and New York.

Birbeck, M. S. C., and Mercer, E. H. (1957). *J. Biophys. Biochem. Cytol.* **3**, 203.
Birshtein, T. M., and Ptitsyn, O. B. (1967). *Biopolymers* **5**, 785.
Blackburn, J. A. (1970). *In* "Spectral Analysis" (J. A. Blackburn, ed.), p. 35. Dekker, New York.
Blake, C. C. F., Mair, G. A., North, A. C. T., Phillips, D. C., and Sarma, V. R. (1967). *Proc. Roy. Soc. Ser. B* **167**, 365.
Blaurock, A. E., and Worthington, C. R. (1966). *Biophys. J.* **6**, 305.
Block, H., and Kay, J. A. (1967). *Biopolymers* **5**, 243.
Blombäck, B., and Yamashina, I. (1958). *Arkiv. Kemi* **12**, 299.
Bloom, S. M., Fasman, G. D., de Lozé, C., and Blout, E. R. (1962). *J. Amer. Chem. Soc.* **84**, 458.
Blout, E. R. (1962). *In* "Polyamino Acids, Polypeptides and Proteins" (M. A. Stahmann, ed.), p. 275. Univ. of Wisconsin Press, Madison, Wisconsin.
Blout, E. R., and Asadourian, A. (1956). *J. Amer. Chem. Soc.* **78**, 955.
Blout, E. R., and Fasman, G. D. (1958). *In* "Advances in Gelatin and Glue Research" (G. Stainsby, ed.), p. 122. Pergamon, Oxford.
Blout, E. R., and Karlson, R. H. (1958). *J. Amer. Chem. Soc.* **80**, 1259.
Blout, E. R., and Lenormant, H. (1957). *Nature (London)* **179**, 960.
Blout, E. R., de Lozé, C., Bloom, S. M., and Fasman, G. D. (1960). *J. Amer. Chem. Soc.* **82**, 3787.
Blundell, T. L., Dodson, E., Dodson, G., and Vijayan, M. (1971). *Contemp. Phys.* **12**, 209.
Bohak, E., and Katchalski, E. (1963). *Biochemistry* **2**, 228.
Bolduan, O. E. A., and Bear, R. S. (1949). *J. Appl. Phys.* **20**, 983.
Bolduan, O. E. A., and Bear, R. S. (1950). *J. Polym. Sci.* **5**, 159.
Bolduan, O. E. A., and Bear, R. S. (1951). *J. Polym. Sci.* **6**, 271.
Bolduan, O. E. A., Salo, T. P., and Bear, R. S. (1951). *J. Amer. Leather Chem. Ass.* **46**, 124.
Bonart, R., and Spei, M. (1972). *Kolloid-Z. Z. Polym.* **250**, 385.
Bondi, A. (1964). *J. Phys. Chem.* **68**, 441.
Born, M., and Wolf, E. (1965). "Principles of Optics." Pergamon, Oxford.
Bornstein, P., and Piez, K. A. (1964). *J. Clin. Invest.* **43**, 1813.
Bornstein, P., and Piez, K. A. (1966). *Biochemistry* **5**, 3460.
Borysko, E. (1963). *In* "Ultrastructure of Protein Fibers" (R. Borasky, ed.), p. 19. Academic Press, New York.
Bouteille, M., and Pease, D. C. (1971). *J. Ultrastruct. Res.* **35**, 314.
Box, M. J. (1966). *Comput. J.* **9**, 67.
Bracewell, R. (1965). "The Fourier Transform and Its Applications." McGraw-Hill, New York.
Brack, A., and Spach, G. (1968). *In* "Peptides" (E. Bricas, ed.), p. 45. North-Holland Publ., Amsterdam.
Bradbury, E. M. (1961). Quoted by Burge (1961).
Bradbury, E. M., and Elliott, A. (1963a). *Spectrochim. Acta* **19**, 995.
Bradbury, E. M., and Elliott, A. (1963b). *Polymer* **4**, 47.
Bradbury, E. M., Burge, R. E., Randall, J. T., and Wilkinson, G. R. (1958). *Discuss. Faraday Soc.* **25**, 173.
Bradbury, E. M., Brown, L., Downie, A. R., Elliott, A., Hanby, W. E., and McDonald, T. R. R. (1959). *Nature (London)* **183**, 1736.
Bradbury, E. M., Downie, A. R., Elliott, A., and Hanby, W. E. (1960a). *Proc. Roy. Soc. Ser. A* **259**, 110.
Bradbury, E. M. *et al.* (1960b). *J. Mol. Biol.* **2**, 276.

Bradbury, E. M., Brown, L., Downie, A. R., Elliott, A., Fraser, R. D. B., and Hanby, W. E. (1962a). *J. Mol. Biol.* **5**, 230.
Bradbury, E. M., Elliott, A., and Hanby, W. E. (1962b). *J. Mol. Biol.* **5**, 487.
Bradbury, E. M., Brown, L., Elliott, A., and Parry, D. A. D. (1965). *Polymer* **6**, 465.
Bradbury, E. M., Carpenter, B. G., and Goldman, H. (1968a). *Biopolymers* **6**, 837.
Bradbury, E. M., Carpenter, B. G., and Stephens, R. M. (1968b). *Biopolymers* **6**, 905.
Bradbury, E. M., Carpenter, B. G., and Stephens, R. M. (1972). *Macromolecules* **5**, 8.
Bradbury, J. H. (1970). *In* "Physical Principles and Techniques of Protein Chemistry" (S. J. Leach, ed.), Part B, p. 99. Academic Press, New York.
Bradbury, J. H., Chapman, G. V., Hambly, A. N., and King, N. L. R. (1966). *Nature (London)* **210**, 1333.
Bradbury, J. H., Chapman, G. V., King, N. L. R., and O'Shea, J. M. (1970). *Aust. J. Biol. Sci.* **23**, 637.
Bradley, D. E. (1965a). *In* "Techniques for Electron Microscopy" (D. H. Kay, ed.), p. 58. Blackwell, Oxford.
Bradley, D. E. (1965b). *In* "Techniques for Electron Microscopy" (D. H. Kay, ed.), p. 75. Blackwell, Oxford.
Bradley, D. E. (1965c). *In* "Techniques for Electron Microscopy" (D. H. Kay, ed.), p. 96. Blackwell, Oxford.
Bragg, W. L. (1939). *Nature (London)* **143**, 678.
Bragg, W. L. (1942). *Nature (London)* **149**, 470.
Bragg, W. L., and Stokes, A. R. (1945). *Nature (London)* **156**, 332.
Bragg, W. L., Kendrew, J. C., and Perutz, M. F. (1950). *Proc. Roy. Soc. Ser. A* **203**, 321.
Brand, J. C. D., and Eglinton, G. (1965). "Application of Spectroscopy to Organic Chemistry." Oldbourne Press, London.
Brant, D. A. (1968). *Macromolecules* **1**, 291.
Brant, D. A., and Flory, P. J. (1965). *J. Amer. Chem. Soc.* **87**, 2791.
Brant, D. A., Miller, W. G., and Flory, P. J. (1967). *J. Mol. Biol.* **23**, 47.
Breedlove, J. R., and Trammell, G. T. (1970). *Science* **170**, 1310.
Brill, R. (1923). *Justus Liebigs Ann. Chem.* **434**, 204.
Brill, R. (1943). *Z. Phys. Chem.* **B53**, 61.
Brindley, G. W. (1955). *In* "X-Ray Diffraction by Polycrystalline Materials" (H. S. Peiser, H. P. Rooksby, and A. J. C. Wilson, eds.), p. 122. Inst. of Phys., London.
Brody, I. (1959a). *J. Ultrastruct. Res.* **2**, 482.
Brody, I. (1959b). *J. Ultrastruct. Res.* **3**, 84.
Brody, I. (1960). *J. Ultrastruct. Res.* **4**, 264.
Brody, I. (1964). *In* "The Epidermis" (W. Montagna and W. C. Lobitz, eds.), p. 251. Academic Press, New York.
Brown, F. R., Corato, A. D., Lorenzi, G. P., and Blout, E. R. (1972). *J. Mol. Biol.* **63**, 85.
Brown, L. (1956). Quoted by Bamford *et al.* (1956).
Brown, L., and Trotter, I. F. (1956). *Trans. Faraday Soc.* **52**, 537.
Brown, T. L. (1958). *Chem. Rev.* **58**, 581.
Brumberger, H. (1967). "Small-Angle X-Ray Scattering." Gordon and Breach, New York.
Buerger, M. J. (1959). "Vector Space." Wiley, New York.
Buerger, M. J. (1960). "Crystal-Structure Analysis." Wiley, New York.
Bunn, C. W. (1961). "Chemical Crystallography." Oxford Univ. Press (Clarendon), London and New York.
Bunn, C. W., and Daubeny, R. P. (1954). *Trans. Faraday Soc.* **50**, 1261.
Burge, R. E. (1959). *Acta Crystallogr.* **12**, 285.
Burge, R. E. (1961). *Proc. Roy. Soc. Ser. A* **260**, 558.

Burge, R. E. (1963). *J. Mol. Biol.* **7**, 213.

Burge, R. E. (1965). *In* "Structure and Function of Connective and Skeletal Tissue" (S. Fitton Jackson *et al.*, eds.), p. 2. Butterworths, London.

Burge, R. E., and Draper, J. C. (1971). *J. Mol. Biol.* **56**, 21.

Burge, R. E., and Hynes, R. D. (1959). *J. Mol. Biol.* **1**, 155.

Burge, R. E., and Randall, J. T. (1955). *Proc. Roy. Soc. Ser. A* **233**, 1.

Burge, R. E., Cowan, P. M., and McGavin, S. (1958). *In* "Recent Advances in Gelatin and Glue Research" (G. Stainsby, ed.), p. 25. Pergamon, Oxford.

Burke, M. J. (1969). Cross-β Structure in Fibrils of Feather Keratin and of Insulin. Ph. D. Thesis, Iowa State Univ.

Burke, M. J., and Rougvie, M. A. (1972). *Biochemistry* **11**, 2435.

Burnstock, G. (1970). *In* "Smooth Muscle" (E. Bülbring, A. F. Brading, A. W. Jones, and T. Tomita, eds.), p. 1. Arnold, London.

Burton, A. C. (1954). *Physiol. Rev.* **34**, 619.

Burton, P. R. (1970). *J. Cell Biol.* **44**, 693.

Cain, G. D. (1970). *Arch. Biochem. Biophys.* **141**, 264.

Carlsen, F., Knappeis, G. G., and Buchthal, F. (1961). *J. Biophys. Biochem. Cytol.* **11**, 95.

Carr, S. H., Walton, A. G., and Baer, E. (1968). *Biopolymers* **6**, 469.

Carver, J. P., and Blout, E. R. (1967). *In* "Treatise on Collagen." (G. N. Ramachandran, ed.), Vol. I, p. 441. Academic Press, New York.

Caspar, D. L. D. (1963). *Advan. Protein Chem.* **18**, 37.

Caspar, D. L. D. (1966a). *In* "Molecular Architecture in Cell Physiology" (T. Hayashi and A. G. Szent-Györgyi, eds.), p. 191. Prentice-Hall, Englewood Cliffs, New Jersey.

Caspar, D. L. D. (1966b). *In* "Principles of Biomolecular Organization" (G. E. W. Wolstenholme and M. O'Connor, eds.), p. 7. Churchill, London.

Caspar, D. L. D., and Cohen, C. (1969). *In* "Symmetry and Function of Biological Systems at the Macromolecular Level" (A. Engström and B. Strandberg, eds.), p. 393. Wiley, New York.

Caspar, D. L. D., and Holmes, K. C. (1969). *J. Mol. Biol.* **46**, 99.

Caspar, D. L. D., and Klug, A. (1962). *Cold Spring Harbor Symp. Quant. Biol.* **27**, 1.

Caspar, D. L. D., Cohen, C., and Longley, W. (1969). *J. Mol. Biol.* **41**, 87.

Cassel, J. M. (1966). *Biopolymers* **4**, 989.

Cassim, J. Y., Tobias, P. S., and Taylor, E. W. (1968). *Biochim. Biophys. Acta* **168**, 463.

Cella, R. J., Byungkook, L., and Hughes, R. E. (1970). *Acta Crystallogr.* **A26**, 118.

Champness, J. N. (1971). *J. Mol. Biol.* **56**, 295.

Champness, J. N., and Lowy, J. (1968). *In* "Symposium on Fibrous Proteins, Australia 1967" (W. G. Crewther, ed.), p. 106. Butterworths, London.

Chandrasekharan, R., and Balasubramanian R. (1969). *Biochim. Biophys. Acta* **188**, 1.

Chapman, B. M. (1969a). *Textile Res. J.* **39**, 1102.

Chapman, B. M. (1969b). *J. Textile Inst.* **60**, 181.

Chapman, G. E., and McLauchlan, K. A. (1969). *Proc. Roy. Soc. Ser. B* **173**, 223.

Chapman, G. E., Campbell, I. D., and McLauchlan, K. A. (1970). *Nature (London)* **225**, 639.

Chapman, G. E., Danyluk, S. S., and McLauchlan, K. A. (1971). *Proc. Roy. Soc. Ser. B* **178**, 465.

Chen, Y., and Yang, J. T. (1971). *Biochem. Biophys. Res. Commun.* **44**, 1285.

Chen, Y., Yang, J. T., and Martinez, H. M. (1972). *Biochemistry* **11**, 4120.

Chiba, A., Hasegawa, A., Hikichi, K., and Furuichi, J. (1966). *J. Phys. Soc. Japan.* **21**, 1777.

Chidambaram, R., Balasubramanian, R., and Ramachandran, G. N. (1970). *Biochem. Biophys. Acta* **221**, 182.

Chirgadze, Y. N. (1965). *Biofizika* **10**, 201.

Chirgadze, Y. N., Venyaminov, S. U., and Lobachev, V. M. (1971). *Biopolymers* **10**, 809.

Clark, E. S., and Muus, L. T. (1962). *Z. Kristallogr.* **117**, 108.

Clark, G. L., and Schaad, J. A. (1936). *Radiology* **27**, 339.

Clark, G. L., Parker, E. A., Schaad, J. A., and Warren, W. J. (1935). *J. Amer. Chem. Soc.* **57**, 1509.

Clegg, J. B., and Bailey, K. (1962). *Biochem. Biophys. Acta* **63**, 525.

Clifford, J., and Sheard, B. (1966). *Biopolymers* **4**, 1057.

Coates, J. H. (1970). *In* "Physical Principles and Techniques of Protein Chemistry" (S. J. Leach, ed.), Part B, p. 1. Academic Press, New York.

Cochran, W., Crick, F. H. C., and Vand, V. (1952). *Acta Crystallogr.* **5**, 581.

Cohen, C. (1961). *J. Polym. Sci.* **49**, 144.

Cohen, C. (1966a). *In* "Principles of Biomolecular Organization" (G. E. W. Wolstenholme and M. O'Connor, eds.), p. 101. Churchill, London.

Cohen, C. (1966b). *In* "Molecular Architecture in Cell Physiology" (T. Hayashi and A. G. Szent-Györgyi, eds.), p. 169. Prentice-Hall, Englewood Cliffs, New Jersey.

Cohen, C., and Bear, R. S. (1953). *J. Amer. Chem. Soc.* **75**, 2783.

Cohen, C., and Hanson, J. (1956). *Biochim. Biophys. Acta* **21**, 177.

Cohen, C., and Holmes, K. C. (1963). *J. Mol. Biol.* **6**, 423.

Cohen, C., and Longley, W. (1966). *Science* **152**, 794.

Cohen, C., and Szent-Györgyi, A. G. (1957). *J. Amer. Chem. Soc.* **79**, 248.

Cohen, C., and Szent-Györgyi, A. G. (1960). *Int. Congr. Biochem. 4th* **8**, 108. Pergamon, Oxford.

Cohen, C., Revel, J. P., and Kucera, J. (1963). *Science* **141**, 436.

Cohen, C., Slayter, H., Goldstein, L., Kucera, J., and Hall, C. (1966). *J. Mol. Biol.* **22**, 385.

Cohen, C., Lowey, S., Harrison, R. G., Kendrick-Jones, J., and Szent-Györgyi, A. G. (1970). *J. Mol. Biol.* **47**, 605.

Cohen, C., Szent-Györgyi, A. G., and Kendrick-Jones, J. (1971a). *J. Mol. Biol.* **56**, 223.

Cohen, C., Caspar, D. L. D., Parry, D. A. D., and Lucas, R. M. (1971b). *Sympos. Quant. Biol.* **36**, 205.

Cohen, C., Harrison, S. C., and Stephens, R. E. (1971c). *J. Mol. Biol.* **59**, 375.

Coleman, D., and Howitt, F. O. (1947). *Proc. Roy. Soc. Ser. A* **190**, 147.

Colthup, N. B., Daly, L. H., and Wiberley, S. E. (1964). "Introduction to Infrared and Raman Spectroscopy." Academic Press, New York.

Cook, D. A. (1967). *J. Mol. Biol.* **29**, 167.

Cooke, P. H., and Fay, F. S. (1972). *Exp. Cell Res.* **71**, 265.

Cooper, A. (1971). *J. Mol. Biol.* **55**, 123.

Cooper, L., and Steinberg, D. (1970). "Introduction to Methods of Optimization." Saunders, Philadelphia, Pennsylvania.

Cope, F. W. (1969). *Biophys. J.* **9**, 303.

Corey, R. B., and Pauling, L. (1953a). *Proc. Roy. Soc. Ser. B* **141**, 10.

Corey, R. B., and Pauling, L. (1953b). *Rev. Sci. Instrum.* **24**, 621.

Corey, R. B., and Wyckoff, R. W. G. (1936). *J. Biol. Chem.* **114**, 407.

Corfield, M. C., Fletcher, J. C., and Robson, A. (1968). *In* "Symposium on Fibrous Proteins, Australia 1967" (W. G. Crewther, ed.), p. 289. Butterworths, London.

Cornell, S. W., and Koenig, J. L. (1968). *J. Appl. Phys.* **39**, 4883.

Cosslett, V. E. (1970). *Ber. Bunsenges. Phys. Chem.* **74**, 1171.

Cosslett, V. E. (1971). *In* "Physical Techniques in Biological Research" (G. Oster, ed.), Vol. 1, Part A, p. 72. Academic Press, New York.

Cosslett, V. E., and Nixon, W. C. (1960). "X-Ray Microscopy." Cambridge Univ. Press, London and New York.

Cowan, P. M., and McGavin, S. (1955). *Nature (London)* **176**, 501.

Cowan, P. M., North, A. C. T., and Randall, J. T. (1953). *In* "Nature and Structure of Collagen" (J. T. Randall, ed.), p. 241. Butterworths, London.

Cowan, P. M., North, A. C. T., and Randall, J. T. (1955a). *Symp. Soc. Exp. Biol.* **9**, 115.

Cowan, P. M., McGavin, S., and North, A. C. T. (1955b). *Nature (London)* **176**, 1062.

Cowley, J. M. (1968). *In* "Progress in Materials Science" (B. Chalmers and W. Hume-Rothery, eds.), Vol. 13, No. 6. p. 267. Pergamon, Oxford.

Cox, R. W., Grant, R. A., and Horne, R. W. (1967). *J. Roy. Microsc. Soc.* **87**, 123.

Coyle, B. A. (1972). *Acta Crystallogr.* **A28**, 231.

Coyle, B. A., and Schroeder, L. W. (1971). *Acta Crystallogr.* **A27**, 291.

Crane, H. R. (1950). *Sci. Mon.* **70**, 376.

Crewther, W. G. (1972). *Textile Res. J.* **42**, 251.

Crewther, W. G., and Dowling, L. M. (1960). *J. Textile Inst.* **51**, T775.

Crewther, W. G., and Dowling, L. M. (1971). *Appl. Polym. Symp. No. 18*, 1.

Crewther, W. G., and Harrap, B. S. (1967). *J. Biol. Chem.* **242**, 4310.

Crewther, W. G., Fraser, R. D. B., Lennox, F. G., and Lindley, H. (1965). *Advan. Protein Chem.* **20**, 191.

Crewther, W. G., Dobb, M. G., Dowling, L. M., and Harrap, B. S. (1968). *In* "Symposium on Fibrous Proteins, Australia 1967" (W. G. Crewther, ed.), p. 329. Butterworths, London.

Crewther, W. G., Dobb, M. G., and Millward, G. R. (1973). *J. Textile Inst.* **64**.

Crick, F. H. C. (1952). *Nature (London)* **170**, 882.

Crick, F. H. C. (1953a). *Acta Crystallogr.* **6**, 685.

Crick, F. H. C. (1953b). *Acta Crystallogr.* **6**, 689.

Crick, F. H. C. (1954). *J. Chem. Phys.* **22**, 347.

Crick, F. H. C., and Rich, A. (1955). *Nature (London)* **176**, 780.

Crick, R. A., and Misell, D. L. (1971). *J. Phys. D* **4**, 1.

Crighton, J. S., and Happey, F. (1968). *In* "Symposium on Fibrous Proteins, Australia 1967" (W. G. Crewther, ed.), p. 409. Butterworths, London.

Crighton, J. S., Happey, F., and Ball, J. T. (1967). *In* "Conformation of Biopolymers" (G. N. Ramachadran ed.), Vol. 2, p. 623. Academic Press, New York.

Crippen, G. M., and Scheraga, H. A. (1969). *Proc. Nat. Acad. Sci. U.S.* **64**, 42.

Crippen, G. M., and Scheraga, H. A. (1971a). *Arch. Biochem. Biophys.* **144**, 453.

Crippen, G. M., and Scheraga, H. A. (1971b). *Arch. Biochem. Biophys.* **144**, 462.

Crowther, R. A. (1971). *Phil. Trans. Roy. Soc. London Ser. B* **261**, 221.

Crowther, R. A., and Amos, L. A. (1971). *J. Mol. Biol.* **60**, 123.

Crowther, R. A., and Klug, A. (1971). *J. Theor. Biol.* **32**, 199.

Crowther, R. A., Amos, L. A., Finch, J. T., DeRosier, D. J., and Klug, A. (1970a). *Nature (London)* **226**, 421.

Crowther, R. A., DeRosier, D. J., and Klug, A. (1970b). *Proc. Roy. Soc. Ser. A* **317**, 319.

Cruickshank, D. W. J. (1965). *In* "Computing Methods in Crystallography" (J. S. Rollett, ed.), p. 112. Pergamon, Oxford.

Cruickshank, D. W. J., Pilling, D. E., Bujosa, A., Lovell, F. M., and Truter, M. R. (1961). *In* "Computing Methods and the Phase Problem in X-Ray Crystal Analysis" (R. Pepinsky, J. M. Robertson and J. C. Speakman, eds.), p. 32. Pergamon, Oxford.

Dalgleish, D. G. (1972). *FEBS Lett.* **24**, 134.

Damiani, A., De Santis, P., and Pizzi, A. (1970). *Nature (London)* **226**, 542.

Damle, V. N. (1970). *Biopolymers* **9**, 937.

Darskus, R. L., and Gillespie, J. M. (1971). *Aust. J. Biol. Sci.* **24**, 515.

Darskus, R. L., Gillespie, J. M. and Lindley, H. (1969). *Aust. J. Biol. Sci.* **22**, 1197.

Davidon, W. C. (1959). AEC Res. and Develop. Rep., ANL–5990.

Davidson, G., Tooney, N., and Fasman, G. D. (1966). *Biochem. Biophys. Res. Commun.* **23**, 156.

Davies, D. R. (1964). *J. Mol. Biol.* **9**, 605.

Davies, D. R. (1965). *Progr. Biophys. Mol. Biol.* **15**, 189.

Davison, P. F., and Huneeus, F. C. (1970). *J. Mol. Biol.* **52**, 429.

DeDeurwaerder, R. A., Dobb, M. G., and Sweetman, B. J. (1964). *Nature (London)* **203**, 48.

Dehl, R. E., and Hoeve, C. A. J. (1969). *J. Chem. Phys.* **50**, 3245.

de Lozé, C., and Josien, M. L. (1969). *Biopolymers* **8**, 449.

Denbigh, K. G. (1940). *Trans. Faraday Soc.* **36**, 936.

Depue, R. H., and Rice, R. V. (1965). *J. Mol. Biol.* **12**, 302.

Derksen, J. C., and Heringa, G. C. (1936). *Polska Gaz. Lekarska* **15**, 532.

Derksen, J. C., Heringa, G. C., and Weidinger, A. (1937). *Acta Neerl. Morphol.* **1**, 31.

DeRosier, D. J., and Klug, A. (1968). *Nature (London)* **217**, 130.

DeRosier, D. J., and Klug, A. (1972). *J. Mol. Biol.* **65**, 469.

DeRosier, D. J., and Moore, P. B. (1970). *J. Mol. Biol.* **52**, 355.

De Santis, P., and Liquori, A. M. (1971). *Biopolymers* **10**, 699.

De Santis, P., Giglio, E., Liquori, A. M., and Ripamonti, A. (1965). *Nature (London)* **206**, 456.

Desper, C. R., and Stein, R. S. (1966). *J. Appl. Phys.* **37**, 3990.

Deveney, M. J., Walton, A. G., and Koenig, J. L. (1971). *Biopolymers* **10**, 615.

Dickerson, R. E. (1964). *In* "The Proteins" (H. Neurath, ed.), Vol. II, p. 603. Academic Press, New York.

Dobb, M. G. (1964a). *Nature (London)* **202**, 804.

Dobb, M. G. (1964b). *J. Mol. Biol.* **10**, 156.

Dobb, M. G. (1970). *J. Textile Inst. Trans.* **61**, 232.

Dobb, M. G., Johnson, D. J., and Sikorski, J. (1962). *Int. Congr. Elect. Microsc., 5th, Philadelphia.* **2**, T4. Academic Press, New York.

Dobb, M. G., Fraser, R. D. B., and MacRae, T. P. (1965). *Proc. Int. Wool Text. Res. Conf. 3rd* **1**, 95. Inst. Textile de France, Paris.

Dobb, M. G., Fraser, R. D. B., and MacRae, T. P. (1967). *J. Cell Biol.* **32**, 289.

Donohue, J. (1953). *Proc. Nat. Acad. Sci. U.S.* **39**, 470.

Donohue, J. (1968). *In* "Structural Chemistry and Molecular Biology" (A. Rich and N. Davidson, eds.), p. 443. Freeman, San Francisco, California.

Donovan, J. W. (1969). *In* "Physical Principles and Techniques of Protein Chemistry" (S. J. Leach, ed.), Part A, p. 101. Academic Press, New York.

Dopheide, T. (1973). *Eur. J. Biochem.* **34**, 120.

Doty, P., Wada, A., Yang, J. T., and Blout, E. R. (1957). *J. Polym. Sci.* **23**, 851.

Downes, A. M., Ferguson, K. A., Gillespie, J. M., and Harrap, B. S. (1966). *Aust. J. Biol. Sci.* **19**, 319.

Downie, A. R., and Randall, A. A. (1959). *Trans. Faraday Soc.* **55**, 2132.

Doyle, B. B., Traub, W., Lorenzi, G. P., Brown, F. R., and Blout, E. R. (1970). *J. Mol. Biol.* **51**, 47.

Doyle, B. B., Traub, W., Lorenzi, G. P., and Blout, E. R. (1971). *Biochemistry* **10**, 3052.

Doyle, P. A., and Turner, P. S. (1968). *Acta Crystallogr.* **A24**, 390.

Draper, M. H., and Hodge, A. J. (1949). *Aust. J. Exp. Biol. Med. Sci.* **27**, 465.

Drenth, J., Jansonius, J. N., Koekoek, R., and Wolthers, B. G. (1971). *Advan. Protein Chem.* **25**, 79.

Dubochet, J., Ducommon, M., Zollinger, M., and Kellenberger, E. (1971). *J. Ultrastruct. Res.* **35**, 147.

Duke, B. J., and Gibb, T. C. (1967). *J. Chem. Soc.* **A** 1478.

Dunn, M. S., Camien, M. N., Rockland, L. S., Shankman, S., and Goldberg, S. C. (1944). *J. Biol. Chem.* **155**, 591.

Dunnill, P. (1968). *Biophys. J.* **8**, 865.

Dupouy, G., Perrier, F., and Durrieu, L. (1960). *C. R. Acad. Sci. Paris* **251**, 2836.

Dupouy, G., Perrier, F., and Verdier, P. (1966). *J. Microsc.* **5**, 655.

Dupouy, G., Perrier, F., Enjalbert, L., Lapchine, L., and Verdier, P. (1969). *C. R. Acad. Sci. Ser. D* **268**, 1341.

Dweltz, N. E. (1962). *Proc. Indian. Acad. Sci.* **56A**, 329.

Eastoe, J. E. (1967). *In* "Treatise on Collagen" (G. N. Ramachandran, ed.), Vol. 1, p. 1. Academic Press, New York.

Ederer, D. (1969). *Appl. Opt.* **8**, 2315.

Edsall, J. T. *et al.* (1966). *Biopolymers* **4**, 121.

Ehrenberg, W. (1949). *J. Opt. Soc. Amer.* **39**, 741.

Ehrenberg, W., and Spear, W. E. (1951). *Proc. Phys. Soc. London, Sect. B* **64**, 67.

Elleman, T. C. (1971). *Nature (London) New Biol.* **234**, 148.

Elleman, T. C. (1972a). *Biochem. J.* **128**, 1229.

Elleman, T. C. (1972b). *Biochem. J.* **130**, 833.

Elleman, T. C. and Dopheide, T. (1972). *J. Biol. Chem.* **247**, 3900.

Elliott, A. (1952a). *Textile Res. J.* **22**, 783.

Elliott, A. (1952b). *Proc. Roy. Soc. Ser. A* **211**, 490.

Elliott, A. (1953a). *Proc. Roy. Soc. Ser. A* **221**, 104.

Elliott, A. (1953b). *Nature (London)* **172**, 359.

Elliott, A. (1953c). Quoted by Bamford *et al.* (1956).

Elliott, A. (1954). *Proc. Roy. Soc. Ser. A* **226**, 408.

Elliott, A. (1962). *In* "Polyamino Acids, Polypeptides and Proteins" (M. A. Stahmann, ed.), p. 218. Univ. of Wisconsin Press, Madison, Wisconsin.

Elliott, A. (1963). *In* "Aspects of Protein Structure" (G. N. Ramachandran, ed.), p. 54. Academic Press, New York.

Elliott, A. (1965). *J. Sci. Instrum.* **42**, 312.

Elliott, A. (1967). *In* "Poly-α-Amino Acids" (G. D. Fasman, ed.), p. 1. Dekker, New York.

Elliott, A. (1968). *In* "Symposium on Fibrous Proteins, Australia 1967" (W. G. Crewther, ed.), p. 115. Butterworths, London.

Elliott, A. (1969). "Infra-Red Spectra and Structure of Organic Long-Chain Polymers." Arnold, London.

Elliott, A. (1970). *J. Photogr. Sci.* **18**, 68.

Elliott, A. (1971). *Phil. Trans. Roy. Soc. London Ser. B* **261**, 197.

Elliott, A., and Ambrose, E. J. (1950). *Nature (London)* **165**, 921.

Elliott, A., and Bradbury, E. M. (1962). *J. Mol. Biol.* **5**, 574.

Elliott, A., and Lowy, J. (1970). *J. Mol. Biol.* **53**, 181.

Elliott, A., and Malcolm, B. R. (1956a). *Trans. Faraday Soc.* **52**, 528.

Elliott, A., and Malcolm, B. R. (1956b). *Nature (London)* **178**, 912.

Elliott, A., and Malcolm, B. R. (1958). *Proc. Roy. Soc. Ser. A* **249**, 30.

Elliott, A., and Robertson, P. (1955). *Acta Crystallogr.* **8**, 736.

Elliott, A., Hanby, W. E., and Malcolm, B. R. (1954). *Brit. J. Appl. Phys.* **5**, 377.

Elliott, A., Malcolm, B. R., and Hanby, W. E. (1957a). *Nature (London)* **179**, 963.

Elliott, A., Hanby, W. E., and Malcolm, B. R. (1957b). *Nature (London)* **180**, 1340.

Elliott, A., Hanby, W. E., and Malcolm, B. R. (1958). *Discuss. Faraday Soc.* **25**, 167.

Elliott, A., Bradbury, E. M., and Zubay, G. (1962). *J. Mol. Biol.* **4**, 61.

Elliott, A., Fraser, R. D. B., MacRae, T. P., Stapleton, I. W., and Suzuki, E. (1964). *J. Mol. Biol.* **9**, 10.

Elliott, A., Fraser, R. D. B., and MacRae, T. P. (1965). *J. Mol. Biol.* **11**, 821.

Elliott, A., Lowy, J., Parry, D. A. D., and Vibert, P. J. (1968a). *Nature (London)* **218**, 656.

Elliott, A., Lowy, J., and Squire, J. M. (1968b). *Nature (London)* **219**, 1224.

Elliott, G. F. (1963). *Acta Crystallogr.* **16**, A81.

Elliott, G. F. (1964a). *Proc. Roy. Soc. Ser. B* **160**, 467.

Elliott, G. F. (1964b). *J. Mol. Biol.* **10**, 89.

Elliott, G. F., and Lowy, J. (1961). *J. Mol. Biol.* **3**, 41.

Elliott, G. F., Huxley, A. F., and Weis-Fogh, T. (1965). *J. Mol. Biol.* **13**, 791.

Ellis, D. O., and McGavin, S. (1970). *J. Ultrastruct. Res.* **32**, 191.

Ellis, J. W., and Bath, J. (1938). *J. Chem. Phys.* **6**, 723.

Engel, J. (1967). *In* "Conformation of Biopolymers" (G. N. Ramachandran, ed.), Vol. 2, p. 483. Academic Press, New York.

Engel, J., Kurtz, J., Katchalski, E., and Berger, A. (1966). *J. Mol. Biol.* **17**, 255.

Engel, J., Liehl, E., and Sorg, C. (1971). *Eur. J. Biochem.* **21**, 22.

Englander, S. W., and von Hippel, P. H. (1962). Abstr. 142, Nat. Meeting, Amer. Chem. Soc., Atlantic City, New Jersey.

Epand, R. F., and Scheraga, H. A. (1968). *Biopolymers* **6**, 1551.

Epstein, E. H., Scott, R. D., Miller, E. J., and Piez, K. A. (1971). *J. Biol. Chem.* **246**, 1718.

Erenrich, E. H., Andretta, R. H., and Scheraga, H. A. (1970). *J. Amer. Chem. Soc.* **92**, 1116.

Erickson, H. P., and Klug, A. (1971). *Phil. Trans. Roy. Soc. London Ser. B* **261**, 105.

Ericson, L. G., and Tomlin, S. G. (1959). *Proc. Roy. Soc. Ser. A* **252**, 197.

Erlander, S. R., Koffler, H., and Foster, J. F. (1960). *Arch. Biochem. Biophys.* **90**, 139.

Esipova, N. G., Andreeva, N. S., and Gatovskaia, T. V. (1958). *Biofizika* **3**, 529.

Fanconi, B., Tomlinson, B., Nafie, L. A., Small, W., and Peticolas, W. L. (1969). *J. Chem. Phys.* **51**, 3993.

Fantl, P., and Ward, H. A. (1965). *Biochem. J.* **96**, 886.

Fasman, G. D. (1967a). "Poly-α-Amino Acids." Dekker, New York.

Fasman, G. D. (1967b). *In* "Poly-α-Amino Acids" (G. D. Fasman, ed.), p. 499. Dekker, New York.

Fasman, G. D., and Blout, E. R. (1960). *J. Amer. Chem. Soc.* **82**, 2262.

Fasman, G. D., and Potter, J. (1967). *Biochem. Biophys. Res. Commun.* **27**, 209.

Fasman, G. D., Idelson, M., and Blout, E. R. (1961). *J. Amer. Chem. Soc.* **83**, 709.

Fasman, G. D., Landsberg, M., and Buchwald, M. (1965). *Can. J. Chem.* **43**, 1588.

Fasman, G. D., Hoving, H., and Timasheff, S. N. (1970). *Biochemistry* **9**, 3316.

Felix, W. D., McDowall, M. A., and Eyring, H. (1963). *Textile Res. J.* **33**, 465.

Ferrier, R. P. (1969). *In* "Advances in Optical and Electron Microscopy" (R. Barer and V. E. Cosslett, eds.), Vol. III, p. 155. Academic Press, London.

Feughelman, M. (1959). *Textile Res. J.* **39**, 229.

Feughelman, M. (1966). *Nature (London)* **212**, 497.

Feughelman, M. (1968). *In* "Symposium on Fibrous Proteins, Australia 1967" (W. G. Crewther, ed.), p. 397. Butterworths, London.

Feughelman, M., Haly, A. R., and Snaith, J. W. (1962). *Textile Res. J.* **32**, 913.

Filshie, B. K., and Rogers, G. E. (1961). *J. Mol. Biol.* **3**, 784.

Filshie, B. K., and Rogers, G. E. (1962). *J. Cell Biol.* **13**, 1.

Filshie, B. K., Fraser, R. D. B., MacRae, T. P., and Rogers, G. E. (1964). *Biochem. J.* **92**, 19.

Finch, J. T., and Holmes, K. C. (1967). *In* "Methods in Virology" (K. Maramorosch and H. Koprowski, eds.), Vol. 3, p. 351. Academic Press, New York.

Finch, J. T., and Klug, A. (1971). *Phil. Trans. Roy. Soc. London Ser. B* **261**, 211.

Finch, J. T., Klug, A., and Nermut, M. V. (1967). *J. Cell Sci.* **2**, 587.

Fischer, E. W. (1947). *Ann. N.Y. Acad. Sci.* **47**, 783.

Fischer, E. W. (1964). *In* "Newer Methods of Polymer Characterization" (B. Ke, ed.), p. 279. Wiley (Interscience), New York.

Fisher, H. F. (1964). *Proc. Nat. Acad. Sci. U.S.* **51**, 1285.

Finkelstein, A. V., and Ptitsyn, O. B. (1971). *J. Mol. Biol.* **62**, 613.

Fitton Jackson, S. (1968). *In* "Treatise on Collagen" (B. S. Gould, ed.), Vol. 2. Part B, p. 1. Academic Press, New York.

Fitton Jackson, S. *et al.* (1953). *In* "Nature and Structure of Collagen" (J. T. Randall, ed.), p. 106. Butterworths, London.

Fletcher, R. (1965). *Comput. J.* **8**, 33.

Fletcher, R., and Powell, M. J. D. (1963). *Comput. J.* **6**, 163.

Flett, M. S. C. (1963). "Characteristic Frequencies of Chemical Groups in the Infrared." Elsevier, Amsterdam.

Florkin, M., and Stotz, E. H. (1968). "Comprehensive Biochemistry" Vol. 26B. Elsevier, Amsterdam.

Florkin, M., and Stotz, E. H. (1971). "Comprehensive Biochemistry" Vol. 26C, Elsevier, Amsterdam.

Flory, P. J. (1967). *In* "Conformation of Biopolymers" (G. N. Ramachandran, ed.), Vol. I, p. 339. Academic Press, New York.

Flower, N. E. (1968). Quoted by Lucas and Rudall (1968a), p. 535.

Flower, N. E., and Kenchington, W. (1967). *J. Roy. Microsc. Soc.* **86**, 297.

Flower, N. E., Geddes, A. J., and Rudall, K. M. (1969). *J. Ultrastruct. Res.* **26**, 262.

Foà, C. (1912). *Kolloid Z.* **10**, 7.

Franklin, R. E., and Gosling, R. G. (1953). *Acta Crystallogr.* **6**, 678.

Franklin, R. E., and Klug, A. (1955). *Acta Crystallogr.* **8**, 777.

Franks, A. (1955). *Proc. Phys. Soc. London, Sect. B* **68**, 1054.

Franks, A. (1958). *Brit. J. Appl. Phys.* **9**, 349.

Franzblau, C. (1971). *In* "Comprehensive Biochemistry" (M. Florkin and E. H. Stotz, eds.), Vol. 26C, p. 659. Elsevier, Amsterdam.

Franzblau, C., Sinex, F. M., and Faris, B. (1965). *Nature (London)* **205**, 802.

Franzini-Armstrong, C., and Porter, K. R. (1964). *J. Cell Biol.* **22**, 675.

Fraser, I. E. B. (1969). *Aust. J. Biol. Sci.* **22**, 213.

Fraser, R. D. B. (1950a). *Discuss. Faraday Soc.* **9**, 378.

Fraser, R. D. B. (1950b). *Discuss. Faraday Soc.* **9**, 398.

Fraser, R. D. B. (1951). The Application of Infra-Red Spectroscopy to Biological Problems. Ph. D. Thesis, Univ. of London.

Fraser, R. D. B. (1953a). *J. Chem. Phys.* **21**, 1511.

Fraser, R. D. B. (1953b). *Nature (London)* **172**, 675.

Fraser, R. D. B. (1955). *Nature (London)* **176**, 358.

Fraser, R. D. B. (1956a). *J. Chem. Phys.* **24**, 89.

Fraser, R. D. B. (1956b). *Proc. Int. Wool Text. Res. Conf., 1st, Australia 1955 F* 120.

Fraser, R. D. B. (1956c). *Proc. Int. Wool Text. Res. Conf., 1st, Australia 1955 F* 130.

Fraser, R. D. B. (1958a). *J. Chem. Phys.* **28**, 1113.

Fraser, R. D. B. (1958b). *J. Chem. Phys.* **29**, 1428.

Fraser, R. D. B. (1960). *In* "A Laboratory Manual of Analytical Methods of Protein Chemistry" (P. Alexander and R. J. Block, eds.), Vol. 2, p. 285. Pergamon, Oxford.

Fraser, R. D. B. (1961). *In* "Structure de la Laine," p. 25. Inst. Textile de France, Paris.

Fraser, R. D. B. (1969). *Sci. Amer.* **221**, No. 2, 86.

Fraser, R. D. B., and MacRae, T. P. (1957a). *Nature (London)* **179**, 732.

Fraser, R. D. B., and MacRae, T. P. (1957b). *Textile Res. J.* **27**, 384.

Fraser, R. D. B., and MacRae, T. P. (1958a). *Biochim. Biophys. Acta* **29**, 229.

Fraser, R. D. B., and MacRae, T. P. (1958b). *J. Chem. Phys.* **28**, 1120.

Fraser, R. D. B., and MacRae, T. P. (1958c). *J. Chem. Phys.* **29**, 1024.

Fraser, R. D. B., and MacRae, T. P. (1959a). *J. Mol. Biol.* **1**, 387.

Fraser, R. D. B., and MacRae, T. P. (1959b). *Nature (London)* **183**, 179.

Fraser, R. D. B., and MacRae, T. P. (1959c). *J. Chem. Phys.* **31**, 122.

Fraser, R. D. B., and MacRae, T. P. (1961a). *Nature (London)* **189**, 572.

Fraser, R. D. B., and MacRae, T. P. (1961b). *J. Mol. Biol.* **3**, 640.

Fraser, R. D. B., and MacRae, T. P. (1962a). *J. Mol. Biol.* **5**, 457.

Fraser, R. D. B., and MacRae, T. P. (1962b). *Nature (London)* **195**, 1167.

Fraser, R. D. B., and MacRae, T. P. (1963). *J. Mol. Biol.* **7**, 272.

Fraser, R. D. B., and MacRae, T. P. (1969). *In* "Physical Principles and Techniques of Protein Chemistry" (S. J. Leach, ed.), Part A, p. 59. Academic Press, New York.

Fraser, R. D. B., and MacRae, T. P. (1971). *Nature (London)* **233**, 138.

Fraser, R. D. B., and MacRae, T. P. (1973). *Polymer* **14**, 61.

Fraser, R. D. B., and Millward, G. R. (1970). *J. Ultrastruct. Res.* **31**, 203.

Fraser, R. D. B., and Price, W. C. (1952). *Nature (London)* **170**, 490.

Fraser, R. D. B., and Suzuki, E. (1964). *J. Mol. Biol.* **9**, 829.

Fraser, R. D. B., and Suzuki, E. (1965). *J. Mol. Biol.* **14**, 279.

Fraser, R. D. B., and Suzuki, E. (1966). *Anal. Chem.* **38**, 1770.

Fraser, R. D. B., and Suzuki, E. (1969). *Anal. Chem.* **41**, 37.

Fraser, R. D. B., and Suzuki, E. (1970a). *In* "The Physical Principles and Techniques of Protein Chemistry" (S. J. Leach, ed.), Vol. B, p. 213. Academic Press, New York.

Fraser, R. D. B., and Suzuki, E. (1970b). *In* "Spectral Analysis" (J. A. Blackburn, ed.), p. 171. Dekker, New York.

Fraser, R. D. B., and Suzuki, E. (1970c). *Spectrochim. Acta* **26A**, 423.

Fraser, R. D. B., and Suzuki, E. (1973). *In* "The Physical Principles and Techniques of Protein Chemistry" (S. J. Leach, ed.), Vol. C, p. 301. Academic Press, New York.

Fraser, R. D. B., MacRae, T. P., and Rogers, G. E. (1960). *J. Textile Inst.* **51**, T497.

Fraser, R. D. B., MacRae, T. P., and Stapleton, I. W. (1962a). *Nature (London)* **193**, 573.

Fraser, R. D. B., MacRae, T. P., and Rogers, G. E. (1962b). *Nature (London)* **193**, 1052.

Fraser, R. D. B., MacRae, T. P., and Rogers, G. E. (1962c). Unpublished observations.

Fraser, R. D. B., MacRae, T. P., Rogers, G. E., and Filshie, B. K. (1963). *J. Mol. Biol.* **7**, 90.

Fraser, R. D. B., MacRae, T. P., and Miller, A. (1964a). *Acta Crystallogr.* **17**, 813.

Fraser, R. D. B., MacRae, T. P., and Miller, A. (1964b). *Acta Crystallogr.* **17**, 769.

Fraser, R. D. B., MacRae, T. P., Miller, A., and Suzuki, E. (1964c). *J. Mol. Biol.* **9**, 250.

Fraser, R. D. B., MacRae, T. P., and Miller, A. (1964d). *Nature (London)* **203**, 1231.

Fraser, R. D. B., MacRae, T. P., and Miller, A. (1964e). *J. Mol. Biol.* **10**, 147.

Fraser, R. D. B., MacRae, T. P., Stewart, F. H. C., and Suzuki, E. (1965a). *J. Mol. Biol.* **11**, 706.

Fraser, R. D. B., Harrap, B. S., MacRae, T. P., Stewart, F. H. C., and Suzuki, E. (1965b). *J. Mol. Biol.* **12**, 482.

Fraser, R. D. B., MacRae, T. P., and Stewart, F. H. C. (1965c). *J. Mol. Biol.* **13**, 949.

Fraser, R. D. B., Harrap, B. S., MacRae, T. P., Stewart, F. H. C., and Suzuki, E. (1965d). *J. Mol. Biol.* **14**, 423.

Fraser, R. D. B., MacRae, T. P., and Miller, A. (1965e). *J. Mol. Biol.* **14**, 432.

Fraser, R. D. B., MacRae, T. P., Miller, A., Stewart, F. H. C., Suzuki, E. (1965f). *Proc. Int. Wool Text. Res. Conf., 3rd* **1**, 85. Inst. Textile de France, Paris.

Fraser, R. D. B., MacRae, T. P., and Miller, A. (1965g). *Acta Crystallogr.* **18**, 1087.

Fraser, R. D. B., MacRae, T. P., Miller, A., Stewart, F. H. C., and Suzuki, E. (1965h). Unpublished observations.

Fraser, R. D. B., MacRae, T. P., and Stewart, F. H. C. (1966). *J. Mol. Biol.* **19**, 580.

Fraser, R. D. B., Harrap, B. S., MacRae, T. P., Stewart, F. H. C., and Suzuki, E. (1967a). *Biopolymers* **5**, 251.

Fraser, R. D. B., Harrap, B. S., Ledger, R., MacRae, T. P., Stewart, F. H. C., and Suzuki, E. (1967b). *Biopolymers* **5**, 797.

Fraser, R. D. B., Harrap, B. S., Ledger, R., MacRae, T. P., Stewart, F. H. C., and Suzuki, E. (1968a). *In* "Symposium on Fibrous Proteins, Australia 1967" (W. G. Crewther, ed.), p. 57. Butterworths, London.

Fraser, R. D. B., MacRae, T. P., and Parry, D. A. D. (1968b). *In* "Symposium on Fibrous Proteins, Australia 1967" (W. G. Crewther, ed.), p. 279. Butterworths, London.

Fraser, R. D. B., MacRae, T. P., Parry, D. A. D., and Suzuki, E. (1969a). *Polymer* **10**, 810.

Fraser, R. D. B., MacRae, T. P., and Millward, G. R. (1969b). *J. Textile Inst.* **60**, 343.

Fraser, R. D. B., MacRae, T. P., and Millward, G. R. (1969c). *J. Textile Inst.* **60**, 498.

Fraser, R. D. B., MacRae, T. P., Parry, D. A. D., and Suzuki, E. (1971a). *Polymer* **12**, 35.

Fraser, R. D. B., MacRae, T. P., Millward, G. R., Parry, D. A. D., Suzuki, E., and Tulloch, P. A. (1971b). *Appl. Polym. Symp. No. 18*, 65.

Fraser, R. D. B., MacRae, T. P., and Rogers, G. E. (1972). "Keratins—Their Composition, Structure, and Biosynthesis." Thomas, Springfield, Illinois.

Fraser, R. D. B., Gillespie, J. M., and MacRae, T. P. (1973). *Comp. Biochem. Physiol.* **44B**, 943.

Frater, R. (1966). *Aust. J. Biol. Sci.* **19**, 699.

Frieder, G., and Herman, G. T. (1971). *J. Theor. Biol.* **33**, 189.

Friedman, M. H. (1970). *J. Ultrastruct. Res.* **32**, 226.

Friedrich-Freksa, H., Kratky, O., and Sekora, A. (1944). *Naturwissenschaften* **32**, 78.

Frisch, D. (1969). *J. Ultrastruct. Res.* **29**, 357.

Fujime-Higashi, S., and Ooi, T. (1969). *J. Microsc.* **8**, 535.

Fukami, A., and Adachi, K. (1965). *J. Electron Microsc.* **14**, 112.

Furedi, H., and Walton, A. G. (1968). *Appl. Spectrosc.* **22**, 23.

Furneaux, P. J. S., and MacKay, A. L. (1972). *J. Ultrastruct. Res.* **38**, 343.

Gall, M. J., Hendra, P. J., Watson, D. S., and Peacock, C. J. (1971). *Appl. Spectrosc.* **25**, 423.

Gates, R. E., and Fisher, H. F. (1971). *Proc. Nat. Acad. Sci. U.S.* **68**, 2928.

Geddes, A. J., Parker, K. D., Atkins, E. D. T., and Beighton, E. (1968). *J. Mol. Biol.* **32**, 343.

Geil, P. H. (1963). "Polymer Single Crystals." Wiley, New York.

Gerasimov, V. I. (1970). *Kristallografia* **15**, 156 (*English transl.: Sov. Phys. Crystallogr.* **15**, 122).

Gergely, J. (1966). *Annu. Rev. Biochem.* **35**, 691.

Gerngross, O., and Katz, J. R. (1926). *Kolloid-Z.* **39**, 181.

Gerngross, O., Herrmann, K., and Abitz, W. (1930). *Biochem. Z.* **228**, 409.

Gerngross, O., Herrmann, K., and Abitz, W. (1931). *Collegium* **53**.

Gibson, K. D., and Scheraga, H. A. (1966). *Biopolymers* **4**, 709.
Gibson, K. D., and Scheraga, H. A. (1967). *Proc. Nat. Acad. Sci. U.S.* **58**, 420.
Gibson, K. D., and Scheraga, H. A. (1969a). *Proc. Nat. Acad. Sci. U.S.* **63**, 9.
Gibson, K. D., and Scheraga, H. A. (1969b). *Physiol. Chem. Phys.* **1**, 109.
Gibson, K. D., and Scheraga, H. A. (1970). *Comput. Biomed. Res.* **3**, 375.
Gilbert, P. F. C. (1972a). *Proc. Roy. Soc. Ser. B* **182**, 89.
Gilbert, P. (1972b). *J. Theor. Biol.* **36**, 105.
Gilëv, V. P. (1966a). *In* "Electron Microscopy 1966" (R. Uyeda, ed.), Vol. 2, p. 689. Maruzen, Tokyo.
Gilëv, V. P. (1966b). *Biochim. Biophys. Acta* **112**, 340.
Gilëv, V. P. (1970). *In* "Microscopie Électronique, 1970" (P. Favard, ed.), Vol. 3, p. 785. Société Francaise de Microscopie Électronique, Paris.
Gillespie, J. M. (1963). *Aust. J. Biol. Sci.* **16**, 259.
Gillespie, J. M. (1965). *In* "Biology of Skin and Hair Growth" (A. G. Lyne and B. F. Short eds.). Angus and Robertson, Sydney.
Gillespie, J. M. (1972). *Comp. Biochem. Physiol.* **41B**, 723.
Gillespie, J. M., and Broad, A. (1972). *Aust. J. Biol. Sci.* **25**, 139.
Gillespie, J. M., and Darskus, R. L. (1971). *Aust. J. Biol. Sci.* **24**, 1189.
Gillespie, J. M., and Harrap, B. S. (1963). *Aust. J. Biol. Sci.* **16**, 252.
Gillespie, J. M., Reis, P. J., and Schinkel, P. G. (1964). *Aust. J. Biol. Sci.* **17**, 548.
Giroud, A., and Champetier, G. (1936). *Bull. Soc. Chim. Biol.* **18**, 656.
Giroud, A., and Leblond, C. P. (1951). *Ann. N.Y. Acad. Sci.* **53**, 613.
Giroud, A., Bulliard, H., and Leblond, C. P. (1934). *Bull. Histol. Appl. Physiol. Pathol. Tech. Microsc.* **11**, 129.
Glaeser, R. M., and Thomas, G. (1969). *Biophys. J.* **9**, 1073.
Glatt, L., and Ellis, J. W. (1948). *J. Chem. Phys.* **16**, 551.
Glauert, A. M. (1965). *In* "Techniques for Electron Microscopy" (D. H. Kay, ed.), p. 254. Blackwell, Oxford.
Glauert, A. M., and Phillips, R. (1965). *In* "Techniques for Electron Microscopy" (D. H. Kay, ed.), p. 213. Blackwell, Oxford.
Glimcher, M. J., and Krane, S. M. (1968). *In* "Treatise on Collagen" (B. S. Gould, ed.), Vol. 2B, p. 68. Academic Press, New York.
Gō, N., and Scheraga, H. A. (1969). *J. Chem. Phys.* **51**, 4751.
Gō, N., and Scheraga, H. A. (1970). *Macromolecules* **3**, 188.
Go, Y., Noguchi, J., Asai, M., and Hayakawa, T. (1956). *J. Polym. Sci.* **21**, 147.
Go, Y., Ejiri, S., and Fukada, E. (1969). *Biochim. Biophys. Acta* **175**, 454.
Goddard, D. R., and Michaelis, L. (1934). *J. Biol. Chem.* **106**, 605.
Goddard, D. R., and Michaelis, L. (1935). *J. Biol. Chem.* **112**, 361.
Godfrey, J. E., and Harrington, W. F. (1970). *Biochemistry* **9**, 894.
Goldsack, D. E. (1969). *Biopolymers* **7**, 299.
Goldstein, M., and Halford, R. S. (1949). *J. Amer. Chem. Soc.* **71**, 3854.
Goodman, M., and Listowsky, I. (1962). *J. Amer. Chem. Soc.* **84**, 3770.
Goodman, M., and Rosen, I. G. (1964). *Biopolymers* **2**, 537.
Goodman, M., Boardman, F., and Listowsky, I. (1963a). *J. Amer. Chem. Soc.* **85**, 2491.
Goodman, M., Felix, A. M., Deber, C. M., Brause, A. R., and Schwartz, G. (1963b). *Biopolymers* **1**, 371.
Goodman, M., Listowsky, I., Masuda, Y., and Boardman, F. (1963c). *Biopolymers* **1**, 33.
Goodman, M., Masuda, Y., and Verdini, A. S. (1971). *Biopolymers* **10**, 1031.
Gordon, R., Bender, R., and Herman, G. T. (1970). *J. Theor. Biol.* **29**, 471.

Gotte, L., Mammi, M., and Pezzin, G. (1968). *In* "Symposium on Fibrous Proteins, Australia 1967" (W. G. Crewther, ed.), p. 236. Butterworths, London.

Grant, R. A., Horne, R. W., and Cox, R. W. (1965). *Nature (London)* **207**, 2822.

Grant, R. A., Cox, R. W., and Horne, R. W. (1967). *J. Roy. Microsc. Soc.* **87**, 143.

Gratzer, W. B. (1967). *In* "Poly-α-Amino Acids" (G. D. Fasman, ed.), p. 177. Dekker, New York.

Gratzer, W. B., Holzwarth, G. M., and Doty, P. (1961). *Proc. Nat. Acad. Sci. U.S.* **47**, 1785.

Greenberg, J., Fishman, L., and Levy, M. (1964). *Biochemistry* **3**, 1826.

Greenfield, N., and Fasman, G. D. (1969). *Biochemistry* **8**, 4108.

Greenfield, N., Davidson, B., and Fasman, G. D. (1967). *Biochemistry* **6**, 1630.

Gregory, J., and Holmes, K. C. (1965). *J. Mol. Biol.* **13**, 796.

Grimstone, A. V. (1968). *Symp. Int. Soc. Cell Biol.* **6**, 219.

Grimstone, A. V., and Klug, A. (1966). *J. Cell Sci.* **1**, 351.

Gross, J. (1956). *J. Biophys. Biochem. Cytol., Suppl.* **2**, 261.

Gross, J. (1963). *Comp. Biochem.* **5**, 307.

Gross, J., and Nagai, Y. (1965). *Proc. Nat. Acad. Sci. U.S.* **54**, 1197.

Gross, J., and Schmitt, F. O. (1948). *J. Exp. Med.* **88**, 555.

Gross, J., Highberger, J. H., and Schmitt, F. O. (1952). *Proc. Soc. Exp. Biol. Med.* **80**, 462.

Gross, J., Highberger, J. H., and Schmitt, F. O. (1954). *Proc. Nat. Acad. Sci. U.S.* **40**, 679.

Grover, N. B. (1965). *J. Ultrastruct. Res.* **12**, 574.

Guzzo, A. V. (1965). *Biophys. J.* **5**, 809.

Hall, C. E. (1948). *J. Appl. Phys.* **19**, 198.

Hall, C. E. (1949). *J. Biol. Chem.* **179**, 857.

Hall, C. E. (1956). *Proc. Nat. Acad. Sci. U.S.* **42**, 801.

Hall, C. E. (1960). *J. Biophys. Biochem. Cytol.* **7**, 613.

Hall, C. E. (1963). *Lab. Invest.* **12**, 998.

Hall, C. E., and Slayter, H. S. (1959). *J. Biophys. Biochem. Cytol.* **5**, 11.

Hall, C. E., Jakus, M. A., and Schmitt, F. O. (1942). *J. Amer. Chem. Soc.* **64**, 1234.

Hall, C. E., Jakus, M. A., and Schmitt, F. O. (1945). *J. Appl. Phys.* **16**, 459.

Halvarson, M., and Afzelius, B. A. (1969). *J. Ultrastruct. Res.* **26**, 289.

Haly, A. R., and Snaith, J. W. (1967). *Textile Res. J.* **37**, 898.

Haly, A. R., and Swanepoel, O. A. (1961). *Textile Res. J.* **31**, 966.

Hamilton, W. C. (1964). "Statistics in Physical Science." Ronald Press, New York.

Hamilton, W. C. (1965). *Acta Crystallogr.* **18**, 502.

Hanson, J. (1967). *Nature (London)* **213**, 353.

Hanson, J. (1968). *In Symp. Muscle* (E. Ernst and F. B. Straub, eds.), p. 93. Akad. Kiado, Budapest.

Hanson, J. (1969). *Quart. Rev. Biophys.* **1**, 177.

Hanson, J., and Huxley, H. E. (1953). *Nature (London)* **172**, 530.

Hanson, J., and Huxley, H. E. (1955). *In Symp. Soc. Exp. Biol.* (R. Brown and J. F. Danielli, eds.), No. 9, p. 228. Cambridge Univ. Press, London and New York.

Hanson, J., and Lowy, J. (1963). *J. Mol. Biol.* **6**, 46.

Hanson, J., and Lowy, J. (1964a). *In* "Biochemistry of Muscle Contraction" (J. Gergely, ed.), p. 141. Little, Brown, Boston, Massachusetts.

Hanson, J., and Lowy, J. (1964b). *In* "Biochemistry of Muscle Contraction" (J. Gergely, ed.), p. 400. Little, Brown, Boston, Massachusetts.

Hanson, J., Lowy, J., Huxley, H. E., Bailey, K., Kay, C. M., and Rüegg, J. C. (1957). *Nature (London)* **180**, 1134.

Hanson, J., O'Brien, E. J., and Bennett, P. M. (1971). *J. Mol. Biol.* **58**, 865.

Hanszen, K. J. (1971). *In* "Advances in Optical and Electron Microscopy" (R. Barer, ed.), Vol. 4, p. 1. Academic Press, New York.

Happey, F. (1950). *Nature (London)* **166**, 396.

Happey, F. (1955). *In* "X-Ray Diffraction by Polycrystalline Materials" (H. S. Peiser, H. P. Rooksby, and A. J. C. Wilson, eds.), p. 533. Inst. of Phys., London.

Harada, J., and Kashiwase, Y. (1962). *J. Phys. Soc. Japan* **17**, 829.

Harburn, G. (1972). *In* "Optical Transforms" (H. S. Lipson, ed.), p. 189. Academic Press, New York.

Harkness, R. D. (1968). *In* "Treatise on Collagen" (B. S. Gould, ed.), Vol. 2A, p. 247. Academic Press, New York.

Harrap, B. S. (1969). *Int. J. Protein Res.* **1**, 245.

Harrap, B. S. (1970). Personal communication.

Harrap, B. S., and Stapleton, I. W. (1963). *Biochim. Biophys. Acta* **75**, 31.

Harrap, B. S., and Woods, E. F. (1964a). *Biochem. J.* **92**, 8.

Harrap, B. S., and Woods, E. F. (1964b). *Biochem. J.* **92**, 19.

Harrap, B. S., and Woods, E. F. (1967). *Comp. Biochem. Physiol.* **20**, 449.

Harrington, W. F. (1964). *J. Mol. Biol.* **9**, 613.

Harrington, W. F., and Rao, N. V. (1967). *In* "Conformation of Biopolymers" (G. N. Ramachandran, ed.), Vol. 2, p. 513. Academic Press, New York.

Harrington, W. F., and Sela, M. (1958). *Biochim. Biophys. Acta* **27**, 24.

Harrington, W. F., and von Hippel, P. H. (1961). *Advan. Protein Chem.* **16**, 1.

Harrison, R. G., Lowey, S., and Cohen, C. (1971). *J. Mol. Biol.* **59**, 531.

Hartley, G. S., and Robinson, C. (1952). *Trans. Faraday Soc.* **48**, 847.

Hashimoto, M. (1966). *Bull. Chem. Soc. Japan* **39**, 2713.

Hashimoto, M., and Arakawa, S. (1967). *Bull. Chem. Soc. Japan* **40**, 1698.

Hashimoto, M., and Aritomi, J. (1966). *Bull. Chem. Soc. Japan* **39**, 2707.

Havsteen, B. H. (1966). *J. Theor. Biol.* **10**, 1.

Hawes, R. C., George, K. P., Nelson, D. C., and Beckwith, R. (1966). *Anal. Chem.* **38**, 1842.

Hawn, C. V. Z., and Porter, K. R. (1947). *J. Exp. Med.* **86**, 285.

Hayakawa, T., Kondo, Y., Yamamoto, S., and Noguchi, J. (1970). *Kobunshi Kagaku* **27**, 229.

Haydon, G. B. (1968). *J. Ultrastruct. Res.* **25**, 349.

Haydon, G. B. (1969). *J. Microsc.* **89**, 73.

Haylett, T., and Swart, L. S. (1969). *Textile Res. J.* **39**, 917.

Haylett, T., Swart, L. S., Parris, D., and Joubert, F. J. (1971). *Appl. Polym. Symp. No. 18*, 37.

Hayes, D., Huang, M., and Zobel, C. R. (1971). *J. Ultrastruct. Res.* **37**, 17.

Hazlewood, C. F., Nichols, B. L., and Chamberlain, N. F. (1969). *Nature (London)* **222**, 747.

Hearle, J. W. S., and Greer, R. (1970). *In* "Textile Progress" (P. W. Harrison ed.), Vol. 2, No. 4, p. 1. Textile Inst., Manchester.

Hecht, K. T., and Wood, D. L. (1956). *Proc. Roy. Soc. Ser. A* **235**, 174.

Heffelfinger, C. J., and Burton, R. L. (1960). *J. Polym. Sci.* **47**, 289.

Hegetschweiler, R. (1949). *Makromol. Chem.* **4**, 156.

Heidemann, G., and Halboth, H. (1967). *Nature (London)* **213**, 71.

Heidemann, G., and Halboth, H. (1970). *Textile Res. J.* **40**, 861.

Heidenreich, R. D. (1964). "Fundamentals of Transmission Electron Microscopy." Wiley (Interscience), New York.

Hendra, P. J. (1969). *Advan. Polym. Sci.* **6**, 151.

Henry, N. F. M., and Lonsdale, K. (1952). "International Tables for X-Ray Crystallography," Vol. I. Kynoch Press, Birmingham.

Herbstein, F. H., Boonstra, E. G., Dunn, H. M., Chipman, D. R., Boldrini, P., and Loopstra, B. O. (1967). "Methods of Obtaining Monochromatic X-Rays and Neutrons," I.U.C. Bibliography. Oosthoek, Utrecht.

Herman, G. T., and Rowland, S. (1971). *J. Theor. Biol.* **33**, 213.

Hermans, J., and Puett, D. (1971). *Biopolymers* **10**, 895.

Hermans, P. H. (1946). "Contribution to the Physics of Cellulose Fibres." Elsevier, Amsterdam.

Herriott, J. R., Sieker, L. C., and Jensen, L. H. (1970). *J. Mol. Biol.* **50**, 391.

Herrmann, K., Gerngross, O., and Abitz, W. (1930). *Z. Phys. Chem.* **B10**, 371.

Herzberg, G. (1945). "Molecular Spectra and Molecular Structure," Vol. II. Van Nostrand-Reinhold, Princeton, New Jersey.

Herzog, R. O., and Gonnell, H. W. (1931). *Z. Phys. Chem.* **B12**, 228.

Herzog, R. O., and Jancke, W. (1920). *Ber. Deut. Chem. Ges.* **B53**, 2162.

Herzog, R. O., and Jancke, W. (1921). *In* "Festschrift Kaiser-Wilhelm Gesellschaft" p. 118. Springer, Berlin.

Herzog, R. O., and Jancke, W. (1926). *Ber. Deut. Chem. Ges.* **B59**, 2487.

Herzog, R. O., and Jancke, W. (1929). *Z. Phys.* **52**, 755.

Higgs, P. W. (1953). *Proc. Roy. Soc. Ser. A* **220**, 472.

Highberger, J. H., Gross, J., and Schmitt, F. O. (1950). *J. Amer. Chem. Soc.* **72**, 3321.

Highberger, J. H., Gross, J., and Schmitt, F. O. (1951). *Proc. Nat. Acad. Sci. U.S.* **37**, 286.

Highberger, J. H., Kang, A. H., and Gross, J. (1971). *Biochemistry* **10**, 610.

Hirabayashi, K., Uchiyama, K., Ishikawa, H., and Go, Y. (1967). *Sen -i Gakkaishi* **23**, 538.

Hirabayashi, K., Ishikawa, H., Kakudo, M., and Go, Y. (1968). *Sen -i Gakkaishi* **24**, 397.

Hirabayashi, K., Ishikawa, H., Kasai, N., and Kakudo, M. (1970). *Rep. Prog. Phys. Japan* **13**, 223.

Hirsch, P. B. (1955). *In* "X-Ray Diffraction by Polycrystalline Materials" (H. S. Peiser, H. P. Rooksby, and A. J. C. Wilson eds.), p. 278. Inst. of Phys., London.

Hitchborn, J. H., and Hills, G. J. (1967). *Science* **157**, 705.

Hodge, A. J. (1952). *Proc. Nat. Acad. Sci. U.S.* **38**, 850.

Hodge, A. J. (1959). *Rev. Mod. Phys.* **31**, 409.

Hodge, A. J. (1967). *In* "Treatise on Collagen" (G. N. Ramachandran, ed.), Vol. 1, p. 185. Academic Press, New York.

Hodge, A. J., and Petruska, J. A. (1963). *In* "Aspects of Protein Structure" (G. N. Ramachandran, ed.), p. 289. Academic Press, New York.

Hodge, A. J., and Schmitt, F. O. (1960). *Proc. Nat. Acad. Sci. U.S.* **46**, 186.

Hodge, A. J., Huxley, H. E., and Spiro, D. (1954). *J. Exp. Med.* **99**, 201.

Hodges, R. S., and Smillie, L. B. (1970). *Biochem. Biophys. Res. Commun.* **41**, 987.

Hodges, R. S., and Smillie, L. B. (1972a). *Can. J. Biochem.* **50**, 312.

Hodges, R. S., and Smillie, L. B. (1972b). *Can. J. Biochem.* **50**, 330.

Hodges, R. S., Soder, J., Smillie, L. B., and Jurasek, L. (1973). *Sympos. Quant. Biol.* **37**, 299.

Holmes, K. C., and Blow, D. M. (1965). *In* "Methods of Biochemical Analysis" (D. Glick, ed.), Vol. 13, p. 113. Wiley (Interscience), New York.

Holtzer, A., Clark, R., and Lowey, S. (1965). *Biochemistry* **4**, 2401.

Hooke, R., and Jeeves, T. A. (1961). *J. Ass. Comput. Mach.* **8**, 212.

Hooper, C. W., Seeds, W. E., and Stokes, A. R. (1955). *Nature (London)* **175**, 679.

Hopfinger, A. J., and Walton, A. G. (1970a). *J. Macromol. Sci., Part B* **4**, 185.

Hopfinger, A. J., and Walton, A. G. (1970b). *Biopolymers* **9**, 29.

Hopfinger, A. J., and Walton, A. G. (1970c). *Biopolymers* **9**, 433.

Horne, R. W. (1965a). *In* "Techniques for Electron Microscopy" (D. H. Kay, ed.), p. 311. Blackwell, Oxford.

Horne, R. W. (1965b). *In* "Techniques for Electron Microscopy" (D. H. Kay, ed.), p. 328. Blackwell, Oxford.

Horne, R. W., and Markham, R. (1972). *In* "Practical Methods in Electron Microscopy" (A. M. Glauert, ed.), Vol. I, p. 327. North-Holland, Amsterdam.

Hosemann, R. (1951). *Acta Crystallogr.* **4**, 520.

Hosemann, R. (1962). *Polymer* **3**, 349.

Hosemann, R., and Bagchi, S. N. (1962). "Direct Analysis of Diffraction by Matter." North-Holland, Amsterdam.

Hosken, R., Moss, B. A., O'Donnell, I. J., and Thompson, E. O. P. (1968). *Aust. J. Biol. Sci.* **21**, 593.

Huc, A., and Sanejouand, J. (1968). *Biochim. Biophys. Acta* **154**, 408.

Huggins, M. L. (1943). *Chem. Rev.* **32**, 195.

Huggins, M. L. (1945). *J. Chem. Phys.* **13**, 37.

Huggins, M. L. (1952). *J. Amer. Chem. Soc.* **74**, 3963.

Huggins, M. L. (1954). *J. Amer. Chem. Soc.* **76**, 4045.

Huggins, M. L. (1957). *Proc. Nat. Acad. Sci. U.S.* **43**, 209.

Huggins, M. L. (1958). *J. Polym. Sci.* **30**, 5.

Huggins, M. L. (1966). *Makromol. Chem.* **92**, 260.

Huneeus, F. C., and Davison, P. F. (1970). *J. Mol. Biol.* **52**, 415.

Hunt, S. (1970). *Biochim. Biophys. Acta* **207**, 347.

Hunt, S. (1971). *Comp. Biochem. Physiol.* **40B**, 715.

Huszar, G., and Elzinger, M. (1971). *Biochemistry* **10**, 229.

Huxley, H. E. (1953). *Acta Crystallogr.* **6**, 457.

Huxley, H. E. (1957). *J. Biophys. Biochem. Cytol.* **3**, 631.

Huxley, H. E. (1963). *J. Mol. Biol.* **7**, 281.

Huxley, H. E. (1966). "Harvey Lectures," Ser. 60, p. 85. Academic Press, New York.

Huxley, H. E. (1967). *J. Gen. Physiol.* **50**, 71.

Huxley, H. E. (1968). *J. Mol. Biol.* **37**, 507.

Huxley, H. E. (1971). *Proc. Roy. Soc. Ser. B* **178**, 131.

Huxley, H. E., and Brown, W. (1967). *J. Mol. Biol.* **30**, 383.

Huxley, H. E., and Hanson, J. (1960). *In* "Structure and Function of Muscle" (G. H. Bourne, ed.), Vol. 1, p. 183. Academic Press, New York.

Huxley, H. E., and Klug, A. (1971). A Discussion on New Developments in Electron Microscopy with Special Emphasis on Their Application in Biology. *Phil. Trans. Roy. Soc. London Ser. B* **261**, pp. 1–230. Roy. Soc., London.

Huxley, H. E., and Perutz, M. F. (1951). *Nature (London)* **167**, 1054.

Huxley, H. E., and Zubay, G. (1960). *J. Mol. Biol.* **2**, 10.

Huxley, H. E., Holmes, K. C., and Brown, W. (1966). *In* "Principles of Biomolecular Organization" (G. E. W. Wolstenholme and M. O'Connor, eds.), p. 259. Churchill, London.

Hyde, A. J., and Wippler, C. (1962). *J. Polym. Sci.* **58**, 1083.

Ibers, J. A. (1962). *In* "International Tables for X-Ray Crystallography" (C. H. MacGillavry and G. D. Rieck, eds.), Vol. III, p. 201. Kynoch Press, Birmingham.

Iio, T., and Takahashi, S. (1970). *Bull. Chem. Soc. Japan* **43**, 515.

Iizuka, E. (1963). *J. Soc. Text. Ind. Japan* **19**, 911.

Iizuka, E. (1966). *Biorheology* **3**, 141.

Iizuka, E. (1968). *Biochim. Biophys. Acta* **160**, 454.

Iizuka, E., and Yang, J. T. (1966). *Proc. Nat. Acad. Sci. U.S.* **55**, 1175.

Ikeda, S., Maeda, H., and Isemura, T. (1964). *J. Mol. Biol.* **10**, 223.

Imahori, K. (1960). *Biochim. Biophys. Acta* **37**, 336.

Imahori, K., and Inouye, H. (1967). *Biopolymers* **5**, 639.

Imahori, K., and Tanaka, J. (1959). *J. Mol. Biol.* **1**, 359.

Imahori, K., and Yahara, I. (1964). *Biopolym. Symp.* No. 1, 421.

Ingwall, R. T., and Flory, P. J. (1972). *Biopolymers* **11**, 1527.

Itoh, K., Shimanouchi, T., and Oya, M. (1969). *Biopolymers* **7**, 649.

IUPAC-IUB Commission on Biochemical Nomenclature (1970). *J. Mol. Biol.* **52**, 1.

Jahnke, E., and Emde, F. (1945). "Tables of Functions." Dover, New York.

Jakus, M. A., and Hall, C. E. (1947). *J. Biol. Chem.* **167**, 705.

Jakus, M. A., Hall, C. E., and Schmitt, F. O. (1944). *J. Amer. Chem. Soc.* **66**, 313.

James, R. W. (1954). "The Optical Principles of the Diffraction of X-Rays." Bell and Sons, London.

Jeffery, J. W. (1971). "Methods in X-Ray Crystallography." Academic Press, New York.

Jeffrey, P. D. (1972). *J. Textile Inst.* **63**, 91.

Jellinek, G. (1972). *Messtechnik* **80**, 179.

Jennison, R. C. (1961). "Fourier Transforms and Convolutions." Pergamon, Oxford.

Jensen, M., and Weis-Fogh, T. (1962). *Phil. Trans. Roy. Soc. Ser. B* **245**, 137.

Jirgensons, B. (1969). "Optical Rotatory Dispersion of Proteins and Other Macromolecules." Springer-Verlag, New York.

Jirgensons, B. (1970). *Biochim. Biophys. Acta* **200**, 9.

Johnson, C. K. (1959). X-Ray Diffraction Studies of Polypeptide Packing. Ph. D. Thesis, Massachusetts Inst. of Technol., Cambridge, Massachusetts.

Johnson, D. J., and Sikorski, J. (1962). *Nature (London)* **194**, 31.

Johnson, D. J., and Sikorski, J. (1964). *Proc. Eur. Reg. Conf. Elect. Microsc., 3rd, Prague* **B**, 63.

Johnson, D. J., and Sikorski, J. (1965). *Proc. Int. Wool Text. Res. Conf., 3rd* **1**, 147. Inst. Textile de France, Paris.

Johnson, D. J., and Speakman, P. T. (1965a). *Nature (London)* **205**, 268.

Johnson, D. J., and Speakman, P. T. (1965b). *Proc. Int. Wool Text. Res. Conf., 3rd* **1**, 173. Inst. Textile de France, Paris.

Johnson, H. M., and Parsons, D. F. (1969). *J. Microsc.* **90**, 199.

Johnson, P., and Perry, S. V. (1968). *Biochem. J.* **110**, 207.

Johnson, P., Harris, C. I., and Perry, S. V. (1967). *Biochem. J.* **105**, 361.

Jones, C. B., and Mecham, D. K. (1943). *Arch. Biochem.* **3**, 193.

Jones, R. L. (1966). *Spectrochim. Acta* **22**, 1555.

Jones, R. N. (1952). *J. Amer. Chem. Soc.* **74**, 2681.

Jordan, B. J., and Speakman, P. T. (1965). *Proc. Int. Wool Text. Res. Conf., 3rd* **1**, 229. Inst. Textile de France, Paris.

Josse, J., and Harrington, W. F. (1964). *J. Mol. Biol.* **9**, 269.

Joubert, F. J., de Jager, P. J., and Swart, L. S. (1968). *In* "Symposium on Fibrous Proteins, Australia 1967" (W. G. Crewther, ed.), p. 343. Butterworths, London.

Kaesberg, P., and Shurman, M. M. (1953). *Biochim. Biophys. Acta* **11**, 1.

Kaesberg, P., Ritland, H. N., and Beeman, W. W. (1948). *Phys. Rev.* **74**, 71.

Kakiuti, Y., Shimozawa, T., and Suzuki, R. (1967). *J. Mol. Spectrosc.* **23**, 383.

Karges, H. E., and Kühn, K. (1970). *Eur. J. Biochem.* **14**, 94.

Karlson, R. H., Norland, K. S., Fasman, G. D., and Blout, E. R. (1960). *J. Amer. Chem. Soc.* **82**, 2268.

Kartha, G., Bello, J., and Harker, D. (1967). *Nature (London)* **213**, 862.

Kasper, J. S., and Lonsdale, K. (1959). "International Tables for X-Ray Crystallography," Vol. II. Kynoch Press, Birmingham.

Katchalski, E., Berger, A., and Kurtz, J. (1963). *In* "Aspects of Protein Structure" (G. N. Ramachandran, ed.), p. 205. Academic Press, New York.

Katchalski, E., Sela, M., Silman, H. I., and Berger, A. (1964). *In* "The Proteins" (H. Neurath, ed.), Vol. 2, p. 405. Academic Press, New York.

Katz, E. P. (1970). *Biopolymers* **9**, 745.

Katz, J. R. (1934). Die Röntgenspektrographie als Untersuchungsmethode. Abderhalden's Handbuch der biologischen Arbeitsmethoden, Abt. 2, Teil 3, p. 3589. Urban and Schwarzenberg, Berlin.

Kawamura, M., and Maruyama, K. (1970). *J. Biochem.* **68**, 885.

Kawasaki, H., Sato, H., and Suzuki, M. (1971). *Insect. Biochem.* **1**, 130.

Kay, D., and Cuddigan, B. J. (1967). *Brit. J. Haemat.* **13**, 341.

Kay, D. H. (1965). "Techniques for Electron Microscopy." Blackwell, Oxford.

Kay, L. M., Schroeder, W. A., Munger, N., and Burt, N. (1956). *J. Amer. Chem. Soc.* **78**, 2430.

Ke, B. (1964a). *In* "Newer Methods of Polymer Characterization" (B. Ke, ed.), p. 347. Wiley (Interscience), New York.

Ke, B. (1964b). *In* "Newer Methods of Polymer Characterization" (B. Ke, ed.), p. 421. Wiley (Interscience), New York.

Keith, H. D., Giannoni, G., and Padden, F. J. (1969a). *Biopolymers* **7**, 775.

Keith, H. D., Padden, F. J., and Giannoni, G. (1969b). *J. Mol. Biol.* **43**, 423.

Kellenberger, E. (1969). *In* "Symmetry and Function of Biological Systems at the Macromolecular Level" (A. Engström and B. Strandberg, eds.), p. 349. Wiley, New York.

Keller, A. (1968). *Rep. Progr. Phys.* **31**, 623.

Keller, W. D., Lusebrink, T. R., and Sederholm, C. H. (1966). *J. Chem. Phys.* **44**, 782.

Kenchington, W., and Flower, N. E. (1969). *J. Microsc.* **89**, 263.

Kendrew, J. C. (1954). *In* "The Proteins" (H. Neurath and K. Bailey, eds.), 1st ed., Vol. 2, Part B, p. 845. Academic Press, New York.

Kendrew, J. C. (1962). *Brookhaven Symp. Biol.* **15**, 216.

Kendrew, J. C., Dickerson, R. E., Strandberg, B. E., Hart, R. G., Davies, D. R., Phillips, D. C., and Shore, V. C. (1960). *Nature (London)* **185**, 422.

Kendrick-Jones, J., Szent-Györgyi, A. G., and Cohen, C. (1971). *J. Mol. Biol.* **59**, 527.

Kerridge, D., Horne, R. W., and Glauert, A. M. (1962). *J. Mol. Biol.* **4**, 227.

Kessler, H. K., and Sutherland, G. B. B. M. (1953). *J. Chem. Phys.* **21**, 570.

Khanagov, A. A. (1971). *Biopolymers* **10**, 789.

Khanagov, A. A., and Gabuda, S. P. (1969). *Biofizika* **14**, 796.

King, M. V., and Young, M. (1970). *J. Mol. Biol.* **50**, 491.

King, M. V., and Young, M. (1972a). *J. Mol. Biol.* **63**, 539.

King, M. V., and Young, M. (1972b). *J. Mol. Biol.* **65**, 519.

Kingham, O. J., and Brisbin, D. A. (1968). *Can. J. Biochem.* **46**, 1199.

Kirimura, J. (1962). *Bull. Sericult. Exp. Sta. Tokyo* **17**, 447.

Kiselev, N. A., and Klug, A. (1969). *J. Mol. Biol.* **40**, 155.

Kiselev, N. A., DeRosier, D. J., and Klug, A. (1968). *J. Mol. Biol.* **35**, 561.

Kleinschmidt, A. K. (1970). *Ber. Bunsenges. Phys. Chem.* **74**, 1190.

Klotz, I. M., and Farnham, S. B. (1968). *Biochemistry* **7**, 3879.

Klotz, I. M., and Franzen, J. S. (1962). *J. Amer. Chem. Soc.* **84**, 3461.

Klug, A. (1968). *Symp. Int. Soc. Cell Biol.* **6**, 1.

Klug, A. (1969). *In* "Symmetry and Function of Biological Systems at the Macromolecular Level" (A. Engström and B. Strandberg eds.), p. 425. Wiley, New York.

Klug, A. (1971). *Phil. Trans. Roy. Soc. London Ser. B* **261**, 173.

Klug, A., and Berger, J. E. (1964). *J. Mol. Biol.* **10**, 565.

Klug, A., and Crowther, R. A. (1972). *Nature (London)* **238**, 435.

Klug, A., and DeRosier, D. J. (1966). *Nature (London)* **212**, 29.

Klug, A., and Franklin, R. E. (1958). *Discuss. Faraday Soc.* **25**, 104.

Klug, A., Crick, F. H. C., and Wyckoff, H. W. (1958). *Acta Crystallogr.* **11**, 199.

Klug, H. P., and Alexander, L. E. (1954). "X-Ray Diffraction Procedures." Wiley, New York.

Knappeis, G. G., and Carlsen, F. (1968). *J. Cell Biol.* **38**, 202.

Kobayashi, K., and Sakaoku, K. (1965). *Lab. Invest.* **14**, 1097.

Kobayashi, Y., Sakai, R., Kakiuchi, K., and Isemura, T. (1970). *Biopolymers* **9**, 415.

Koenig, J. L. (1971). *Appl. Spectrosc. Rev.* **4**, 233.

Koenig, J. L., and Sutton, P. L. (1969). *Biopolymers* **8**, 167.

Koenig, J. L., and Sutton, P. L. (1970). *Biopolymers* **9**, 1229.

Koenig, J. L., and Sutton, P. L. (1971). *Biopolymers* **10**, 89.

Koenig, J. L., Cornell, S. W., and Witenhafer, D. E. (1967). *J. Polym. Sci. Part A2* **5**, 301.

Koffler, H. (1957). *Bacteriol. Revs.* **21**, 227.

Koisi, S. (1941). *Mitt. Med. Akad. Kioto.* **31**, 809.

Koltun, W. L. (1965). *Biopolymers* **3**, 665.

Kominz, D. R. (1965). *In* "Molecular Biology of Muscle Contraction" (S. Ebashi, F. Oosawa, T. Sekine and Y. Tonomura, eds.), p. 90. Elsevier, Amsterdam.

Konishi, T., and Kurokawa, M. (1968). *Sen -i Gakkaishi* **24**, 550.

Konishi, T., Kondo, M., and Kurokawa, M. (1967). *Sen -i Gakkaishi* **23**, 64.

Köppel, G. (1958). *Z. Zellforsch.* **47**, 401.

Köppel, G. (1967). *Z. Zellforsch.* **77**, 443.

Kotelchuck, D., and Scheraga, H. A. (1968). *Proc. Nat. Acad. Sci. U.S.* **61**, 1163.

Kotelchuck, D., and Scheraga, H. A. (1969). *Proc. Nat. Acad. Sci. U.S.* **62**, 14.

Kotelchuck, D., Duggert, M., and Scheraga, H. A. (1969). *Proc. Nat. Acad. Sci. U.S.* **63**, 615.

Kowalik, J., and Osborne, M. R. (1968). "Methods for Unconstrained Optimization Problems." Elsevier, New York.

Krakow, W., Endres, G. F., Siegel, B. M., and Scheraga, H. A. (1972). *J. Mol. Biol.* **71**, 95.

Kratky, O. (1929). *Z. Phys. Chem. (Leipzig)* **B5**, 297.

Kratky, O. (1933). *Kolloid Z.* **64**, 213.

Kratky, O. (1947). *J. Polym. Sci.* **3**, 195.

Kratky, O. (1951). *Z. Naturforsch.* **6b**, 173.

Kratky, O. (1956). *Trans. Faraday Soc.* **52**, 558.

Kratky, O. (1967). *In* "Small-Angle X-Ray Scattering" (H. Brumberger, ed.), p. 63. Gordon and Breach, New York.

Kratky, O., and Kuriyama, S. (1931). *Z. Phys. Chem. (Leipzig)* **B11**, 363.

Kratky, O., and Porod, G. (1955). *Phys. Hochpolym.* **3**, 119.

Kratky, O., and Schauenstein, E. (1951). *Discuss. Faraday Soc.* No. 11, 171.

Kratky, O., and Sekora, A. (1943). *J. Makromol. Chem.* **1**, 113.

Kratky, O., Schauenstein, E., and Sekora, A. (1950). *Nature (London)* **165**, 319.

Kratky, O., Sekora, A., Zahn, H., and Fritze, E. R. (1955). *Z. Naturforsch.* **10b**, 68.

Kraut, J. (1954). A Study of the Molecular Properties of Rat-Tail Tendon Collagen and an Investigation of the Structure of Feather Keratin. Ph. D. Thesis, California Inst. of Technol., Pasadena.

Kresheck, G. C., and Klotz, I. M. (1969). *Biochemistry* **8**, 8.

Krimm, S. (1960a). *Advan. Polym. Sci.* **2**, 51.

Krimm, S. (1960b). *J. Chem. Phys.* **32**, 313.

Krimm, S. (1962). *J. Mol. Biol.* **4**, 528.

Krimm, S. (1966). *Nature (London)* **212**, 1482.

Krimm, S., and Abe, Y. (1972). *Proc. Nat. Acad. Sci. U.S.* **69**, 2788.

Krimm, S., and Kuroiwa, K. (1968). *Biopolymers* **6**, 401.

Krimm, S., Kuroiwa, K., and Rebane, T. (1967). *In* "Conformation of Biopolymers" (G. N. Ramachandran, ed.), Vol. 2, p. 436. Academic Press, New York.

Kühn, K. (1967). *Naturwissenschaften* **54**, 101.

Kühn, K. (1969). *In* "Essays in Biochemistry" (P. N. Campbell and G. D. Greville, eds.), Vol. 5, p. 59. Academic Press, London.

Kühn, K., and Zimmer, E. (1961). *Z. Naturforsch.* **16b**, 648.

Kühn, K., Grassmann, W., and Hofmann, U. (1960). *Naturwissenschaften* **47**, 258.

Kühn, K., Tkocz, C., Zimmerman, B., and Beier, G. (1965). *Nature (London)* **208**, 685.

Kühn, K., Rauterberg, J., Zimmerman, B., and Tkocz, C. (1968). *In* "Symposium on Fibrous Proteins, Australia 1967" (W. G. Crewther, ed.), p. 181. Butterworths, London.

Kuhn, W., and Grün, F. (1942). *Kolloid Z.* **101**, 248.

Kulkarni, R. K., Fasman, G. D., and Blout, E. R. (1967). Quoted by Fasman (1967b).

Kulonen, E., and Pikkarainen, J. (1970). *In* "Chemistry and Molecular Biology of the Intercellular Matrix" (E. A. Balazs, ed.), Vol. 1, p. 81. Academic Press, New York.

Küntzel, A., and Prakke, F. (1933). *Biochem. Z.* **267**, 243.

Kurtz, J., Berger, A., and Katchalski, E. (1956). *Nature (London)* **178**, 1066.

Kurtz, J., Fasman, G. D., Berger, A., and Katchalski, E. (1958). *J. Amer. Chem. Soc.* **80**, 393.

Kushner, D. J. (1969). *Bacteriol. Rev.* **33**, 302.

Labaw, L. W., and Rossmann, M. G. (1969). *J. Ultrastruct. Res.* **27**, 105.

Lake, J. A. (1972a). *In* "Optical Transforms" (H. S. Lipson, ed.), p. 153. Academic Press, New York.

Lake, J. A. (1972b). *J. Mol. Biol.* **66**, 255.

Lake, J. A., and Slayter, H. S. (1970). *Nature (London)* **227**, 1032.

Lakshmanan, B. R., Ramakrishnan, C., Sasisekharan, V., and Thathachari, Y. T. (1962). *In* "Collagen" (N. Ramanathan, ed.), p. 117. Wiley, New York.

Lakshminarayanan, A. V., Sasisekharan, V., and Ramachandran, G. N. (1967). *In* "Conformation of Biopolymers" (G. N. Ramachandran, ed.), Vol. 1, p. 61. Academic Press, New York.

Lang, A. R. (1956a). *Acta Crystallogr.* **9**, 436.

Lang, A. R. (1956b). *Acta Crystallogr.* **9**, 446.

Langridge, R., Wilson, H. R., Hooper, C. W., Wilkins, M. H. F., and Hamilton, L. D. (1960). *J. Mol. Biol.* **2**, 19.

Laskowski, M. (1970). *In* "Spectroscopic Approaches to Biomolecular Conformation" (D. W. Urry, ed.), p. 1. Amer. Med. Ass., Chicago, Illinois.

Lauffer, M. A. (1971). *In* "Subunits in Biological Systems" (S. N. Timasheff and G. D. Fasman, eds.), Part A., p. 149. Dekker, New York.

Leach, S. J., Hill, J., and Holt, L. A. (1964). *Biochemistry* **3**, 737.

Leach, S. J., Némethy, G., and Scheraga, H. A. (1966a). *Biopolymers* **4**, 369.

Leach, S. J., Némethy, G., and Scheraga, H. A. (1966b). *Biopolymers* **4**, 887.

Ledger, R., and Stewart, F. H. C. (1965). *Aust. J. Chem.* **18**, 1477.

Ledger, R., and Stewart, F. H. C. (1966). *Aust. J. Chem.* **19**, 1729.

Lenormant, H. (1956). *Trans. Faraday Soc.* **52**, 549.

Lenormant, H., Baudras, A., and Blout, E. R. (1958). *J. Amer. Chem. Soc.* **80**, 6191.

Leung, Y. C., and Marsh, R. E. (1958). *Acta Crystallogr.* **11**, 17.

Levenberg, K. (1944). *Quart. Appl. Math.* **2**, 164.

Lewis, M. S., and Piez, K. A. (1964). *J. Biol. Chem.* **239**, 3336.

Lewis, P. N., and Scheraga, H. A. (1971). *Arch. Biochem. Biophys.* **144**, 576.

Lewis, P. N., Gō, N., Gō, M., Kotelchuck, D., and Scheraga, H. A. (1970). *Proc. Nat. Acad. Sci. U.S.* **65**, 810.

Liang, C. Y. (1964). *In* "Newer Methods of Polymer Characterizations" (B. Ke, ed.), p. 33. Wiley (Interscience), New York.

Liang, C. Y., and Krimm, S. (1956). *J. Chem. Phys.* **25**, 563.

Liang, C. Y., Krimm, S., and Sutherland, G. B. B. M. (1956). *J. Chem. Phys.* **25**, 543.

Liljas, A. *et. al.* (1972). *Nature (London) New Biol.* **235**, 131.

Lindley, H. (1955). *Biochim. Biophys. Acta* **18**, 194.

Lindley, H., and Haylett, T. (1967). *J. Mol. Biol.* **30**, 63.

Lindley, H., and Rollett, J. S. (1955). *Biochim. Biophys. Acta* **18**, 183.

Lindley, H., Gillespie, J. M., and Rowlands, R. J. (1970). *J. Text. Inst.* **61**, 157.

Lindley, H. *et al.* (1971). *Appl. Polym. Symp. No. 18*, 21.

Lipscomb, W. N. *et al.* (1968). *Brookhaven Symp. Biol.* **21**, 24.

Lipson, H. S. (1959). *In* "International Tables for X-Ray Crystallography" (J. S. Kasper and K. Lonsdale eds.), Vol. II. Kynoch Press, Birmingham.

Lipson, H. S. (1972). "Optical Transforms." Academic Press, London.

Lipson, H. S., and Taylor, C. A. (1958). "Fourier Transforms and X-Ray Diffraction." Bell, London.

Liquori, A. M. (1966). *J. Polym. Sci. Part C* **12**, 209.

Liquori, A. M. (1969). *In* "Symmetry and Function of Biological Systems at the Macromolecular Level" (A. Engström and B. Strandberg, eds.), p. 101. Wiley, New York.

Locker, R. H., and Schmitt, F. O. (1957). *J. Biophys. Biochem. Cytol.* **3**, 889.

Loeb, G. I. (1969). *J. Colloid Interface Sci.* **31**, 572.

Lord, R. C., and Yu, N. (1970a). *J. Mol. Biol.* **50**, 509.

Lord, R. C., and Yu, N. (1970b). *J. Mol. Biol.* **51**, 203.

Lotan, N., Momany, F. A., Yan, J. F., Vanderkooi, G., and Scheraga, H. A. (1969). *Biopolymers* **8**, 21.

Lotz, B., and Keith, H. D. (1971). *J. Mol. Biol.* **61**, 201.

Low, B. W., Lovell, F. M., and Rudko, A. D. (1968). *Proc. Nat. Acad. Sci. U.S.* **60**, 1519.

Lowey, S. (1971). *In* "Subunits in Biological Systems" (S. N. Timasheff and G. D. Fasman, eds.), Part A, p. 201. Dekker, New York.

Lowey, S., and Cohen, C. (1962). *J. Mol. Biol.* **4**, 293.

Lowey, S., Kucera, J., and Holtzer, A. (1963). *J. Mol. Biol.* **7**, 234.

Lowey, S., Goldstein, L., Cohen, C., and Luck, S. M. (1967). *J. Mol. Biol.* **23**, 287.

Lowey, S., Slayter, H. S., Weeds, A. G., and Baker, H. (1969). *J. Mol. Biol.* **42**, 1.

Lowy, J., and Hanson, E. J. (1964). *Nature (London)* **202**, 538.

Lowy, J., and Hanson, E. J. (1965). *J. Mol. Biol.* **11**, 293.

Lowy, J., and McDonough, M. W. (1964). *Nature (London)* **204**, 125.

Lowy, J., and Small, J. V. (1970). *Nature (London)* **227**, 46.

Lowy, J., and Spencer, M. (1968). *Symp. Soc. Exp. Biol.* **22**, 215.

Lowy, J., and Vibert, P. J. (1967). *Nature (London)* **215**, 1254.

Lowy, J., and Vibert, P. J. (1969). *Acta Crystallogr.* **A25**, S199.

Lowy, J., Hanson, J., Elliott, G. F., Millman, B. M., and McDonough, M. W. (1966). *In* "Principles of Biomolecular Organization" (G. E. W. Wolstenholme and M. O'Connor, eds.), p. 229. Churchill, London.

Lowy, J., Vibert, P. J., Haselgrove, J. C., and Poulsen, F. R. (1973). *Proc. Roy. Soc. Ser. B.* (in press).

Lucas, F. (1964). *Discovery* **25**, 20.

Lucas, F. (1966). *Nature (London)* **210**, 952.

Lucas, F., and Rudall, K. M. (1968a). *In* "Comprehensive Biochemistry" (M. Florkin and E. H. Stotz, eds.), Vol. 26B, p. 475. Elsevier, Amsterdam.

Lucas, F., and Rudall, K. M. (1968b). *In* "Symposium on Fibrous Proteins, Australia 1967" (W. G. Crewther, ed.), p. 45. Butterworths, London.

Lucas, F., Shaw, J. T. B., and Smith, S. G. (1955). *J. Textile Inst.* **46**, T440.

Lucas, F., Shaw, J. T. B., and Smith, S. G. (1957). *Nature (London)* **179**, 906.

Lucas, F., Shaw, J. T. B., and Smith, S. G. (1958). *Advan. Protein Chem.* **13**, 107.

Lucas, F., Shaw, J. T. B., and Smith, S. G. (1960). *J. Mol. Biol.* **2**, 339.

Luenberger, D. G., and Dennis, U. E. (1966). *Anal. Chem.* **38**, 715.

Lui, K., and Anderson, J. E. (1970). *J. Macromol. Sci.—Rev. Macromol. Chem.* **C5(1)**, 1.

Lumry, R., and Biltonen, R. (1969). *In* "Structure and Stability of Biological Macro-molecules" (S. N. Timasheff and G. D. Fasman, eds.), p. 65. Dekker, New York.

Lundgren, H. P., and Ward, W. H. (1962). *Arch. Biochem. Suppl.* **1**, 78.

Luzzati, V., Cesari, M., Spach, G., Masson, F., and Vincent, J. M. (1961). *J. Mol. Biol.* **3**, 566.

Luzzati, V., Cesari, G. S., Masson, F., and Vincent, J. M. (1962). *In* "Polyamino Acids, Polypeptides, and Proteins" (M. A. Stahmann, ed.), p. 121. Univ. of Wisconsin Press, Madison, Wisconsin.

Luzzati, V., Reiss-Husson, F., and Saludjian, P. (1966). *In* "Principles of Biomolecular Organization" (G. E. W. Wolstenholme and M. O'Connor, eds.), p. 69. Churchill, London.

MacArthur, I. (1943). *Nature (London)* **152**, 38.

McBride, O. W., and Harrington, W. F. (1967a). *Biochemistry* **6**, 1484.

McBride, O. W., and Harrington, W. F. (1967b). *Biochemistry* **6**, 1499.

McCubbin, W. D., and Kay, C. M. (1968). *Biochim. Biophys. Acta* **154**, 239.

McCubbin, W. D., and Kay, C. M. (1969). *Can. J. Biochem.* **47**, 411.

McCubbin, W. D., Kay, C. M., and Oikawa, K. (1966). *Biochim. Biophys. Acta* **126**, 600.

McDiarmid, R., and Doty, P. (1966). *J. Phys. Chem.* **70**, 2620.

McDonough, M. W. (1965). *J. Mol. Biol.* **12**, 342.

McFarlane, E. F. (1971). *Search* **2**, 171.

MacFarlane, R. G. (1960). *In* "The Plasma Proteins" (F. W. Putnam, ed.), Vol. II, p. 137. Academic Press, New York.

McGavin, S. (1962). *J. Mol. Biol.* **5**, 275.

McGavin, S. (1964). *J. Mol. Biol.* **9**, 601.

MacGillavry, C. H., and Bruins, E. M. (1948). *Acta Crystallogr.* **1**, 156.

MacGillavry, C. H., and Rieck, G. D. (1962). "International Tables for X-Ray Crystal-lography," Vol. III. Kynoch Press, Birmingham.

McGuire, R. F. *et al.* (1971). *Macromolecules* **4**, 112.

McGuire, R. F., Momany, F. A., and Scheraga, H. A. (1972). *J. Phys. Chem.* **76**, 375.

McKee, P. A., Rogers, L. A., Marler, E., and Hill, R. L. (1966). *Arch. Biochem. Biophys.* **116**, 271.

McLachlan, D. (1958). *Proc. Nat. Acad. Sci. U.S.* **44**, 948.

Madison, V., and Schellman, J. (1972). *Biopolymers* **11**, 1041.

Mahl, H., and Weitsch, W. (1960). *Z. Naturforsch* **15a**, 1051.

Maigret, B., Pullman, B., and Dreyfus, M. (1970). *J. Theor. Biol.* **26**, 321.

Malcolm, B. R. (1955). *In Symp. Soc. Exp. Biol.* (R. Brown and J. F. Danielli, eds.), No. 9, p. 265. Cambridge Univ. Press, London and New York.

Malcolm, B. R. (1968a). *Proc. Roy. Soc. Ser. A* **305**, 363.

Malcolm, B. R. (1968b). *Nature (London)* **219**, 929.

Malcolm, B. R. (1968c). *Biochem. J.* **110**, 733.

Malcolm, B. R. (1970). *Biopolymers* **9**, 911.

Mammi, M., Gotte, L., and Pezzin, G. (1970). *Nature (London)* **225**, 380.

Mandelkern, L. (1964). "Crystallization of Polymers." McGraw-Hill, New York.

Mandelkern, L. (1967). *In* "Poly-α-Amino Acids" (G. D. Fasman, ed.), p. 675. Dekker, New York.

Mandelkern, L. (1970). *Progr. Polym. Sci.* **2**, 163.

Markham, R. (1968). *In* "Methods of Virology" (K. Maramorosch and H. Koprowski, eds.), Vol. 4, p. 503. Academic Press, New York.

Markham, R., Frey, S., and Hills, G. J. (1963). *Virology* **20**, 88.

Markham, R., Hitchborn, J. H., Hills, G. J., and Frey, S. (1964). *Virology* **22**, 342.

Marks, M. H., Bear, R. S., and Blake, C. H. (1949). *J. Exp. Zool.* **111**, 55.

Marquardt, D. W. (1963). *J. Soc. Ind. Appl. Math.* **11**, 431.

Marquardt, D. W., Bennett, R. G., and Burrell, E. J. (1961). *J. Mol. Spectrosc.* **7**, 269.

Marsh, R. E., Corey, R. B., and Pauling, L. (1955a). *Acta Crystallogr.* **8**, 710.

Marsh, R. E., Corey, R. B., and Pauling, L. (1955b). *Biochim. Biophys. Acta* **16**, 1.

Marsh, R. E., Corey, R. B., and Pauling, L. (1956). *Proc. Int. Wool Text. Res. Conf., 1st, Australia 1955 B* 176.

Marvin, D. A. (1966). *J. Mol. Biol.* **15**, 8.

Marvin, D. A., Spencer, M., Wilkins, M. H. F., and Hamilton, L. D. (1961). *J. Mol. Biol.* **3**, 547.

Marwick, T. C. (1931). *J. Text. Sci.* **4**, 31.

Masuda, Y., and Miyazawa, T. (1967). *Makromol. Chem.* **103**, 261.

Masuda, Y., Fukushima, K., Fujii, T., and Miyazawa, T. (1969). *Biopolymers* **8**, 91.

Mathews, F. S., Levine, M., and Argos, P. (1972). *J. Mol. Biol.* **64**, 449.

Matoltsy, A. G. (1965). *In* "Biology of Skin and Hair Growth" (A. G. Lyne and B. F. Short, eds.), p. 291. Angus and Robertson, Sydney.

Matoltsy, A. G., and Parakkal, P. F. (1967). *In* "Ultrastructure of Normal and Abnormal Skin" (A. S. Zelickson, ed.), p. 76. Lea and Febiger, Philadelphia, Pennsylvania.

Matsuzaki, T., and Iitaka, Y. (1971). *Acta Crystallogr.* **B27**, 507.

Matthews, B. W., Klopfenstein, C. E., and Colman, P. M. (1972). *J. Phys. (E)* **5**, 353.

Meggy, A. B., and Sikorski, J. (1956). *Nature (London)* **177**, 326.

Meiron, J. (1965). *J. Opt. Soc. Amer.* **55**, 1105.

Menefee, E., and Yee, G. (1965). *Textile Res. J.* **35**, 801.

Mercer, E. H. (1949). *Nature (London)* **163**, 18.

Mercer, E. H. (1952). *Aust. J. Sci. Res.* **B5**, 366.

Mercer, E. H. (1954). *Textile Res. J.* **24**, 135.

Mercer, E. H. (1961). "Keratin and Keratinization." Pergamon, Oxford.

Mercer, E. H., and Matoltsy, A. G. (1969). *In* "Advances in Biology of Skin; Hair Growth" (W. Montagna and R. L. Dobson, eds.), Vol. 9, p. 556, Pergamon, Oxford.

Mercer, E. H., and Olofsson, B. (1951). *J. Polym. Sci.* **6**, 261.

Metcalfe, J. C. (1970). *In* "Physical Principles and Techniques of Protein Chemistry" (S. J. Leach, ed.), Part B, p. 275. Academic Press, New York.

Meyer, K. H. (1929). *Biochem. Z.* **214**, 265.

Meyer, K. H., and Go, Y. (1934). *Helv. Chim. Acta* **17**, 1488.

Meyer, K. H., and Jeannerat, J. (1939). *Helv. Chim. Acta* **22**, 22.

Meyer, K. H., and Mark, H. (1928). *Ber. Deut. Chem. Ges.* **B61**, 1932.

Migchelsen, C., and Berendsen, H. J. C. (1967). *In* "Magnetic Resonance and Relaxation" (R. Blinc, ed.), p. 761. North-Holland, Amsterdam.

Mihalyi, E. (1965). *Biochim. Biophys. Acta* **102**, 487.

Mihalyi, E. (1968). *In* "Fibrinogen" (K. Laki, ed.), p. 61. Dekker, New York.

Mihalyi, E. (1970). *Thromb. Diath. Haemorrh., Suppl.* **39**, 43.

Mikhailov, A. M. (1971). *Kristallografiya* **15**, 818 (*English transl.: Sov. Phys. Crystallogr.* **15**, 701).

Milledge, H. J., and Graeme-Barber, A. (1973). The Joyce-Loebl Review.

Miller, A. (1965). *J. Mol. Biol.* **12**, 280.

Miller, A. (1968). *J. Mol. Biol.* **32**, 687.

Miller, A., and Parry, D. A. D. (1973). *J. Mol. Biol.* **75**, 441.

Miller, A., and Tregear, R. T. (1972). *J. Mol. Biol.* **70**, 85.

Miller, A., and Wray, J. S. (1971). *Nature (London)* **230**, 437.

Miller, C. S., Parsons, F. G., and Kofsky, I. L. (1964). *Nature (London)* **202**, 1196.

Miller, E. J., and Matukas, V. J. (1969). *Proc. Nat. Acad. Sci. U.S.* **64**, 1264.

Miller, R. G. J., and Willis, H. A. (1961). "Molecular Spectroscopy." Heywood, London.

Millionova, M. I., and Andreeva, N. S. (1958). *Biophysics (U.S.S.R.)* **3**, 259.

Millionova, M. I., and Andreeva, N. S. (1959). *Biophysics (U.S.S.R.)* **4**, 138.

Millionova, M. I., Andreeva, N. S., and Lebedev, L. A. (1963). *Biofizika* **8**, 430.

Millman, B. M., and Elliott, G. F. (1965). *Nature (London)* **206**, 824.

Millman, B. M., Elliott, G. F., and Lowy, J. (1967). *Nature (London)* **213**, 356.

Millward, G. R. (1969). *J. Cell Biol.* **42**, 317.

Millward, G. R. (1970). *J. Ultrastruct. Res.* **31**, 349.

Millward, G. R., and Woods, E. F. (1970). *J. Mol. Biol.* **52**, 585.

Mitsui, Y. (1966). *Acta Crystallogr.* **20**, 694.

Mitsui, Y. (1970). *Acta Crystallogr.* **A26**, 658.

Mitsui, Y., Iitaka, Y., and Tsuboi, M. (1967). *J. Mol. Biol.* **24**, 15.

Miyazawa, T. (1960a). *J. Chem. Phys.* **32**, 1647.

Miyazawa, T. (1960b). *J. Mol. Spectrosc.* **4**, 155.

Miyazawa, T. (1960c). *J. Mol. Spectrosc.* **4**, 168.

Miyazawa, T. (1961a). *J. Polym. Sci.* **55**, 215.

Miyazawa, T. (1961b). *Bull. Chem. Soc. Japan* **34**, 691.

Miyazawa, T. (1962). *In* "Polyamino Acids, Polypeptides and Proteins" (M. A. Stahmann, ed.), p. 201. Univ. of Wisconsin Press, Madison, Wisconsin.

Miyazawa, T. (1963). *In* "Aspects of Protein Structure" (G. N. Ramachandran, ed.), p. 257. Academic Press, New York.

Miyazawa, T. (1967). *In* "Poly-α-Amino Acids" (G. D. Fasman, ed.), p. 69. Arnold, London.

Miyazawa, T., and Blout, E. R. (1961). *J. Amer. Chem. Soc.* **83**, 712.

Miyazawa, T., Shimanouchi, T., and Mizushima, S. (1956). *J. Chem. Phys.* **24**, 408.

Miyazawa, T., Shimanouchi, T., and Mizushima, S. (1958). *J. Chem. Phys.* **29**, 611.

Miyazawa, T., Masuda, Y., and Fukushima, K. (1962). *J. Polym. Sci.* **62**, S62.

Miyazawa, T., Fukushima, K., Sugano, S., and Masuda, Y. (1967). *In* "Conformation of Biopolymers" (G. N. Ramachandran, ed.), Vol. 2, p. 557. Academic Press, New York.

Moffitt, W. (1956). *Proc. Nat. Acad. Sci. U.S.* **42**, 736.

Moffitt, W., and Moscowitz, A. (1959). *J. Chem. Phys.* **30**, 648.

Moffitt, W., and Yang, J. T. (1956). *Proc. Nat. Acad. Sci. U.S.* **42**, 596.

Momany, F. A., Vanderkooi, G., Tuttle, R. W., and Scheraga, H. A. (1969). *Biochemistry* **8**, 744.

Momii, R. K., and Urry, D. W. (1968). *Macromolecules* **1**, 372.

Mommaerts, W. F. H. M. (1950). "Muscular Contraction." Wiley (Interscience), New York.

Moody, M. F. (1967). *J. Mol. Biol.* **25**, 201.

Moore, P. B., Huxley, H. E., and DeRosier, D. J. (1970). *J. Mol. Biol.* **50**, 279.

Morales, M. F., and Cecchini, L. P. (1951). *J. Cell. Comp. Physiol.* **37**, 107.

Morales, M. F., Laki, K., Gergely, J., and Cecchini, L. P. (1951). *J. Cell. Comp. Physiol.* **37**, 477.

Morita, H. (1970). *In* "Physical Principles and Techniques of Protein Chemistry" (S. J. Leach, ed.), Part B, p. 437. Academic Press, New York.

Moritani, M., Hayashi, N., Utsuo, A., and Kawai, H. (1971). *Polymer J.* **2**, 74.

Morton, W. E., and Hearle, J. W. S. (1962). "Physical Properties of Textile Fibres." Butterworths, London.

Moscowitz, A. (1960). *In* "Optical Rotatory Dispersion, Applications to Organic Chemistry" (C. Djerassi, ed.), p. 150. McGraw-Hill, New York.

Mukhopadhyay, U., and Taylor, C. A. (1971). *J. Appl. Crystallogr.* **4**, 20.

Murphy, A. J. (1971). *Biochemistry* **10**, 3723.

Murray, R. T., and Ferrier, R. P. (1968). *J. Ultrastruct. Res.* **21**, 361.

Nagai, Y., Piez, K. A., and Gross, J. (1965). *Proc. Conf. Struct. Proteins, 16th, Fukuoka, Japan* p. 94. Jap. Chem. Soc.

Nagatoshi, F., and Arakawa, T. (1970). *Polymer J.* **1**, 685.

Naghski, J., Wisnewski, A., Harris, E. H., and Witnauer, L. P. (1966). *J. Amer. Leather Chem. Ass.* **61**, 64.

Nagy, B. (1966). *Biochim. Biophys. Acta* **115**, 498.

Nagy, B., and Jencks, W. P. (1962). *Biochemistry* **1**, 987.

Nakamura, A., Sreter, F., and Gergely, J. (1971). *J. Cell Biol.* **49**, 883.

Nakanishi, K. (1962). "Infra-Red Absorption Spectroscopy." Nankodo Co., Tokyo.

Nathan, R. (1971). *In* "Advances in Optical and Electron Microscopy" (R. Barer, ed.), Vol. 4, p. 85. Academic Press, New York.

Nelder, J. A., and Mead, R. (1965). *Comput. J.* **7**, 308.

Nemetschek, T. (1968). *Naturwissenschaften* **55**, 346.

Nemetschek, T., Grassmann, W., and Hofmann, U. (1955). *Z. Naturforsch.* **B10**, 61.

Némethy, G., and Scheraga, H. A. (1965). *Biopolymers* **3**, 155.

Némethy, G., Leach, S. J., and Scheraga, H. A. (1966). *J. Phys. Chem.* **70**, 998.

Newman, R., and Halford, R. S. (1950). *J. Chem. Phys.* **18**, 1276.

Nishikawa, K., and Ooi, T. (1971). *Prog. Theor. Phys.* **46**, 670.

Nishikawa, S., and Ono, S. (1913). *Proc. Tokyo Math. Phys. Soc.* [2]**7**, No. 8, 131.

Nockolds, C. E., and Kretsinger, R. H. (1970). *J. Phys. E.* **3**, 842.

Nold, J. G., Kang, A. H., and Gross, J. (1970). *Science* **170**, 1096.

Nordwig, A., and Hayduk, U. (1969). *J. Mol. Biol.* **44**, 161.

Norman, R. S. (1966). *Science* **152**, 1238.

North, A. C. T., Cowan, P. M., and Randall, J. T. (1954). *Nature (London)* **174**, 1142.

O'Brien, E. J., and MacEwan, A. W. (1970). *J. Mol. Biol.* **48**, 243.

O'Brien, E. J., Bennett, P. M., and Hanson, J. (1971). *Phil. Trans. Roy. Soc. London Ser. B* **261**, 837.

O'Donnell, I. J. (1969). *Aust. J. Biol. Sci.* **22**, 471.

O'Donnell, I. J. (1971). *Aust. J. Biol. Sci,* **24**, 1219.

O'Donnell, I. J. (1973). *Aust. J. Biol. Sci.* **26**, 415.

Offer, G. W. (1965). *Biochim. Biophys. Acta* **111**, 191.

Offer, G. W., and Starr, R. L. (1968). *Int. Un. Pure Appl. Biophys., Commun. Mol. Biophys., Cambridge, England,* p. 25.

Ohtsuki, I., and Wakabayashi, K. (1972). *J. Biochem.* **72**, 369.

Olander, J. (1971). *Biochemistry* **10**, 601.

Olander, J., Emerson, M. F., and Holtzer, A. (1967). *J. Amer. Chem. Soc.* **89**, 3058.

Olsen, B. R. (1963a). *Z. Zellforsch. Mikrosk. Anat.* **59**, 184.

Olsen, B. R. (1963b). *Z. Zellforsch. Mikrosk. Anat.* **59**, 199.

Olsen, B. R., Berg, R. A., Sakakibara, S., Kishida, Y., and Prockop, D. J. (1971). *J. Mol. Biol.* **57**, 589.

Onions, W. J., Woods, H. J., and Woods, P. B. (1960). *Nature (London)* **185**, 157.

Ooi, T., and Fujime-Higashi, S. (1971). *Advan. Biophys.* **2**, 113.

Ooi, T., Scott, R. A., Vanderkooi, G., Epand, R. F., and Scheraga, H. A. (1966). *J. Amer. Chem. Soc.* **88**, 5680.

Ooi, T., Scott, R. A., Vanderkooi, G., and Scheraga, H. A. (1967). *J. Chem. Phys.* **46**, 4410.

Oosawa, F., and Kasai, M. (1971). *In* "Subunits in Biological Systems" (S. N. Timasheff and G. D. Fasman, eds.), Part A, p. 261. Dekker, New York.

Oosawa, F., Kasai, M., Hatano, S., and Asakura, S. (1966). *In* "Principles of Biomolecular Organization" (G. E. W. Wolstenholme and M. O'Connor, eds.), Churchill, London.

Oster, G., and Riley, D. P. (1952). *Acta Crystallogr.* **5**, 272.

Ottensmeyer, F. P. (1969). *Biophys. J.* **9**, 1144.

Ottensmeyer, F. P., Schmidt, E. E., Jack, T., and Powell, J. (1972). *J. Ultrastruct. Res.* **40**, 546.

Overend, J. (1963). *In* "Infrared Spectroscopy and Molecular Structure" (M. Davies, ed.), p. 345. Elsevier, Amsterdam.

Padden, F. J., and Keith, H. D. (1965). *J. Appl. Phys.* **36**, 2987.

Padden, F. J., Keith, H. D., and Giannoni, G. (1969). *Biopolymers* **7**, 793.

Page, S. G., and Huxley, H. E. (1963). *J. Cell Biol.* **19**, 369.

Pain, R. H., and Robson, B. (1970). *Nature (London)* **227**, 62.

Pande, A. (1965). *Lab. Pract.* **14**, 1048.

Pardon, J. F. (1967). *Acta Crystallogr.* **23**, 937.

Parker, K. D. (1955). *Biochim. Biophys. Acta* **17**, 148.

Parker, K. D. (1956). Quoted by Bamford *et al.* (1956).

Parker, K. D. (1969). Quoted by Elliott (1969).

Parker, K. D., and Rudall, K. M. (1957). *Nature (London)* **179**, 905.

Parry, D. A. D. (1969). *J. Theor. Biol.* **24**, 73.

Parry, D. A. D. (1970). *J. Theor. Biol.* **26**, 429.

Parry, D. A. D., and Elliott, A. (1965). *Nature (London)* **206**, 616.

Parry, D. A. D., and Elliott, A. (1967). *J. Mol. Biol.* **25**, 1.

Parry, D. A. D., and Squire, J. M. (1973). *J. Mol. Biol.* **75**, 33.

Parry, D. A. D., and Suzuki, E. (1969a). *Biopolymers* **7**, 189.

Parry, D. A. D., and Suzuki, E. (1969b). *Biopolymers* **7**, 199.

Parsons, D. F. (1968). *In* "International Review of Experimental Pathology" (B. W. Richter and M. A. Epstein, eds.), Vol. 6, p. 1. Academic Press, New York.

Parsons, D. F. (1970). *In* "Some Biological Techniques in Electron Microscopy" (D. F. Parsons, ed.), p. 1. Academic Press, New York.

Parsons, D. F., and Martius, U. (1964). *J. Mol. Biol.* **10**, 530.

Parsons, F. G., Miller, C. S., and Kofsky, I. L. (1965). *Joyce-Loebl Rev.* **2**, 2.

Partridge, S. M. (1968). *In* "Symposium on Fibrous Proteins, Australia 1967" (W. G. Crewther, ed.), p. 246. Butterworths, London.

Partridge, S. M., Elsden, D. F., and Thomas, J. (1963). *Nature (London)* **197**, 1297.

Patrizi, G., and Munger, B. L. (1966). *J. Ultrastruct. Res.* **14**, 329.

Pauling, L. (1953). *Discuss. Faraday Soc.* **13**, 170.

Pauling, L., and Corey, R. B. (1950). *J. Amer. Chem. Soc.* **71**, 5349.

Pauling, L., and Corey, R. B. (1951a). *Proc. Nat. Acad. Sci. U.S.* **37**, 235.

Pauling, L., and Corey, R. B. (1951b). *Proc. Nat. Acad. Sci. U.S.* **37**, 241.

Pauling, L., and Corey, R. B. (1951c). *Proc. Nat. Acad. Sci. U.S.* **37**, 251.

Pauling, L., and Corey, R. B. (1951d). *Proc. Nat. Acad. Sci. U.S.* **37**, 256.

Pauling, L., and Corey, R. B. (1951e). *Proc. Nat. Acad. Sci. U.S.* **37**, 261.

Pauling, L., and Corey, R. B. (1951f). *Proc. Nat. Acad. Sci. U.S.* **37**, 272.

Pauling, L., and Corey, R. B. (1951g). *Proc. Nat. Acad. Sci. U.S.* **37**, 729.

Pauling, L., and Corey, R. B. (1953a). *Proc. Nat. Acad. Sci. U.S.* **39**, 253.

Pauling, L., and Corey, R. B. (1953b). *Nature (London)* **171**, 59.

Pauling, L., and Wilson, E. B. (1935). "Introduction to Quantum Mechanics." McGraw-Hill, New York.

Pauling, L., Corey, R. B., and Branson, H. R. (1951). *Proc. Nat. Acad. Sci. U.S.* **37**, 205.

Peacock, N. (1959). *Biochim. Biophys. Acta* **32**, 220.

Pearson, J. D. (1969). *Comput. J.* **12**, 171.

Peggion, E., Cosani, A., Verdini, A. S., Del Pra, A., and Mammi, M. (1968). *Biopolymers* **6**, 1477.

Peiser, H. S., Rooksby, H. P., and Wilson, A. J. C. (1955). "X-Ray Diffraction by Polycrystalline Materials." Inst. of Phys., London.

Pepe, F. A. (1966). *J. Cell Biol.* **28**, 505.

Pepe, F. A. (1967). *J. Mol. Biol.* **27**, 203.

Pepe, F. A. (1971a). *In* "Subunits in Biological Systems" (S. N. Timasheff and G. D. Fasman, eds.), Part A, p. 323. Dekker, New York.

Pepe, F. A. (1971b). *Progr. Biophys. Mol. Biol.* **22**, 77.

Pepe, F. A., and Drucker, B. (1972). *J. Cell Biol.* **52**, 255.

Periti, P. F., Quagliarotti, G., and Liquori, A. M. (1967). *J. Mol. Biol.* **24**, 313.

Perry, S. V. (1967). *Progr. Biophys. Mol. Biol.* **17**, 325.

Perutz, M. F. (1951a). *Nature (London)* **167**, 1053.

Perutz, M. F. (1951b). *Nature (London)* **168**, 653.

Perutz, M. F. (1951c). *Rep. Progr. Chem.* **48**, 361.

Perutz, M. F., Kendrew, J. C., and Watson, H. C. (1965). *J. Mol. Biol.* **13**, 669.

Phillips, D. C. (1966). *In* "Advances in Structure Research by Diffraction Methods" (R. Brill and R. Mason, eds.), Vol. 2, p. 75. Wiley (Interscience), New York.

Phillips, D. C. (1967). *Proc. Nat. Acad. Sci. U.S.* **57**, 484.

Philpott, D. E., and Szent-Györgyi, A. G. (1954). *Biochim. Biophys. Acta* **15**, 165.

Piez, K. A. (1965). *Biochemistry* **4**, 2590.

Piez, K. A. (1967). *In* "Treatise on Collagen" (G. N. Ramachandran, ed.), Vol. 1, p. 207. Academic Press, New York.

Piez, K. A., and Gross, J. (1960). *J. Biol. Chem.* **235**, 995.

Piez, K. A., Eigner, E., and Lewis, M. S. (1963). *Biochemistry* **2**, 58.

Pikkarainen, J., Rantanen, J., Vastamäki, M., Lampiaho, K., Kari, A., and Kulonen, E. (1968). *Eur. J. Biochem.* **4**, 555.

Pilz, I. (1973). *In* "Physical Principles and Techniques of Protein Chemistry," Part C, p. 141. Academic Press, New York.

Pimentel, G. C., and McClellan, A. L. (1960). "The Hydrogen Bond." Freeman, San Francisco, California.

Pimentel, G. C., and Sederholm, C. H. (1956). *J. Chem. Phys.* **24**, 639.

Pitha, J., and Jones, R. N. (1966). *Can. J. Chem.* **44**, 3031.

Pitha, J., and Jones, R. N. (1967). *Can. J. Chem.* **45**, 2347.

Pohl, F. M. (1971). *Nature (London) New Biol.* **234**, 277.

Poland, D., and Scheraga, H. A. (1967). *Biochemistry* **6**, 3791.

Pomeroy, C. D., and Mitton, R. G. (1951). *J. Soc. Leather Trades' Chem.* **35**, 360.

Ponnuswamy, P. K., and Sasisekharan, V. (1971a). *Int. J. Protein Res.* **3**, 1.
Ponnuswamy, P. K., and Sasisekharan, V. (1971b). *Int. J. Protein Res.* **3**, 9.
Powell, M. J. D. (1964). *Comput. J.* **7**, 155.
Predecki, P., and Statton, W. O. (1965). *In* "Small-Angle X-Ray Scattering" (H. Brumberger, ed.), p. 131. Gordon and Breach, New York.
Prothero, J. W. (1966). *Biophys. J.* **6**, 367.
Ptitsyn, O. B. (1969). *J. Mol. Biol.* **42**, 501.
Pullman, B. (1971). *Int. J. Quantum Chem., Symp. No. 4*, 319.
Pysh, E. S. (1966). *Proc. Nat. Acad. Sci. U.S.* **56**, 825.
Quiocho, F. A., and Lipscomb, W. N. (1971). *Advan. Protein Chem.* **25**, 1.
Rainford, P., and Rice, R. V. (1970). *Biopolymers* **9**, 1.
Ramachandran, G. N. (1956). *Nature (London)* **177**, 710.
Ramachandran, G. N. (1960). *Proc. Indian Acad. Sci.* **A52**, 240.
Ramachandran, G. N. (1963a). *Int. Rev. Connect. Tissue Res.* **1**, 127.
Ramachandran, G. N. (1963b). *In* "Aspects of Protein Structure" (G. N. Ramachandran, ed.), p. 39. Academic Press, New York.
Ramachandran, G. N. (1964). *In* "Advanced Methods of Crystallography" (G. N. Ramachandran, ed.), p. 25. Academic Press, New York.
Ramachandran, G. N. (1967). *In* "Treatise on Collagen" (G. N. Ramachandran, ed.), Vol. 1, p. 103. Academic Press, New York.
Ramachandran, G. N. (1969). *Int. J. Protein Res.* **1**, 5.
Ramachandran, G. N., and Ambady, G. K. (1954). *Curr. Sci.* **23**, 349.
Ramachandran, G. N., and Chandrasekharan, R. (1968). *Biopolymers* **6**, 1649.
Ramachandran, G. N., and Dweltz, N. E. (1962). *In* "Collagen" (N. Ramanathan, ed.), p. 147. Wiley (Interscience), New York.
Ramachandran, G. N., and Kartha, G. (1954). *Nature (London)* **174**, 269.
Ramachandran, G. N., and Kartha, G. (1955a). *Nature (London)* **176**, 593.
Ramachandran, G. N., and Kartha, G. (1955b). *Proc. Indian Acad. Sci.* **A42**, 215.
Ramachandran, G. N., and Santhanam, M. S. (1957). *Proc. Indian Acad. Sci.* **A45**, 124.
Ramachandran, G. N., and Sasisekharan, V. (1956). *Arch. Biochem. Biophys.* **63**, 255.
Ramachandran, G. N., and Sasisekharan, V. (1961). *Nature (London)* **190**, 1004.
Ramachandran, G. N., and Sasisekharan, V. (1965). *Biochim. Biophys. Acta* **109**, 314.
Ramachandran, G. N., and Sasisekharan, V. (1968). *Advan. Protein Chem.* **23**, 283.
Ramachandran, G. N., and Srinivasan, R. (1970). "Fourier Methods in Crystallography." Wiley (Interscience), New York.
Ramachandran, G. N., and Venkatachalam, C. M. (1966). *Biochim. Biophys. Acta* **120**, 459.
Ramachandran, G. N., Sasisekharan, V., and Thathachari, Y. T. (1962). *In* "Collagen" (N. Ramanathan, ed.), p. 81. Wiley (Interscience), New York.
Ramachandran, G. N., Ramakrishnan, C., and Sasisekharan, V. (1963). *J. Mol. Biol.* **7**, 95.
Ramachandran, G. N., Sasisekharan, V., and Ramakrishnan, C. (1966a). *Biochim. Biophys. Acta* **112**, 168.
Ramachandran, G. N., Venkatachalam, C. M., and Krimm, S. (1966b). *Biophys. J.* **6**, 849.
Ramachandran, G. N., Ramakrishnan, C., and Venkatachalam, C. M. (1967). *In* "Conformation of Biopolymers" (G. N. Ramachandran, ed.), Vol. 2, p. 429. Academic Press, New York.
Ramachandran, G. N., Doyle, B. B., and Blout, E. R. (1968). *Biopolymers* **6**, 1771.
Ramakrishnan, C. (1964). *Proc. Indian Acad. Sci.* **59**, 317.
Ramakrishnan, C., and Balasubramanian, R. (1972). *Int. J. Peptide Protein Res.* **4**, 79.
Ramakrishnan, C., and Prasad, N. (1971). *Int. J. Protein Res.* **3**, 209.
Ramakrishnan, C., and Ramachandran, G. N. (1965). *Biophys. J.* **5**, 909.

Ramanathan, N., and Nayudamma, Y. (1962). *In* "Collagen" (N. Ramanathan, ed.), p. 453. Wiley (Interscience), New York.

Ramsden, W. (1938). *Nature (London)* **142**, 1120.

Randall, J. T. (1954). *J. Soc. Leather Trades' Chem.* **38**, 362.

Randall, J. T., Fraser, R. D. B., Jackson, S., Martin, A. V. W., and North, A. C. T. (1952). *Nature (London)* **169**, 1029.

Randall, J. T. *et al.* (1953). *In* "Nature and Structure of Collagen" (J. T. Randall, ed.), p. 213. Butterworths, London.

Rao, N. V., and Harrington, W. F. (1966). *J. Mol. Biol.* **21**, 577.

Reedy, M. K. (1968). *J. Mol. Biol.* **31**, 155.

Reedy, M. K., Holmes, K. C., and Tregear, R. T. (1965). *Nature (London)* **207**, 1276.

Rees, M. K., and Young, M. (1967). *J. Biol. Chem.* **242**, 4449.

Rice, R. V. (1961). *Biochim. Biophys. Acta* **53**, 29.

Rice, R. V., McManus, G. M., Devine, C. E., and Somlyo, A. P. (1971). *Nature (London) New Biol.* **231**, 242.

Rich, A., and Crick, F. H. C. (1955). *Nature (London)* **176**, 915.

Rich, A., and Crick, F. H. C. (1958). *In* "Recent Advances in Gelatin and Glue Research" (G. Stainsby, ed.), p. 20. Pergamon, Oxford.

Rich, A., and Crick, F. H. C. (1961). *J. Mol. Biol.* **3**, 483.

Riddiford, L. M. (1966). *J. Biol. Chem.* **241**, 2792.

Rigby, B. J. (1968a). *Biochim. Biophys. Acta* **133**, 272.

Rigby, B. J. (1968b). *In* "Symposium on Fibrous Proteins, Australia 1967" (W. G. Crewther, ed.), p. 219. Butterworths, London.

Riley, D. P. (1955). *In* "X-Ray Diffraction by Polycrystalline Materials" (H. S. Peiser, H. P. Rooksby and A. J. C. Wilson, eds.), p. 232. Inst. of Phys., London.

Riley, D. P., and Arndt, U. W. (1952). *Nature (London)* **169**, 138.

Rill, R. L. (1972). *Biopolymers* **11**, 1929.

Rippon, W. B., Koenig, J. L., and Walton, A. G. (1970). *J. Amer. Chem. Soc.* **92**, 7455.

Rippon, W. B., Anderson, J. M., and Walton, A. G. (1971). *J. Mol. Biol.* **56**, 507.

Roberts, B. W., and Parrish, W. (1962). *In* "International Tables for X-Ray Crystallography" (C. H. MacGillavry and G. D. Rieck, eds.), Vol. III, p. 73. Kynoch Press, Birmingham.

Roberts, G. C. K., and Jardetzky, O. (1970). *Advan. Protein Chem.* **24**, 447.

Robinson, C. (1966). *Mol. Cryst.* **1**, 467.

Robinson, C., and Ambrose, E. J. (1952). *Trans. Faraday Soc.* **48**, 854.

Robinson, C., and Bott, M. J. (1951). *Nature (London)* **168**, 325.

Robson, A., Woodhouse, J. M., and Zaidi, Z. H. (1970). *Int. J. Protein Res.* **2**, 181.

Robson, B., and Pain, R. H. (1971). *J. Mol. Biol.* **58**, 237.

Rogers, G. E. (1959a). *Ann. N.Y. Acad. Sci.* **83**, 378.

Rogers, G. E. (1959b). *Ann. N.Y. Acad. Sci.* **83**, 408.

Rogers, G. E. (1959c). *J. Ultrastruct. Res.* **2**, 309.

Rogers, G. E., and Clarke, R. M. (1965). *Nature (London)* **205**, 77.

Rogers, G. E., and Filshie, B. K. (1963). *In* "Ultrastructure of Protein Fibers" (R. Borasky, ed.), p. 123. Academic Press, New York.

Rogulenkova, V. N., Millionova, M. I., and Andreeva, N. S. (1964). *J. Mol. Biol.* **9**, 253.

Rosenbrock, H. H. (1960). *Comput. J.* **3**, 175.

Rosenheck, K., and Doty, P. (1961). *Proc. Nat. Acad. Sci. U.S.* **47**, 1775.

Rosenheck, K., and Sommer, B. (1967). *J. Chem. Phys.* **46**, 532.

Rosenheck, K., Miller, H., and Zakaria, A. (1969). *Biopolymers* **7**, 614.

Ross, R., and Benditt, E. P. (1961). *J. Biophys. Biochem. Cytol.* **11**, 677.

Ross, R., and Bornstein, P. (1971). *Sci. Amer.* **224**, No. 6, 44.

Roth, S. I., and Jones, W. A. (1970). *J. Ultrastruct. Res.* **32**, 69.

Rougvie, M. A. (1954). Studies on a Monomer of Feather Keratin. Ph.D. Thesis, Massachusetts Inst. of Technol., Cambridge, Massachusetts.

Rougvie, M. A., and Bear, R. S. (1953). *J. Amer. Leather Chem. Ass.* **48**, 735.

Rowe, A. J. (1964). *Proc. Roy. Soc. Ser. B* **160**, 437.

Rowe, A. J., and Rowe, H. J. (1970). *J. Microsc.* **91**, 31.

Rozsa, G., Szent-Györgyi, A. G., and Wyckoff, R. W. G. (1949). *Biochim. Biophys. Acta* **3**, 561.

Rudall, K. M. (1946). *In* "Fibrous Proteins Symposium" (C. L. Bird, ed.), p. 15. Chorley and Pickersgill, Leeds.

Rudall, K. M. (1947). *Biochim. Biophys. Acta* **1**, 549.

Rudall, K. M. (1952). *Advan. Protein Chem.* **7**, 253.

Rudall, K. M. (1956). *In* "Lectures on the Scientific Basis of Medicine," Vol. 5, p. 217. British Postgraduate Med. Fed., Univ. of London.

Rudall, K. M. (1962). *In* "Comparative Biochemistry" (M. Florkin and H. S. Mason, eds.), Vol. 4, p. 397. Academic Press, New York.

Rudall, K. M. (1965). *In* "Aspects of Insect Biochemistry" (T. W. Goodwin, ed.), p. 83. Academic Press, New York.

Rudall, K. M. (1968a). *In* "Treatise on Collagen" (B. S. Gould, ed.), Vol. 2, Part A, p. 83. Academic Press, New York.

Rudall, K. M. (1968b). *In* "Comprehensive Biochemistry" (M. Florkin and E. H. Stotz, eds.), Vol. 26B, p. 559. Elsevier, Amsterdam.

Rudall, K. M., and Kenchington, W. (1971). *Annu. Rev. Entomol.* **16**, 73.

Rugo, H. J. (1949). The Effect of Heat and Moisture on the Small Angle X-Ray Diffraction Diagram of Feather Keratin. M. Sc. Thesis, Massachusetts Inst. of Technol., Cambridge, Massachusetts.

Rugo, H. J. (1950). The Effects of Physical and Chemical Treatments on the Small Angle X-Ray Diffraction Pattern of Feather Keratin. Ph. D. Thesis, Massachusetts Inst. of Technol., Cambridge, Massachusetts.

Rumplestiltskin, R. (1957). Hides, Skins and Leather under the Microscope. British Leather Manufacturers' Res. Ass., London.

Ruska, E. (1966). *In* "Advances in Optical and Electron Microscopy" (R. Barer and V. E. Cosslett, eds.), Vol. I, p. 115. Academic Press, New York.

Ruska, H., and Wolpers, C. (1940). *Klin. Wochenschr.* **18**, 695.

Rymer, T. B. (1970). "Electron Diffraction." Methuen, London.

Sakakibara, S., Kishida, Y., Kikuchi, Y., Sakai, R., and Kakiuchi, K. (1968). *Bull. Chem. Soc. Japan* **41**, 1273.

Sakakibara, S., Kishida, Y.,Okuyama, K., Tanaka, N., Ashida, T., and Kakudo, M. (1972). *J. Mol. Biol.* **65**, 371.

Sakar, P. K., and Doty, P. (1966). *Proc. Nat. Acad. Sci. U.S.* **55**, 981.

Saludjian, P., and Luzzati, V. (1966). *J. Mol. Biol.* **15**, 681.

Saludjian, P., and Luzzati, V. (1967). *In* "Poly-α-Amino Acids" (G. D. Fasman, ed.), p. 157. Dekker, New York.

Saludjian, P., de Lozé, C., and Luzzati, V. (1963a). *C.R. Acad. Sci. Paris* **256**, 4297.

Saludjian, P., de Lozé, C., and Luzzati, V. (1963b). *C.R. Acad. Sci. Paris* **256**, 4514.

Samulski, E. T., and Tobolsky, A. V. (1968). *Macromolecules* **1**, 555.

Samulski, E. T., and Tobolsky, A. V. (1971). *Biopolymers* **10**, 1013.

Sandeman, I. (1955). *Proc. Roy. Soc. Ser. A* **232**, 105.

Sandeman, T., and Keller, A. (1956). *J. Polym. Sci.* **19**, 401.

Sarathy, K. P., and Ramachandran, G. N. (1968). *Biopolymers* **6**, 461.

Sasisekharan, V. (1959a). *Acta Crystallogr.* **12**, 897.

Sasisekharan, V. (1959b). *Acta Crystallogr.* **12**, 903.

Sasisekharan, V. (1960). *J. Polym. Sci.* **47**, 373.

Sasisekharan, V., and Ramachandran, G. N. (1957). *Proc. Indian Acad. Sci.* **A45**, 363.

Saxena, V. P., and Wetlaufer, D. B. (1971). *Proc. Nat. Acad. Sci. U.S.* **68**, 969.

Scatturin, A., Del Pra, A., Tamburro, A. M., and Scoffone, E. (1967). *Chem. Ind. M.* **49**, 970.

Schellman, J. A., and Schellman, C. (1964). *In* "The Proteins" (H. Neurath, ed.), Vol. 2, p. 1. Academic Press, New York.

Scheraga, H. A. (1968a). *Advan. Phys. Org. Chem.* **6**, 103.

Scheraga, H. A. (1968b). *Harvey Lect.* **63**, 99.

Scheraga, H. A. (1971). *Chem. Rev.* **71**, 195.

Scheraga, H. A., and Laskowski, M. (1957). *Advan. Protein Chem.* **12**, 1.

Schiffer, M., and Edmundson, A. B. (1967). *Biophys. J.* **7**, 121.

Schimmel, P. R., and Flory, P. J. (1967). *Proc. Nat. Acad. Sci. U.S.* **58**, 52.

Schimmel, P. R., and Flory, P. J. (1968). *J. Mol. Biol.* **34**, 105.

Schmitt, F. O. (1939). *Physiol. Rev.* **19**, 270.

Schmitt, F. O. (1959). *Rev. Mod. Phys.* **31**, 349.

Schmitt, F. O. (1968). *Proc. Nat. Acad. Sci. U.S.* **60**, 1092.

Schmitt, F. O., Hall, C. E., and Jakus, M. A. (1942). *J. Cell. Comp. Physiol.* **20**, 11.

Schmitt, F. O., Gross, J., and Highberger, J. H. (1953). *Proc. Nat. Acad. Sci. U.S.* **39**, 459.

Schmitt, F. O., Gross, J., and Highberger, J. H. (1955a). *Symp. Soc. Exp. Biol.* **9**, 148.

Schmitt, F. O., Gross, J., and Highberger, J. H. (1955b). *Exp. Cell Res. Suppl.* **3**, 326.

Schnabel, E., and Zahn, H. (1958). *Justus Liebigs Ann. Chem.* **614**, 141.

Schor, R., and Krimm. S. (1961a). *Biophys. J.* **1**, 467.

Schor, R., and Krimm, S. (1961b). *Biophys. J.* **1**, 489.

Schroeder, W. A., and Kay, L. M. (1955). *J. Amer. Chem. Soc.* **77**, 3908.

Schwartz, A., Andrics, J. C., and Walton, A. G. (1970). *Nature (London)* **226**, 161.

Schwenker, R. F., and Dusenbury, J. H. (1960). *Textile Res. J.* **30**, 800.

Scott, R. A., and Scheraga, H. A. (1966). *J. Chem. Phys.* **45**, 2091.

Seeds, W. E. (1953). *In* "Nature and Structure of Collagen" (J. T. Randall, ed.), p. 250. Butterworths, London.

Segal, D. M. (1969). *J. Mol. Biol.* **43**, 497.

Segal, D. M., and Traub, W. (1969). *J. Mol. Biol.* **43**, 487.

Segal, D. M., Traub, W., and Yonath, A. (1969). *J. Mol. Biol.* **43**, 519.

Segrest, J. P., and Cunningham, L. W. (1971). *Nature (London) New Biol.* **234**, 26.

Seifter, S., and Gallop, P. M. (1966). *In* "The Proteins" (H. Neurath, ed.), 2nd ed., Vol. 4, p. 153. Academic Press, New York.

Selby, C. C., and Bear, R. S. (1956). *J. Biophys. Biochem. Cytol.* **2**, 71.

Senti, F. R., Eddy, C. R., and Nutting, G. C. (1943). *J. Amer. Chem. Soc.* **65**, 2473.

Seshadri, K. S., and Jones, R. N. (1963). *Spectrochim. Acta* **19**, 1013.

Shaw, J. T. B. (1964). *Biochem. J.* **93**, 45.

Shaw, J. T. B., and Smith, S. G. (1961). *Biochim. Biophys. Acta* **52**, 305.

Sheard, B., and Bradbury, E. M. (1971). *In* "Progress in Biophysics and Molecular Biology" (J. A. V. Butler and D. Noble, eds.), Vol. 20, p. 187. Pergamon, Oxford.

Shibnev, V. A., Rogulenkova, V. N., and Andreeva, N. S. (1965). *Biofizika* **10**, 164.

Shimizu, M. (1941). *Bull. Imp. Sericult. Exp. Sta. Japan* **10**, 475.

Shmueli, U., and Traub, W. (1965). *J. Mol. Biol.* **12**, 205.

Shulman, S. (1953). *J. Amer. Chem. Soc.* **75**, 5846.

Signer, R., and Strässle, R. (1947). *Helv. Chim. Acta* **30**, 155.

Sikorski, J. (1963). *In* "Fibre Structure" (J. W. S. Hearle and R. H. Peters, eds.), p. 269. Butterworths, London.

Sikorski, J., and Woods, H. J. (1960). *J. Textile Inst.* **51**, T506.

Simmons, N. S., Cohen, C., Szent-Györgyi, A. G., Wetlaufer, D. B., and Blout, E. R. (1961). *J. Amer. Chem. Soc.* **83**, 4766.

Simpson, W. S., and Woods, H. J. (1960). *Nature (London)* **185**, 157.

Sisson, W. A. (1936). *J. Phys. Chem.* **40**, 343.

Sjöstrand, F. S. (1967). "Electron Microscopy of Cells and Tissues," Vol. I. Academic Press, New York.

Sjöstrand, F. S., and Andersson, E. (1956). *Exp. Cell Res.* **11**, 493.

Skertchly, A. R. B. (1957). *Acta Crystallogr.* **10**, 535.

Skertchly, A. R. B., and Woods, H. J. (1960). *J. Textile Inst.* **51**, T517.

Slayter, E. M. (1969). *In* "Physical Principles and Techniques of Protein Chemistry" (S. J. Leach, ed.), Part A, p. 1. Academic Press, New York.

Slayter, H. S., and Hall, C. E. (1962). *Proc. Ann. Meeting Biophys. Soc., 6th. Washington D.C., 1962* Paper TB–10.

Slayter, H. S., and Lowey, S. (1967). *Proc. Nat. Acad. Sci. U.S.* **58**, 1611.

Small, E. W., Fanconi, B., and Peticolas, W. L. (1970). *J. Chem. Phys.* **52**, 4369.

Small, J. V., Lowy, J., and Squire, J. M. (1971). *Eur. Biophys. Congr., 1st, Vienna* **5**, 419.

Smith, J. W. (1965). *Nature (London)* **205**, 356.

Smith, J. W. (1968). *Nature (London)* **219**, 157.

Smith, J. W., and Frame, J. (1969). *J. Cell Sci.* **4**, 421.

Smith, M., Walton, A. G., and Koenig, J. L. (1969). *Biopolymers* **8**, 29.

Snyder, R. G. (1971). *J. Mol. Spectrosc.* **37**, 353.

Sobajima, S. (1967). *J. Phys. Soc. Japan* **23**, 1070.

Somlyo, A. P., Somlyo, A. V., Devine, C. E., and Rice, R. V. (1971a). *Nature (London) New Biol.* **231**, 243.

Somlyo, A. P., Devine, C. E., and Somlyo, A. V. (1971b). *Nature (London) New Biol.* **233**, 218.

Spadaro, J. A. (1970). *Nature (London)* **228**, 78.

Spei, M. (1970a). *Z. Naturforsch.* **25b**, 420.

Spei, M. (1970b). *Z. Naturforsch.* **25b**, 421.

Spei, M. (1970c). *Kolloid-Z. Z. Polym.* **238**, 436.

Spei, M. (1971). *Appl. Polym. Symp.* No. 18, 659.

Spei, M. (1972a). *Kolloid-Z. Z. Polym.* **250**, 207.

Spei, M. (1972b). *Kolloid-Z. Z. Polym.* **250**, 214.

Spei, M., and Zahn, H. (1971). *Monatsh. Chem.* **102**, 1163.

Spei, M., Heidemann, G., and Zahn, H. (1968a). *Naturwissenschaften* **55**, 346.

Spei, M., Heidemann, G., and Halboth, H. (1968b). *Nature (London)* **217**, 247.

Spei, M., Stein, W., and Zahn, H. (1970). *Kolloid-Z. Z. Polym.* **238**, 447.

Spencer, M. (1969). *Nature (London)* **223**, 1361.

Squire, J. M. (1971). *Nature (London)* **233**, 457.

Squire, J. M., and Elliott, A. (1969). *Mol. Cryst. Liquid Cryst.* **7**, 457.

Squire, J. M., and Elliott, A. (1972). *J. Mol. Biol.* **65**, 291.

Srinivasan, R. (1969). Quoted by Ramachandran (1969).

Stein, R. S. (1958). *J. Polym. Sci.* **31**, 327.

Stein, R. S. (1961). *J. Polym. Sci.* **50**, 339.

Stein, R. S. (1964). *In* "Newer Methods of Polymer Characterization" (B. Ke, ed.), p. 155. Wiley (Interscience), New York.

Stein, R. S. (1969). *Ann. N.Y. Acad. Sci.* **155**, 566.

Steinberg, I. Z., Berger, A., and Katchalski, E. (1958). *Biochim. Biophys. Acta* **28**, 647.

Steinberg, I. Z., Harrington, W. F., Berger, A., Sela, M., and Katchalski, E. (1960). *J. Amer. Chem. Soc.* **82**, 5263.

Stenn, K., and Bahr, G. F. (1970). *J. Ultrastruct. Res.* **31**, 526.

Stephens, R. E. (1971). *In* "Subunits in Biological Systems" (S. N. Timasheff and G. D. Fasman, eds.), Part A, p. 355. Dekker, New York.

Sternlieb, I., and Berger, J. E. (1969). *J. Cell Biol.* **43**, 448.

Steven, F. S. (1970). *In* "Chemistry and Molecular Biology of the Intercellular Matrix" (E. A. Balazs, ed.), Vol. 1, p. 511. Academic Press, New York.

Stevens, L., Townend, R., Timasheff, S. N., Fasman, G. D., and Potter, J. (1968). *Biochemistry* **7**, 3717.

Stokes, A. R. (1955). *Acta Crystallogr.* **8**, 27.

Stoyanova, I. G., and Mikhailovskii, G. A. (1959). *Biofizika* **4**, 483.

Straub, F. B. (1942). *Stud. Inst. Med. Chem. Univ. Szeged.* **2**, 3.

Straub, F. B. (1943). *Stud. Inst. Med. Chem. Univ. Szeged.* **3**, 23.

Stryer, L., Cohen, C., and Langridge, R. (1963). *Nature (London)* **197**, 793.

Sugeta, H., and Miyazawa, T. (1967). *Biopolymers* **5**, 673.

Sugeta, H., and Miyazawa, T. (1968). *Biopolymers* **6**, 1387.

Susi, H. (1961). *Spectrochim. Acta* **17**, 1257.

Susi, H. (1969). *In* "Structure and Stability of Biological Macromolecules" (S. N. Timasheff and G. D. Fasman, eds.), p. 575. Dekker, New York.

Susi, H., Ard, J. S., and Carroll, R. J. (1971). *Biopolymers* **10**, 1597.

Sutherland, G. B. B. M. (1955). *Rend. Ist. Lombardo Sci.* **89**, 67.

Sutherland, G. B. B. M., Tanner, K. N., and Wood, D. L. (1954). *J. Chem. Phys.* **22**, 1621.

Sutor, D. J. (1963). *J. Chem. Soc.* 1105.

Sutton, P., and Koenig, J. L. (1970). *Biopolymers* **9**, 615.

Suwalsky, M., and Traub, W. (1972). *Biopolymers* **11**, 623.

Suzuki, E. (1967). *Spectrochim. Acta* **23A**, 2303.

Suzuki, E. (1971). Unpublished observations.

Suzuki, E. (1972). Unpublished data.

Suzuki, E. (1973). *Aust. J. Biol. Sci.* **26**, 435.

Suzuki, E., Crewther, W. G., Fraser, R. D. B., MacRae, T. P., and McKern, N. M. (1973). *J. Mol. Biol.* **73**, 275.

Suzuki, S., Iwashita, Y., and Shimanouchi, T. (1966). *Biopolymers* **4**, 337.

Swanbeck, G. (1959). *Acta Dermato-Venereol. Suppl. 43*, **39**, 1.

Swanbeck, G. (1961). *Exp. Cell Res.* **23**, 420.

Swanbeck, G. (1963). *In* "Aspects of Protein Structure" (G. N. Ramachandran, ed.), p. 93. Academic Press, New York.

Swanbeck, G., and Forslind, B. (1964). *Biochim. Biophys. Acta* **88**, 422.

Swart, L. S., Joubert, F. J., and Strydom, A. J. C. (1969). *Textile Res. J.* **39**, 273.

Swift, J. A. (1969). *Histochemie* **19**, 88.

Szent-Györgyi, A. G., and Cohen, C. (1957). *Science* **126**, 697.

Szent-Györgyi, A. G., Cohen, C., and Philpott, D. E. (1960). *J. Mol. Biol.* **2**, 133.

Szent-Györgyi, A. G., Cohen, C., and Kendrick-Jones, J. (1971). *J. Mol. Biol.* **56**, 239.

Tait, M. J., and Franks, F. (1971). *Nature (London)* **230**, 91.

Takahashi, M., Yasuda, M., Yabu, H., Suzuki, K., and Inadome, H. (1962). *Sapporo Igaku Zasshi* **21**, 7.

Takeda, Y., Iitaka, Y., and Tsuboi, M. (1970). *J. Mol. Biol.* **51**, 101.

Tanaka, S., and Naya, S. (1969). *J. Phys. Soc. Japan* **26**, 982.

Tashiro, Y., and Otsuki, E. (1970a). *Biochim. Biophys. Acta* **214**, 265.

Tashiro, Y., and Otsuki, E. (1970b). *J. Cell Biol.* **46**, 1.

Taylor, C. A. (1972). *In* "Optical Transforms" (H. S. Lipson, ed.), p. 115. Academic Press, New York.

Taylor, C. A., and Lipson, H. (1964). "Optical Transforms." Bell, London.

Taylor, H. S. (1941). *Proc. Amer. Phil. Soc.* **85**, 1.

Thomas, L. E., Humphreys, C. J., Duff, W. R., and Grubb, D. T. (1970). *Radiat. Eff.* **3**, 89.

Thompson, B. J. (1972). *In* "Optical Transforms" (H. S. Lipson, ed.), p. 267. Academic Press, New York.

Thompson, E. O. P., and O'Donnell, I. J. (1965). *Aust. J. Biol. Sci.* **18**, 1207.

Thon, F. (1966a). *Z. Naturforsch.* **21**, 476.

Thon, F. (1966b). *In* "Electron Microscopy 1966" (R. Uyeda, ed.), Vol. 1, p. 23. Maruzen, Tokyo.

Thon, F., and Siegel, B. M. (1970). *Ber. Bunsenges. Phys. Chem.* **74**, 1116.

Timasheff, S. N., and Fasman, G. D. (1969). "Structure and Stability of Biological Macromolecules." Dekker, New York.

Timasheff, S. N., and Townend, R. (1970). *In* "Physical Principles and Techniques of Protein Chemistry" (S. J. Leach, ed.), Part B, p. 147. Academic Press, New York.

Tinoco, I., and Cantor, C. R. (1970). *In* "Methods of Biochemical Analysis" (D. Glick, ed.), Vol. 18, p. 81. Wiley (Interscience), New York.

Tinoco, I., Halpern, A., and Simpson, W. T. (1962). *In* "Polyamino Acids, Polypeptides Proteins" (M. H. Stahmann, ed.), p. 147. Univ. Wisconsin Press, Madison, Wisconsin.

Tkocz, C., and Kühn, K. (1969). *Eur. J. Biochem.* **7**, 454.

Tobin, M. C. (1968). *J. Opt. Soc. Amer.* **58**, 1057.

Tobin, M. C. (1971). "Laser Raman Spectroscopy." Wiley, New York.

Tomita, K., Rich, A., de Lozé, C., and Blout, E. R. (1962). *J. Mol. Biol.* **4**, 83.

Tomlin, S. G. (1956). *In Proc. Int. Wool Text. Res. Conf., Australia 1955* (W. G. Crewther, ed.), Vol. B, p. 187. C.S.I.R.O., Australia.

Tomlin, S. G., and Ericson, L. G. (1960). *Acta Crystallogr.* **13**, 395.

Tomlin, S. G., and Worthington, C. R. (1956). *Proc. Roy. Soc. Ser. A* **235**, 189.

Toniolo, C., Falxa, M. L., and Goodman, M. (1968). *Biopolymers* **6**, 1579.

Tooney, N. M., and Cohen, C. (1972). *Nature (London)* **237**, 23.

Torchia, D. A., and Bovey, F. A. (1971). *Macromolecules* **4**, 246.

Traub, W. (1969). *J. Mol. Biol.* **43**, 479.

Traub, W., and Piez, K. A. (1971). *Advan. Protein Chem.* **25**, 243.

Traub, W., and Shmueli, U. (1963a). *Nature (London)* **198**, 1165.

Traub, W., and Shmueli, U. (1963b). *In* "Aspects of Protein Structure" (G. N. Ramachandran, ed.), p. 81. Academic Press, New York.

Traub, W., and Yonath, A. (1966). *J. Mol. Biol.* **16**, 404.

Traub, W., and Yonath, A. (1967). *J. Mol. Biol.* **25**, 351.

Traub, W., Shmueli, U., Suwalsky, M., and Yonath, A. (1967). *In* "Conformation of Biopolymers" (G. N. Ramachandran, ed.), Vol. 2, p. 449. Academic Press, New York.

Traub, W., Yonath, A., and Segal, D. M. (1969). *Nature (London)* **221**, 914.

Trelstad, R. L., Kang, A. H., Igarashi, S., and Gross, J. (1970). *Biochemistry* **9**, 4993.

Trillat, J. J. (1930). *C. R. Acad. Sci. Paris* **190**, 265.

Trogus, C., and Hess, K. (1933). *Biochem. Z.* **260**, 376.

Tromans, W. J., Horne, R. W., Gresham, G. A., and Bailey, A. J. (1963). *Z. Zellforsch. Mikrosk. Anat.* **58**, 798.

Trombka, J. I., and Schmadebeck, R. L. (1970). *In* "Spectral Analysis" (J. A. Blackburn, ed.), p. 121. Dekker, New York.

Trotter, I. F., and Brown, L. (1956). As quoted by Bamford *et al.* (1956), p. 409.

Tsao, T. C., Kung, T. H., Peng, C. M., Chang, Y. S., and Tsou, Y. S. (1965). *Scientia Sinica* **14**, 91.

Tsuboi, K. K. (1968). *Biochim. Biophys. Acta* **160**, 420.

Tsuboi, M. (1962). *J. Polym. Sci.* **59**, 139.

Tsuboi, M. (1964). *Biopolym. Symp. No. 1*, 527.

Tsuboi, M., and Wada, A. (1961). *J. Mol. Biol.* **3**, 480.

Tsuboi, M., Wada, A., and Nagashima, N. (1961). *J. Mol. Biol.* **3**, 705.

Tubomura, T. (1956). *J. Chem. Phys.* **24**, 927.

Tulloch, P. A. (1971). The Application of Electron Diffraction to Biological Materials. Ph. D. Thesis, Univ. of Melbourne.

Tumanyan, V. G. (1970). *Biopolymers* **9**, 955.

Tyson, C. N., and Woods, H. J. (1964). *J. Mol. Biol.* **9**, 266.

Ueki, T., Ashida, T., Kakudo, M., Sasada, Y., and Katsube, Y. (1969). *Acta Crystallogr.* **B25**, 1840.

Vainshtein, B. K. (1964). "Structure Analysis by Electron Diffraction." Pergamon, Oxford.

Vainshtein, B. K. (1966). "Diffraction of X-Rays by Chain Molecules." Elsevier, London.

Vainshtein, B. K., and Tatarinova, L. I. (1967a). *In* "Conformation of Biopolymers" (G. N. Ramachandran, ed.), Vol. 2, p. 569. Academic Press, New York.

Vainshtein, B. K., and Tatarinova, L. I. (1967b). *Sov. Phys. Crystallogr.* **11**, 494.

Valentine, R. C. (1964). *Nature (London)* **204**, 1262.

Valentine, R. C. (1966). *In* "Advances in Optical and Electron Microscopy" (R. Barer and V. E. Cosslett, eds.), Vol. I, p. 180. Academic Press, London.

Vazina, A., Frank, G., and Lemazhikhin, B. (1965). *J. Mol. Biol.* **14**, 373.

Veis, A. (1967). *In* "Treatise on Collagen" (G. N. Ramachandran, ed.), Vol. 1, p. 367. Academic Press, New York.

Veis, A., Anesey, J., and Mussell, S. (1967a). *Nature (London)* **215**, 931.

Veis, A., Kaufman, E., and Chao, C. C. W. (1967b). *In* "Conformation of Biopolymers" (G. N. Ramachandran, ed.), Vol. 2, p. 499. Academic Press, New York.

Veis, A., Bhatnagar, R. S., Shuttleworth, C. A., and Mussell, S. (1970). *Biochim. Biophys. Acta* **200**, 97.

Venkatachalam, C. M. (1968a). *Biochim. Biophys. Acta* **168**, 411.

Venkatachalam, C. M. (1968b). *Biopolymers* **6**, 1425.

Venkatachalam, C. M., and Ramachandran, G. N. (1967). *In* "Conformation of Biopolymers" (G. N. Ramachandran, ed.), Vol. I, p. 83. Academic Press, New York.

Vibert, P. J., Haselgrove, J. C., Lowy, J., and Poulsen, F. R. (1972). *J. Mol. Biol.* **71**, 757.

von der Mark, K., Wendt, P., Rexrodt, F., and Kühn, K. (1970). *FEBS Lett.* **11**, 105.

von Hippel, P. H. (1967). *In* "Treatise on Collagen" (G. N. Ramachandran, ed.), Vol. 1, p. 253. Academic Press, New York.

von Weimarn, P. P. (1932). *In* "Colloid Chemistry" (J. Alexander, ed.), p. 363. Reinhold, New York.

Wada, A., Tsuboi, M., and Konishi, E. (1961). *J. Phys. Chem.* **65**, 1119.

Wakabayashi, K., and Mitsui, T. (1970). *J. Mol. Biol.* **53**, 567.

Waldschmidtt-Leitz, E., and Zeiss, O. (1955). *Z. Physiol. Chem.* **300**, 49.

Walton, A. (1969). *In* "Progress in Stereochemistry" (B. J. Aylett and M. M. Harris, ed.), Vol. 4, p. 335. Butterworths, London.

Wang, S. N. (1939). *J. Chem. Phys.* **7**, 1012.

Ward, J. C. (1955). *Proc. Roy. Soc. Ser. A* **228**, 205.
Ward, W. H., High, L. M., and Lundgren, H. P. (1946). *J. Polym. Sci.* **1**, 22.
Warme, P. K., Gō, N., and Scheraga, H. A. (1972). *J. Comput. Phys.* **9**, 303.
Warren, R. C., and Hicks, R. M. (1971). *J. Ultrastruct. Res.* **36**, 861.
Warwicker, J. O. (1954). *Acta Crystallogr.* **7**, 565.
Warwicker, J. O. (1956). *Trans. Faraday Soc.* **52**, 554.
Warwicker, J. O. (1960a). *J. Mol. Biol.* **2**, 350.
Warwicker, J. O. (1960b). *Brit. J. Appl. Phys.* **11**, 92.
Warwicker, J. O. (1961). *Biochim. Biophys. Acta* **52**, 319.
Warwicker, J. O., and Ellis, K. C. (1965). *J. Polym. Sci. Part A*, **3**, 4159.
Watanabe, M., Okazaki, I., Honjo, G., and Mihama, K. (1960). *In Int. Conf. Electron Microsc., 4th, Berlin, 1958* (G. Möllenstedt, H. Niehrs, and E. Ruska, eds.), **1**, 90. Springer-Verlag, Berlin.
Watson, H. C. (1970). *In* "Progress in Stereochemistry" (B. J. Aylett and M. M. Harris, eds.), Vol. 4, p. 299. Butterworths, London.
Watson, M. R., and Silvester, N. R. (1959). *Biochem. J.* **71**, 578.
Waugh, D. F. (1954). *Advan. Protein Chem.* **9**, 325.
Weber, K., and Osborn, M. (1969). *J. Biol. Chem.* **244**, 4406.
Weibull, C. (1950). *Nature (London)* **165**, 482.
Weis-Fogh, T. (1961a). *J. Mol. Biol.* **3**, 520.
Weis-Fogh, T. (1961b). *J. Mol. Biol.* **3**, 648.
Welscher, H. D. (1969). *Int. J. Protein Res.* **1**, 267.
Werner, P. E. (1970). *Acta Crystallogr.* **A26**, 489.
Whitmore, P. G. (1971). Unpublished observations.
Whitmore, P. G. (1972). *J. Cell Biol.* **52**, 174.
Whittaker, E. J. W. (1955). *Acta Crystallogr.* **8**, 571.
Wiener, O. (1912). *Abh. Sächs. Akad. Wiss.* **32**, 507.
Wiener, O. (1927). *Kolloidchem. Beihefte* **23**, 189.
Wilkes, G. L. (1971). *Advan. Polym. Sci.* **8**, 91.
Wilson, A. J. C. (1962). "X-Ray Optics." Methuen, London.
Wilson, A. J. C. (1970). "Elements of X-Ray Crystallography." Addison-Wesley, Reading, Massachusetts.
Wilson, E. B., Decius, J. C., and Cross, P. C. (1955). "Molecular Vibrations; The Theory of Infra-Red and Raman Vibrational Spectra." McGraw-Hill, New York.
Wilson, G. A. (1972a). *Polymer* **13**, 63.
Wilson, G. A. (1972b). *Biophys. Biochim. Acta* **278**, 440.
Wilson, H. R. (1963). *J. Mol. Biol.* **6**, 474.
Wilson, H. R. (1966). "Diffraction of X-Rays by Proteins, Nucleic Acids and Viruses." Arnold, London.
Winkler, F. K., and Dunitz, J. D. (1971). *J. Mol. Biol.* **59**, 169.
Winklmair, D., Engle, J., and Ganser, V. (1971). *Biopolymers* **10**, 721.
Witnauer, L. P., and Wisnewski, A. (1964). *J. Amer. Leather Chem. Ass.* **59**, 598.
Witz, J. (1969). *Acta Crystallogr.* **A25**, 30.
Wolpers, C. (1943). *Klin. Wochschr.* **32**, 624.
Woodin, A. M. (1954). *Biochem. J.* **57**, 99.
Woods, E. F. (1967). *J. Biol. Chem.* **242**, 2859.
Woods, E. F. (1969a). *Biochemistry* **8**, 4336.
Woods, E. F. (1969b). *Biochem. J.* **113**, 39.
Woods, E. F. (1971). *Comp. Biochem. Physiol.* **39A**, 325.
Woods, E. F., and Pont, M. J. (1971). *Biochemistry* **10**, 270.

Woods, H. J. (1938). *Proc. Roy. Soc. Ser. A* **166**, 76.

Woods, H. J. (1960). *Brit. J. Appl. Phys.* **11**, 94.

Woods, H. J. (1967). *J. Polym. Sci. Pt. C* **20**, 37.

Woods, H. J., and Tyson, C. N. (1966). *Nature (London)* **209**, 399.

Woody, R. W. (1969). *Biopolymers* **8**, 669.

Woolfson, M. M. (1970). "An Introduction to X-Ray Crystallography." Cambridge Univ. Press, London and New York.

Wooster, W. A. (1962). "Diffuse X-Ray Reflections from Crystals." Oxford Univ. Press (Clarendon).

Worthington, C. R. (1959). *J. Mol. Biol.* **1**, 398.

Worthington, C. R. (1961). *J. Mol. Biol.* **3**, 618.

Wright, B. A. (1948). *Nature (London)* **162**, 23.

Wright, C. S., Alden, R. A., and Kraut, J. (1969). *Nature (London)* **221**, 235.

Wu, J. Y., and Yang, J. T. (1970). *J. Biol. Chem.* **245**, 212.

Wyckoff, H. W., Bear, R. S., Morgan, R. S., and Carlstrom, D. (1957). *J. Opt. Soc. Amer.* **47**, 1061.

Wyckoff, H. W., Tsernglou, D., Hanson, A. W., Knox, J. R., Lee, B., and Richards, F. M. (1970). *J. Biol. Chem.* **245**, 305.

Wyckoff, R. W. G., and Corey, R. B. (1936). *Proc. Soc. Exp. Biol. Med.* **34**, 285.

Wyckoff, R. W. G., Corey, R. B., and Biscoe, J. (1935). *Science* **82**, 175.

Yaguchi, M., Foster, J. F., and Koffler, H. (1964). *Proc. Int. Congr. Biochem., 6th, II* **210**, 189.

Yahara, I., and Imahori, K. (1963). *J. Amer. Chem. Soc.* **85**, 230.

Yakel, H. L. (1953). *Acta Crystallogr.* **6**, 724.

Yakel, H. L., and Schatz, P. N. (1955). *Acta Crystallogr.* **8**, 22.

Yakel, H. L., Pauling, L., and Corey, R. B. (1952). *Nature (London)* **169**, 920.

Yamaguchi, M. (1965). *In* "Handbook of Organic Structural Analysis" (Y. Yukawa, ed.), p. 265. Benjamin, New York.

Yamaguchi, T., Hayashi, M., Wakabayashi, K., and Higashi-Fujime, S. (1972). *Biochim. Biophys. Acta* **257**, 30.

Yan, J. F., Vanderkooi, G., and Scheraga, H. A. (1968). *J. Chem. Phys.* **49**, 2713.

Yan, J. F., Momany, F. A., Hoffman, R., and Scheraga, H. A. (1970). *J. Phys. Chem.* **74**, 420.

Yang, J. T. (1967). *In* "Poly-α-Amino Acids" (G. D. Fasman, ed.), p. 239. Dekker, New York.

Yang, J. T. (1969). *In* "Analytical Methods of Protein Chemistry" (P. Alexander and H. P. Lundgren, eds.), Vol. 5, p. 23. Pergamon, Oxford.

Yonath, A., and Traub, W. (1969). *J. Mol. Biol.* **43**, 461.

Young, M. (1969). *Annu. Rev. Biochem.* **38**, 913.

Young, M., Blanchard, M. H., and Brown, D. (1968). *Proc. Nat. Acad. Sci. U.S.* **61**, 1087.

Young, R. P., and Jones, R. N. (1971). *Chem. Rev.* **71**, 219.

Zahn, H. (1948). *Kolloid-Z.* **111**, 96.

Zahn, H. (1949). *Kolloid-Z.* **112**, 91.

Zahn, H. (1952). *Melliand Textilber.* **33**, 1076.

Zahn, H., and Biela, M. (1968a). *Textile Praxis* **2**, 103.

Zahn, H., and Biela, M. (1968b). *Eur. J. Biochem.* **5**, 567.

Zahn, H., Kratky, O., and Sekora, A. (1951). *Z. Naturforsch.* **6b**, 9.

Zbinden, R. (1964). "Infrared Spectroscopy of High Polymers." Academic Press, New York.

Zernike, F., and Prins, J. A. (1927). *Z. Phys.* **41**, 184.

Zobel, C. R., and Carlson, F. D. (1963). *J. Mol. Biol.* **7**, 78.

Author Index

Numbers in italics refer to the pages on which the complete references are listed.

C

Subject Index

A

Acidic residues, effect on conformation, 287–288

Actin, 404–418
  amino acid composition of, 404, 406
  chain conformation in, 406
  electron microscope studies of, 406–409
  G → F transformation in, 405–406
  infrared spectrum of, 406
  molecular weight of, 405
  ordered aggregates of, 406–412
    electron microscope studies of, 406–409
    X-ray pattern of, 409–412
  in thin filaments, 413–418
    electron microscopy of, 410, 413, 418
    X-ray pattern of, 413–418

Alanyl residues
  effect on conformation of, 289
  steric map for, 139–142

Alpha helix, *see also* individual materials
  coordinates for poly(L-alanine), 185
  coordinates, for standard, 181
  hydrogen bond length in, 188, 205–206
  parameters of
    interrelationship of, 212–217
    in lysozyme, 215
    in myoglobin, 215
    in polypeptides, 214

screw sense, factors affecting, 179, 210–212

stability of, 276–290
  effect of acidic residues on, 287
  effect of basic residues on, 287
  effect of cysteinyl residues on, 281–283
  effect of glycyl residues on, 285–286
  effect of prolyl residues on, 276, 286–287
  effect of seryl residues on, 283–285
  effect of valyl residues on, 279–281

Alpha-helix favoring residues, 276–278

Alpha-keratin, *see* α-Keratin

Amide group
  dimensions of standard, 138
  planarity of, 128, 134, 137, 360
  transition moment direction in, 97–98, 206–207
  vibrations of, 96

Atomic scattering factor, definition of, 6

B

Basic residues, effect on conformation, 287

Beta conformation, *see also* individual materials
  in [Ala]$_n$, 228–232
    coordinates of model for, 230
  in globular proteins, 219
  in β-keratin, 521–525

619

Molecular Biology

An International Series of Monographs and Textbooks

Editors

BERNARD HORECKER

Department of Molecular Biology
Albert Einstein College of Medicine
Yeshiva University
Bronx, New York

NATHAN O. KAPLAN

Department of Chemistry
University of California
At San Diego
La Jolla, California

JULIUS MARMUR

Department of Biochemistry
Albert Einstein College of Medicine
Yeshiva University
Bronx, New York

HAROLD A. SCHERAGA

Department of Chemistry
Cornell University
Ithaca, New York

WALTER W. WAINIO. The Mammalian Mitochondrial Respiratory Chain. 1970

LAWRENCE I. ROTHFIELD (Editor). Structure and Function of Biological Membranes. 1971

ALAN G. WALTON AND JOHN BLACKWELL. Biopolymers. 1973

WALTER LOVENBERG (Editor). Iron-Sulfur Proteins. Volume I, Biological Properties — 1973. Volume II, Molecular Properties — 1973

A. J. HOPFINGER. Conformational Properties of Macromolecules. 1973

R. D. B. FRASER AND T. P. MacRae. Conformation in Fibrous Proteins. 1973

In preparation

OSAMU HAYAISHI (Editor). Molecular Mechanisms of Oxygen Activation